Advances in Crop Production and Climate Change

About the Editors

Dr A.S. Yadav, Ph.D. (Agronomy) from Acharya Narendra Deva University of Agriculture & Technology, Kumarganj Ayodhyay, India. He is scientific officer at U.P. Council of Agricultural Research, Lucknow, Uttar Pradesh, India. He has an experience for over 10 years in research and extension in the field of agriculture. He is active member of various academic and professional societies/associations. He is author of 122 peer-reviewed research papers and technical articles that has been published in various national and international journals. He has published 8 book chapters. He has experience of handling five externally funded research/consultancy projects and he is recipient of Young Scientist Award 2014, Young Fellow Award 2014, Young Scientist Award 2015, Distinguished Service Award 2015, Dr Masood Ali Vishishth Yuva Vaigyanik Puruskar 2015, Outstanding Scientist in Agriculture Award 2016, SCSI Leadership Award 2018, Biodiversity Conservation Award 2019 and Fellow Award 2019. He is member of editorial board, Crop Research: An International Journal, Indian Journal of Waste Management and Indian Journal of Plant and Soil.

Dr Narendra Kumar, completed his Ph.D. from Indian Agricultural Research Institute, New Delhi and joined Agricultural Research Service (ARS) of ICAR in 2003. Dr. Kumar received Junior Research Fellowship and Senior Research Fellowship of ICAR for his post-graduation and doctoral degree. Dr. Kumar started his career as scientist at ICAR-VPKAS, Almora and developed technologies for crop diversification with off-season vegetables and pulses, intercropping, relay cropping, protected agriculture, weed management and resource conservation practices. He is associated in release of 18 varieties of pulses and oilseeds for different zones of country. He is recipient of Dr. P.S. Deshmukh Young Agronomist Award 2009 and Associateship Award 2016 by Indian Society of Agronomy. He is Fellow of Indian Society of Pulses Research. He has published more than 80 research papers in international as well as national journals. Dr. Kumar is presently working as Principal Scientist at ICAR-Indian Institute of Pulses Research, Kanpur and involved in research activities related to crop diversification, resource conservation technology and weed management in pulse based cropping systems.

Dr Sanjay Arora, Ph.D. in Soils science from Punjab Agricultural University, Ludhiana, Punjab, India. He is Principal Scientist (Soil Science) at Regional Research Station of ICAR-Central Soil Salinity Research Institute at Lucknow, Uttar Pradesh, India. He has made significant contributions in sustainable management of soil and water resources. He has developed relative soil quality and relative production efficiency indices based expert system using soil parameters and climatic variables for different agro-climatic zones. He was successful in his pioneering work of isolation and characterization of halophilic soil microbes having potential for bio-remediation of salt affected inland and coastal soils. The database of the dominant halophytes of economic use for bio-saline agriculture has been generated from coastal Gujarat. He has been instrumental in reporting emergence of salt affected and water logged soils in Ravi-Tawi command of Jammu through remote sensing and GIS approach. He is active member of various academic and professional societies/associations. He is author of 142 peer-reviewed research papers and technical articles that has been published in various national and international journals. He has edited nine books and contributed 42 book chapters. He has guided 13 M.Sc. students for research related to soil health management. He has experience of handling 5 externally funded research/ consultancy projects and he is recipient of Young Scientist Award, SCSI Leadership Award, Groundwater Augmentation Award, 2009 (MOWR), Dr. R.N. Dwivedi Medal, National Fellow, SCSI and Fellow, NABS, Fellow, SES, J.S. Bali Award, 2016. He is chief editor, Journal of Soil and Water Conservation as well as editor of International Journal of Forest, Soil and Erosion, Iran and associate editor, International Journal of Microbiology Research.

Dr D.S. Srivastava is presently working as scientist (plant protection) at KVK-II, Sitapur, U.P and Ph.D. in the field of nematology. He has more than 15 years research and extension experience in the field of agriculture. He has developed new concept of IPM ambassadors for conservation of natural enemy and ecosystem and promotion of IPM practices and development of organic farming and demonstrated ITKs of rodent, termite, sucking pest, blue bull and wild boar management in more than 100 villages of Sitapur district, Uttar Pradesh. He has introduced Lac culture technology with established IFS models first time in Sitapur district of Uttar Pradesh. He has prepared blue prints for on-line open access journals of I.C.A.R at DKMA portal at I.C.A.R, New Delhi. He has completed 4 projects as co-investigator of reputed agencies *viz* DST, ICAR and NABARD. He has published 32 research papers and 13 book chapters and guided 3 M.Sc students as major guide. He has been awarded with Young Scientist Award 2007, Distinguished Service Award 2009, Young Scientist Associate Award 2016, Best Agriculture Technology Promotion Award 2016, Best Agri-communication award 2016 and Global IPR Award 2017 at various national and international forums.

Dr Hemlata Pant (FISEP, FBPS, FSFLS, FLSc, FATDS, FSPPS, FESW) graduated from the University of Allahabad. She is the assistant professor in Department of Zoology, CMP PG College (University of Allahabad), Prayagraj, U.P. She has more than 63 peer reviewed national and international publications and has delivered numerous oral and posters presentations in numerous national and international meetings. She has published 120 popular articles and 20 book chapters. She has edited/author of 15 important books of emerging issues. She is editor of journal of Natural resource and development and Gramin vikas sandesh (Prayagraj). Dr. Pant is co-editor of journal of fisheries sciences (Manglore). She is in editorial board of Annals of plant protection sciences (New Delhi) and in advisory board member of Al Da Beatz magazine, Prayagraj. She was awarded of Dr. Jagdish Chandra Bose Hindi Granth Lekhan Award from DBT, Govt. of India, New Delhi. She got 35 other awards in the field of environmental sciences, popularization of science, hindi writing and in different emerging issues.

Advances in Crop Production and Climate Change

A.S. Yadav
Scientific Officer (Agronomy)
U.P. Council of Agricultural Research
8th, Floor Kisan Mandi Bhawan, Vibhuti Khand
Gomti Nagar, Lucknow-226 010, Uttar Pradesh

Narendra Kumar
Principal Scientist (Agronomy)
Crop Production Division
ICAR-Indian Institute of Pulses Research
Kanpur-208 024, Uttar Pradesh

Sanjay Arora
Principal Scientist (Soil Science)
ICAR-Central Soil Salinity Research Institute, Regional Research Station
Post Office Dilkusha-226 002, Lucknow, Uttar Pradesh

D.S. Srivastava
Scientist (Plant Protection)
Krishi Vigyan Kendra-II, Village Katia
Vill. Katia, P.O. Manpur-261 145, Sitapur, Uttar Pradesh

Hemlata Pant
Assistant Professor
Department of Zoology, CMP PG College
University of Allahabad
Mahatma Gandhi Marg, George Town, Prayagraj
Allahabad -211 001, Uttar Pradesh

CRC Press
Taylor & Francis Group
Boca Raton London New York

CRC Press is an imprint of the
Taylor & Francis Group, an **informa** business

NEW INDIA PUBLISHING AGENCY
New Delhi-110 034

First published 2021
by CRC Press
2 Park Square, Milton Park, Abingdon, Oxon, OX14 4RN

and by CRC Press
6000 Broken Sound Parkway NW, Suite 300, Boca Raton, FL 33487-2742

CRC Press is an imprint of Informa UK Limited

British Library Cataloguing-in-Publication Data
A catalogue record for this book is available from the British Library

Library of Congress Cataloging-in-Publication Data
A catalog record has been requested

ISBN: 978-1-032-02429-5 (hbk)

त्रिलोचन महापात्र, पीएच.डी.
सचिव, एवं महानिदेशक
TRILOCHAN MOHAPATRA, Ph.D.
SECRETARY & DIRECTOR GENERAL

भारत सरकार
कृषि अनुसंधान और शिक्षा विभाग एवं
भारतीय कृषि अनुसंधान परिषद
कृषि एवं किसान कल्याण मंत्रालय, कृषि भवन, नई दिल्ली 110 001
GOVERNMENT OF INDIA
DEPARTMENT OF AGRICULTURAL RESEARCH & EDUCATION
AND
INDIAN COUNCIL OF AGRICULTURAL RESEARCH
MINISTRY OF AGRICULTURE AND FARMERS WELFARE
KRISHI BHAVAN, NEW DELHI 110 001
Tel.: 23382629; 23386711 Fax: 91-11-23384773
E-mail: dg.icar@nic.in

Foreword

Agricultural development is one of the most powerful tools to end extreme poverty, boost shared prosperity and feed a projected 9.7 billion people by 2050. Growth in the agriculture sector is 2-4 times more effective in raising incomes among the poorest compared to other sectors. Climate change is already impacting crop yields, especially in the world's most food-insecure regions. In present era, shocks related to climate change, conflict, dwindling resources, emerging pests and diseases are hampering food production, disrupting supply chains and stressing people's ability to access nutritious and affordable food. Climate change mitigation in the agriculture sector is also need of the hour.

Agriculture is highly dependent on climatic, soil, water, seed and diverse range of variables for varied crops and cultivation. Unforeseen impacts of Climate Change on natural resources as well as agriculture and environment resulted in substantial risks and compel for advancement of techniques involving blend of scientific and traditional approaches to mitigate adversities. Adoption of advanced and innovative crop production techniques including quality seed to proper management of resources and mechanized operations needs widespread dissemination for achieving livelihood and food security.

Towards this goal, the authors have brought out an publication on "*Advances in Crop Production and Climate Change*". I hope the book would be an invaluable resource helping researchers, policymakers, academician, students, as well as farmers to develop strategic action plan to combat the effect of climate change.

I compliment the editors and contributors for collating the advanced technologies in the form of this useful publication and hope that it will serve as technology treasure useful for the stakeholders.

10th August, 2020, New Delhi

(T. MOHAPATRA)

Preface

India is the world's 7th largest country (by area) with a human population of about 1.387 billion as on 1st January, 2020 and is expected to touch 1.705 billion by 2050. It is characterized by an immense diversity in climate, topography, flora, fauna, land use and socio-economic conditions. Agriculture, as the backbone of Indian economy, plays the most crucial role in the socio-economic sphere of the country. Indian agriculture is a diverse and extensive sector involving a large number of stakeholders. As per 2018, agriculture employed more than 50% of the Indian work force and contributed 17–18% to country's GDP. During last 70 years, Indian agriculture has experienced remarkable change in terms of production as well as productivity. It has been one of the remarkable success stories of the post independence era through the association of Green Revolution technologies. Although one side the Green Revolution driven technologies like HYVs, fertilizers and pesticides contributed to the Indian economy by providing food self-sufficiency in the country, while the other side it has led to the serious issue of natural resources degradation. The declining factor productivity and plateauing yield of major crops since last 3-4 decades has threatened the sustainability of agricultural production system. Therefore, a paradigm shift is required for enhancing the system's productivity and sustainability.

This book has comprehensive coverage and advances in agriculture for sustainable development and is expected to provide a valuable sources book for scholars and researchers, as well as guide book to famer's community and development agencies. Contents in the book are organized in 20 chapters, which includes advances in production technologies of crops e.g. rice, wheat, barley, maize, pearl millet, pulses and oilseeds; sugarcane; medicinal and aromatic plants; vegetable crops; fodder crops; resource conservation technologies; management of degraded and sodic lands; soil biodiversity; farm mechanization etc. The text is adequately illustrated with tables, figures and photographs to bring out the significant findings. The book provides cutting edge scientific knowledge as well as solid background information that are accessible for those who have a strong interest in agricultural research and development and want to have insight on the challenges faced by global agricultural production systems. Editors are grateful to all the authors for their valuable contributions. The valuable contributions of scientists and researchers across the country involved in developing the crop production technologies is also gratefully acknowledged. It is hoped that the students, teachers, researchers, development managers, extension workers and policy makers of agricultural and allied disciplines find this publication informative and useful.

A.S. Yadav
Narendra Kumar
Sanjay Arora
D.S. Srivastava
Hemlata Pant

Contents

List of Contributors

Chapter 1: Advances in Rice Production Technologies

R.D. Jat, Assistant Professor (Agronomy), Department of Agronomy, Ch. Charan Singh Haryana Agricultural University, Hisar-125 004, Haryana

S.K. Kakraliya, Senior Research Fellow, ICAR-Central Soil Salinity Research Institute, Zarifa Farm, Kachhwa Road, Karnal-132 001, Haryana

K.K. Choudhary, Scientist (Soil Science), Regional Research Station, Pali Marwar, Rajasthan

P. Kapoor, Assistant Professor (Agronomy), Department of Agronomy, Ch. Charan Singh Haryana Agricultural University, Hisar-125 004, Haryana

Sardar Singh Kakraliya, Ph.D. Scholar, Department of Plant Pathology, College of Agriculture, Sher-e-Kashmir University of Agricultural Sciences and Technology, Jammu-180009, Jammu & Kashmir

Hardev Ram, Scientist (Agronomy), ICAR-National Dairy Research Institute, Karnal-132 001, Haryana

Chapter 2: Advances in Wheat and Barley Production Technologies

S. K. Singh, Principal Scientist (Plant Breeding), ICAR-Indian Institute of Wheat and Barley Research, Karnal-132 001, Haryana

Satish Kumar, Senior Scientist (Plant Breeding), ICAR-Indian Institute of Wheat and Barley Research, Karnal-132 001, Haryana

Chapter 3: Advances in Maize Production Technologies

Hardev Ram, Scientist (Agronomy), ICAR-National Dairy Research Institute, Karnal-132 001, Haryana

Thomas Abraham, Professor, Department of Agronomy, Sam Higginbottom University of Agriculture, Technology and Sciences, Prayagraj-211 007, Uttar Pradesh

Shailesh Marker, Director Research, Sam Higginbottom University of Agriculture, Technology and Sciences, Prayagraj-211 007, Uttar Pradesh

Surgyan Rundla, Research Scholar, Department of Agronomy, Ch. Charan Singh Haryana Agricultural University, Hisar-125 004, Haryana

Chapter 4: Improved Technologies for Pearl Millet Cultivation

Rajesh C. Jeeterwal, Senior Research Fellow ICAR-AICRP on Pearl Millet, Agricultural Research Station, Mandor, Jodhpur-342 304, Rajasthan

Anju Nehra, Research Scholar, Sri Karan Narendra University of Agriculture, Jobner-303 329, Jaipur, Rajasthan

Rupa Ram Jakhar, Senior Research Fellow, Swami Keshwanand Rajasthan Agricultural University, Bikaner-334 006, Rajasthan

Chapter 5: Advances in Pulses Production Technologies: A Holistic Approach for New Millennium

Narendra Kumar, Principal Scientist (Agronomy), Crop production Division, ICAR-Indian Institute of Pulses Research, Kanpur-208 024, Uttar Pradesh

Chapter 6: Advances in Oilseeds Production Technologies

Kartikeya Srivastava, Professor, Department of Genetics and Plant Breeding, Institute of Agricultural Sciences, Banaras Hindu University, Varanasi-221 005, Uttar Pradesh

Ayushi Srivastava, Research Scholar, Department of Genetics and Plant Breeding, Institute of Agricultural Sciences, Banaras Hindu University, Varanasi-221 005, Uttar Pradesh

Akanksha, Research Scholar, Department of Genetics and Plant Breeding, Institute of Agricultural Sciences, Banaras Hindu University, Varanasi-221 005, Uttar Pradesh

Chapter 7: Advance Production Technologies of Sugarcane: A Step Towards Higher Productivity

A.K. Mall, Senior Scientist, Crop Improvement Division, ICAR-Indian Institute of Sugarcane Research, Raibareli Road, Post Office Dilkusha-226 002, Lucknow, Uttar Pradesh

Varucha Misra, Research Associate, Crop Improvement Division, ICAR-Indian Institute of Sugarcane Research, Raibareli Road, Post Office Dilkusha-226 002, Lucknow, Uttar Pradesh

A.D. Pathak, Director, ICAR-Indian Institute of Sugarcane Research, Raibareli Road, Post Office Dilkusha-226 002, Lucknow, Uttar Pradesh

B.D. Singh, Technical Officer, Crop Improvement Division, ICAR-Indian Institute of Sugarcane Research, Raibareli Road, Post Office Dilkusha-226 002, Lucknow, Uttar Pradesh

Rajan Bhatt, Scientist (Soil Science), PAU Regional Research Station, Kapurthala-144601, Punjab

Chapter 8: Advances in Vegetable Production Technologies

Hari Har Ram, Former-Professor and Head, Vegetable Science and Nodal Officer, Centre for Plant Genetic Resources, Govind Ballabh Pant University of Agriculture and Technology, Pantnagar, Udham Singh Nagar-263 153, Uttarakhand

Chapter 9: Advances in Medicinal and Aromatic Crop Production Technologies

Neha Singh, Senior Research Fellow, CSIR-National Botanical Research Institute, Rana Pratap Marg, Lucknow-226 001, Uttar Pradesh\

Hemant Kumar Yadav, Senior Scientist, CSIR-National Botanical Research Institute, Rana Pratap Marg, Lucknow-226 001, Uttar Pradesh

Sujit Kumar Yadav, Scientific Officer (Plant Breeding), UP Council of Agricultural Research, 8th Floor, Kisan Mandi Bhawan, Vibhutikhand, Gomtinagar-226 010, Lucknow, Uttar Pradesh

Chapter 10: Advances in Forage Crop Production Technologies

D. Vijay, Principal Scientist, ICAR- Division of Seed Science and Technology, ICAR-Indian Agricultural Research Institute (IARI), Pusa Campus-110 012, New Delhi

N. Manjunatha, Scientist, ICAR-Indian Grassland and Fodder Research Institute, Near Pahuj Dam, Gwalior Road, Jhansi-2840 03, Uttar Pradesh

Sanjay Kumar, Scientist, ICAR-Indian Grassland and Fodder Research Institute, Near Pahuj, Dam, Gwalior Road, Jhansi- 284 003, Uttar Pradesh

Chapter 11: Restoration of Degraded Sodic Lands Through Agroforestry Practices

Y.P. Singh, Principal Scientist (Agronomy), ICAR-Central Soil Salinity Research Institute, Regional Research Station, Post Office Dilkusha-226 002, Lucknow, Uttar Pradesh

Chapter 12: Advances in Farm Mechanization in India

Sanjay K. Patel, Associate Professor (Farm Machinery & Power), College of Agricultural Engineering, Dr. Rajendra Prasad Central Agriculture University, Pusa-848 125, Samastipur, Bihar

B.K. Yaduvanshi, Assistant Professor, Polytechnic in Agricultural Engineering, Anand Agricultural University, Muvaliya Farm, Dahod-389 151, Anand, Gujarat

Prem K. Sundaram, Scientist (SS), Division of Land and Water Management, ICAR-Research Complex for Eastern Region, ICAR Parisar, Post Office Bihar Veterinary College, Patna-800 014, Bihar

Chapter 13: Resource Conservation Techniques for Sustaining Crop Production in Rainfed Foothills Under Changing Climate

Sanjay Arora, Principal Scientist (Soil Science), ICAR-Central Soil Salinity Research Institute, Regional Research Station, Post Office Dilkusha-226 002, Lucknow, Uttar Pradesh

Rajan Bhatt, Scientist (Soil Science), PAU Regional Research Station, Kapurthala-144601, Punjab

Chapter 14: Advances in Reclamation and Management of Salt Affected Soils for Sustainable Crop Production

Sanjay Arora, Principal Scientist (Soil Science), ICAR-Central Soil Salinity Research Institute, Regional Research Station, Post Office Dilkusha-226 002, Lucknow, Uttar Pradesh

Y.P. Singh, Principal Scientist (Agronomy), ICAR-Central Soil Salinity Research Institute, Regional Research Station, Post Office Dilkusha-226 002, Lucknow, Uttar Pradesh

Atul K. Singh, Principal Scientist (SWCE), ICAR-Central Soil Salinity Research Institute, Regional Research Station, Post Office Dilkusha-226 002, Lucknow, Uttar Pradesh

Chapter 15: Physio-molecular Mechanisms of Drought Tolerance in Crop

Shambhoo Prasad, Associate Professor, Department of Plant Molecular Biology and Genetic Engineering, Acharya Narendra Dev University of Agricultural & Technology, Kumarganj-224 229, Ayodhya (Faizabad), Uttar Pradesh

Chapter 16: Soil Biodiversity and Its Management for Sustainable Agriculture

Sanjay Arora, Principal Scientist (Soil Science), ICAR-Central Soil Salinity Research Institute, Regional Research Station, Post Office Dilkusha-226 002, Lucknow, Uttar Pradesh

Chapter 17: Impact of Climate Change in Crop Protection

Mukesh Sehgal, Principal Scientist, ICAR-National Centre for Integrated Pest Management, L.B.S. Centre, Pusa Campus-110 012, New Delhi

D.S. Srivastava, Scientist (Plant Protection), Krishi Vigyan Kendra-II, Village Katia, Post Office Manpur-261 145, Sitapur, Uttar Pradesh

H. Ravindra, Assistant Director Research, Department of Plant Pathology, Z.A.R.S., University of Agriculture and Horticulture Sciences, Navile- 577 204, Shivamogga, Karnataka

Chapter 18: Analysis of Field Experimental Data Using Statistical Calculator

D.S. Dhakre, Assistant Professor, Institute of Agriculture, Visva-Bharati, Sriniketan-731 236, Birbhum, West Bengal

D. Bhattacharya, Assistant Professor, Institute of Agriculture, Visva-Bharati Sriniketan-731 236, Birbhum, West Bengal

1

Advances in Rice Production Technologies

R.D. Jat, S.K. Kakraliya, K.K. Choudhary, P. Kapoor, Sardar Singh Kakraliya and Hardev Ram*

Introduction

Rice is the staple food for over half the world's population. China and India alone account for ~50% of the rice grown and consumed. Rice provides up to 50% of the dietary caloric supply for millions living in poverty in Asia and is, therefore, critical for food security. Globally, share of Asia continent in rice production is more than 90% followed by America (5.2%), Africa (3.4%) and Europe (0.6%). It is becoming an important food staple in both Latin America and Africa. Record increases in rice production have been observed since the start of the 'Green Revolution'. However, rice remains one of the most protected food commodities in world trade. Rice is a poor source of vitamins and minerals, and losses occur during the milling process. Populations that subsist on rice are at high risk of vitamin and mineral deficiency.

The Indian population of 1.32 billion is projected to reach 1.53 billion by 2030 A.D. Goyal and Singh (2002) estimated a demand of 146 million tonnes of rice (taken as 50% of total cereals) by the year 2030. On the other end of the scale, 3rd projection puts rice demand to be 156 million tonnes by 2030 (ICAR, 2010). To meet this, it is immensely necessary to increase the productivity levels without adversely affecting the natural resource base. Achievement of the targeted production would be an uphill task in the coming decades with the shrinking natural resource base, deteriorating soil health and soil productivity, declining input use efficiency, plateauing of yields in irrigated ecologies and lack of a major yield breakthrough in rainfed ecologies. Further, slowdown in the growth rate of cereal production and growing population pressure have emerged as formidable challenges for the future food and nutritional security in India. These challenges will be more extreme under emerging situations of

**Corresponding author email: kakraliyask@gmail.com*

natural resource degradation, high energy demands, volatile markets and risks associated with global climate change (Jat *et al*., 2016; Lal, 2016). Since, rice is the staple food for our country's food security primarily depends on this wonderful crop. Globally rice is cultivated now in 167 million ha with annual production of around 769 million tonnes (499.3 million tonnes, milled rice) of rough rice and average productivity of 4.6 tonnes/ha (3.62 t/ha, milled rice) of rough rice (FAO STAT, 2017). In Asian countries, >90% of the rice is produced and consumed. The other continents in which rice is grown are Africa (7.78% of the global area), South America (6.4%) and North America (1.4%). In India during the period 2016-17, rice was cultivated in an area of 43.4 million ha with a production of 115.6 million tonnes of paddy (nearly 44% of the total food grain production) with an average productivity of 2.65 t/ha, milled rice or 3.52 t/ha rough rice (Anonymous, 2018). Although, rice production growth trend had kept in pace with population growth rate during last five decades, signs of decreasing growth rate are evident. This has been a cause of concern. In green revolution period the semi-dwarf, more irrigation demand, fertilizer responsive, high yielding genotypes of rice and wheat were implemented, which led to remarkable increase in the production and productivity of these crops (Kakraliya *et al.,* 2018). It seems that the technology implemented during the green revolution period has touched a phase of diminishing returns. Hence, it is very pertinent to critically consider whether the rice production can be further increased to keep pace with population growth with the current green revolution technologies. It is projected that by 2030 at least 140-150 million tonnes of milled rice is to be produced to maintain the present level of self-sufficiency of country's peoples. Is there urgent need for a paradigm shift in rice research to meet the challenges of the future decades for ensuring food security? Do we need to adopt the gene revolution technologies? The possibilities and prospects of utilizing the new technologies for enhancing rice productivity in order to achieve food and nutritional security are examined in this chapter.

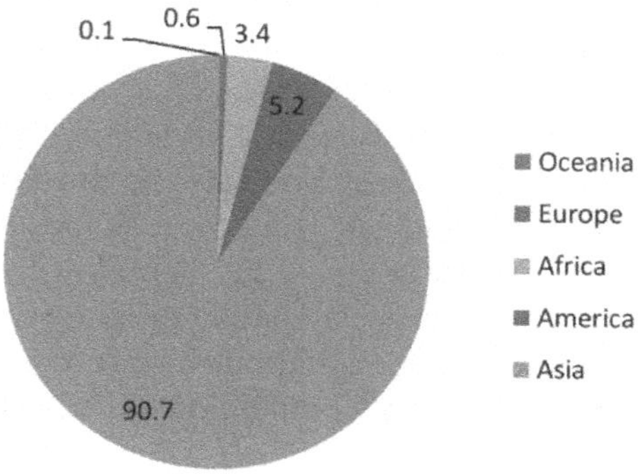

Fig. 1: Global share of rice production (FAO STAT, 2020)

Major emerging issues of paddy production system

Despite these practices, rice is suffering from a number of ecological, production and social issues are listed in table 1.

Table 1. Major issues of rice production

Production issues	Ecological and social issues
• Climatic issues	• Environment pollution
• Decreased land and water productivity	• Falling ground water
• Declining soil health	• Increasing ground water pollution
• Big management yield gaps	• New diverse weed flora
• Crop residue management	• Emerging new insect-pests and diseases
• Labour shortage	• Shrinking biodiversity
• Lessening crop response	• Population expansion
• Land holding- getting smaller and fragmented	• Farmers mindset with traditional practices
• High energy requirement	• Youth moving away from farming
• Poor incomes of farmers	• Less mechanization
	• Lack of appropriate seeders
	• Lack of adoption of new technology

Production issues

The declining soil health at an alarming rate results in deficiencies of various macro and micro nutrients in one and another part of the country (Kakraliya *et al.,* 2018). Intensive tillage practices break the larger soil aggregates further coupled with poor contact with seed, there by reduces potential yields. Therefore, puddling in rice is a water, labour and energy consuming process which deteriorates the soil physical structure (Gathala *et al.*, 2016). Traditional rice-wheat (R-W) system is a labour intensive system, therefore, the labour shortage is an emerging issue in the prevailing R-W system due to narrow window period and legal binding to transplant paddy after 10th of June. This labour shortage is more prevails from last few years due to assured working days offered under MANREGA scheme of govt. of India and thus flow of labour to region is decreased to a remarkable extent (Bhatt *et al.*, 2016). Lower labour availability is responsible for higher wage rates. In R-W system, the on-farm residue management is a very serious issue. Due to higher silica content, rice straw is not used for animal consumption but the wheat residue is used for animal consumption. Similarly, the rice residue enhance N immobilization when it incorporate in the soil due to its wider C:N ratio which in turn severely decreases crop yield (Singh *et al.,* 2010). Presently, rice residue burning is serious issues which causes environmental pollution, global warming, killing the beneficial insects, create multiple nutrient deficiencies and also degrade the soil, reduces organic matter levels and finally results in the soil health deterioration (Lohan *et al.*, 2018). Before green revolution, only the major plant nutrients was provided through the use of inorganic fertilizers, although with time micro-nutrients also being provided through chemically formed fertilizers due to deficiency of micro-nutrient availability in several regions under R-W cropping system (Chauhan *et al.,* 2012). In Punjab, wheat grown after rice suffered from manganese deficiency in coarse textured soils. In R-W system, decline of wheat yields due to boron deficiency was found in some regions soils of West Bengal. Selenium toxicity is also an emerging issue coming out of intensive R-W system in the region, however, under maize-wheat cropping rotation, no selenium toxicity was observed (Mahajan *et al.*, 2012; Bhatt *et al.*, 2016). Currently, nitrogen fertilizers gives low use efficiency in rice where it is only 30-40% of applied N due to surface runoff, ammonia volatilisation, leaching and denitrification (Fageria and Baligar, 2005).

Under traditional rice cultivation, the decline of underground water levels associated with replacement in centrifugal pumps by submersible pumps which lifts up water from the deeper depths are serious concern in rice production system. In some states of India, the free electricity provided by the state government to the farmers in agricultural sector further compounded this problem as farmers are now having no interest in saving irrigation water. With an effect the water levels below the ground declining at an alarming rate. As per the recent condition, by 2023, in Punjab state (India), the electricity cost for pumping ground water from a depth of 43 metre would increase by 93% compared with that for 2006 (Humphreys *et al.,* 2010). In the time of ever increasing population and hungry stomachs, decreasing land productivity might be quite dangerous. Approximately 2.5% growth in cereal production will be required to provide sufficient food for the future population (Hobbs and Morris, 1996). Since three decades, production goal has been able to keep pace with population growth and this is only possible through an increase in the area and yield growth. Amplified demand by urban areas and industry creates a competition for land allocation to different sectors including the agriculture. Some studies reported stagnant yields in R-W system because of different factors and their compounded effects (Dawe, 2000; Duxbury *et al.,* 2000). Continuous decreasing water productivity of rice is a major cause of concern as reported by many workers throughout the region under different agro-climatic conditions (Humphreys *et al.,* 2010; Bhatt, 2013). As per the above described issues, there is an urgent need to improve the declining rice water productivity. Efficiency of water use is different from the productivity in terms that the former deals with total water discharged from the tube well up to the field and include the conveyance losses while the latter is the grain yield obtained from a particular volume of applied water to a particular field and it does not include the conveyance losses. It was reported that around 35 to

55% of applied water for the main field was lost before reaching the main field. Besides, degradation of the soil structure, formation of hard pan and declining underground water table along with outbreak of in various biotic and abiotic stresses usually result in lowered land productivity. Lower land productivity means lower grain yield produced per piece of land. Land holdings are already shrinking with generation after generations. Therefore, by keeping in mind these facts, the issue of rice sustainability is of great concern throughout the rice growing areas.

Ecological and social issues

In India, annually, more than 501 million tonnes crop residues are produced as about 90% the acreage under R-W cultivation is harvested by combine, which left back loose straw in wind rows. The burning of rice loose residues is mostly adapted by Indian farmers owing to its easiness and quickness in disposing which cases considerable, air pollution (Gupta *et al.*, 2003), loss of soil nutrients, death in beneficial soil microbes in surface few centimetre soil, and loss in soil physical and biological health (Singh and Sidhu, 2014). In each year, cereals contribute 70% of the total crop residues (352 million tonnes) comprising 34% by rice and 22% by wheat crops. About 25-30% of N and P, 35-40% of S, and 70-75% of K uptake are retained in wheat residue. In general, the conventional method of rice production needs about 1500 mm of water. This huge water demands by rice cultivation greatly depleted the ground water level since last four decades. In Punjab alone, every year water shortage of 1.2 million ha meter has been reported (Hira *et al.*, 1998; Jain and Kumar, 2007). The declining ground water table is become an alarming task which needs an immediate attention of scientists as well as the policy makers. Similarly, over use of agro-chemicals including synthetic fertilizers in R-W system deteriorate the ground water quality owing to runoff and leaching of chemical residues and thus dissolving in ground water (Bhatt, 2013). Consequently, the use of this contaminated poor quality ground water for irrigation purpose invites several diseases of crops and livestock which ultimately affect the human health and livelihood. This indiscriminate and unbalanced nitrogenous fertilizer application makes groundwater nitrate rich which a serious concern to human health (Bhatt *et al.*, 2016). The rise of diverse weed flora also became a matter of concern in R-W system threatening sustainability of agricultural system. The intensive R-W system brings the emergences of more grassy weeds in the field which compete with crop plants for light, water, space, and nutrients which in turn decreases the overall productivity of the system. This cause's serious yield reduction as it is major biotic constraints in sustainable agricultural production (Chauhan *et al.*, 2012). Both the crops in the system are grown in lavish environment and so both these need higher amount of fertilizers and irrigation water which increases the insect-pests outbreak. Outbreak of the diseases and insect-pests attack is mainly responsible for the lower water and land productivity and is considered a serious issue in the way to sustainable agriculture.

Production technologies of rice

Climatic requirements

Rice is best performed under high temperature, high humidity, prolonged sunshine and assured water-supply. A temperature range of 20 to 37.5 °C is required for its optimum growth. The crop requires a higher temperature at tillering than that during early growth. The temperature requirement for blossoming ranges between 26.5 and 29.5 °C.

Soil type

Generally, loam to clay loam soil, which becomes soft on wetting but develops cracks on drying, is suitable for rice crop. Rice can grow well on soils with low permeability and over a wide range of soil reaction *viz*. pH 5 to 9.

Application of organic manure

After harvesting wheat or any other preceding crops, apply 15-20 tonnes of farmyard manure (FYM) per hectare and save 40-45 kg N/ha. Since, FYM is not available in needed quantities, green manuring by dhaincha/cowpea/ moong bean is a very practicable alternative and sow 40-50 kg dhaincha seed pre-soaked in water for 8 hours or 25-30 kg/ha of cowpea or 20 kg/ha of moong bean up to the first week of May. Flowering stage or 45 to 55 days old dhaincha/cowpea/moong bean should be buried one day before transplanting of paddy (Fig. 2). Generally, dhaincha should be preferred in alkali (also known as *kallar*) and recently reclaimed soils. This practice results in a saving of 60 kg of N/ha. Dhaincha also ameliorates iron deficiency in rice. Apply 10 tonnes of pressmud or 6 tonnes of poultry manure per ha to rice and reduce the N fertilizer dose by half and omit the application of P_2O_5 fertilizer to rice even on soils testing low in available P.

Fig. 2. Green manuring in rice field

Field preparation

Use laser land leveller for precision land levelling before puddling to enhance water use efficiency. Rice field is prepared primarily by ploughing with soil turning plough, followed by harrowing. Before transplanting, the main field is filled with water and is puddled twice by paddy puddler or once by rotavator to reduce water loss through percolation, to maintain good seedling vigour and to control weeds. Puddle the field by tractor drawn rotavator if green manure crop like dhaincha or moong bean has been taken so that it can be incorporated well in soil. Puddling should be followed by planking.

Seed selection and treatment

Dip the seed into a 10% salt solution in a tub/bucket. Stir the seed and remove immature grains which float at the top. For this purpose, dissolve 1 kg salt in 10 litres of water for 10 kg of seed. Collect heavy seeds settled at bottom of the solution and wash them 2-3 times by fresh water before treating them. Dip the heavy seed in 10 litres of water (for 10 kg seed) containing 20 g Bavistin 50 WP (carbendazim) or MEMC (Emisan 6 Hg) and 1g Streptocycline (streptomycin + tetracycline) for 8 to 10 hours before sowing. Take out the treated seed from the solution and cover it with moist gunny bags and allow it to sprout by sprinkling water frequently on the gunny bags.

Seed rate

Kilogram per ha of the area to be transplanted:

- High yielding (Non-scented medium, mid early and early) varieties: 25-30 kg/ha
- Hybrids varieties: 15-20 kg/ha
- Scented varieties: 20 kg/ha

Nursery raising

For getting healthy seedlings, the time and method of nursery sowing are very important practices time of nursery sowing: 15th to 30th May is the optimum time of sowing for all the recommended varieties.

Mix 10-15 tonnes/ha of well-rotten FYM in the nursery area at least 25-30 days before sowing and mix it well in the soil by ploughing 2-3 times. Irrigate the field to permit the germination of weeds followed by ploughing the field twice after about a week to kill germinated weeds. Thereafter, add water to the nursery field and puddle it 1-2 times by paddy puddler or rotavator followed by planking. For raising healthy seedlings broadcast N, P_2O_5 and zinc sulphate each at 25 kg/ha of nursery area before planking. Broadcast the sprouted seeds uniformly at 1 kg in 20 m^2 of the nursery area after at least 4-5 hours of planking (after the silt has settled). Apply Pretilachlor 30 EC+safener @ 1500 ml/ha of nursery area by mixing it with sand at 1-3 days after sowing. Keep the soil moist by irrigating the plot frequently. To check the damage from birds, broadcast a thin layer of well decomposed FYM immediately after broadcasting rice seed. Apply another dose of 26 kg urea per acre about a fortnight after sowing so as to get the seedlings ready for transplanting in 25-30 days. If the seedlings in the nursery show the yellowing of new leaves, spray 0.5 per cent ferrous sulphate solution at weekly intervals.

Preparation of mat-type nursery for paddy transplantation

Select the mat type paddy nursery location having fertile soil preferably of medium type. Irrigate the field to permit the germination of weeds followed by tilling the field with rotavator/tiller and one to two planking to destroy the germinated weeds and to pulverize the soil. Make two raised beds (10-15 cm high) of size 1.0 × 1.2 m and make channels around them for irrigation and drainage. Cover the beds with a polythene sheet (100 μ) made perforated by making holes in the sheet at 15-22.5 cm distance with the help of sharp needle so that the nursery remains in contact with the soil. Prepare a 4-5 quintals mixture of field soil (sieved) and manure (well rotten FYM/vermi-compost/ pressmud) in 4:1 ratio. Spread this mixture in 1.25-1.87 cm thick layer uniformly on the polythene sheet. 20-25 kg seed is sufficient to sow about 500 mats required for transplanting one ha area. Seeds should be cover by a thin layer of soil and apply fine sprinkle water by hand sprayer for proper setting of the soil and maintain optimum moisture for the next 4-5 days. After that apply water in the surrounding channels so that it reaches up to upper surface of the beds. In this way, maintain optimum moisture up to 15 days or until the nursery is ready. If zinc deficiency appears, spray the nursery with a solution of 0.5% zinc sulphate. In case of iron deficiency, spray 0.5% ferrous sulphate. Drain the water from the nursery field a few hours before nursery uprooting after that cutting of the nursery into cakes of 60 × 20 cm size with a sharp knife or sickle. Put the nursery cakes in plastic tray or basket to carry them to the transplanting site.

Transplanting

Time of transplanting: Time of transplanting is the most important factor influencing the yield of crop. Performance of the variety depends on the planting time. According to duration of varieties the planting time are as under.

- Medium and mid early duration varieties and hybrids (non-scented): 15 June to 7 July
- Early varieties (non-scented): 15 June to 30 July
- Scented (Basmati) varieties: 1 July to 15 July

Method of transplanting: Following methods of rice transplanting are practised in India.

Manual transplanting: Start uprooting the nursery when the seedlings become 25 to 30 days old and wash them with water to eliminate mud from their roots. Transplant 2-3 seedlings at a place (hill) lines at 20 × 15 cm (33 hills/m^2) for normal and 15 × 15 cm (44 hills/m^2) for the late transplanting. Put two seedlings per hill. The seedlings should be transplanted upright and about 2-3 cm deep. This practice ensures good establishment of seedlings and early tillering, which are essential for good tiller development and synchronous flowering.

In salt affected soils, transplant the crop one week earlier using old age seedlings (35- 40 days old) with 3-4 seedlings per hill.

Bed transplanting: Transplant 25-30 days old seedlings on the middle of the slopes of beds prepared with wheat bed planter on heavy textured soils. After field preparation (without puddling), apply basal dose of fertilizer and make beds. Further, irrigate the furrows and immediately transplant seedlings by maintaining a plant to plant distance of 9 cm to ensure 33 seedlings/m^2. During the first week after transplanting irrigation water should be allowed to pass over the beds once in a day. Thereafter apply irrigation in furrows only 2-3 days after the ponded water has infiltrated into the soil. All care should be taken so that field does not develop cracks in the furrows. With the above method about 25% of total applied irrigation water can be saved without affecting the grain yield. For controlling broad and narrow leaf weeds spray bispyribac sodium 10 SC at 250 ml/ha in 300-400 litres of water as post emergence at 20-25 days after transplanting.

Mechanical transplanting: Mat type rice nursery is required for the mechanical transplanting. By self-propelled paddy transplanter can be used for paddy transplanting both in zero till and unpuddled fields. Before transplanting in unpuddled fields, plough up the field once or twice followed by planking in dry or vattar condition. Later apply light irrigation and remove excess water to allow the soil to settle for 12-24 hours and transplant the nursery using mechanical transplanter in flooded field (2-3 cm water).

Before flooding the field should be levelled with a laser land leveller to save water, equal distribution of water and enhance nutrient use efficiency. For transplanting under zero tillage conditions, apply 1.0% glyphosate 41 SL solution + 0.1% surfactant for control of weeds in field. Apply irrigation 7-10 days after the spray and transplant with paddy transplanter in standing water (2-3 cm) 12 hours after weedicide spray.

For transplanting, put the nursery cakes on the platform of transplanter. The line to line distance is already fixed at 23.5 cm and the plant to plant distance can be fixed at 12-14 cm with the help of the lever. The number of seedlings/ hill can be adjusted to 2-3 by rotating the screw of the fingers. The depth of transplanting can be increased or decreased with the help of lever fixed on float board of the machine. Start transplanting with machine by leaving one round around the field to be transplanted. Apply light irrigation in the transplanted field during initial five days of transplanting. The other recommended practices are the same as already mentioned under conventionally transplanted rice.

Manures and fertilizer management

To get higher yield of rice and to maintain soil health particularly when exhaustive cropping systems like rice-wheat are followed, use of FYM or green manure along with chemical fertilizers is necessary.

- Before 25-30 days of transplanting, apply FYM at 10-15 tonnes/ha and incorporate it into soil by ploughing.
- For green manuring in manual transplanting, sow dhaincha using 25-30 kg seed/ha by first week of May and incorporate it into the soil preferably

with the help of rotavator during puddling. Method of green manuring for mechanical transplanting has already been explained.

- The fertilizers should be applied as per soil test report. However, in its absence, apply fertilizers according to the following general schedule (Table 2).

Table 2. Fertilizer application rate

Varieties	N kg/ha	P_2O_5 kg/ha	K_2O kg/ha	$ZnSO_4$ kg/ha
Medium and mid-early duration (non- scented) varieties and hybrids	150	60	60	25
Early (non-scented) varieties	120	60	60	25
Scented semi dwarf varieties	90	30	-	25
Scented tall varieties	60	30	-	25

Apply full dose of phosphorus, potash and zinc at transplanting

- N should be applied in ammonical form through splits for higher nitrogen use efficiency. Therefore, apply N in 3 equal splits at 0 (transplanting), 21 and 42 days after transplanting (DAT) in semi dwarf (non-scented and scented) varieties and in 2 equal splits at 21 and 42 DAT in tall scented varieties.
- Avoid standing water in the field at the time of application of N fertilizer.
- Apply 2/3rd of NPK in non-scented varieties if FYM or green manures have been added but avoid any fertilizer in scented rice if FYM or green manure has been incorporated.
- Spray the crop with a solution of 0.5% zinc sulphate and 2.5% urea if zinc deficiency symptoms appear. For this purpose, 1 kg zinc sulphate and 5 kg urea dissolved in 200 litres of water is sufficient for spraying one acre of the crop. Repeat the spray if needed.

Advance tools and techniques for precision nutrient management

The SSNM (site specific nutrient management) provides an approach for need based feeding of rice crop with nutrients while recognizing the inherent spatial variability. It avoids indiscriminate use of fertilizers and enables the farmer to dynamically adjust the fertilizer use to fill the deficit optimally between nutrient needs of the variety and nutrient supply from natural resources, organic sources, irrigation water etc. it aims at nutrient supply at optimal rates and times to achieve high yield and efficiency of nutrient use in the rice crop. (Pampolino *et al*., 2012; Sapkota *et al*., 2014; Jat *et al*., 2014).

Nutrient expert tool: Nutrient expert® is an interactive, computer-based decision-support tool that enables small holder farmers to rapidly implementation of SSNM in their individual fields with or without soil test data. The software estimates the attainable yield in farmer's field based on the growing conditions determines the nutrient balance in the cropping system based on yield and fertilizer/manure applied in the previous crop. Combination of such information was used for expected N, P and K response in target fields to generate location-specific nutrient recommendations. The software also does a simple profit analysis comparing costs and benefits between farmers' current practices and recommended alternative practices. The algorithm for calculating fertilizer requirements was developed from on-farm research data and validated over 5 years of testing. The software is currently available without charge for wheat, rice and maize systems in South Asia (http://software.ipni.net/article/nutrient-expert).

Precision nitrogen management: Improving the match between crop nitrogen demand and the nitrogen supply from soil and/or the applied nitrogen fertilizer is likely to be the most promising strategy to increase nitrogen use efficiency (NUE). As the size of more than 50% of operational land holdings in South Asia is < 2 ha, fertilizer NUE can be improved through field-specific fertilizer N management that takes care of both spatial and temporal variability in soil N supply (Jat *et al.*, 2014). Recent advances in N management for transplanted as well as direct seeded rice consist of reduced early N application to match the relatively low demand of young rice plants and varying rates and distribution of fertilizer N within the growing season as per crop demand. Both photosynthetic rate, which is closely related to leaf N status and biomass production serve as sensitive indicators of the crop demand for N during the growing season.

Leaf colour chart (LCC) and chlorophyll meter: As rice leaf colour is a good indicator of leaf N content, the leaf colour chart (LCC) and chlorophyll meter (SPAD meter) are increasingly being used as tools for evaluating colour of the leaves (Fig. 2). These gadgets are becoming popular, easy-to-use tools for managing fertilizer N in rice in Asia.

Green seeker: Use of optical sensors which measure visible and near-infrared (NIR) spectral response from plant canopies to detect N stress is rapidly increasing (Fig. 3). The green seeker TM canopy sensor (N Tech. Industries, Inc., Ukiah, California, USA) is a tool to guide nitrogen application in rice which is based on vegetative index known as NDVI (Normalized difference vegetative index) (Singh *et al.*, 2015). The device measures the fraction of the emitted light in the sensed area that is returned to the sensor (reflectance). These fractions are used within the sensor to compute NDVI according to the following formula:

$$\text{NDVI} = \frac{\text{NIR} - \text{VIS}}{\text{NIR} + \text{VIS}}$$

Where, NIR is the fraction of emitted NIR radiation returned from the sensed area (reflectance) and VIS is the fraction of emitted visible red radiation returned from the sensed area (reflectance).

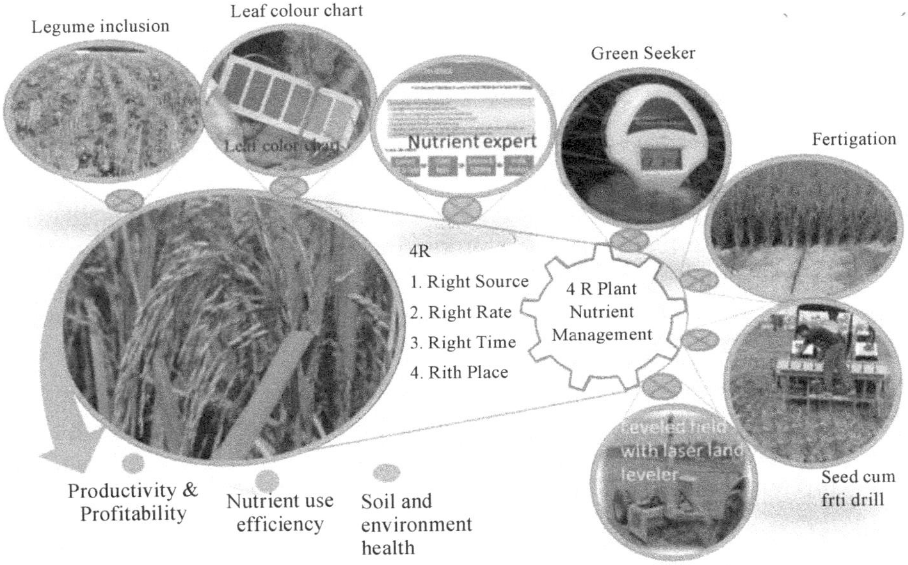

Fig. 3. Advanced tools and techniques for precision nutrient management in rice

Weed management

Rice is infested with broadleaf, grassy weeds and annual sedges which can be controlled effectively by applying following recommended herbicides. For grassy weeds: Apply Butachlor 50 EC at 3000 ml/ha or Anilophos 30 EC at 1325 ml or Pretilachlor 50 EC at 2000 ml/ha by mixing with 120 kg sand after 2-3 days of transplanting in 4-5 cm standing water and maintain standing water for 4-5 days after application.

For broad leaf weeds spray of Metsulfuron + Chlorimuron 20 WP at 20 g/ha or Ethoxysulfuron 15 WDG at 125 g/ha or 2,4-D at 1000 ml/ha in 400-500 litres of water after 20-25 days of transplanting. For mixed weed flora spray of Bispyribac sodium 10 SL at 250 ml/ha in 400-500 litres of water after 15-25 days of transplanting.

Water management

Maintain standing water (3-7 cm) in the field during vegetative growth period but thereafter, keep the soil wet by frequent irrigation. Withdraw water from the field after 6-10 days of transplanting for better establishment of the seedlings.

Drain out the water at the time of N fertilizer application and weeding/hoeing or spray of herbicides. Withdraw irrigation one week before harvesting to ensure timely harvest of paddy and sowing of next crop. In case of limited water availability, keep the soil wet by frequent irrigations throughout the season.

Plant protection

Diseases

Foot rot and bakanae (*Fusarium moniliforme*): Seed should be treated with fungicide (as described earlier). Uproot the young seedling or nursery in standing water and treated with Carbendazim (Bavistin) mix with sand and apply in nursery at the rate of 1 g/m^2 nursery area 7 days before uprooting. Remove and destroy the diseased plants from field to reduce infection

Blast (*Pyricularia grisea*): Treat the seed as described earlier. Crop should be transplanted before 15th July to minimize disease incidence. When symptoms appear, spray the crop (two times) with Tricyclazole 300 g or Carbendazim (Bavistin) 500 g in 400-500 litres of water.

Bacterial leaf blight (*Xanthomonas oryzae pv. oryzae*): Seed treatment as described earlier. Avoid excessive use of nitrogen, remove and destroy disease affected plants.

False smut (*Ustilaginoidea virens*): Apply balanced dose of fertilizers. Avoid late application of nitrogen (after 6 weeks of transplanting). Spray the crop with 1250 g Copper oxychloride per ha in 400-500 litres water at 50% panicle emergence. Spraying should be done with knapsack sprayer.

Brown spot (*Drechslera oryzae*): Treat the seed as described earlier. Spray the crop with Mancozeb @ 1500 g/ha in 400-500 litres of water twice, first at disease initiation and second after 15 days of first sprayed.

Stem rot (*Sclerotium oryzae*): Avoid continuous stagnation of water. Avoid movement of water from diseased to healthy field. Remove and destroy the stubbles and fungal sclerotia floating on water surface before transplanting. Burn the diseased stubbles after harvesting of crop.

Sheath blight (*Rhizoctonia solani*) : Keep the bunds and field free of weeds particularly doob grass (*Cynodon dactylon*) and avoid excessive use of nitrogen fertilizer. Burn the diseased stubbles after harvesting of crop. Spray the crop with Carbendazim 25% + Flusilazole12.5% SE @ 1000 ml/ha at the time of disease initiation and after 15 days of first spray.

Insect-pests

Hoppers (WBPH and BPH): Spray the crop with Monocrotophos @ 625 ml or Dichlorvos @ 312.5 ml or Carbaryl @ 1000 g in 400-600 litres of water per ha or dusting of Methyl parathion @ 25 kg/ha and broadcasting of Dichlorvos @ 625 ml diluted in 3.75 litres of water mixed in 40-50 kg sand in standing water.

Leaf folder: Spray the crop with Monocrotophos @ 500 ml or Quinalphos @ 1000 ml in 400-600 litres of water per ha or dusting with methyl parathion (2%) @ 25 kg/ha.

Stem borer: Spray the crop with Methyl parathion @ 1250 ml or Monocrotophos @ 1250 ml or Chloropyriphos @ 1 litre in 400-600 litres water per acre at 30, 50 and 70 DAT or broadcasting of Cartap hydrochloride 4G @ 18.75 kg or Fipronil 0.3 G @ 18.75 kg or Phorate @ 12.5 kg mixed with 25 kg sand per acre in standing water at 30, 50 and 70 days after transplanting.

Root weevil: Apply Carbaryl 4G @ 25 kg or Savidol 4G @ 25 kg or Carbofuran 3G @ 25 kg or Phorate 10 G @10 kg per ha.

Harvesting and threshing

Crop should be harvested whenever the panicles are mature and the plants have turned considerably yellow. It can be harvested manually by sickles or by combine harvester. The manually harvested crop, tied into small bundles, is hit against hard surface to obtain the grains from the straw, after that winnowing operation done.

Storage and marketing

After threshing and winnowing, shift the produce directly to the grain market for its sale. For safe storage, dry the produce in the sun for about one week and transfer it into the fumigated store. The moisture content should be about 12% for safe storage.

Other advanced technologies for rice production

Direct Seeded Rice (DSR)

Direct seeding of rice refers to the process of establishing a rice crop from seeds sown in the field with the define row pattern or broadcasting rather than by transplanting seedlings from the nursery. Rice can be successfully grown as direct seeded rice (DSR) under unpuddled conditions with proper management practices. It will help in improving soil health and saving of water resources along with labour and energy. Methodology of the DSR is as under.

Soil type: Medium textured soils are suitable for DSR.

Laser levelling: The field should be laser levelled for better germination and saving of water. It is first most important step to adaptation of DSR, so before direct sowing rice land should be levelled through automatic laser guided machine (laser land leveller), is well known for achieving higher level of accuracy in land levelling. Because undulate condition of land often leads to poor establishments of DSR due to uneven depth of seedling as well as uneven water distribution in irrigation. Laser land levelling technology offers a great potential for water saving, uniform crop establishment, uniform distribution of irrigation, improve nutrient use efficiency, better environmental quality and higher grain yields.

Planting time: The best time of planting *kharif* DSR is about 10-15 days before onset of the monsoon. Historical onset of monsoon indicate that monsoon arrives near 20-25 June in Uttar Pradesh, Haryana and Punjab, therefore best seedling time in this region is around 15-20 June, farmers and research experience with DSR in participatory field and research trails across Indo-Gangetic Plains (IGPs) suggests that the seedling after onset of monsoon is difficult due to high rainfall during early onset monsoon so poor crop establishments and also problem of field access for machinery in wet soil condition resulting poor germination and crop establishment.

Selection of varieties: In general, high yielding varieties should be selected according local environment/ region for higher production of DSR. Varieties can select for DSR which are famous in local region. Basmati type and hybrids cultivar have proven to be suited for DSR than others. List of suitable varieties are given in table 3.

Table 3. Suitable varieties for directed seeded rice

Basmati type	Coarse and Hybrid type
Pusa 1121, CSR-30, Pusa Basmati-1, Pusa 2511, Basmati super, Basmati 385, PB 1509	Arize-6129, PR-113, PR-114, PR-124, PRH-10, RH-664, Pusa-44, RH-2014, HKRH-1, HKRH-401, Sahyadri

Sowing machine: For precise seeding, rice should be drilled with a multi-crop planter fitted with inclined plates seed metering systems and inverted T-type tynes. For ZT-DSR, when only anchored residues are retained, then same multi-crop planter can be used for seeding. However, if loose crop residues are available on the soil surface, specialized machines are needed for drilling rice. In such situations, rice can be drilled using any of the following machines: turbo happy seeder, rotary disc drill.

Methods of sowing: Direct seeding can be done by adopting any of the following two methods.

Wet seedling: Before seedling, field should be prepared by 2-3 ploughing (2-3 times) at optimum moisture (vattar condition) followed by planking. Just after that, sowing should be done with drill at a row to row spacing of 20 cm and a depth of 3-5 cm.

Dry seedling: Before seedling, field should be prepared by ploughing (2-3 times) followed by planking. Sowing in the ploughed dry field is done with drill at a row to row spacing of 20 cm and a depth of 2-3 cm. Irrigation should be applied just after sowing.

Water seeding: Prepare the field by ploughing (2-3 times) followed by planking. Apply irrigation just after ploughing and seeds sown into standing water.

Seed rate and seed treatment: The optimum seed rate is 20-25 kg/ ha. Seed should be treated by soaking in a solution of Bavistin @ 10 g + Streptocycline @ 1 g per 10 kg seed for 24 hours (as mentioned for transplanted rice). After soaking, seed should be dried in shade for 1-2 hours to make it friable for proper seeding with seed-cum-fertilizer drill.

Weed management

Weed flora in DSR fields is quite different from that of the puddled transplanted rice due to difference in moisture status in the soil. The grasses (*Leptochloa, Eragrostis, Dactyloctenium* etc.) other than *Echinochloa crusgalli* and *E. colonum*, along with sedges like *Cyperus rotundus* may dominate the field of direct seeded rice. For broad spectrum weed control, apply the following herbicides with knapsack sprayer fitted with flat fan nozzle. Weeds are the major constraint towards the success of DSR, so weed management is a very important practices to success DSR instead transplanting rice, if anybody fail to control of weeds in DSR, yield penalty will be sure even all practices done very precisely. So, if we able to control of weeds at initial stage means we can get equal or more yield than transplanted rice.

Agronomic/cultural methods

Stale seedbed method: In this method, weeds are push to germinate by giving one irrigation and then killed by a non-selective herbicide (Paraquat or Glyphosate) or by deep tillage prior to sowing of rice. This method has great potential in suppressing weeds and is feasible under Dry-DSR because of 45-60 days window of fallow period after wheat harvest and sowing of succeeding rice crop.

Sesbania co-culture: It involves seeding rice and *Sesbania* crops (*Sesbania rostrata*) together and then killing *Sesbania* with 2,4-D ester about 25-30 DAS. In DSR field 10-12 kg *Sesbania* seed are sufficient to sowing one ha field, *Sesbania* grows rapidly and suppresses weed. This practice is found more effective in suppressing broad leaf weeds than grasses and therefore if combined with pre-emergence application of Oxadiargyl 80% WP, 100 g/ha or Pendimethalin 2.6-3.75 litre/ha, its performance in suppressing weeds increases.

Chemical method: Herbicide should be applied by flat fan nozzle to uniform weeds control in DSR field, for the effective management of weeds should be used Pendimethalin/Oxadiargyl as pre-emergence followed by post-emergence application of Bispyribac sodium or Azimsulfuron or Bispyribac sodium + Azimsulfuron or Bispyribac sodium + Pyrizosulfuron at 1-3 *fb* 20-30 DAS. Herbicides, alone or in combination, which have been found effective against different weed species are summarized in table 4.

Table 4. Recommended herbicides to manage weeds in paddy

Herbicides	Dose/ha	Time of application	Remarks
Pendimethalin 30% EC	2500-3075ml (750-1125 g ai/ha)	Pre-emergence	Control of most grasses, some broadleaves and annual sedges. But good moisture is needed for its activity.
Oxadiargyl 80% WP	125 g (100 g ai/ha)	Pre-emergence	Control of most grasses, some broadleaves and annual sedges. But good moisture is needed for its activity.
Bispyribac sodium 10% SL	250 ml (75 g ai/ha)	Post emergence 20-30 DAS	Broad-spectrum weed control of grasses, broadleaves and annual sedges. Good control of *Echinochloa* spp.
Pyrizosulfuron 10% WP	150 g (15 g ai/ha)	Post emergence	Control of broadleaves weeds
Azimosulfuron 50% WGG	62.5-75 g (31.3-37.5 g ai/ha)	Post emergence 20-30 DAS	Controls wide variety of weeds, including broadleaved and sedges like *Cyperus rotundus* and *Dactyloctenium* spp.
Ethoxysulfuron 15% WGG	132.5 g (20 g ai/ha)	Post emergence 20-25 DAS	Effective on broadleaves and annual sedges such as motha (*Cypers rotundus*).
Fenoxaprop ethyl + safner 6.7 EC	1117.5 ml (75 g ai/ha)	Post emergence 20-25 DAS	Excellent control of annual grassy weeds, safe on rice at early stage, good control of Barnyard grass. Does not control broadleaves and sedges.
Helosulfuron 75% WG	90 g (67.5 g ai/ha)	Post emergence 15-20 DAS	Effective on broadleaves and annual sedges such as motha (*Cypers rotundus*).

Cyhallofop 10 EC	1,000 ml (100 g ai/ha)	Post emergence 15-20 DAS	Good control of annual grassy weeds, safe on rice at early stage.
Metsulfuron + Chlorimuron 20% WP	20 g (2.0+2.0 g ai/ha)	Post emergence 20-25 DAS	Effective on broadleaves and annual sedges.
Bispyribac sodium + Azimsulfuron	250 ml+35 ml (25+17.5 g ai/ha)	Post emergence 20-25 DAS	Broad-spectrum weed control of grasses, broadleaves and sedges, including *C. rotundus*
Bispyribac sodium + Pyrazosulfuron	250 ml+150 g (62.5 g+15 g ai/ha)	Post emergence 20-25 DAS	Broad-spectrum weed control of grasses, broadleaves and sedges, including *C. rotundus.*

Note: More than 90% weed control could be achieved by using aforesaid herbicidal combinations. However, one manual weeding (easy in line sown DSR) or spray of 2,4-D ester @ 1250 g/ha (against broad leaf weeds) or Ethoxysulfuron 15% WDG @ 125 g/ or metsulfuron + chlorimuron @ 15 g/ha (against broad leaf weeds and sedges) at 25-30 days after sowing may be required to manage additional weed infestation, if any (similar to as done in puddle transplanted rice).

Water management

DSR is a major opportunity to change production practices to attain optimal plant density and high-water productivity in water scarce areas. Traditionally, rice is grown by transplanting one month old seedlings into puddled and continuously flooded soil. In dry DSR, after sowing of rice, a light irrigation should be applied to maintain proper soil moisture for proper seed germination. But at the the time of germination, water should not stagnate for long time, In wet (Vatar) DSR, after sowing of rice, a light irrigation should be apply after full germination to maintain proper soil moisture for excellent crop establishment. After crop establishments irrigation should be apply at an interval of 5 to 6 days (depending on weather conditions) or on the basis of light hair cracks in soil (Fig. 4).

Fig. 4. Irrigation based on soil hair cracking **(a)** and tensiometer **(b)** in DSR field (*Source*: Research platform climate smart village Anjanthali, Karnal)

Drip irrigation in rice crop: A majority of the farmers of South Asia, rice is grown during summer season by manually transplanting into puddled soil and use traditional surface (flood) irrigation method. Most of this flood irrigation water goes as deep drainage component, which is considered to be an energy consuming process when irrigation water source is ground water. In case of canal-irrigated areas, it could result in waterlogging and related salinization. To switch these conditions, drip irrigation (surface and subsurface) system is the best option. Drip irrigation systems (DIS) are much more water-efficient than continuous flooding of transplanted rice. DIS have 100% conveyance efficiency and 70-90% application efficiency, while the corresponding figures for traditional irrigation method (basin) are 40–70% and 60–70%, respectively (Von *et al.*, 2004; Narayanamoorthy, 2006). Drip irrigation have the potential to rise irrigation water productivity by providing water to match crop desires, reducing deep drainage losses, and generally keeping the soil drier, thereby decreasing soil evaporation (Camp, 1998). Drip irrigation not only results in savings in water usage in rice production, but also increases the rice yield (Dhawan, 2002; Tiwari *et al.*, 2003; Yuan, 2003). Adoption of drip irrigation for suitable crops in the potential areas may lead to reduction in crop water requirements to the level of 44.46 BCM in India (Sharma *et al.*, 2009). A properly designed and managed drip irrigation system gives producers the best uniformity and application efficiency available, consequently saving the time, energy and water, all while maximizing rice yields (Fig. 5). In drip irrigation, the water pumped out from a well is first sent through sand separators and media/screen filters to remove silt and impurities such as algae or dead plant matter. This filtered water is, then, applied to the crop via a network of mainline and sub-mainline pipes, valves (that turn on or off the water flow) and smaller diameter poly-tubes or 'laterals', which have pre-installed emitters at spaces corresponding with the placement of each plant. These ensure delivery of water directly to each plant's root zone and at discharge rates as low as one litre per hour.

Surface drip irrigation system: Surface drip irrigation is a method of applying small amounts of water uniformly across a specific area. It is a specialized subset of drip irrigation where drip line or drip tape "lateral lines" (tubes keep near the crop rows) and supply and flushing "sub-mains" (pipes supplying water to the lateral lines) are retained above the soil surface for seasonal use. The irrigation water is delivered directly to the rice root zone, eliminating runoff, evaporation and drift from rice field. Surface drip irrigation systems also have provision for 'fertigation' application of fertilizer, in liquefied form from a separate tank along with the water. "Surface drip irrigation works well in rice crop, where only less water per plant is needed for rice growth period.

Fig. 5. (a) Surface and **(b)** Sub-surface drip irrigation system in direct seeded rice (*Source*: BISA farm Ludhiana)

Subsurface drip irrigation: Subsurface drip irrigation (SDI) is the irrigation of crops through buried plastic tubes containing embedded emitters located at regular spacing. It is a specialized subset of drip irrigation where drip line or drip tape "lateral lines" (tubes buried beneath the crop rows) and supply and flushing "sub-mains" (pipes supplying water to the lateral lines) are buried beneath the soil surface for multiyear use. There are wide varieties of configurations and equipment used, however drip tubes are adjusted accordance with the root zone of crops below the soil surface (Fig. 4). The SDI applies water directly to the crop root zone using buried polyethylene tubing, also known as a drip line, dripper line, or drip tape. For, rice crop, the optimum depth of drip line is 10-15 cm for placing the subsurface drip irrigation system. Subsurface drip irrigation provides the ultimate in water use efficiency for rice crop, often resulting in water savings of more than 50% compared to flood irrigation. The use of SDI offers many other advantages for rice production, including less nitrate leaching compared to surface irrigation, higher yields, a dry soil surface for improved weed control and crop health, the ability to apply water and nutrients to the most active part of the root zone, protection of drip lines from damage due to cultivation and other operations, and the ability to safely irrigate with waste water while preventing human contact.

System of Rice Intensification (SRI)

The SRI is an advanced method comprising uncomplicated management practices that allow rice-growers to get higher productivity of rice by changing the management of plants, soil, water and nutrients. Similar to the central principle of sustainable agriculture which seeks to make optimal use of naturally available resources as functional inputs. SRI is not a technology, but a set of simple ideas and principles that help produce more productive and robust plants. The ideas are:

(1) Raising seedlings in a carefully managed, garden-like nursery.

(2) Transplant young seedlings to preserve growth potential (ideally 8–15 days old, or 2-3 leaf stage).

(3) Provide plants wider spacing-one plant per hill and in square pattern (typically 25 × 25 cm and possibly wider).

(4) Keeping soil moist but not flooded-soil should be mostly aerobic, not continuously saturated.

(5) Early and regular weeding, ideally using a mechanical rotary weeder which churns and aerates the soil or by hand.

(6) Fertilization, preferably using organic sources (compost, FYM and green manure) (Uphoff, 1999; Stoop *et al.*, 2002; Ram *et al.*, 2014; Ram *et al.*, 2015).

Methodology of the SRI

Varieties: In SRI, long-duration rice varieties perform better with wider spacing than the short-duration varieties of rice. Stoop (2005) explained that this is because of the extended crop growth period of long- duration varieties of rice. Latif *et al.* (2005) also reported that long-duration varieties are more suitable for SRI practices, as they led to a higher grain yield.

Seed rate: In SRI, a single seedling is planted per hill and the hills are more widely spaced. Thus, the number of seedlings required for the planting unit area is reduced to a great magnitude. Only 5-7.5 kg/ha of seeds are required.

Nursery: The nursery area is required 100 m^2, with the use of SRI management techniques. Additionally, active maintenance of the nursery is only required for about 14 days.

Water management: Water management in SRI method does not require intermittent irrigation or AWD (alternate wetting and drying). Irrigation is given to maintain soil moisture near saturation initially and water is let in when surface soil develops hairline cracks. This method also helps in better growth and spread of roots. The field should be irrigated again when the soil develops hair line cracks. Depending upon the soil and the environment conditions, the frequency of irrigation should be decided. After the panicle initiation stage until maturity, shallow water should be maintained in the field. The water in the paddy field should be drained at 20 days before harvest.

Nutrient management: Nutrient management for SRI is the same as for TPR and DSR.

Weed management: As there is no standing water in SRI method, weeds would be more. In SRI, the weeds are incorporated by operating cono-weeder between rows at the right time. First weeding is to be done 10-12 days after planting. Further weeding may be undertaken depending on the necessity at 10-15 days interval until crop reaches panicle stage. Other chemical weed management for SRI is the same as for TPR and DSR. Other management practices for SRI are the same as for TPR and DSR. The difference between transplanted rice (TPR) and SRI management are given in table 5.

Table 5. Difference between transplanted rice and SRI management systems

TPR	SRI
Transplant older seedlings, 20-30 days old, or even 40 days old	Transplant young seedlings, 8-12 days old, and certainly less than 15 days old to preserve subsequent growth potential
Required high seed rate (60-80 kg/ha)	Required relatively low seed rate (5-6 kg/ha)
Maintain soil continuously flooded, with standing water throughout the growth cycle	Keep soil moist, but not continuously saturated, so that mostly aerobic soil conditions prevail
Use water to control weeds, supplemented by hand weeding or use herbicides	Control weeds with frequent weeding by a mechanical hand weeder (rotating hoe or cono-weeder) that also aerates the soil
Use chemical fertilizers to enhance soil nutrients	Apply as much organic matter to the soil as possible; can use chemical fertilizer, but best results from compost, mulch etc.

Conclusion

In current scenario, increasing rice production in India is not only important from the point of view of food security but also its immense role in poverty alleviation. However, the current progress in rice production does not really cater to the demand of the burgeoning human population. Consequently, this puts global food and nutritional security at a great risk. This challenge calls for concerted efforts of all stakeholders to produce required quantity and quality of assured rice production for ensuring food security. The scope of area expansion in rice has been exhausted and future production growth in rice has to come from yield growth through advance approach/technological improvement. Meanwhile, the yield growth has been slowed down in recent years in dominated rice producing regions, because of these rice production regions are using still traditional practices. Resulted, the rice production in the country is passing through serious constraints like plateauing of yield, water scarcity, increased use of agro inputs, multiple nutrient deficiency, labour shortage, weed resistance, invasive pests & diseases etc. These challenges will be more intense under emerging scenarios of natural resource degradation, energy crisis, volatile markets and risks associated with global climate change. In view of the constraints, it is imperative that rice production and productivity need to be enhanced through application of modern tools of science. It is necessary to focus on proper utilization of resources by using novel rice cultivation practices such, Smart crop establishments (Direct Seeded Rice, SRI, mechanical trans-planter), Water smart (DSR, AWD, Tensiometer based irrigation, Surface and sub-surface drip irrigation), Nutrient smart (SSNM-Nutrient expert tool, Leaf colour chart, Green-Seeker) etc. Further, there is need to focus on proper utilization of resources in diverse agro-climatic zones in the country by providing quality seeds, developing high yielding varieties/ hybrid rice, effective natural resource management, developing strategies on biotic and abiotic stress management, cost effective mechanization and promoting agricultural stewardship which will help in enhancing production and productivity of rice. Anticipatory, strategic and basic research on rice needs to be strengthened with financial and policy support to meet the future challenges of climate change, water crisis and land and labour shortages. It is also equally important to make rice cultivation more profitable and less labour dependent. Paradigm shift is needed in the way we grow rice in the backdrop of declining resources, escalating labour cost and deteriorating soil health. 'Grow more with lesser inputs' would be the way forward for sustainable rice production in the coming decades.

References

Anonymous. 2018. Economic survey 2018-19. Website: https://www.indiabudget.gov.in/economicsurvey/

Bhatt, R., Kukal, S.S., Arora, S., Busari, M.A. and Yadav, M. 2016. Sustainability issues on rice-wheat cropping system. *Int. Soil and Water Cons. Res.* **4**(1): 68-83.

Bhatt, R. 2013. Soil test based fertilization to improve production of oilseed crops in Kapurthala district of Punjab. *Int. J. of Sci., Env. and Tech.* **2**(3): 521–526.

Chauhan, B.S., Mahajany, G., Sardanay, V., Timsina, J. and Jat, M.L. 2012. Productivity and sustainability of the rice-wheat cropping system in the Indo-Gangetic Plains of the Indian subcontinent: Problems opportunities and strategies. *Adv. Agron.* **117**: 315–369.

Dawe, D. 2005. Increasing water productivity in rice-based systems in Asia past trends, current problems and future prospects. *Pl. Prod. Sci.* **8**: 221-230.

Duxbury, J.M., Abrol, I.P., Gupta, R.K. and Bronson, K.F. 2000. Analysis of long term fertility experiments with rice-wheat rotations in South Asia. In: Abrol, I.P., Bronson, K.F., Duxbury, J.M., Gupta, R.K. (Eds.), Long term soil fertility experiments in rice-wheat cropping systems. Rice-wheat consortium paper series 6. *RWC*, New Delhi, India, vii-xxii.

FAO STAT. 2020. http://www.fao.org/faostat/en/#data/QC/visualize.

Fageria, N.K. and Baligar, V.C. 2005. Enhancing nitrogen use efficiency in crop plants. *Adv. in Agron.* **88**: 97-185.

Food and Agriculture Organization. 2016. FAO STAT, website: www.fao.org/corp/statistics.

Gathala, M.K., Timsinaa, J., Islama, Md.S., Krupnika, T.J., Bose, T.R., Islamh, N., Rahman, Md.M., Hossain, Md.I., Rashid, Md. H.A., Ghosh, A.K., Hasan, Md. M.K., Khayer, Md. A., Islam Md. Z., Tiwari, P.T. and McDonald, A. 2016. Productivity, profitability and energetics: a multi-criteria assessment of farmers' tillage and crop establishment options for maize in intensively cultivated environments of South Asia. *Field Crops Res.* **186**: 32-46.

Goyal, S.K. and Singh, J.P. 2002. Demand versus supply of foodgrains in India: Implications to food security. Paper presented in 13th International Farm Management Congress, Wageningen, The Netherlands, July 7-12, 2002, pp. 20.

Gupta, R.K., Hobbs, P.R., Jiaguo, J. and Ladha, J.K. 2003. Sustainability of post-green revolution agriculture. In: J.K. Ladha, et al., (Eds.), Improving the productivity and sustainability of rice-wheat systems: issues and impacts, ASA Spec. Publ. ASA, CSSA, and SSSA, Madison, WI. **65**: 1-25.

Indian Council of Agricultural Research (ICAR) 2010. Vision 2030. Indian Council of Agricultural Research, New Delhi. pp.24.

Hira, G.S., Gupta, P.K. and Josan, A.S. 1998. Waterlogging causes and remedial measures in South-West Punjab. Research bulletin no. 1/98. Ludhiana: Department of Soils, Punjab Agricultural University.

Hobbs, P.R. and Morris, M.L. 1996. Meeting South Asia's future food requirements from rice-wheat cropping systems: Priority issues facing researchers in the post green revolution era. NRG Mexico, D.F.: CIMMYT, 96-101.

Humphreys, E., Kukal, S.S., Christen, E.W., Hira, G.S., Singh, B., Yadav, S. and Sharma, R.K. 2010. Halting the groundwater decline in north-west India-which crop technologies will be winners? *Adv. in Agron.* **109**: 156–199.

Jain, A.K. and Kumar, R. 2007. Water management issues–Punjab, North-West India. In Indo-U.S. workshop on innovative E-technologies for distance education and extension/ outreach for efficient water management. ICRISAT, Hyderabad.

Jat, M.L., Dagar, J.C., Sapkota, T.B., Singh, Y., Govaerts, B., Ridaura, S.L., Saharawat, Y.S., Sharma, R.K., Tetarwal, J.P., Jat, R.K., Hobbs, H. and Stirling, C. 2016. Climate change and agriculture: Adaptation strategies and mitigation opportunities for food security in South Asia and Latin America. *Adv. in Agron.* **137**: 127–236.

Jat, M.L., Singh, B. and Gerard, B. 2014. Nutrient management and use efficiency in wheat systems of South Asia. *Adv. in Agron.* **125**:171-259.

Kakraliya, S.K., Jat, H.S., Singh, I., Sapkota, T.B., Singh, L.K., Sutaliya, J.M., Sharma, P.C., Jat, R.D., Lopez-Ridaura, S. and Jat, M.L. 2018a. Performance of portfolios of climate smart agriculture practices in a rice-wheat system of western Indo-gangetic plains. *Agril. Water Manag.* **202**: 122–133.

Kakraliya, S.K., Kumar, S., Kakraliya, S.S., Choudhary, K.K. and Singh, L.K. 2018. Remedial options for the sustainability of rice-wheat cropping system. *J. of Pharm. and Phytochem.* **7**(2): 163-171.

Lal, R. 2016. Feeding 11 billion on 0.5 billion hectare of area under cereal crops. *Food Energy Sec.* **5**: 239–251.

Lohan, S.K., Jat, H.S., Yadav, A.K., Sidhu, H.S., Jat, M.L., Choudhary, M., Peter, J.K. and Sharma, P.C. 2018. Burning issues of paddy residue management in north-west states of India. *Renew. Sustain. Energy Rev.* **81**: 693–706.

Mahajana, G., Chauhan, B.S., Timsina, J., Singh, P.P. and Singh, K. 2012. Crop performance and water and nitrogen use sciences in dry seeded rice in response to irrigation and fertilizer amounts in northwest India. *Field Crops Res.* **134**: 59–70.

Pampolino, M., Majumdar, K., Jat, M.L., Satyanarayana, T., Kumar, A. and Shahi, V.B. 2012. Development and evaluation of nutrient expert for wheat in South Asia. *Better Crops*. **96**(3): 29-31.

Pandian, B.J., Rajkumar, D. and Chellamuthu, S. 2011. System of rice intensification (SRI), a synthesis of scientific experiments and experiences, national consortium of SRI, Tamil Nadu Agricultural University, pages 1-100.

Ram, H., Singh, J.P., Bohra, J.S., Singh, R.K. and Sutaliya, J.M. 2014. Effect of seedlings age and plant spacing on growth, yield, nutrient uptake and economics of rice (*Oryza sativa*) genotypes under system of rice intensification. *Indian J. of Agron.* **59**(2): 256-260.

Ram, H., Singh, J.P., Bhora, J.S., Yadav, A.S. and Sutaliya, J.M. 2015. Assessment of productivity, profitability and quality of rice (*Oryza sativa*) under system of rice intensification in eastern Uttar Pradesh. *Indian J. of Agril. Sci.* **85**(1): 38-42.

Rao, K.S. 2013. Strategies for enhancing production and productivity of rice in India. Innovations in rice production. *Editors* P.K., Shetty; M.R., Hegde; M., Mahadevappa. National Institute of Advanced Studies. *Indian Institute of Scicence Campus, Bangalore.* ISBN: 978-81-87663-70-6.

Sapkota, T.B., Majumdar, K., Jat, M.L., Kumara, A., Bishnoi, D.K. and McDonaldd, A.J. 2014. Precision nutrient management in conservation agriculture based wheat production of Northwest India: Profitability, nutrient use efficiency and environmental footprint. *Field Crops Res.* **155**: 233-244.

Sidhu, H.S., Singh, M., Singh, Y., Blackwell, J., Lohan, S.K., Humphreys, E., Jat, M.L., Singh, V. and Singh, S. 2015. Development and evaluation of the turbo happy seeder for sowing wheat into heavy rice residues in N-W India. *Field Crops Res.* **184**: 201–212.

Singh, B., Singh, V., Purba, J., Sharma, R.K., Jat, M.L., Singh, Y. and Gupta, R. 2015. Site-specific fertilizer nitrogen management in irrigated transplanted rice (*Oryza sativa*) using an optical sensor. *Precision Agri.* **16**(4): 455-475.

Singh, Y. and Sidhu, H.S. 2014. Management of cereal crop residues for sustainable rice-wheat production system in the Indo-gangetic plains of India. (*In*) *Proc. of the Indian National Sci. Acad.* **80**: 95-114.

Singh, Y., Gupta, R.K., Jagmohan, S., Gurpreet, S., Gobinder, S. and Ladha, J.K. 2010. Placement effects on rice residue decomposition and nutrient dynamics on two soil types during wheat cropping in rice-wheat system in north-western India. *Nutr. Cycl. Agroecosyst.* **88**: 471-480.

Viraktamath, B.C. 2013. Key research inputs and technologies in rice production in pre and post green revolution era. Innovations in rice production. *Edt.* P.K., Shetty; M. R., Hegde; M., Mahadevappa. National institute of advanced studies, *Indian Institute of Science Campus, Bangalore.* ISBN: 978-81-87663-70-6.

2

Advances in Wheat and Barley Production Technologies

S.K. Singh and Satish Kumar*

Introduction

Wheat is pre-eminent both in regard to its antiquity and its importance as a food of mankind in the world. In prehistoric times, it was cultivated throughout Europe and was one of the most valuable cereals of ancient Persia, Greece and Egypt. Wheat has been cultivated for several thousand years in India with the evidence of presence of wheat grains in the Mohen-Jo-Daro excavations. Till 1947, the total wheat production was only about 6 million tonnes and was not sufficient to meet the demand, leading to large scale importation of wheat. Subsequently production increased due to an increase in both total cropped and irrigated area. Wheat crop has exhibited a robust growth trend since the onset of the 'green revolution' in 1968. India, one of the greatest success stories of 'green revolution', is the second largest producer of wheat in the world after China. During 2018-19 (Fig. 1) India produced a record breaking 101.20 million tonnes of wheat from 29.62 million ha area indicating 3.42 tonnes/ha wheat yield (ICAR-IIWBR, 2019). On the other hand, India is also the second largest wheat consumer after China. Thus, wheat and its various products play an increasingly important role in managing India's food security and India became the wheat surplus nation as against the wheat deficient nation during 1960's. The tremendous progress in area, production and productivity of wheat to the tune of 2.9, 12.2 and 4.2 times, respectively as compared to 1950 has made India (Table 1) the member of elite group of wheat exporting countries (Workman, 2018).

**Corresponding author email: sksingh.dwr@gmail.com*

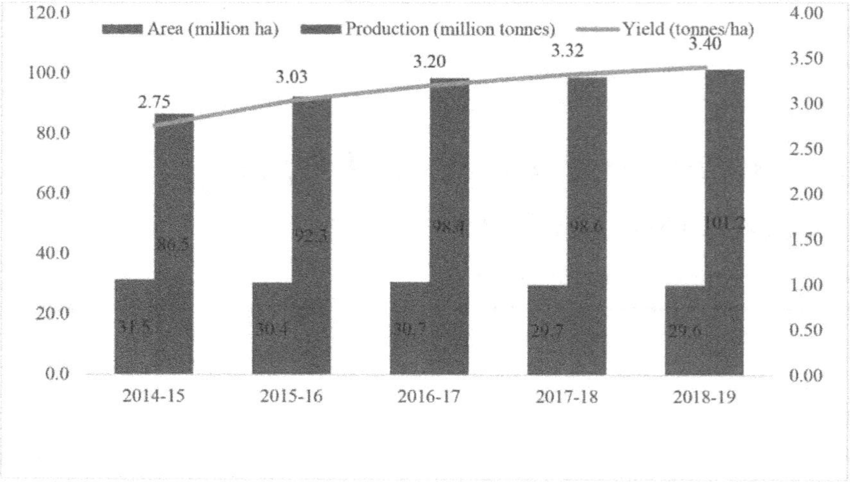

Fig. 1. Quinquennial change in area, production and yield of wheat in India

Barley is a temperate crop and has been traditionally used as a grain crop for human consumption and animal food in India. Based on its use, it can be divided as barley for feed, fodder and malt. Some varieties are used for dual purpose, i.e. feed and fodder both. Barley occupies 0.46% of the total cropped area, 0.62% of the food grains and 0.76% of the cereals in the country. Similarly, it contributes 0.86% of the total production of cereals and 0.81% of the food grains in India. There has been steady shortfall in the area and production of the crop since 1960-61 onward with the beginning of the 'green revolution'. Its area has decreased from 32.23 lakh ha in 1960-61 to 6.93 lakh ha in 2002-03 recording an average annual decline of 1.87 per cent. Similarly, the production has fallen down from 28.66 lakh tonnes in 1960-61 to 14.06 lakh tonnes in 2002-03 at an average annual rate of 1.21 per cent. This decline is mainly due to the transfer of the barley area to wheat cultivation. Area and production of barley have been stabilized in last 15 years. During 2018-19 (Fig. 2), national barley production was 1.73 million tonnes from an area of 0.63 million ha with an average productivity of 2.61tonnes/ha (ICAR-IIWBR, 2019).

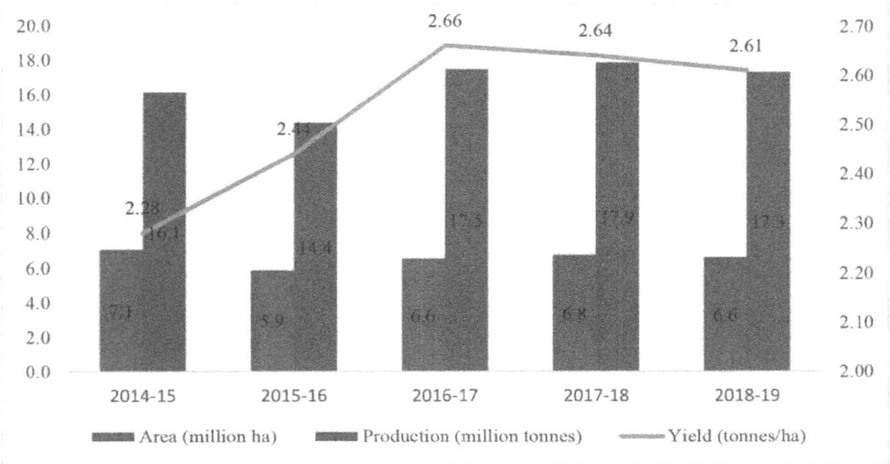

Fig. 2. Quinquennial change in area, production and yield of barley in India

Wheat and barley production scenario

About 91.5% of the wheat is produced in six states *viz.* Uttar Pradesh, Punjab, Haryana, Madhya Pradesh, Rajasthan and Bihar. On an average basis of last 5 years, Uttar Pradesh continues to be the highest producer of wheat followed by Punjab and Haryana. The contribution from Haryana and Punjab is mainly attributed to their high productivity whereas the contribution of other states such as Uttar Pradesh and Madhya Pradesh is due to relatively large area (approx. 50% of the total area) sown to wheat. The states of low wheat productivity have shown an increasing trend in grain yield per unit area in recent past but the stagnated productivity levels in highly productive states of Punjab and Haryana is posing a major concern for future production targets.

Barley is basically a crop of north India. Uttar Pradesh and Rajasthan are the two major states of barley producers in the country. These two states together provide 64 per cent of the total area and 72 per cent of the total production of barley in India. Besides these two states barley is mainly grown in Punjab, Madhya Pradesh, Haryana, Bihar, Himachal Pradesh, West Bengal and Jammu & Kashmir.

Table 1. Area, production and yield of wheat and barley in India (last 10 years)

Year	Wheat			Barley		
	Area (m ha)	Production (m tonnes)	Yield (t/ha)	Area (m ha)	Production (m tonnes)	Yield (t/ha)
2008-09	27.8	80.7	2.90	7.80	15.40	1.97
2009-10	28.5	80.7	2.80	6.24	13.55	2.17
2010-11	29.2	85.9	2.90	7.05	16.63	2.35
2011-12	29.9	93.9	3.14	6.43	16.19	2.51
2012-13	30.0	93.5	3.12	6.95	17.52	2.52
2013-14	30.5	95.9	3.15	6.74	18.31	2.71
2014-15	31.5	86.5	2.75	7.07	16.13	2.28
2015-16	30.4	92.3	3.03	5.89	14.38	2.44
2016-17	30.7	98.4	3.20	6.56	17.47	2.66
2017-18	29.7	98.6	3.32	6.77	17.88	2.64
2018-19	29.6	101.2	3.40	6.63	17.34	2.61

Source: ICAR–IIWBR annual progress report (2019)

The growing zones, seasons and cultural conditions

Wheat in India is cultivated in almost every state except Kerala, thus representing diverse crop growing conditions and situations. Based on the agro-climatic conditions and varying agro-ecological production conditions, the country is broadly divided into six wheat growing zones namely Northern Hills Zone (NHZ), North Western Plain Zone (NWPZ), North Eastern Plain Zone (NEPZ), Central Zone (CZ), Peninsular Zone (PZ) and Southern Hill Zone (SHZ). The barley is also grown in all these zones except SHZ. The growing period is variable from one agro-climatic zone to another that affect the vegetative and grain filling duration leading to differences in attainable yield. The maximum growing duration is in NHZ and minimum in PZ.

Both the crops are cultivated during winter season from mid October to April (except in higher hills of north India where harvesting is done in the month of May). Sowings are initiated when the average of day-night temperatures equal to 23 °C. The months of December and January remain to be coldest followed by comparatively warmer and higher temperatures in the months of March-April coinciding to later grown stages of the crop till maturity. These crops are mainly grown under three production conditions, viz., timely sown, medium to good fertility, irrigated; late sown, medium fertility, irrigated and timely sown, low fertility, rainfed. In recent years, a new situation of timely sown, restricted/ limited irrigation has emerged in some of the wheat growing areas of the central and peninsular parts where water for irrigation is not available in sufficient quantity and thus, the wheat crop is grown with one to two irrigations only.

Available irrigation for crop

Nearly 93% of the wheat area in the country is irrigated and most of it lies in north India. Wheat cultivation in major wheat producing states is supported by >95% area under irrigation. Haryana (99.5%) and Rajasthan (99.2%) have almost all the wheat area under irrigation and U.P. and Punjab have covered most of the wheat area under irrigation. West Bengal, Bihar and Gujarat also have >90% wheat area under irrigation. The central, peninsular and hilly areas of northern and southern regions grow mostly rainfed wheat whose success largely depends on residual moisture build up from monsoon, limited availability of water for irrigation and casual winter rains. Madhya Pradesh (89.3%), Jharkhand (89.1%), Maharashtra (73.9%), Uttarakhand (57.6%) and Karnataka (55.9%) have comparatively lower coverage of area under irrigation and rest of the area depends on rains. In the areas of less rainfall, barley is preferred crop in *rabi* season as compared to wheat.

Wheat species cultivated in India

The three species of wheat namely, *Triticum aestivum* (Bread wheat), *Triticum durum* (Macaroni wheat) and *Triticum dicoccum* (Emmer or Khapli) grown on commercial basis in India are of spring type but cultivated during winter season. Bread wheat represents hexaploid group with 2n=6x=42 chromosomes whereas macaroni and *khapli* wheat represents tetraploid group with 2n=4x=28 chromosomes. Of these species, *T. aestivum* continues to be the most important species accounting about 90-95% of total wheat area of the country and is grown in almost all the wheat growing states. *T. durum* is next in importance with approximately 5% of total wheat area and confined mostly to central and southern parts of India. It is primarily grown under rainfed condition but in recent past, its cultivation under high fertility and irrigated conditions has extended its cultivation in new areas of Punjab state. The cultivation of *T. dicoccum* is confined largely to the southern region mainly Karnataka and southern Maharashtra. Another wheat species *T. sphaerococcum (*Indian dwarf wheat) has now almost vanished and cultivated in some pockets in Gujarat.

Challenges in wheat production

Stagnating yield potential: After realizing the benefits of green revolution, a steep growth in wheat productivity in frontline states of north western India was achieved from 1975 to 1995 through churning of the gene pool and deployment of rust resistance genes in better agronomic backgrounds that resulted in release of some of the landmark varieties namely, Kalyan Sona,

WL 711, Arjun, HD 2329, CPAN 3004, UP 2338 and PBW 343. The increasing trend of wheat productivity in NWPZ has now reached to the saturation level which is a major concern for enhanced wheat production. In order to fulfil our domestic requirements, India need to produce 109 million tonnes of wheat by 2020 for which at least 1.6% annual growth rate in production is required (Mishra *et al.*, 2014). Progress made in irrigated and high fertility wheat regions is significantly higher than that in marginal rainfed area which is often associated with abiotic stresses. Since significant scope exists for improvement in these new areas, one of the major challenges is to develop high yielding varieties having tolerance to abiotic stresses especially heat, drought, salinity and waterlogging.

***Unavailability of quality seeds and low seed replacement**:* Physical and genetic purities of seed are of utmost importance for realizing the actual yield potential of the variety. The rate of seed replacement of newer varieties to the older ones is also important factor to put the yield levels in high momentum. The major constraints in most of the area are unavailability of pure quality seeds and substantially lower seed replacement rate. An enhanced rate of seed replacement will certainly result in significant increase in total wheat production in the country. Thus, there is an urgent need to strengthen the seed production system and efficient distribution channel to bring more area under new wheat varieties.

***Global climate change**:* An impact of climate change on crop production is expected for various latitude limits for all the crop seasons and the wheat crop is the most affected during winter season. Significant increases in area, accompanied by crop yield reductions are expected in the mid to higher latitudes. Contrary to this, significant areas will become unsuitable for wheat and yield may decline at lower latitudes with increasing temperature. Wheat is sensitive to high temperature (both early and late heat) but magnitude of damage depends on the existing ambient temperature, stage of crop development and variety. Adverse weather has frequently affected the wheat productivity and production during recent years. The rise in temperature during December, the period of tillering and subsequently higher temperatures above 30 °C during the February and March at the stages of anthesis, grain formation and filling has led to decreased productivity during the last 4-5 years. Effects of increased CO_2 on wheat yields will normally be positive but the benefits vary with the prevailing temperature regime and availability of other inputs (water and nutrients). It is predicted that with the doubling of CO_2, ambient temperature in India would increase by 3 °C and will affect both the area and productivity of wheat. The encounter of negative effects of varying temperature regime with the benefits of increased CO_2 activity is therefore, a critical issue for any assessment of wheat production under changing climate. It has been observed that increase in temperature (about 2 ^{0}C) reduced potential wheat grain yields at most places. Regions with higher potential productivity were relatively less affected than the areas with lower potential productivity. A net reduction in wheat production is anticipated due to reduction in growth period as a result of increased temperature. Besides, increased water requirements may be anticipated in all regions, which will highlight the importance of irrigation management in mitigating climate change.

Sustainability of rice-wheat system: Rice-wheat cropping system prevalent in north India that occupies about 11.0 million ha area is the most intensively adopted system which consumes substantially high amount of agricultural inputs. Excessive and imbalanced use of these inputs and similar crop growth pattern over the years is affecting the soil physical environment and the productivity of the system. Sustainability, therefore, is the major issue and needs to be ensured through efficient management of resources. The declining response to inputs has been perceived to be the major issue challenging the sustainability of wheat based cropping system across the Indo-Gangetic plains.

Reduced total factor productivity and imbalanced use of fertilizers: The intensive tillage and over-exploitation of the natural resources i.e. soil and water, resulted in the situation where the benefits per unit input used in wheat cultivation is continuously declining as evident from experimental as well as on farm trials. There is an indication that fertilizer recommendation which was 120:60:30 kg NPK/ha at present has to be amended to 150:60:40 or 150:60:60 kg NPK/ha in some soils. Over mining of essential plant nutrients and burning of crop residues have increased wide spread incidences of nutrients deficiencies. The deficiency of Zn in rice-wheat system and S, Fe, Mn and B from various pockets in the intensive cropping areas is severely reported. Farmers in general are not applying potash and in some areas soil status has dropped to such an extent that further nutrient mining may change soil nutrient status substantially. Potash and to some extent zinc is going to play a major role in sustaining and enhancing production and productivity. Further increase in fertilizer requirement in future to get same productivity level is also predicted in case of continuing unbalanced fertilisation. The depleting soil organic carbon due to intensive tillage is also a very crucial factor at present that has reduced the water and nutrient holding capacities of the soils. As a consequence, frequency of irrigation has increased in those areas where water is not a limiting factor. This is further leading to deep percolation of water that often leaches down plant nutrients such as nitrogen and ultimately pollutes ground water.

Limited irrigation availability: The non-judicious use of water is another major issue related with sustainability of wheat production system. Over the years, the per cent area irrigated by canal water has decreased in the states of Punjab and Haryana and the area under diesel operated tube wells has increased. There are nearly a million tube wells that exploit the ground water for agricultural purposes in the NWPZ. Since, the water recharge due to monsoon rains is not proportionate to its utilisation, water table is going down. Though wheat requires only 4-6 irrigations the consequences of water shortage on production can be substantially high. Taking advantage of the farmers' friendly state policy like flat electricity charges, farmers continue to irrigate even during rainy days. Consequently, water level in northern Haryana and Punjab is declining. On the other hand, in southern Punjab and Haryana, water level and salinity is increasing which makes the land unsuitable for wheat crop. The need of the hour is to adopt integrated water management (IWM) practices which require the judicious use of good quality water and conjunctive use of brackish water.

Restricted access to exotic germplasm: The success of green revolution in early sixties and the later varietal improvement programmes were dependent on the exchange of germplasm lines from exotic sources especially International Maize and Wheat Improvement Centre (CIMMYT) that were used directly as a variety or as a donor parent in wheat improvement programmes. The exotic germplasm in the form of international wheat trials and nurseries from CIMMYT continued to serve as one of the major sources of genetic material for yield potential. However, the emergence of Intellectual Property Rights (IPRs) as one of the major global issues, germplasm exchange between the countries is likely to be restricted and this may influence the pace of genetic improvement in wheat. In the post General Agreement on Tariffs and Trade (GATT) scenario the issue of IPRs gained significant importance and has become a major hurdle in germplasm exchange. Therefore, the Indian wheat programme has to focus on pre-breeding activities by utilising unexploited elite sources like land races, synthetics and other available wheat genotypes for broadening the genetic diversity for higher yield potential.

Yield gaps at farm level: The yield gaps have been observed between the cluster frontline demonstrations (CFLDs) and the farmer's practice at their own field. The zone wise analysis indicated the maximum yield gap (1.5 t/ha) in southern hills zone followed by central zone (1.4 t/ha), north eastern plains zone (1.2 t/ha) and northern hills zone (1.0 t/ha). The state-wise scenario indicated the highest yield gap in Madhya Pradesh (2.1 t/ha) followed by Rajasthan (1.6 t/ha), Uttar Pradesh (1.3 t/ha), Bihar (1.2 t/ha) and Karnataka (1.0 t/ha). This gap may be due to many related factors to seed, water and technological adoptions which can be bridged for enhanced overall production.

Breeding objectives for wheat

High yield: Perhaps the foremost among the problems and objective of wheat breeding, is the production of varieties or strains that are superior in yield performance when compared to the best among the previously existing varieties or strains. The main objective is to create new genotypes that contribute to greater yield potential and improved product quality. Yield potential in wheat refers to the ability of the plant to manufacture, translocate and store food materials in the wheat grain. Emphasis is now being given to the breeding of high yielding wheat cultivars. The strategy for turning the 'green revolution' which started during 1960's with the introduction of Mexican dwarf wheat varieties such as Sonora 64 and Lerma Rojo into an ever-green revolution –includes following four objectives: a) Collection, evaluation and utilization of germplasm from diverse sources. b) Selection of appropriate parents using biometrical approaches. c) Evaluation of segregating generations for target oriented selections from chosen crosses. d) Bi-parental mating for releasing locked up genetic variability.

Resistance to biotic stresses: The wheat crop in India is affected by a number of diseases and insect-pests. It is estimated that, on an average, 10 per cent of the crop is lost due to infestation by diseases annually. The rust diseases like the black or stem rust *(Puccimagraminis tritici)*, the brown or leaf rust *(P. recondita),* and the yellow or stripe rust *(P. striiformis)* are among the most important diseases that affect the crop globally. Next in importance is loose smut *(Ustilago nudatritici)*. The powdery mildew *(Erisyphe graministritici)* is found in the hills both in northern hills as well as the hills of southern India. The Karnal bunt *(Neovossia indica)* and the flag smut *(Urocystis tritici),* are found to attack the wheat crop in southern Punjab, northern Rajasthan, Delhi and western U.P. The leaf spot diseases including blight caused by species of *Helmithosporium* and *Alternaria* are responsible for considerable losses in the states of U.P., Bihar, West Bengal, Maharashtra and Gujarat. The *tundu* disease, in which a bacterium (*Corynebacterium tritici)* and a nematode (*Anguina tritici*) are associated, also takes a toll of the crop in western U.P, Delhi, southern Punjab and northern Rajasthan. Every year on an average 20-25% of wheat is lost due to insect-pests and diseases. The situation is further aggravated on storage. The termites, aphids, armyworm, American pod borer and brown mite are the major pests of wheat. In view of the above, it is necessary to breed for combined resistance for all the important diseases and insect-pests.

Adaptability against abiotic stress conditions (salinity, water and temperature): Breeding for adaptability implies that variety is adaptable over a wide range of environments with consistency in performance. For a given agro-climate, it is also necessary to develop varieties/ hybrids that have high yield with wider adaptation. One of the primary considerations in wheat breeding in India had been the development of early-maturing wheat, or varieties maturing within the optimum duration of time which is characteristics of a wheat growing zone. Further, a substantial part of wheat growing area in India is not covered under irrigation. For such areas, the breeding of early maturing, drought tolerant or drought resistant varieties is of big importance. On the other hand, the irrigated tracts offer great potentialities for the attainment of maximum crop yields and they, therefore, deserve to be fully exploited by the development of varieties which would respond very favourably, in terms of yield, to intensive irrigation and fertilization. An important factor to reckon with, when the crop are intensively irrigated and manured, is the possibility of damage to crop due to lodging. While adjustment culture and manurial practices, especially in respect of nitrogen application, can be helpful in minimizing or preventing lodging, the breeding of lodging resistant varieties would offer distinct advantage. In the existing agricultural economy, the yield of straw per unit area is considered as important as yield of grain therefore the breeding of dwarf or short strawed and, therefore, lodging resistant varieties was not considered promising in the past. Therefore, in future breeding attempts to develop such varieties, care should be taken that the decrease in plant height is compensated for, as much as possible, by such factors as increased tillering, which would secure a good yield of straw per unit area.

Quality improvement: Wheat is cultivated primarily for its grain that is mainly processed into flour and is utilized for numerous end products. The quality of end product is of utmost consideration to the wheat consumers. Broadly the wheat grain quality criteria include features like physical appearance (like colour, vitreousness, texture/hardness, appearance, grain weight, test weight), processing qualities (functionality of proteins, starch, amylose and amylopectin content as per desirable end product such as noodles, pasta, thickness, binding agents, bread etc.), nutritional values (lysine and mineral nutrient content like iron, zinc, copper, manganese, etc.) and biological properties each of these is composed of several components influenced by genetic makeup of the variety. The industrial quality parameters that includes physical characteristics, flour recovery, milling quality, dough quality as well as gluten content useful for specific product development are also of utmost importance for breeding new varieties.

Improved wheat production technologies

Improvement in wheat production technologies are being made through continuous experimentations at multilocations in targeted environments. For the purpose a number of special trials are being conducted under AICRP-Wheat and Barley at funded centres in U.P. namely, Kanpur, Ayodhya (Faizabad), Varanasi, Modipuram, Meerut and more than 25 voluntary centres under SAUs across the state. The latest improved technologies to harness the maximum yield potential are detailed as below.

Selection of variety: Most of the farmers are still growing old varieties having low productivity like Sonalika, Raj 3765, UP 262, WH 147, and HUW 234. The older varieties are having very less yield levels and more susceptibility to diseases. In order to increase production, replacement of older varieties is essential. In this situation, farmers have to grow recommended new varieties. A list of varieties (Table 2) recommended for different production condition is given below.

Table 2. Wheat varieties recommended by CVRC for different production conditions of India

Production condition	Agro-climatic zone			
	NWPZ	NEPZ	CZ	PZ
Irrigated (timely sown)	DBW 222, DBW 187, HD 3226, PBW 723, HPBW 01, WB 2, DBW 88, HD 3086, WH 1105, HD 2967, DPW 621-50, PBW 550, DBW 17, WHD 943(d), PDW 314 (d), PDW 291(d)	HD 3249, DBW 187, NW 5054, K 1006, HD 2967, DBW 39, HD 2733	HI 1544, GW 366, HI 8759(d), HI 8737(d), HD 4728(d), HI 8713 (d), MPO 1215 (d)	DBW 168, MACS 6478, UAS 304, MACS 6222, NIAW 917, MACS 3949(d), UAS 428 (d), UAS 415 (d), MACS 2971(dic), DDK 1029 (dic), DDK 1025(dic)
Irrigated (late sown)	PBW 771, HI 1621 (VLS), HD 3271 (VLS),PBW 752, PBW 757(VLS), DBW 173, DBW 90, WH 1124, DBW 71, HD 3059	HI 1621 (VLS), HD 3271 (VLS), DBW 107, HD 3118, HI 1563, DBW 14	MP 3336, MP 1203, HD 2932	HD 3090, AKAW 4627, HD 2932, Raj 4083
Restricted irrigation/ Rainfed (timely sown)	HI 1628, HD 3237, HI 1620, PBW 660, WH 1142, PBW 644, WH 1080, HD 3043	DBW 252, HI 1612, K 1317, HD 3171, HD 2888	Raj 4238, DBW 110, MP 3288, MP 3173, HI 1531, DDW 47 (d), UAS 466 (d), HD 4672(d)	UAS 375, HI 1605, UAS 347, DBW 93, GW 1346 (d), HI 8805(d), HI8802(d), MACS 4028(d), HI 8777(d), UAS 446 (d)

Source: ICAR–IIWBR annual progress report (2019)

Climatic conditions: Wheat is cultivated in all types of climatic conditions i.e. tropical, sub-tropical and temperate. In India, major wheat area is under sub-tropical region. The cool and sunny winters are very conducive for growth of wheat crop. The temperature required for optimum growth and development of wheat differs for different growth stages. At the time of germination temperature must be around 20-25 °C, later on for growth of wheat 20-23 °C (mean daily) is the favourable range. For better tillering in plants, the temperature should be around 16-20 °C and it should be 23-25 °C at the time of grain filling.

Soil texture: Wheat can be grown on all kinds of soils, except the highly deteriorated alkaline and water logged soils. Soils with clay loam or loam texture, good structure and moderate water holding capacity are ideal for wheat cultivation. Durum wheat should preferably be sown on medium to fine textured soils.

Land preparation: Wheat cultivation requires a well pulverized but compact seed bed for uniform germination. One deep ploughing with soil turning plough followed by two harrowing and planking is desirable. To protect the crop from termites and white ants particularly in rainfed area, treatment of seed is suggested with Chlorpyriphos 20 EC @ 700 ml per 100 kg of seed by mixing in 5 litres of water and spraying over the seed followed by seed drying overnight before sowing. For the control of diseases like bunts and smuts, seed treatment is suggested with Bavistin or Thiram @ 2.5 g/kg of seed.

Depth of sowing and spacing: The depth of sowing should be around 5±2 cm and row spacing for timely sown should be 20-22.5 cm. Under late sown condition row to row spacing should be 15-18 cm. For rainfed areas, line spacing should be 20-25 cm depending upon soil moisture condition (Table 3).

Table 3. Time of sowing, seed rate and fertilizer application

Cultural condition	Time of sowing	Seed rate (kg/ha)	Fertilizer dose (kg/ha)
Irrigated (timely sown)	First fortnight of November	100	N150:P60:K40, 1/3rd N and full P and K at sowing time. Remaining 2/3rd N in two equal splits at first and second irrigation.
Irrigated (late sown)	December	125	N120:P60:K40, 1/3rd N and full P and K at sowing time. Remaining 2/3rd N in two equal splits at first and second irrigation.
Restricted Irrigation, Rainfed	Last fortnight of October	100	N60:P30:K20, Full dose at the time of sowing as basal.

Integrated nutrient management: The improvement in productivity as well as quality was observed when bio-fertilizers and farm yard manure (FYM) were used along with recommended doses of fertilizers. If we supply 50% of nitrogen through organic manures and 50% nitrogen through inorganic fertilizers, then the productivity was significantly higher than application of recommended inorganic fertilizers alone. Besides micronutrient deficiencies like zinc, magnesium, sulphur, boron, iron, etc. is also being observed in some areas for which following measures should be adopted.

i. Zinc deficiency appears on younger leaves indicated by interveinal chlorosis, resetting and mottled leaf, reduced shoot growth, shortened internode and stunted growth. Zinc sulphate at the rate of 25 kg/ha once in a year or 10 kg/ha to each crop during *rabi* and *kharif* season should be applied in areas where its deficiency is prevalent. If zinc deficiency symptoms are visible in the crop then spray 0.5% zinc sulphate solution. For this, dissolve 2.5 kg zinc sulphate and 1.25 kg unslaked lime or 12.5 kg urea in 500 litres of water for one ha and spray it at 30-35 days after sowing (DAS) and if needed, repeat 2-3 times at 15 days interval.

ii. Magnesium deficiency appears on the tip and margin of the leaf blade and indicated by interveinal yellowing and premature defoliation of leaves. In areas having manganese deficiency, spray 0.5% solution of $MnSO_4$ for which dissolve 2.5 kg $MnSO_4$ and 1.25 kg unslaked lime in 500 litres of water. First spray should be done at 30-35 DAS and additional 2-3 sprays at 15 days interval if required. Spray should be done during clear and bright sunny days.

iii. Some of the wheat areas have sulphur deficiency where symptoms appear first at younger leaves which turn yellow due to chlorosis, root and stem becomes abnormally long and develop woodiness. Thus, sulphur should be applied at the time of sowing as basal dose. For optimum sulphur availability, apply 250 kg gypsum/ ha or Reap (90WDG)/ Cosavet (80WDG) @10 kg/ ha. It can also be applied at first irrigation. If symptoms appear at later stages, then 1.0% solution of Reap (90WDG)/ Cosavet (80WDG)+ 0.5% unslaked lime may be sprayed in one ha which can be made through dissolving 5.0 kg Reap/Cosavet and 2.5 kg unslaked lime in 500 litres of water.

iv. Boron deficiency show yellowing and resetting of leaves, abortion of buds, low pollination and fertilization, cracking of stem and fruits. It is rare but in affected areas, apply borax @10 kg/ ha as basal dose. If basal dose is not applied, spraying of 0.5% boron+0.25% unslaked lime solution (dissolve 2.5 kg borax and 1.25 kg unslaked lime in 500 litres of water) in one area at 35 DAS and subsequent two spray at 15 days interval.

v. Iron deficiency leads to chlorosis of young leaves in general but interveinal chlorosis is seen under acute deficiency conditions. It can be corrected by spraying of 0.5% solution of iron sulphate ($FeSO_4$) and 0.25% unslaked lime solution (dissolve 2.5 kg $FeSO_4$ and 1.25 kg unslaked lime in 500 litres of water) in one ha area at 35 DAS and may be repeated at 15 days interval if deficiency persists.

vi. If the deficiency of more than one micronutrient appears, then two micronutrients can be mixed and sprayed at an interval of 15 days starting at 35 DAS. For mixture of two or more micronutrients, dose of unslaked lime should also be increase, e.g., for correction of iron and zinc deficiency, dissolve 2.5 kg $FeSO_4$+2.5 kg $ZnSO_4$ and 2.5 (1.25+1.25) kg of unslaked lime in 500 litres of water for one ha crop.

Irrigation management: Irrigation can be applied depending upon availability of water and requirement of crop. Generally, wheat crop requires 3-6 irrigations. Crown root initiation (CRI) i.e., 21 DAS and heading are the two most sensitive stages where, moisture stress causes maximum damage to this crop. In zero-tillage sown wheat crop also, irrigation need to be managed similar to conventional method of wheat cultivation. In raised bed sown crop, if moisture is less at sowing time then apply light irrigation immediately after sowing for better germination and crop stand. If, irrigation water is available in limited amount, irrigation may be applied as give in table 4.

Table 4. Critical stages for wheat crop for irrigation under limited water condition

Number of irrigations	Irrigation at days after sowing	Stages (DAS)
One	21	Crown root initiation (CRI) (21)
Two	21, 85	Tiller completion (40-45)
Three	21, 65, 105	Booting/ Late jointing (60-65)
Four	21, 45, 85, 105	Heading/ Flowering (80-85)
Five	21, 45, 65, 85, 105	Milking (100-105)
Six	21, 45, 65, 85, 105, 120	Dough (115-120)

Source: Nand *et al.* (2011)

Weed management: Weed infestation is one of the major biotic constraints in wheat production. Both grassy and broad leaf weed infests wheat crop. For control of weeds, farmers prefer chemical weed control due to cost and timely effectiveness. Depending on the weed flora infesting the field, herbicide should be selected. The major weeds in wheat crop are *Phalaris minor* (mandusi/ kanki), *Avena ludoviciana* (jangli jai), *Chenopodium album* (bathua), *C. murale* (kharbathua,), *Cyperus rotundus* (montha), *Medicago sativa* (maina), *Melilotus alba* (senji), *Malwa parviflora* (chughra), *Convolvulus arvensis* (hirankhuri), *Cirsium arvense* (kandai), *Anagalis arvensis* (krishnanil), *Argemone mexicana* (satyanashi), *Rumex dentatus* (jangali palak), etc. Manual weeding with hand hoe is beneficial after first irrigation at optimum moisture. Weed free seed should be used for sowing. Early sowing and reduced spacing also helps in weed control. Chemical measures should be adopted as mentioned below in case of heavy infestation.

i. For control of broad leaved weeds, spray Metsulfuran @ 4 g or Carfentrazon @ 20 g or 2,4-D @ 500 g/ha dissolved in 250-300 litres of water at 30-35 DAS.
ii. For control of narrow leaved weeds, spray Isoproturon @ 1.0 kg or Clodinafop @ 60 g or Pinoxaden (5 EC) @ 40 g or Fenoxaprop @ 100 g or Sulfosulfuron @ 25 g/ha dissolved in 250-300 litres water at 30-35 DAS.
iii. For control of both narrow and broad leaved weeds, spray Isoproturon (75 WP) @ 500 g or Sulfosulfuron @ 13 g or Sulfosulfuron+Metsulfuron @ 16 g or Metsulfuron+Iodosulfuron @ 160 g or Fenoxaprop+Metribuzin @ 500-600 ml dissolved in 250-300 litres of water/ha at 30-35 DAS. Pendimethalin @ 1.0-1.5 litres dissolved in 500 litres of water can also be used for one ha within 3 days of sowing.

Crop protection measures: The recommended plant protection measures against prevalent diseases and insect pests are detailed as under.

Stripe and leaf rusts: In western and central U.P., stripe rust (yellow rust) is very important disease in wheat and for avoiding the losses varieties like HD 3086, DBW 88, WH 1105, HD 2967, DBW 621-50 for timely sown and DBW 71, DBW 16, DBW 90, PBW 590, WH 1021 and HD 3059 for late sown conditions may be preferred. Usually, it is observed that the early infection of stripe rust starts in wheat fields under the poplar trees wherever these are grown having early sown crop (i.e. October). Hence, strict watch is needed by the farmers in such fields. In eastern U.P., leaf rust (brown rust) is more important for which latest cultivars resistant to brown rust should be grown. In the event of severe disease condition, chemical spray of Propiconazole (25 EC @ 0.1%), or Tebuconazole (250 EC @ 0.1%) or Triademefon (25WP @ 0.1%) is recommended to control rust diseases.

Leaf blight: Leaf blight is the main crop health problem in NEPZ. For effective management of the diseases, cultivation of recommended varieties, like NW 5054, K 1006, DBW 39, CBW 38, K0307, DBW 107, HD 3118, HI 1563, HD 2985, DBW 14, NW 2036 should be encouraged.

Loose smut: Loose smut is a seed borne disease. In view of the horizontal distribution of the seed material among the farmers and the use of the carry over seed effective control measures for loose smut should be undertaken. For this, seed treatment with Carboxin (75 WP @ 2.5 g/kg seed) or Carbendazim (50 WP @ 2.5 g/kg seed) or Tebuconazole (2DS @ 1.25 g/kg seed) or a combination of a reduced dosage of Carboxin (75 WP @ 1.25 g/kg seed) and a bio-agent fungus *Trichoderma viride* @ 4 g/kg seed is recommended.

Karnal bunt: Karnal bunt (KB) control is required for seed crop and the produce grown for export purposes. For producing KB free wheat, farmers are advised to grow KB resistant varieties recommended for the respective area. Irrigation at heading time should be avoided to keep KB in control. Zero-tillage also helps in reducing KB incidence. One spray of Propiconazole (25 EC) @ 0.1% or Tebuconazole (250 EC) @ 0.1% be given in mid February for management of the disease.

Powdery mildew: One spray of propiconazole (25 EC) @ 0.1 % at ear head emergence or appearance of disease (whichever is earlier) is recommended for the powdery mildew prone areas.

Termite: In the termite prone areas, seed treatment with chloropyriphos @ 0.9 g a.i /kg seed, be taken up for their management. Seed treatment with thiamethoxam (70WS) @ 0.7 g a.i./kg seed or Fipronil @ 0.3 g a.i./kg seed is also very effective. In the standing crop, the broadcasting of the insecticide treated soil 15 DAS be practiced. For this, chloropyriphos @ 3 litres mixed in 50 kg soil be used for one ha field. Crop planted under FIRBS is more prone to termite attack in the termite-prone areas, while zero-tillage shows less termite damage. Hence, proper attention should be given in crop planted under FIRBS.

Nematodes: Ear cockle is an important disease in eastern parts of India, hence proper precautions be taken, especially in eastern U.P., Bihar and Jharkhand. Wider publicity should be given by extension agencies on the use of gall-free seed, well before the sowings. Farmers should adopt floatation technique for the separation of galls from the infested seed lots. The infested seed lot should be floated in 2% brine solution for this purpose. The galls will float on the surface. These should be separated and destroyed away from the field by burning. The seed should be thoroughly washed to remove the salt solution before sowing.

Integrated pest management (IPM): The IPM module developed and validated in NWPZ can be adopted in parts of north-west plain zone. This involves the seed treatment with *T. viride* (@ 4 g/kg seed) + Carboxin (75WP @ 1.25 g/kg seed) or Tebuconazole (@ 1.0 g/kg seed) for the control of loose smut, followed by broadcast of insecticide treated soil (with chloropyriphos @ 3.0 litres/ha) at 15 DAS for termites. For the management of aphids, foliar spray of Imidacloprid 200 SL @ 20 g a.i./ha on border rows at the start of the aphid colonization be given. This will help in protection of the bio-agent insect, the lady bird beetle inside the field which feeds on aphids. In KB prone areas, the seed crop can be given one spray of Propiconazole or two sprays of *T. viride* at tillering and ear head emergence. For the control of powdery mildew in disease prone areas, one need-based spray of Propiconazole (25 EC @ 0.1%)

can be given at ear head emergence or appearance of disease on flag leaf, whichever is earlier.

Resource conservation technologies (RCTs)

In post green revolution era, India is now on the verge of another agricultural revolution through resource conservation technologies that will help to reduce the cost of cultivation as well as conserve the natural resources. The innovations in resource conservation tillage technologies have helped in reducing the cost of cultivation and now the emphasis is being laid on conservation agriculture through crop residue retention at soil surface (Mishra *et al.*, 2005; Sharma *et al.*, 2007). Some of the technologies for this purpose are elucidated as below.

Laser land levelling: Laser land levelling is the fore most requirements for the adoption of any of the resource conservation technologies as fields are not properly levelled and part of area suffers water stress and the other parts excess of water leading to poor performance of the crop. It is a process of smoothening the field within ± 2 cm from the average elevation of the field using laser equipped bucket, which scraps from higher places and spread onto the low lying areas. After laser levelling the field, yield increase has been observed. The higher yields are due to proper crop stand, uniform fertilizer and water distribution, crop growth, uniform maturity and increased cultivable area by 3 to 6 per cent due to reduction in area under bunds and channels. In addition to higher yield, the savings of water is to the tune of 35-45% due to higher application as well as use efficiencies and increased nutrient use efficiency by 15-25%. In recent years, the adoption of laser levelling by the farmers has gained momentum due to evident benefits of this technology right in the first crop. The major limitation of this technique is, it's higher cost of around Rs. 3.5 lakh which is not within the reach of every farmer and hence has to be used on custom hire basis.

Zero-tillage: It is a profitable resource conservation technique in which wheat seed and fertilizers are directly placed at proper depth into the undisturbed soil after rice harvesting using a specially designed machine that creates narrow slits by the knife type furrow openers of zero-tillage (ZT) ferti-seed drill instead of shovel type furrow opener in conventional ferti-seed drill. The money and time to be spent in field preparation are saved by using this machine and sowing can be advanced by 7-10 days. Both timely and late sowing of wheat is possible by this method and in case of late sown even sowing can be advanced by 7-10 days. Seed rate, fertilizer doses and other package of practices in ZT should be the same as in conventional method to get good yield. Apply 1/3rd or even lesser N, full P and K at sowing and the remaining nitrogen in two equal splits at first and second irrigation. Avoid planking after sowing of wheat by this machine. The cost effectiveness and development of resistance against 'Isoproturon' in *Phalaris minor* was also responsible for ZT adoption in rice-wheat system due to lower incidence. Incidence of KB and termite has also been reported to be less in ZT This machine can sow about two acres (0.81 ha) of wheat in one hour. The carbon dioxide emission due to burning of fuel (assuming 2.6 kg CO_2 production/ litre of diesel burnt) during field preparation is about 208 kg/ha in conventional tillage where as in ZT it is only around 16 kg/ha. This technology provides an opportunity to save > ₹ 3000/ha in cost of cultivation thereby increasing the profit margin of the farmers.

***Rotary tillage**:* An important resource conservation technology which facilitates field preparation and placing of seed and fertilizer in a single pass. The rotary till drill has six L-shaped blades on each gang which completely pulverises soil. Seed rate and fertilizers doses are similar to conventional method. This machine is also very effective for incorporation of *Dhaincha* into soil as green manuring crop. It can also be used for single pass puddling of paddy field after removing the drilling mechanism. This machine can cover about 0.40 ha area in one hour. It saves time, labour and energy and reduces drudgery in field preparation compared to conventional method. There is a saving of > ₹ 2500/ ha in field preparation. Rotary tillage advances sowing of wheat by 3-5 days and gives yield advantage of 5-10% than conventional and ZT technologies.

***Bed planting**:* The field preparations, bed formation, placement of fertilizer and sowing of seed are completed in one go by bed planter. Furrows are used for irrigation as well as for drainage of excess water, if there is heavy rain, during crop season. Generally, 2-3 rows of wheat can be planted on the top of each bed. The top width of each bed is 40-45 cm for a bed of 70-75 cm with furrow of 30-35 cm. Inter cropping of sugarcane can be taken up with wheat by this technique. Crop cultivars are known to vary significantly in their performance on raised beds. The direction of sowing should be North-South so that every plant gets equal sun shine. In situations where sowing is expected to be delayed due to pre-sowing irrigation, dry seeding can be done on raised beds followed by irrigation immediately after seeding. Light irrigation can also be given at grain filling stage, which is generally avoided by the farmers for fear of crop lodging. Growing crops on raised beds can also help in diversification of the rice-wheat system. Raised bed system also provides an option for growing intercrops like sugarcane in furrows and wheat, chickpea, lentil, peas, mustard and various vegetable crops on beds. Since, only $1/3^{rd}$ area is irrigated i.e. irrigation is applied in furrows, there is substantial saving of irrigation water. It saves 20-25 per cent seed i.e. 30-32 kg/acre seed is sufficient for sowing. Although, there is no saving on the cost of land preparation or time but, if same beds are used for seeding next crop; the saving similar to rotary tillage can be made as seeding and shaping of beds can be done in a single operation. It is also suitable for seed production because of production of bold grain and easy rouging by moving in furrows to uproot unwanted plants.

Harvesting, threshing and storage: The crop is harvested when the grains become hard and the straw becomes yellow, dry and brittle. When the moisture level of grain is 20% it is the proper time for manual harvesting. Generally, wheat is manually harvested with sickle but for quick harvesting, combine harvester should be used to avoid losses in grain yield due to shattering and lodging. In eastern U.P., time of harvesting starts from latter part of March and continues until mid-April whereas, wheat is harvested during second fortnight of April in western U.P. Wheat should be harvested 4-5 days before it is dead ripe. Combine harvesting is convenient due to synchronized maturity of wheat crop. Morning is the best time for harvesting. In case of manual harvesting, bundles are made and dried for 3-4 days and threshed by thresher. Before storage, grain should be dried by spreading on tarpoline plastic sheets in bright sunlight to a moisture level below 12%. For storage, use bins and silos made of GI sheets. Now a day, aluminium bins, puma bins, silos and poly lined bags are available for storage. Farmers can store wheat grains in their traditional storage as well. To protect from storage insects-pests, it is necessary to fumigate with FDB 5 g/tonnes and keeping room sealed for 24 hours. Farmers can also apply aluminium phosphide @ 3 g/tonnes.

Barley species cultivated in India

Barley has basic chromosome no. of X= 7 and based on ploidy level, it is grouped as diploid, tetraploid and hexaploid. *Hordeum vulgare* is the only cultivated species of the barley which represents diploid (2n=2x=14) group. A species *H. bulbosum* exists both as diploid and polyploidy and is used to produce haploid barley embryos. *H. vulgare* has two distinct phenotypic forms, viz., two rowed and six rowed types based on ear morphology. These two forms have same chromosome number, intercross freely and produce fertile hybrids. Mostly two rowed barleys are preferred for malt purposes due to their bold grains.

Constraints in barley production and productivity

The major constraints responsible for stagnation in area and production are less coverage of barley under improved varieties, limited use of cash inputs like fertilizers, plant protection measures, inadequate seed rate and low preference to barley sowing as compared to wheat. The major constrains have been discussed below.

Area constraints: Although barley area in India has been more or less stabilized over the last decade, but challenges remain to keep this level and make further improvement in area to meet the increasing demand for industrial utilization, by making availability of high yielding, better quality varieties. Also there will be increasing demand for feed and forage type barley varieties especially in the era of climate change where water availability will become a limiting factor for growing more water requiring crops like rice.

Low demand for quality raw material for malting: In India barley is now becoming more and more important as a commercial crop for industrial raw materials for malting and brewing. Although the proportion in malting utilization is less at present but it is expected to increase with changing scenario in south-east Asia with respect to rapid growth for beer and other malt based products.

Increasing area under problematic soils: The most commonly observed fact is that the farmer will always prefer to grow wheat in resource rich conditions and push barley to problematic soils with minimum input. There is a big challenge to make yield break through in barley varieties for abiotic stresses like salinity, drought and heat tolerance for near future under changing climatic conditions.

Disease and pest management: Incorporation of varietal resistance for prevailing diseases has been the main objective of the barley improvement programmes all over the country. Two major aspects for resistance breeding are rusts and blights. Though the available barley varieties show some amount of resistance against all three rusts, but they lack good resistance to leaf blight. Thus, there is need for identifying sources of blight resistance and incorporating the same into cultivated varieties.

Supplementing the demand for feed/forage: It is visualized that barley is an important crop in semi-arid regions for supplementing grain feed as well as green fodder due to its limited water requirement and faster growth. Thus, there is a need for better biomass and nutritional quality of barley forage.

Resource management in changing scenario: The most important aspect of study in barley agronomy still remains to be covered is the detailed nutrient and moisture utilization pattern under different situations supported by physiologic and economic studies. Agronomic research on malt barley, hulless barley, feed and forage barley, barley saline and coastal lands needs to be undertaken on priority. There is a need to develop lodging resistant varieties as barley is prone to lodging under high input application.

Socio-economic constraints: Poor economic base of the farmers with virtually no support from the government for amelioration of their economic environment is the leading socio-economic factor affecting barley production. There are no incentives for farmers to encourage them to raise barley production. Besides this demand for barley is almost constant due to the biased policies towards the fine cereals (rice and wheat). Change in the lifestyle of population has turned people to fine cereals.

Agro-ecological constraints: Photo-sensitive barley varieties, normally carry vernal and photoperiod sensitivity traits, which induce lateness. Barley is affected by high temperature at various stages in its growth and development. Higher temperatures are known to induce abnormalities and causes reduction in floret number at higher temperature of 24 °C, in comparison to 18 °C. Tillering in barley is affected by the day and night temperature regimes. Frost and low temperature during winter nights damage the crop. Barley crop is highly susceptible to winter injury and is, therefore, less dependable than wheat under severe winter conditions. Barley is mainly cultivated on poor and marginal soils with poor fertility. It is mainly grown as rainfed under conserved moisture conditions. In view of the above constraints, it is the high time to overcome at least agronomic and socio-economic constraints, so that cultivation of barley could be encouraged.

Breeding objectives for barley

Majority of barley produced in India is utilized for cattle and poultry feed. Major emphasis of barley breeding programmes is laid on improvement for grain/feed purposes. A large number of varieties have been released for areas addressing different production conditions. In some parts of the country barley is also cultivated as a fodder for cattle. Thus, breeding for dual purpose barley with higher grain yield and more foliage is one of the most important objectives of the barley improvement programmes.

Malt barley improvement: Due to low demand of malt products in country and low demand from industry research efforts on malt barley improvement in the country did not receive much attention. Recently malt barley improvement has become important in the wake of industrial demand. The Indian barley varieties were never bred for malting purpose and thus, there is a huge need to develop varieties with better malting characteristics.

Breeding barley for abiotic stress tolerance: Barley is known for its inherent tolerance to salinity and alkalinity as compared to other cereals. It has a good potential for problematic soils where otherwise it's difficult to grow crop in winter season. A large area in country comes under this kind of problematic soils. However, it's a higher challenging task to screen materials for such conditions due to very high variability in the field conditions affected by salinity and alkalinity. As a result the field screening for salinity tolerance sometimes cannot be fully reliable, because of non-repetitive performance due to soil heterogeneity. To have more efficient evaluation, there is a need for *in-vitro* screening for salinity/alkalinity tolerance to supplement the regular research efforts.

Breeding for resistance to insect-pests and diseases: Barley is exposed to various diseases and insect-pests that are responsible for heavy reduction in yield and quality. Among the major diseases of barley, rusts and blight are of significant importance. Crop losses due to these diseases especially blight are huge. Since both of these diseases are multiple cycle diseases, the use of chemical control measure may not be sufficient. The use of resistant varieties is highly recommended. Other than these diseases, loose smut and covered smut also cause yield loss in barley. These diseases being spread via seed are easily controlled by seed treatments. Aphid is also a major problem in barley. This insect causes heavy loss to the crop as well as the grain quality. Cereal cyst nematodes are also known to infest the barley crop. They cause heavy losses by reducing the tillering and ear head formation. Use of resistant varieties is generally encouraged to control the damage by nematodes.

Improved barley production technology

Similar to wheat production technologies, experimentations were made on barley agronomic manipulations and based on the recommendations under AICRP on Wheat & Barley, the latest improved technologies to harness the maximum yield potential are detailed as below.

Climatic requirement: Barley may be grown in subtropical climatic conditions. The crop requires around 12-15 °C temperature during growth period and around 30 °C at maturity. It cannot tolerate frost at any stage of the growth and any incidence of frost at flowering is highly detrimental for the yield. The crop posses very high degree of tolerance to drought conditions.

Soil requirement: Sandy to moderately heavy loam soils of Indo-Gangetic plains having neutral to saline reaction and medium fertility are the most suitable type for barley cultivation. However, it may be grown on variety of soil types, *viz.* saline, sodic and lighter soils. Acidic soils are not fit for barley cultivation, as such.

Landpreparation: Two to three ploughing with cultivator followed by planking after every ploughing is required to make good soil tilth. Seed treatment is advisable to save the crop from termite, ants and other insects.

Time of sowing and seed rate: The seed rate, time of sowing and spacing standards are different for different barley production conditions. For irrigated timely sown (1-20 November) and rainfed (25 October-10 November) varieties, the seed rate (100 kg/ha) and spacing (23 cm) is same. For irrigated late sown varieties (10-25 December) preferred seed rate is 125 kg/ha and spacing is 18 cm.

Selection of varieties: The ICAR and the state agriculture universities (SAUs) have developed many varieties suitable for different zones, regions and sowing conditions (Table 5).

Table 5. Barley varieties recommended for different production conditions

Production conditions	Western and central India	Eastern India	Saline and alkaline soils
Irrigated (timely sown)	DWRUB 52*+, RD 2668*+, DWR 28*+, RD 2503+, RD 2035#, RD 2715*, BH 902	BCU 73*+, K551+, Azad#, K 508, PL751, RD 2786, RD 2715#	NDB 1173, RD 2552#, NB1, NB3
Irrigated (late sown)	DWRB 91+	NB2	
Rainfed (timely sown)	RD 2508, RD 2624, RD 2660	K 560, K603, JB 58	

2-row barley, # dual purpose, + malt barley

Method of sowing: The best method of sowing is with a seed drill or dropping seed with a *Chonga* (a tube connected with a funnel like structure) attached to a *desi* plough. Dropping seeds in open furrows behind a *desi* plough and broadcasting are found to be inferior to line sowing with seed drill.

Seed treatment: To control loose smut disease, treatment with Vitavax or Bavistin @ 2 g/kg seed or for the covered smut treatment with 1:1 ratio mixture of Thiram + Bavistin or Vitavax @ 2.5 g/kg seed is used. To avoid crop loss due to termite, seed treatment with 150 ml of Chloropyriphos (20 EC) or 250 ml Formathion (25 EC) in 5 litres of water for 100 kg seed is recommended.

Fertilizer dose and method of application: The nitrogen and phosphorus requirements vary for different production conditions. The recommended fertilizer dose is 60 kg N, 30 kg P_2O_5 and 20 kg K_2O/ha for irrigated timely sown and irrigated late sown barley crop. For malt barley it is 80 kg N, 40 kg P_2O_5 and 20 kg K_2O/ha. In rainfed area, the recommended dose of N:P:K is 30:20:20 kg/ha. In case of irrigated condition, half of the nitrogen and full dose of phosphorous should be applied as basal and remaining half of the nitrogen should be top dressed after first irrigation or 30 DAS. In case of light soils, $1/3^{rd}$ of nitrogen and full dose of phosphorous should be applied as basal, $1/3^{rd}$ of nitrogen after first irrigation and rest $1/3^{rd}$ of nitrogen after second irrigation.

Irrigation: Barley is grown in irrigated as well as rainfed/water scarce areas. Generally barley crop require 2 to 3 irrigations for better yield. Depending upon the water availability, suitable stages for irrigation should be identified.

If water is available for single irrigation, it should be applied at tillering stage (35-40 DAS) if two irrigations are available, 1st irrigation should be given at CRI stage (20-25 DAS) and another at panicle emergence stage (65-70 DAS). If the third irrigation is available, it should be given at grain formation stage (90-95 DAS). The malt barley requires 3-4 irrigations to ensure better yield, grain uniformity and quality.

Weed control: Barley is a fast growing crop and it does not let weeds smoother it, if proper crop stand is maintained. The major weeds in barley crop are *Phalaris minor* (mandusi/kanki), *Avena fatua* (jangli jai), *Chenopodium album* (bathua), *Convolvulus arvensis* (hirankhuri), *Anagalis arvensis* (krishnanil), *Cronopus didymus* (wild carrot) etc. If necessary chemical weed control measures can be applied. For control of broad leaved weeds, spray Metsulfuran @ 4 g or 2,4 D @ 500 g/ha dissolved in 400-500 litres of water at 30-35 DAS. For control of narrow leaved weeds, spray Isoproturon 75% @ 1,000 g or Pinoxaden (5 EC) @ 40 g/ha dissolved in 400-500 litres water at 30-35 DAS. For control of both narrow and broad leaved weeds, spray Isoproturon 75% (1.0 kg) or Isoproturon (750 g) + 2,4-D (500 g) or Isoproturon (750 g) + Metsulfuron (4 g) dissolved in 400-500 litres of water for a one ha at 30-35 DAS. Pendimethalin @ 1.0-1.5 litres dissolved in 500 litres of water can also be used in one ha within 3 days of sowing.

Plant Protection measures: Barley is affected by fungus, virus and different pests that causes significant yield loses. The main diseases infecting barley crop, causal organisms, symptoms and management are given in the following table 6.

Table 6. Insect-pest and diseases, symptoms and their managements

Diseases/ insect-pests	Symptoms	Managements
Diseases		
Covered smut	Smutted heads; grains replaced by black agglutinated spore masses, covered by persistent white, papery membrane.	Treat seeds with Ceresan or Agrosan GN before sowing @ 2-2.5 g/kg of seed; grow resistant varieties; rogue out the smutted plants.
Loose smut	Smutted heads; grains replaced by a black powdery mass of spores; finally only the naked rachis remaining behind.	Solar-heat or hot-water treatment of seed; grow resistant varieties; rogue out the smutted plants; dry seed treatment with Carboxin (0.25%).
Leaf rust	Small, round light-yellowish- brown pustules on leaves later the pustules turn black, covered by the epidermis.	Grow resistant varieties, seed treatment with Benzjmidazole.
Stripe (yellow) rust	Small, yellow, elliptical pustules on leaves, forming stripes which later turn black; sometimes the pustules appear on the leaf sheaths and glumes.	Spray the crop with Zineb @1.70 kg/ha at fortnightly intervals; seed treatment with Oxycarboxin (0.25%) for seedling infection; grow resistant varieties.
Powdery mildew	White to dark powdery masses appears on all aerial parts of the plant.	Dust with finely powdered sulphur @ 15-20 kg/ha; grow resistant varieties; soil treatment with Benomyl (600-1000 g/100kg).
Spot blotch	Dark-brown to black lesions usually occur first on the coleoptiles and progress up wards individual lesions on leaves round to oblong with de margins; spots coalesce to form blotches, heavily Infected leaves urn lure early or dry up.	Seed treatment with organo-mercurials; grow resistant varieties.
Insect-pests		
Termites	Social insects that live underground in colonies ; attack young seedlings as well as grown up plants; the attacked plants wither and ultimately die.	Mix thoroughly 5% Aldrin or Chlordane dust with the soil just at the time of sowing or during preparation of the land for sowing.
Aphids	Nymphs and adults suck sap from leaves, tender shoots and immature rain; extremely fast, forming large colonies.	Spray 0.02% low Phosphamidon or 0.03% Dimethoate or Diazinon.

Harvesting and storage: Barley crop gats ready for harvest by the end of March to first fortnight of April. Since, barley has shattering character; it should be harvested before over ripening to avoid breaking of spikes due to dryness. Barley grain absorbs moisture from the atmosphere and should be stored at an appropriate dry place to avoid storage pest losses.

Short term interventions for enhanced wheat & barley production in India

The wheat production is distributed in major four wheat agro-climatic zones, *viz.* North Western Plains Zone (NWPZ), North Eastern Plains Zone (NEPZ), Central Zone (CZ) and Peninsular Zone (PZ). The yield gap between farmers' fields and frontline demonstration is between 1.5 to 2.0 tonnes/ha. Some short term interventions for enhanced wheat & barley productivity may be as under.

1. New varieties continue to be the primary vehicle of technological advance. This situation is most likely to perpetuate into the future as well.
2. Prioritization of quality aspects remains suboptimal, in deference to overwhelming productivity concerns and a procurement channel aiming almost exclusively at national food security.
3. Biotechnological solutions are now actively sought and their potential is well acknowledged. Molecular markers have emerged as a powerful breeding tool. Its large scale application has however, been thwarted due to low throughput and higher cost. Technologies have evolved rapidly and we are on the cusp of an era of molecular breeding. Transgenic wheat, though largely ignored by public programmes in the present hostile opinion environment is likely to be of great importance.
4. Resource conservation technologies (RCTs) have economic and environmental consequences and are finding increasing adoption at the farm level. A related development is mechanization for precision of farm operations and labour saving. Future evaluation of varietal candidates and management practices is expected to be performed in the context of RCTs.
5. The plant material development has to be on a breeding scale. Consequently, corresponding infrastructural upscaling would be necessary.
6. Wheat improvement can benefit greatly from cutting out the protracted selfing phase. Application of new accelerated breeding methods can strengthen the 'breeding option' in face of sudden biotic and abiotic challenges.

7. Screening systems need to be created with precision as a target but without sacrificing scale and robustness. The system for screening against major biotic stresses such as rusts and KB are well developed and routinely being used. Screening for tolerance to two major abiotic stresses in wheat: heat and drought however need significant technical and infrastructural input.
8. Development and fine-tuning the package of practices and varieties specific recommendations and more emphasis on conservation agriculture to address the issue of climate change and ill-effects of residue burning.
9. To achieve the future projections of wheat production following strategies and framework would be adopted. In order to address the site specific issues, the coordinating centres will be strengthened. Various activities as shown in chart below will be taken in an integrated manner to enhance the wheat production.

Future strategies for enhanced wheat and barley production

Although the problems are daunting, the phenomenal achievements in the food grain production during the past 40 years in the country provides ample encouragement that this challenge can be met through agricultural research. The major thrust for increased wheat & barley production will be in the agro-climatic regions of Indo-Gangetic plains especially the north eastern plains zone (major area in eastern U.P.), the plateau region of central India and the hill regions. An overall growth of 2.50 % in wheat output was shown with increase in production level of 69.7 million tonnes in 2001-02 to 101.20 million tonnes in 2018-19. In potential regions, the production is expected to increase at higher rate. In the remaining regions wheat production is expected to grow with slight increase in yield levels. These expectations are possible with technologies available at present; however, future research efforts would be focused to evolve new and innovative production technologies which can fit into the framework of changing production scenario. Development of better yielding, disease-resistant varieties and their accelerated adoption along with improved production technologies seems the best approach. However, the future yield gains are to be made in the shadow of shrinking land availability, depleting natural resources such as soil health and diminishing water bodies and unpredictable and adverse climatic conditions. The main emphasis of futuristic research for enhancing wheat & barley production can be summed up in following points:

Breaking yield barriers through genetic enhancement: The existing yield plateau in the main wheat growing areas is a major concern and enhanced efforts are being made to break the yield barriers. This has to be accomplished in present production scenario with more emphasis on soil and climate driven environmental challenges. Genetic diversity will continue to be the key factor and new approaches needs to be adopted to enhance yield potential of wheat genotypes. Genotypes with high potential grain yield (a potential of >7.0 tonnes/ha) need to be developed. In this regard, the future strategy will be aimed to introgress desirable genes complexes from unexploited germplasm and wild progenitors for creating new variability. Non-conventional sources can be used in pre-breeding programs for broadening the gene pool for different yield component traits. Development of synthetics and utilization of winter wheat can pave way for wheat genotypes having high yielding ability. Hybrid wheat can also be assessed as an option for breeding for high yield potential. Similarly, in barley, introgression of new genetic variability from wild species may be an innovative approach for enhancing yield potential.

Molecular approach for precision breeding: One of the important goals of molecular biology is to expedite the breeding program with precision and accuracy to develop cultivars with specific traits within a short period of time. Molecular mapping of genes is essential to expedite marker-assisted selection (MAS) in wheat & barley improvement that can complement conventional breeding methodologies for faster and precise breeding. Doubled haploid system can also be utilised to facilitate improvement programmes for obtaining completely homozygous lines in a single step. Besides this, an important area is participation in the global efforts of understanding structural and functional genomics of wheat for harnessing the knowledge thus generated in targeted incorporation/enhanced expression of specific traits, eg., terminal heat tolerance, etc. The transformation approach can also be explored for improving popular varieties for missing traits like disease resistance, abiotic stress tolerance and quality traits.

Tailoring wheat genotypes in cropping system perspective: The wheat based cropping system has many crop sequences among which rice-wheat cropping system is the most extensively followed. The continuity of such sequences over the years shows a tendency for decrease in wheat yield though the total productivity per year has increased. The new RCTs have been developed for phasing out the high output from wheat cultivation in a profitable and sustained manner. This needs to be accomplished with specific genotypes for cultivation under different RCTs. In this regard, the experiments showed significant interaction between varieties and RCTs suggesting differential behaviours of wheat genotypes. The hybridization and handling of segregating populations under different tillage options is the future thrust for developing tillage specific genotypes for different agro-climatic zones.

Improved varieties for abiotic stresses: The wheat & barley crop in the northern plains exposed to higher ambient temperatures at the time of grain filling, which significantly reduces the productivity. The experimental results indicate that each degree rise in ambient temperature reduces wheat yield by 3-4%. Hence, breeding for heat tolerance is one of the major issues under prevailing stress environments. The salinity and alkalinity condition is another threat that severely affects productivity. Although resistant genotypes have been released for these suppressive soils, more efforts are needed to reclaim and provide highly tolerant genotypes for these conditions. It has also been observed that a large area sown to wheat & barley (about 3-6 million ha) is subjected to irregular water logging at early stages of growth particularly where soils are sodic. Genotypes tolerant to water logging have been identified which needs to be incorporated in high yielding background.

Disease resistance breeding: Among biotic stresses, the rust diseases pose major threat to wheat & barley production in India. A limited number of known resistance genes could be used in our breeding programmes so far. Also isolated efforts have been made in characterizing the phenomenon of slow rusting (a trait of durability) in Indian wheat germplasm. At this juncture none of the genes or their compatible combinations in use is capable of providing complete protection from stripe and leaf rusts. Therefore, it becomes imperative to make use of other unexploited genes especially those derived from alien sources as well as those related to adult plant resistance (APR) to keep the rust epidemics under check in Indian sub-continent. Leaf blight and flag smut are other diseases of importance that needs to be taken care. Besides, KB is diseases of importance that need to be addressed for export quality wheat production.

Conservation agriculture: The new RCTs have shifted the thrust to conservation agriculture (CA). These technologies are capable of placing seeds of wheat directly in the presence of residues left after the harvest of rice/wheat by the combines which at present is being burnt. This can tremendously contribute to conserving soil health by preventing loss of nutrients. CA will help to retain the residue on the surface of the soil which will act as biological tillage, conserve the soil moisture, release nutrients to the plant as and when required, reduce soil erosion, keep a check on weeds and provide enabling environment that will ultimately help in sustenance of soil and crop productivity.

Integrated water, nutrient and weed management: Water management is the key to development of sustainable agriculture for both irrigated as well as rainfed areas. In this context the issues like water conservation, watershed management, sprinkler and drip irrigation and Furrow Irrigated Ridge-till and Bed-planting system (FIRBs) of wheat cultivation needs to be addressed with respect to increased nutrients and moisture use efficiency, avoiding lodging, reducing seed rates, minimising weed infestation, better residue management and thus reducing cost of wheat cultivation without affecting the productivity.

Integration of inorganic and organic fertilizers is the key to sustained soil and crop productivity. Attention is also needed for balanced use of chemical fertilizers, use of bio-fertilizers and other sources for improving soil structure and texture. Sustained yield and soil productivity can be accomplished with balanced nutrient addition using animal manures /locally available organic manures with commercial fertilizers. Losses of N through surface drainage, leaching and gaseous emissions need to be minimised through innovative approaches like split application and deep placement of N-fertilizer, irrigation scheduling, introduction of sprinkler/drip irrigation systems and adopting FIRBs. Some of the measures have been suggested to improve micronutrient status of soils which are subjected to intensive cultivation. These include use of leaf colour chart for saving of nitrogen, *in-situ* green manuring without additional water, diversification with pulses, residue incorporation/ retention, balanced application of fertiliser including micronutrients and development of micronutrient efficient cultivars.

Weeds play an important role in realising the yield potential of any crop. Unfortunately, they are more resistant to abiotic stresses and their nutrient absorption capacity is also more than the wheat crop. As most of the area follows the wheat based cropping system, there is a need to focus on integrated weed management and succession of weeds in a cropping sequence. Besides, physical, cultural and chemical means, biological weed control in wheat crop using plant pathogens, especially in the form of myco-herbicides needs to be focused in future that will reduce the hazards of ground water contamination and promote the food safety and protection of endangered species.

Diversification/intercropping/companion cropping: Since, area under wheat is not going to expand further, there is a need to evolve suitable genotypes and production technologies for various synergistic and parallel intercropping/ companion-cropping systems. Under irrigated conditions opportunities for intercropping of wheat exist with autumn planted sugarcane and potato. This could be achieved by establishing inter-institutional linkages and effective extension network. Continuous use of rice-wheat system is depleting the soil health and lowering the water table. Hence, there is also a need to opt for possible replacement of rice by some other remunerative crops i.e, maize, baby corn, soybean etc. Also FIRBs technology can be used in combination to reduce water usage. Alternatively, there is possibility to intensify the system through introduction of leguminous crops like moong bean, which can also improve soil nutrient status. This will reduce the cost of cultivation and enhance crop productivity and water use efficiency. The present efforts to diversify the rice-wheat system need to be enhanced through support from policy makers and extension workers.

***Quality improvement in wheat and barley**:* In changing socio-economic scenario, value added wheat products that can be consumed instantly would dominate the domestic market. With tremendous human resource and emerging food processing technologies India has a large scope to be develop instant food industries, thus, the Indian wheat programme has to strengthen to meet the quality requirement of domestic and international market. Increasing global demand, value addition potential, better price in market and resistance to KB makes durum an export commodity. Although the high yielding varieties with better quality are available, efforts are in progress to create variability for beta-carotene, protein content, semolina recovery and test weight, besides enhancing resistance against yellow, brown and black rust. Barley has its industrial use due to malt quality traits. Today the industrial requirement of barley is about 0.25 million tonnes and is growing annually @ 10% (Singh *et al.*, 2014). The better quality malt barley production is needed to fulfil the industry requirements.

***Post-harvest technology**:* Through an exhaustive exercise involving grain samples from experimental fields and grain markets, it has been convincingly demonstrated that the varieties of wheat grown at present have attributes suitable for the best quality end products such as biscuit and pasta. Food processing industries can take maximum benefit from superior varieties of wheat, if procurement and processing of grains at market level is attended with special care. Grain storage and transport need special attention. Small-scale industries for wheat-based value added products needs to be encouraged in the rural sector to improve livelihood of farmers and especially of rural woman.

***Access to quality seeds**:* The varietal diversity under commercial cultivation plays a barrier for any probable pathogenic hazard. During 1967 to 1977 there were only two varieties namely Sonalika and Kalyan Sona which occupied largest area in the country. Now >160 varieties are in seed chain at national level but farmers are still deprived of quality seed. For quicker replacement of old, susceptible and otherwise uneconomical varieties the Central Research Institutes, SAUs and State Departments of Agriculture should develop a strategy to distribute and popularize newly released cultivars. For providing quality seed to the farmers, extensive breeder seed production programme has been taken up under coordinated project every year and it has been further strengthened through ICAR initiative of mega seed project involving SAUs and ICAR institutes to enhance seed research and production capabilities.

Strengthening linkage between research and extension: Strengthening research capabilities of centers should be one of the priorities to keep pace with advances in science and to adopt technologies. Transfer of technology (ToT) through frontline demonstrations will be the major focus to bridge yield gap between experimental field and farmer's field. If the production gaps are bridged, an additional production of 29.4 million tonnes wheat under irrigated conditions is possible through existing technologies (Aggrawal *et al*., 2008). Cooperation from state agricultural extension units and SAUs would be a key factor in this process. The participatory approach may also be beneficial in selecting suitable varieties for commercial cultivation. Krishi Vigyan Kendras (KVKs) can also play a significant role in rapid ToT and needs to be strengthened and their activities should be strongly linked to national wheat research centres.

Developmental and policy issues: Availability of agricultural inputs in time for timely sowing of wheat can lead to remarkable increase in wheat production. Policy support is needed to popularise and spread eco-friendly production technologies such as ZT and crop diversification, production and marketing of high quality product specific wheat, installation of modern silos to prevent post-harvest loss of quality and quantity of grains and development of wheat based rural industries and cooperatives for producing and marketing value added products. Support for human resource development through training at national and international institutes to acquire expertise in advanced science and technologies are also crucial.

Conclusion

Wheat and barley are important cereal crops in world and contributes significantly in the nutrition of human beings. Barley is also used for malt industries and animal feed. Three types of wheat are grown commercially. Wheat crop is attacked by fungi, bacteria, viruses and nematodes. The fungal diseases are known to cause epidemics in wheat. Wheat and barley are known to be affected due to rust diseases and there had been epidemics of these in some countries. Being a compound interest disease, it is difficult to manage in field in situations of susceptible cultivars, favorable weather and virulent pathotypes. The deployment of resistant varieties against rusts is most practical, simple, effective, ecofriendly and economic method to keep away the rusts. Through coordinated research efforts > 460 wheat varieties suited to different agro-ecological conditions and growing situations have been released so far. These genotypes were very successful in increasing the wheat production from a mere 12.5 million tonnes in 1964 to 101.2 million tonnes during 2018-19. Barley is a temperate crop and has been traditionally used as a grain crop for human consumption and animal food in India. Based on its use, it can be divided as barley for feed, fodder and malt. Some varieties are used for dual purpose, i.e. feed and fodder both. Barley occupies 0.46% of the total cropped area, 0.62% of the food grains and 0.76% of the cereals in the country. Similarly, it contributes 0.86% of the total production of cereals and 0.81% of the food grains in India. There has been steady shortfall in the area and production of the crop since 1960-61 onward with the beginning of the 'green revolution'. A tremendous improvement has been made in the development and production of wheat and barley varieties across the country. These crops, particularly has been playing an important role in the nation's food security.

References

Aggrawal, P.K., Hebber, K.B.,Venugopalan, M.V., Rani, S., Bala, A., Biswal, A. and Wani, S.P. 2008. Quantification of yield gaps in rainfed rice, wheat, cotton and mustard in India. Global theme on Agroecosystems, report no. 43, ICRISAT Hyderabad p 36.

ICAR-IIWBR. 2019. Director's report of AICRP on wheat and barley 2018-19, Ed: G.P. Singh. ICAR-Indian Institute of Wheat and Barley Research, Karnal, Haryana, India.

Mishra, B., Shoran, J., Chatrath, R., Sharma, A.K., Gupta, R.K., Sharma, R.K., Singh, R., Rane, J. and Kumar, A. 2005. Cost effective and sustainable wheat production technologies. Directorate of Wheat Research, Karnal-132001, Technical Bulletin no. **8**: p 36.

Mishra, S.C., Singh, S.K., Patil, R., Bhusal, N., Malik, A. and Sareen, S. 2014. Breeding for heat tolerance in wheat. In: Wheat: Recent trends on production strategies of wheat in India (Eds. R.S. Shukla; P.C., Mishra; R. Chatrath; R.K. Gupta; S.S. Tomar and I. Sharma). JNKVV, Jabalpur & IIWBR, Karnal. Pp 15-29.

Nand, B., Khokhar, R.P.S. and Pyare, R. 2011. Studies on irrigation scheduling of wheat varieties for central Uttar Pradesh, India. *Plant Arch.* **11**(1): 243-248.

Sharma, R.K., Chhokar, R.S., Meena, R.P., Gill, S.C. and Jha, A. 2016. Management strategies to bridge the yield gap. In Souvenir of the 55th All India Wheat & Barley Research Worker's Meet, August 21-24, 2016, CCSHAU, Hisar. Pp 66-71.

Sharma, R.K., Chhokar, R.S. and Gill, S.C. 2007. Resource conservation tillage technologies for sustainability and higher productivity of rice-wheat system. *J. of Wheat Res.* **1**(1&2): 97-102.

Singh, R., Kumar, A., Kharub, A.S., Kumar,V., Chhokar, R.S., Selvakumar, R., Chand, R. and Sharma, I. 2014. Barley cultivation in India. Directorate of Wheat Research, Karnal-132001, Extension Bulletin no.**53**: p18.

Workman, D., 2018. Wheat Exports by Country, May 19, 2018. Website: http://www.worldstopexports.com/wheat-exports-country.

3

Advances in Maize Production Technologies

Hardev Ram, Thomas Abraham, Shailesh Marker and Surgyan Rundla*

Introduction

Maize (*Zea mays L.*) is one of the most versatile promising crops having wider adaptability under diverse agro-climatic conditions. Worldwide, maize is known as "Queen of cereals" because it has the maximum genetic yield potential amongst the cereals. It is cultivated on nearly 197 million ha in about 170 countries having wider diversity of climate, soil, biodiversity and management practices that contributes 36% (1135 million tonnes) of the global grain production. The United States of America (U.S.A.) is the largest producer of maize contributes nearly 33% of the total production in the world and other important growing countries are China, Brazil, India, Argentina, Indonesia, Ukraine and Mexico (Table 1).

Table 1. Area, production and yield of maize in various countries in 2017

Country	Area (million ha)	Production (million tonnes)	Yield (tonnes/ ha)	Production (%)
U.S.A.	33.469	370.960	11.084	32.69
China	42.399	259.071	6.110	22.83
Brazil	17.394	97.722	5.618	8.61
Argentina	6.531	49.476	7.576	4.36
India	9.219	28.720	3.115	2.53
Indonesia	5.375	27.952	5.200	2.46
Mexico	7.328	27.762	3.789	2.45
Ukraine	4.481	24.669	5.506	2.17
World	197.186	1134.747	5.755	100.00

Source: FAO STAT (Assessed on 08.01.2020)

**Corresponding author email: devagron@gmail.com*

Maize is the driver of the U.S.A. economy. The U.S.A. has the highest productivity (>11 tonnes/ha) which is double than the global average (5.7 tonnes/ha). Whereas, the average productivity of India is 3.1 tonnes/ha (Table 1).

In India, maize is the 3rd most important food crops after rice and wheat. The total maize produced during 1950-51 was around 1.73 million tonnes, which has increased to 28.75 million tonnes by 2017-18 which is close to 17 times higher. The average productivity during the period has also increased by 5.6 times to 3065 kg/ha from 547 kg/ha during the period (1950-51 to 2017-18) (Table 2). However, the current yield levels are much lower than U.S.A. and the world average (Table 1). Andhra Pradesh, Bihar, Maharashtra, Karnataka, Telangana, Madhya Pradesh, Rajasthan, Uttar Pradesh, Tamil Nadu and Gujarat are major maize growing states in India. The maize is cultivated throughout the year in all states of the country for various purposes including grain, fodder, green cobs, sweet corn, baby corn, popcorn, etc. in peri-urban areas. The predominant maize growing states that contributes more than 80% of the total maize production are Andhra Pradesh (20.9%), Karnataka (16.5%), Rajasthan (9.9 %), Maharashtra (9.1%), Bihar (8.9%), Uttar Pradesh (6.1%), Madhya Pradesh (5.7%) and Himachal Pradesh (4.4%). Apart from these states maize is also grown in Jammu and Kashmir and North-Eastern states. Hence, the maize has emerged as important crop in the non-traditional regions i.e. peninsular India as the state like Andhra Pradesh which ranks 5th in area (0.79 million ha) has recorded the highest production (4.14 million tonnes) and productivity (5.26 tonnes/ha) in the country although the productivity in some of the districts of Andhra Pradesh is more or equal to the U.S.A.

Table 2. Area, production and productivity of maize in India

Year	Area (million ha)	Production (million tonnes)	Productivity (kg/ha)
2007-08	8.12	18.96	2335
2008-09	8.17	19.73	2414
2009-10	8.26	16.72	2024
2010-11	8.55	21.73	2542
2011-12	8.78	21.76	2478
2012-13	8.67	22.26	2566
2013-14	9.07	24.26	2676
2014-15	9.19	24.17	2632
2015-16	8.81	22.57	2563
2016-17	9.63	25.90	2689
2017-18	9.47	28.75	3065
2018-19$	9.86	27.23	2965

Source: Anonymous (2019), $4th advance estimates

Maize utilization pattern

The multiple utilities of maize as a 'food', 'fodder' and 'feed' makes it further more demand friendly and insulates it against low demand situations. These unique characteristics of maize make the crop a suitable crop candidate for enhancing farmer's income and livelihoods in India.

Over the last two decades, maize is increasingly being used for feed and industrial use (Fig. 1), particularly poultry feed (51%) and starch (12%). Poultry industry has been growing by about 4-5 per cent per annum on higher consumer demand for animal proteins due to a growing economy and expanding middle class. The starch industry, largely catering to textile production, is growing at 3-4 per cent on domestic and export demand. Maize is also used to produce traditional foods, snacks and savories.

The demand of maize depends largely on demand as feed for poultry and livestock, and partially on its direct demand for human food and industrial uses. Per capita starch consumption in India is 1.5 kg as compared to global average of 6.1 kg. The consumption in India is nearly 1/4th in comparison to China, thereby indicating enormous scope of growth.

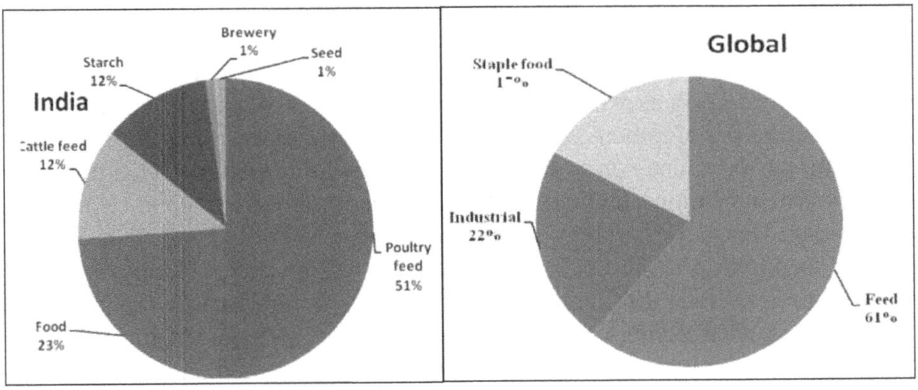

Fig. 1. Utilization pattern of maize in India and world

Soils

Maize can be grown successfully in various soils ranging from clay loam to loamy sand. However, soils having good organic matter content, high water holding capacity and neutral pH are considered excellent for higher productivity. Being a sensitive crop to moisture stress mainly deficit soil moisture and salinity stresses; it is enviable to avoid low lying areas having poor drainage and higher salinity problems. Hence, the fields having stipulation of good drainage should be preferred for cultivation of maize.

Maize research in India

Maize improvement in India has evolved very strongly since 1950. The first phase involved improvement of landraces during 1950-60. The principal breeding efforts were mainly focused towards improvement of local material through mass selection and hence the productivity remained very low (547 kg/ha). In India, organized research on maize improvement started in 1957 under the auspices of the All India Coordinated Maize Improvement Project (AICMIP), which was the first in a series of coordinated projects under the ICAR system. The AICMIP was established after recommendation from renowned maize breeders, Dr. E.J. Wellhausen and Dr. U.J. Grant of Rockefeller Foundation (https://iimr.icar.gov.in/index.php?option=com_content&view=article&id=2&Itemid=114). In 1961, the first set of four double-cross hybrids, viz. Ganga 1, Ganga 101, Ranjit and Deccan were released for commercial cultivation in India. These were followed by series of double-cross hybrids like VL 54, Him 123, Deccan 101, etc. The AICMIP was upgraded to the Directorate of Maize Research in January, 1994 with mandates to organize, conduct, coordinate and generate technology for continuous enhancement in productivity and production of maize.

Late eighties witnessed the launching of single cross hybrid (SCH) breeding programme and adoption of New Seed Policy. SCH breeding activities witnessed many positive changes and accomplishments in generating vital scientific information as well as commercial products. Research efforts were focused on the development of vigorous genetically diverse inbred lines that have good performance *per se* as well as in cross combinations. This resulted in development of high yielding SCHs for different agro-ecological regions of the country. The major strategy therefore, became to evolve and disseminate inbred-based hybrid technology. A total of 212 hybrids and 119 composites of maize have been released till date. These represent a wide range of maturity to cater to the need of farmers in different production ecologies of various states. The improved cultivars have been widely adopted by Indian farmers. The adoption of improved cultivars and production technology had a synergistic effect on crop productivity and provided encouraging results. By 2050, 3.25 times increase in production, 2.2 times increase in productivity and 1.4 times increase in acreage is anticipated (https://iimr.icar.gov.in/attachment/articles/6/Vision%202050.pdf).

Looking to the growing importance of maize and future challenges in Indian agriculture, ICAR strategically upgraded the Directorate of Maize Research to the Indian Institute of Maize Research on 9th February 2015 to further strengthen and fortify the maize research programme of the country for meeting the ever increasing demand of maize in human food, animal feed and industrial raw material.

Varieties

Maize is a cross pollinated crop. Now-a-days plants are quite different from wild species. The maize varieties grown in India comprise the improved selection, varieties, hybrids and composites.

Improved selection: Indigenous (improved local) maize varieties comprise various types of selections arrived at by the farmers, wherein mass selection and its modifications have been used in obtaining a variety to increase frequency of desirable alleles, thus improving the characteristics of populations, being used for seed production. The mass selection is based on maternal parent only and is allowed to open pollinate freely.

Hybrids: Hybridization is the most persuasive technique for breaking yield barriers and this has been used most effectively for maize production in U.S.A. Hybrids are the first generation (F1) from crosses of two pure lines. Hybrids exploits the heterosis vigour. It could be single gene or multiple gene hybrids; or single cross or double cross hybrids. Producing hybrid seed is a costly affair and therefore, hybrid seed is 3-10 times costlier than the recommended common variety seed.

Composites: The term composite is often used as a synonym of synthetic, which is not really true. A composite variety is produced by mixing the seed of several phenotypically outstanding lines and encouraging open pollination to produce crosses in all combinations among themselves. The mixed lines used are rarely tested for their combining ability of strains.

Over the years, many improved varieties, hybrids and composites of maize have been released, suited for cultivation over different agro-ecological regions, growing situations/conditions. These are listed in table 3, 4 and 5.

Table 3: Maize hybrids (H) and composites(C) varieties of different maturity groups for different states for *kharif* season

States	Extra early maturity	Early maturity	Medium maturity	Late maturity
Delhi	H: Vivek 17 & 21, PMH 2	H: PAU 352, PEH 3, Parkash, X 3342	H: HM4, HM 8 10, DK701	H: PMH 3, Buland, NK 61, Pro 311, Bio 9681, Seed Tech 2324
Punjab	H: Vivek 17 & 21, PEEH 5	H: PAU 352, PEH 3, JH 3459, Parkash, PMH 2, X 3342	H: HM4, HM 8 & 10, DK 701	H: PMH 3, PMH-1, Buland, Pro 311, Bio 9681 , NK 61, Pro 311, Seed Tech 2324
Haryana	H: Vivek 17 & 21, PMH 2, PEEH 5	H: HHM 1,PAU 352, Pusa Early Hybrid 3, JH 3459 Parkash, X 3342	H: HM 2, HM 4, 8 & 10 DK 701	H: PMH 3, Buland, ,HM 5, NK 61, Pro 311, Bio 9681, Seed Tech 2324
Uttar Pradesh	H: Vivek 5, 15, 17, 21 & 27 PMH 2	H: JH 3459, Parkash,PEH 2, X 3342 C: Pusa Composite 4,	H: HM 8 & 10, Malviya hybrid Makka 2, Bio 9637, DK 701	H: PMH 3, Buland, Pro Agro 4212, Pro 311, Bio 9681, NK 61, Seed Tech 2324
Rajasthan	H: Pratap hybrid 1, Vivek 4 & 17	H: PEHM 2 ,Parkash ,Pro 368, X 3342 C: PratapMakka 3, AravaliMakka 1, JawaharMakka 8,Amar, Azad Kamal, Pant SankulMakk 3	H: HM 10, NK 21 C: Pratap Makka 5	H: Trishulata, Pro 311, Bio 9681, Seed Tech 2324
Madhya Pradesh	H: Vivek 4 & 17	H: PEHM 2, Parkash, Pro 368, X 3342 C: JawaharMakka 8, Jawahar composite12, Amar, Azad Kamal, Pant SankulMakka 3, Chandramani, PratapMakka 3	H: HM 10, NK 21 C: Pratap Makka 5	H: Trishulata , Pro 311, Bio 9681, Seed Tech 2324
Gujarat	H: Vivek 4 & 17	H: PEHM 2, Parkash, Pro 368, X 3342 C: JawaharMakka 8, Pant Sankul-Makka 3, PratapMakka 3, G M 2,4 & 6 Aravali Makka 1, Narmada Moti	H: HM 10, NK 21 C: Pratap Makka 5	H: Trishulata , Pro 311, Bio 9681, Seed Tech 2324; C: G M 3

Andhra Pradesh	H: Vivek 9, 15, 17& 27, PEEH 5	H: PEHM 1, PEHM 2, DHM 1, BH-2187, Parkash, JKMH 1701, X 3342	H: HM 8 & 10, DHM111,DHM117	H: DHM113,Kargil 900 M, Seed Tech 2324, Pro 311, Bio 9681, Pioneer 30 v 92, Prabal, 30 V 92,
Tamil Nadu	H: Vivek 9, 15, 17, 21& 27, PEEH 5	H: PEHM 2 , Parkash, X 3342 JKMH 1701	H: HM 8 & 10, COHM 4	H: COHM 5, Prabal , Pro 311, Bio 9681, Seed Tech 2324, 30 V 92,
Maharashtra	H: Vivek 9, 15,17, 21&27, PEEH 5	H: PEHM 1& 2, Parkash, X 3342, JKMH 1701	H: HM 8 & 10	H: Prabal, Pro 311, Bio 9681, Seed Tech 2324, 30 V 92,
Karnataka	H: Vivek 9, 15, 21& 27, PEEH 5	H: PEHM 2 , Parkash, X 3342 JKMH 1701C: NAC 6002	H: HM 8 & 10	H: Nithya Shree,EH434042, DMH 1, DMH 2, Bio 9681, Prabal, Pro 311, Seed Tech 2324C: NAC 6004, 30 V 92
Jammu & Kashmir	H: Vivek 15, 21, 25 &33, PEEH 5 C: Pratap Kanchan 2, Shalimar KG 1 & 2,Vivek 35and 37	H: Vivek 33, Parkash, JKMH 1701, X 3342 C: C 8,14 & 15	H: HM 10 C: C 6	-
Uttarakhand	H: Vivek 5, 9, 21& 25 PEEH 5 C: Pratap Kanchan 2,Vivek 35 and 37	H: Vivek hybrid 33, Vivek hybrid 23, Parkash	H: HM 10 C: Bajaura Makka	-
Bihar	H: Vivek 27 C: D 994	H: Parkash, X 3342 C: Dewaki, BirsaVikas Makka 2	H: HM 9, Malviya hybrid makka 2	H: Pro 311, Bio 9681, Seed Tech 2324, 30 V 92, 900 M; C: Hemant, Suwan & Lakshmi
Jharkhand	H: Vivek 27 C: D 994,	H: Parkash, X 3342 C: Dewaki, B V M 2, B M 1	H: HM 9, Malviya hybrid makka 2, DK 701	H: Pro 311, Bio 9681, Seed Tech 2324 C: Suwan
Orissa	H: Vivek 27 C: D 994,	H: Parkash, HIM 129, X 3342	H: HM 9, Malviya hybrid makka 2, DK 701, DMH 115, Pro 345	H: Pro 311, Bio 9681, Seed Tech 2324 , PAC 705
West Bengal	H: Vivek 27	H: Parkash, X 3342	H: Malviya hybrid Makka 2	H: Pro 311, Bio 9681, Seed Tech 2324

Himachal Pradesh	H: Vivek 15, 21& 25, PEEH 5	H: Parkash, X 3342	C: Bajaura Makka, Pratap-Makka 4	H: Pro 311, Bio 9681, Seed Tech 2324
NEH Region	H: Vivek 21& 25, PEEH 5	H: Parkash , JKMH 1701, X 3342	C: Pratap Makka 4	H: Pro 311, Bio 9681, Seed Tech 2324 C: NLD white
Chhattisgarh	H: Vivek 27	H: Parkash, X 3342	C: PratapMakka 5	H: PEHM 1, Pioneer 30 V 92 & 30 R 26, Bio 9681, Pro 4640 & 4642
Assam	-	H: Parkash, X 3342	H: DK 701; C: Pratap Makka 4	C: Vijay, NLD white

H: Hybrids, C: Composites

Table 4. Maize hybrids (H) and composites (C) varieties of late maturity groups for different states for *rabi* season

States	Late maturity
Andhra Pradesh	H: The late maturing hybrids of kharif e.g. Kargil 900 M, Seed Tech 2324, Pro 311, Bio 9681, Pioneer 30 v 92, Prabal, 30V 92, 900 M
Assam	C: NLD white
Bihar	H: Rajendra Hybrid 2, Rajendra Hybrid 1, Pro 311, Bio 9681, Seed Tech 2324, 30 V 92, 900 M C: Hemant, Suwan & Lakshmi
Chhattisgarh	H: PEHM 1, Pioneer 30 V 92 & 30 R 26, Bio 9681, Pro 4640 & 4643, 900 M
Delhi	H: PMH 3, Buland, NK 61, Pro 311, Bio 9681, Seed Tech 2324, HM11, HM8
Gujarat	H: Pro 311, Bio 9681, Seed Tech 2324 C: G M 3, Ganga safed 2
Haryana	H: PMH 3, Buland, ,HM 5, NK 61, Pro 311, Bio 9681, Seed Tech 2324, HM 11, HM 2, HM 1, HM 8
Himachal Pradesh	H: Pro 311, Bio 9681, Seed Tech 2324
Jharkhand	H: Pro 311, Bio 9681, Seed Tech 2324 C: Suwan
Karnataka	H: Nithya Shree, DMH 1, DMH 2, 900 M, Bio 9681, Prabal, Pro 311, Bio 9681, Seed Tech 2324 C: NAC 6004, 30 V 92
Maharashtra	H: Prabal, Pro 311, Bio 9681, Seed Tech 2324, 30 V 92, 900 M
Madhya Pradesh	H: Pro 311, Bio 9681, Seed Tech 2324
NEH Region	H: Pro 311, Bio 9681, Seed Tech 2324 C: NLD white
Odisha	H: Pro 311, Bio 9681, Seed Tech 2324 , PAC 705
Punjab	H: PMH 3, PMH-1 , Buland, Sheetal , Pro 311, Bio 9681 , NK 61, Pro 311, Bio 9681, Seed Tech 2324,HM11,HM8
Rajasthan and U.P.	H: PMH 3, Buland, Pro Agro 4212, Pro 311, Bio 9681, NK 61,Seed Tech 2324, HM 8 H: Pro 311, Bio 9681, Seed Tech 2324, HM 8
Tamil Nadu	H: COHM 5, Prabal , Pro 311, Bio 9681, Seed Tech 2324, 30 V 92, 900 M
West Bengal	H: Pro 311, Bio 9681, Seed Tech 2324

Table 5. Maize hybrids (H) and composites(C) varieties of different maturity groups for different states for *spring* season

States	Extra early maturity	Early maturity
Andhra Pradesh	H: Vivek 9, 15, 17 & 27, PEEH 5	H: PEHM 1, PEHM 2, DHM 1, BH-2187, Parkash, JKMH 1701, X 3342
Assam	-	H: Parkash, X 3342
Bihar	H: Vivek 27 C: D 994, Gujarat Makai 6	H: Parkash, X 3342, C: Dewaki, Birsa Vikas Makka 2
Chhattisgarh	H: Vivek 27	H: Parkash, X 3342

Delhi	H:Vivek 17 & 21, PMH 2	H: PAU 352, PEH 3, Parkash, X 3342
Gujarat	H: Vivek 4 & 17	H: PEHM 2, Parkash, Pro 368, X 3342 C: Jawahar Makka 8,Pant Sankul Makka 3, Pratap Makka 3, G M 2,4 & 6, Aravali Makka 1, Narmada Moti
Haryana	H:Vivek 17 & 21, PMH 2, PEEH 5	H: HHM 1, PAU 352, Pusa Early Hybrid 3, JH 3459 Parkash, X 3342
Himachal Pradesh	H: Vivek 15, 21 & 25, PEEH 5	H: Parkash, X 3342
Jammu & Kashmir	H: Vivek 15, 21, 25 & 33, PEEH 5 C: Pratap Kanchan 2, Shalimar KG 1 & 2	H: Vivek 33, Parkash, JKMH 1701, X 3342, C: C 8,14 & 15
Jharkhand	H: Vivek 27, C: D 994,	H: Parkash, X 3342, C: Dewaki, B V M 2, B M 1
Karnataka	H:Vivek 9, 15, 21 & 27, PEEH 5	H: PEHM 2 , Parkash, X 3342, C: NAC 6002
Maharashtra	H:Vivek 9, 15,17, 21 & 27, PEEH 5	H: PEHM 1 & 2, Parkash, X 3342 , C:, JKMH 1701
Madhya Pradesh	H: Vivek 4 & 17	H: PEHM 2, Parkash, Pro 368, X 3342 C: Jawahar Makka 8, Jawahar composite 12, Amar, Azad Kamal, Pant Sankul, Makk 3, Chandramani, Pratap Makka 3
NEH Region	H: Vivek 21 & 25, PEEH 5	H: Parkash, JKMH 1701, X 3342
Odisha	H: Vivek 27, C: D 994,	H: Parkash, HIM 129, X 3342
Punjab	H: Vivek 17 & 21, PEEH 5	H: PAU 352, PEH 3, JH 3459, Parkash, PMH 2, X 3342
Rajasthan	H: Pratap hybrid 1, Vivek 4 & 17,	H: PEHM 2, Parkash, Pro 368, X 3342 C: Pratap Makka 3, Aravali Makka 1, Jawahar Makka 8, Amar, Azad Kamal, Pant Sankul Makk 3, Mahi Kanchan, Mahi Dhawal
Tamil Nadu	H: Vivek 9, 15, 17, 21 & 27, PEEH 5	H: PEHM 2 , Parkash, X 3342
Uttar Pradesh	H: Vivek 5, 15, 17, 21 & 27 PMH 2,	H: JH 3459, Parkash,PEH 2, X 3342, C: Pusa Composite 4, Gaurav, Azad Uttam, Surya, Kiran, Tarun
Uttarakhand	H: Vivek 5, 9, 21 & 25 PEEH 5 C: Pratap Kanchan 2	H: Vivek hybrid 33, Vivek hybrid 23, Parkash

Time of sowing

Maize can be grown in all seasons viz. *kharif*, post monsoon, *rabi* and spring. During *rabi* and spring seasons, to accomplish higher yield at farmer's field, assured irrigation amenities are essential. In *kharif* season it is desirable to complete the sowing operation 12-15 days before the onset of monsoon. Though, in rainfed areas, the sowing time should be coincided with onset of monsoon. The optimum sowing time for maize is given below (Table 6).

Table 6. Optimum sowing time for maize in Indo-Gangetic Plains

Season	Time of sowing
Kharif	Last week of June to 15th July
Rabi	As inter cropping - Last week of October, Sole crop - Up to 15th November
Spring	Ist week of February

Usually, there are many benefits related to early planting date compared to late planting date and this include a long growth duration that allows a greater choice of hybrid maturities and wider window of opportunities for replant decisions. Early planting date could contribute significantly to higher maize yields. The higher yield is not the only advantage of early planting because other benefits can also be achieved from high plant density and high fertilizer rates. It also allows harvesting earlier in the season when conditions are usually better and field and time losses can be minimized (Varma *et al*., 2014).

Seed rate and plant geometry

To achieve higher productivity and resource-use efficiencies, optimum plant stand is the key factor. The seed rate varies depends on purpose, seed size, plant type, season, sowing methods etc. The following crop geometry and seed rate should be adopted (Table 7).

Table 7. Seed rate and crop geometry of maize

Purpose	Seed rate (kg/ha)	Plant geometry (plant × row, cm)	Plant population (No.)
Grain normal and Quality Protein Maize (QPM)	20.0	60 × 20	83,333
		75 × 20	66,666
Baby corn	25.0	60 × 20	83,333
		60 × 15	1,11,111
Pop corn	12.0	60 × 20	83,333
Sweet corn	8.0	75 × 25	53,333
		75 × 30	44,444
Fodder	50.0	30 × 10	3,33,333

Source: Parihar *et al.* (2011)

It has been reported that the optimal plant population is dependent on rainfall, and that maize grain yield varies significantly across environments with different climatic conditions. Overall, results suggest that plant populations of 90,000 to 1,20,000 plants/ha are optimal to maximize maize grain yield across most rainfall regions. When the effects of plant population on maize grain yield were investigated for conventional tillage (CT) and no-tillage (NT) systems, the optimum plant population was lower for NT than CT systems, but that, at a given plant population, maize grain yields were higher in NT than in CT systems (Haarhof and Swanepoe, 2018).

Seed treatment

To protect the maize crop from soil and seed borne diseases and insect-pests, seed treatment with fungicides and insecticides before sowing is recommended (Table 8). The details are as follows:

Table 8. Important diseases and insect-pests of maize

Diseases/insect-pests	Fungicide/Pesticide	Rate of application (g/kg seed)
Sheath blight, Turcicum leaf blight, Maydis leaf blight and Banded leaf	Bavistin + Captan in 1:1 ratio	2.0
Pythium stalk rot	Captan	2.5
BSMD	Apran 35 SD	4.0
Termite and shoot fly	Imidachlorpid	4.0

Tillage and crop establishment

Tillage and crop establishment is the key for achieving the optimum plant stand that is the main driver of the crop yield. Though the crop establishment is a series of events (seeding, germination, emergence and final establishment) that depends on interactions of seed, seedling depth, soil moisture, method of sowing, machinery etc but, the method of planting plays a vital role for better establishment of crop under a set of growing situation. Maize is mainly sown directly through seed by using different methods of tillage & establishment but during winters where fields are not remain vacant in time (till November), transplanting can be done successfully by raising the nursery. However, the sowing method (establishment) mainly depends on several factors *viz.* the complex interaction over time of seeding, soil, climate, biotic, machinery and management season, cropping system, etc. Recently, resource conservation technologies (RCTs) that include several practices viz. zero tillage, minimum tillage, surface seeding etc. had came in practice in various maize based cropping system and these are cost effective and environment friendly. Therefore, it is very important that different situations require different sowing methods for achieving higher yield as described below:

Raised bed (ridge) planting: Generally the raised bed planting is considered as best planting method for maize during monsoon and winter seasons both under excess moisture as well as limited water availability/rainfed conditions. Sowing/planting should be done on the southern side of the east-west ridges/ beds, which helps in good germination. Planting should be done at proper spacing. Preferably, the raised bed planter having inclined plate, cupping or roller type seed metering systems should be used for planting that facilitates in placement of seed and fertilizers at proper place in one operation that helps in getting good crop stand, higher productivity and resource use efficiency. Using raised bed planting technology, 20-30% irrigation water can be saved with higher productivity.

Moreover, under temporary excess soil moisture/water logging due to heavy rains, the furrows will act as drainage channels and crop can be saved from excess soil moisture stress. For realizing the full potential of the bed planting technology, permanent beds are advisable wherein sowing can be done in a single pass without any preparatory tillage. Permanent beds are more beneficial under excess soil moisture situations as the infiltration rate is much higher and crop can be saved from the temporary water logging injury. Singh and Vashist (2016) reported that crop planted on southern slope of east-west raised beds and ridges recorded 18.0-18.1% higher grain yield, 19.7-21.2% higher irrigation water productivity (IWP) over variable cost of comparison to flat planted crop (Table 9).

Table 9. Grain yield and irrigation water productivity of spring maize as influenced by methods of planting

Treatment	Grain yield (tonnes/ha)	Total irrigation water applied (cm)	IWP (kg/ha-mm)
Methods of planting			
Bed	8.33	58.9	15.2
Ridge	8.34	59.0	15.4
Flat	7.06	60.0	12.7
CD (*P=0.05*)	0.27	-	0.5

Source: Singh and Vashist (2016)

Zero-till planting: Maize can be successfully grown without any primary tillage under no-till situation with less cost of cultivation, higher farm profitability and better resource use efficiency. Under such condition one should ensure good soil moisture at sowing and seed and fertilizers should be placed in band using zero-till seed-cum-fertilizer planter with furrow opener as per the soil texture and field conditions. The technology is in place with large number of farmers particularly under rice-maize and maize-wheat systems in peninsular and eastern India. However, use of appropriate planter having suitable furrow opener and seed metering system is the key of success of the no-till technology.

Conventional planting: Under heavy weed infestation where chemical/ herbicidal weed management is uneconomical in no-till and also for rainfed areas where survival of crop depends on conserved soil moisture, in such situations conventional planting can be done using seed-cum-fertilizer planters.

Furrow planting: To prevent evaporative losses of water during spring season from the soil under flat as well as raised bed planting is higher and hence crop suffers due to moisture stress. Under such situation/condition, it is always advisable to grow maize in furrows for proper growth, seed setting and higher productivity.

Transplanting: Under intensive cropping systems where it is not possible to vacate the field on time for planting of winter maize, the chances of delayed planting exists. Due to delay planting crop establishment is a problem due to low temperature so under such conditions transplanting is an alternative and well established technique for winter maize. Therefore, for the situation where fields are vacated during December-January, it is advisable to grow nursery and transplant the seedlings in furrows and apply irrigation for optimum crop establishment. Use of this technique helps in maintenance of temporal isolation in corn seed production areas for production of pure and good quality seed as well as quality protein maize grain. For planting of one ha, 700 m^2 nursery area is required and the nursery should be raised during second fortnight of November. The age of seedlings for transplanting should be 30-40 days old (depending on the growth) and transplant in the month of December-January in furrows to obtain higher productivity.

Maize based cropping systems in India

As maize has wide adaptability and compatibility under diverse soil and climatic conditions and hence it is cultivated in sequence with different crops under various agro-ecologies. Hence, it is considered as one of the potential driver of crop diversification under different situation. Among different maize based cropping systems, maize-wheat ranks 1st having 1.8 million ha area mainly concentrated in rainfed ecologies. Maize-wheat is the 3rd most important cropping systems after paddy-wheat and paddy-paddy that contributes about 3% in the national food basket. The other major maize systems in India are maize-mustard, maize-chickpea, maize-maize, cotton-maize etc. Recently, due to changing scenario of natural resource base, paddy-maize has emerged a potential maize based cropping system in peninsular and eastern India (Table 10).

Table 10. Maize based sequential cropping systems in different ago-climatic zones of India

Agro-climatic region	Cropping system	
	Irrigated	Rainfed
Western Himalayan region	Maize-wheat, Maize-potato-wheat, Maize-wheat-moong bean, Maize-mustard, Maize-sugarcane	Maize-mustard, Maize-legumes
Eastern Himalayan region	Summer paddy-maize-mustard, Maize-maize, Maize-maize-legumes	Sesame-paddy+maize,
Lower Gangetic plain region	Autumn paddy-maize, Jute-paddy-maize	Paddy-maize
Middle Gangetic plain region	Maize-early potato-wheat-moong bean, Maize-wheat, Maize-wheat-moong bean, Maize-wheat-urdbean, Maize-sugarcane-moong bean	Maize-wheat
Upper Gangetic plain region	Maize-wheat, Maize-wheat-moong bean, Maize-potato-wheat, Maize-potato-onion, Maize-potato-sunflower, Paddy-potato-maize, Maize-potato-sugarcane-ratoon	Maize-wheat, Maize-barley, Maize-safflower
Trans Gangetic plain region	Maize-wheat, Maize-wheat-moong bean, Maize-potato-wheat, Maize-potato-onion, Maize-potato-sunflower, Moong bean-maize-toria-wheat, Maize-potato-moong bean	Maize-wheat
Eastern plateau & hills region	Maize-groundnut-vegetables, Maize-wheat-vegetables	Paddy-potato-maize Jute-maize-cowpea
Central plateau & hills region	Maize-wheat	Maize-groundnut
Western plateau & hills region	Sugarcane + maize	
Southern plateau & hills region	Paddy-maize, Maize-paddy	Sorghum-maize, Maize- sorghum-pulses, Maize-potato - groundnut
East coast plain and hills region	Paddy-maize-pearl millet, Paddy-maize, Maize-Paddy, Paddy-paddy-maize	Maize-maize- pearl millet, Paddy-maize + cowpea

West coast plain and hills region	Maize-pulses, Paddy-maize	Paddy-maize, groundnut- maize
Gujrat plains and hills region	Maize-wheat	Paddy-maize
Western dry region	Maize-mustard, Maize-chickpea	Maize + legumes
Island region	Paddy-maize	Maize-paddy, Paddy-maize + cowpea, Paddy-maize-urdbean, Paddy- paddy -maize

Source: Parihar *et al.* (2011)

In peri-urban interface, maize based high value intercropping systems are also gaining importance due to market driven farming (Table 11). Further, maize has compatibility with several crops of different growth habit that led to development of various intercropping systems. Studies carried out under various soil and climatic conditions under All India Coordinated Research Project on Cropping Systems revealed that compared to existing cropping systems like rice-wheat and rice-rice, maize based cropping systems are better user of available resources and the water use efficiency of maize based cropping systems was about 100 to 200% higher at different locations.

Table 11. Maize based intercropping systems

Suitable area/situation	Intercropping systems
All maize growing areas	Paddy + Maize, Maize + Pigeon pea, Maize + Cowpea, Maize + Soybean, Maize + Mungbean / Urdbean, Maize + Sugarcane
Peri-urban interface	Maize + high value vegetables, Maize + flowers Baby corn + vegetables, Sweet corn + vegetables

Source: Parihar *et al.* (2011)

Maize and climate change

Under the changing climate scenario the limitations of rising temperature during grain filling of wheat particularly in eastern India, and declining yield of boro rice in West Bengal and Orissa, water scarcity areas in peninsular India (AP and Tamil Nadu) affecting yield of *rabi* rice, maize being a photo-insensitive crop has better options for adaptation and mitigation of these climatic changes. Peninsular India is considered to be a neutral environment for maize wherein maize can be cultivated in either of the seasons. Therefore, it is emerging as a potential driving force for diversification i.e. diversification of rice-rice with rice-maize and other maize based high value cropping systems in water scarcity/lowering of water table is a major concern in rice growing belt of India and making rice cultivation non-remunerative. Hence, maize has emerged as a potential as well as profitable crop in these areas. The *rabi* rice in Peninsular India and upland rice in Odisha and North-east hill (NEH) region has low productivity. Therefore, maize is only suitable alternative crop and more area is likely to shift towards maize cultivation in near future in these non-traditional areas. Wheat crop adversely affected with terminal heat due to sudden rise in temperature during crop growth and maturity but this favours maize crop positively.

In view of the changing farming scenario in the country, maize has been emerging as one of the potential crops that addresses several issues like food and nutritional security, climate change, water scarcity, farming systems, bio-fuel etc. Further, a recent study by National Centre for Agricultural Economics and Policy Research (NCAP) has showed that there is an increasing demand for maize in the industry sector which caters to consumer needs like textiles, paper, glue, alcohol, confectionery, food processing and pharmaceutical industry etc., of which the demand keeps on increasing with population pressure.

Nutrient management

Among all the cereals, maize in general and hybrids in particular are responsive to nutrients applied either through organic or inorganic sources. The rate of nutrient application depends mainly on soil nutrient status/balance and cropping system. For obtaining desirable yields, the doses of applied nutrients should be matched with the soil supplying capacity and plant demand (Site specific nutrient management approach) by keeping in view of the preceding crop (cropping system). Response of maize to applied organic manures is notable and hence integrated nutrient management (INM) is very important nutrient management strategy in maize based production systems. Therefore, for higher economic yield of maize, application of 10 tonnes FYM/ha, 10-15 days prior to sowing supplemented with 150-180 kg N, 70-80 kg P_2O_5, 70-80 kg K_2O and 25 kg $ZnSO_4$ per ha is recommended. Full doses of P, K and Zn should be applied as basal preferably drilling of fertilizers in bands along the seed using seed-cum-fertilizer drills. Nitrogen application at grain filling results in better grain filling. Therefore, nitrogen should be applied in 5 splits as per below mentioned crop growth stages for higher productivity and nitrogen use efficiency (Table 12).

Table 12. Nitrogen synchronization with growth stages

S. No.	Crop stage	Nitrogen rate (%)
1	At sowing (basal)	20
2	4 leaf stage (V_4)	25
3	8 leaf stage (V_8)	30
4	Tasseling stage (V_T)	20
5	Grain filling stage (GF)	5

Nutrient deficiencies in crops reduce yields, quality and profits to the farmer. Yield can often be reduced 10-30% by deficiencies of major nutrients before any clear symptoms of deficiency are observed in the field. Application of 150 kg/ha N, 75 kg/ha P and 37.5 kg/ha K along with 25 kg/ha $ZnSO_4$ to maize (Table 13) recorded significantly higher grain (5.42 tonnes/ha) and stover (7.61 tonnes/ha) yields (Hiremath *et al.*, 2016).

Table 13. Grain and stover yield of maize affected by different nutrients

Treatment	Grain yield (tonnes/ha)	Stover yield (tonnes/ha)
Control	1.57	2.29
N (150 kg/ha)	2.98	4.27
NP (150+75 kg/ha)	3.86	5.54
NK (150+37.5 kg/ha)	3.56	5.05
NPK (150+75+37.5 kg/ha)	4.89	6.94
NPK+$ZnSO_4$ (150+75+37.5+25 kg/ha)	5.45	7.61
Farmer practices	3.28	4.67
CD (*P*=*0.05*)	0.099	0.147

Source: Hiremath *et al.* (2016)

Water management

The irrigation water management depends on season as about 80% of maize is cultivated during monsoon season particularly under rainfed conditions. However, in areas with assured irrigation facilities are available, depending upon the rains and moisture holding capacity of the soil, irrigation should be applied as and when required by the crop and first irrigation should be applied very carefully wherein water should not overflow on the ridges/beds. In general, the irrigation should be applied in furrows up to 2/3rd height of the ridges/beds. Young seedlings, knee high stage (V_8), flowering (V_T) and grain filling (GF) are the most sensitive stages for water stress and hence irrigation should ensure at these stages. In raised bed planting system and limited irrigation water availability conditions, the irrigation water can also be applied in alternate furrow to save more irrigation water. In rainfed areas, tied-ridges are helpful in conserving the rainwater for its availability in the root zone for longer period. For winter maize, it is advisable to keep soil wet (frequent and mild irrigation) during 15 December to 15 February to protect the crop from frost injury. Singh and Vashist (2016) observed that the irrigation water productivity was recorded (18.2 kg/ha-mm) highest in IW: CPE ratio of 0.6, while the grain yield and economic returns was the highest with irrigation at 0.9 IW: CPE ratio (Table 14).

Table 14. Grain yield and irrigation water productivity of spring maize as influenced by irrigation regime

Treatment	Grain yield (tonnes/ha)	Total irrigation water applied (cm)	Irrigation water productivity (kg/ha-mm)
Irrigation regime (IW: CPE ratio)			
1.2	8.85	81.8	11.0
0.9	8.50	60.4	14.3
0.6	6.38	36.6	18.2
CD (*P=0.05*)	0.44	-	0.5

Source: Singh and Vashist (2016)

Ramachandiran and Pazhanivelan (2016) conducted a field experiment at the Tamil Nadu Agricultural University, Coimbatore, during the winter (*rabi*) seasons of 2013 and 2014. Irrigation at IW: CPE ratio of 0.80 along with 250 kg N/ha (100% recommended dose of nitrogen) favorable increased the higher growth parameters, nutrient uptake, yield attributes and higher grain yield (7.31 tonnes/ha) of maize.

Weed management

Weeds are the major problem, particularly during monsoon/*kharif* season as they compete with plants for nutrients, water, space and sunlight and causes yield losses up to 35%. Therefore, timely weed management is needed for achieving higher yield. Atrazine being a selective and broad-spectrum herbicide in maize control the emergence of wide spectrum of weeds. Pre-emergence application of Atrazine @ of 1.0-1.5 kg a.i/ha in 600 liters water, Alachlor @ 2-2.5 kg a.i/ha, Metolachlor @ 1.5-2.0 kg a.i/ha, Pendamethalin @ 1-1.5 kg a.i./ha are effective way for control of various annual and broad leaved weeds. Atrazine followed by tembotrione @1,000 and 100 g/ha (resp.) pre-emergence (1-2 days after sowing) + post-emergence (25-30 days after sowing) to controls a broad spectrum of weeds and can render maize field almost free of weeds including sedges.

2,4-D applied @ 750 post-emergence (30-35 days after sowing) to controls broad leaved weeds and few sedges. While spraying, following precautions should be take care by the person during spray, he should move backward so, that the Atrazine film on the soil surface may not be disturbed. Preferably 3 boom flat fan nozzle should be used for suitable coverage of ground and time saving. One to two hoeing are recommended for aeration and uprooting of the remaining weeds, if any. While doing hoeing, the person should move backward to avoid compaction and better aeration. For areas where zero tillage is practiced, pre-plant application (10-15 days prior to seeding) of non-selective herbicides viz., Glyphosate @ 1.0 kg a.i./ha in 400-600 liters water or Paraquat @ 0.5 kg a.i./ha in 600 liters water is recommended to control the weeds. Under heavy weed infestation, post-emergence application of Paraquat can also be done as protected spray using hoods.

Crop Protection

Insect-pest management

Stem borer (*Chilopartellus*): Major pest of maize in India is Stem borer (*Chilopartellus*) popularly known as stalk borer that occurs during monsoon season throughout the country. *Chilo* lays eggs 10-25 days after germination on lower side of the leaves. The larva of the *Chilo* enters in the whorl and cause damage in the leaves.

Pink borer (*Sesamia inference*): Pink borer occurs during winter season particularly in peninsular India. The moth of the *Sesamia* is nocturnal and lays eggs on lower leaf sheath. The larvae of the *Sesamia* enter the plant near the base and cause damage to stem.

Management of stem and pink borer: For control of *Chilo* and *Sesamia*, foliar spray of Carbofuron 3G: 1,000 g a.i./ha or Flubendamide 20WG: 25 g a.i./ha is very effective. The *Chilo* can also be controlled by release of 8 Trichocards (*Trichogramma chilonis*) per ha at10 days after germination. Intercropping of maize with suitable varieties of cowpea is an eco-friendly option for reducing the incidence of *stem borer* on maize.

Shoot fly (*Atherigona* sp.): It is a serious pest in South India, but it also appears on summer and spring maize crop in North India. It attacks mainly at seedling stage of the crop. The tiny maggots creep down under the leaf sheaths till they reach the base of the seedlings. After this they cut the growing point or central shoot which results in to 'dead heart' formation.

Management of shoot fly: Sowing of crop must be completed before first week of February so that the crop will escape shoot fly infestation. For spring sowing, crop seed must be treated with Imidacloprid @ 6 ml/kg seed.

Termites (*Odontotermes obesus*): Termite is also an important pest in many areas. For management of termite fepronil granules should be applied @ 20 kg/ha followed by light irrigation. If the termite incidence is in patches, than spot application of fepronil @ 2-3 granule/plant should be done. Clean cultivation delays termite attack.

Other promising pests: Recently some other non-traditional pests are also causing damage to maize crop viz. larvae of American bollworm (*Helicoverpa armigera*) which causes damage to cob in Southern part of India while the Chaffer beetle (*Chiloloba acuta*) feeds on maize pollen which adversely affects pollination in northern part of India.

Disease management

Across the country several diseases occurs during different seasons, if they are not managed at proper time then they leads to yield loss. Estimated losses due to major diseases of maize in India is about 13.2% of which foliar diseases (5%), stalk rots, root rots, ear rots (5%) cause major yield losses. The major diseases and their management practices are described as below:

Downy mildews (DM): Himachal Pradesh, Sikkim, West Bengal, Meghalaya, Punjab, Haryana, Rajasthan, Delhi, Uttar Pradesh, Bihar, Madhya Pradesh and Gujarat are prone to downy mildews. The main symptoms of downy mildew are legends developing on lower leaves as narrow chlorosis strips. Strips extend in parallel fashion, well defined margined delimited by veins. Downy whitish to creamy growth usually on the ventral surface of the infected leaves appears corresponding to stripes. For control of downy mildew, the infected plants should be rogue out and destroyed. The planting of crop before onset of rains minimizes the incidence of mildew. Seed should be treated with fungicides like Apron 35 WS @ 2.5 g/kg seed. Also the resistant varieties should be used.

Banded leaf and sheath blight (BLSB): This disease mainly occurs in Jammu and Kashmir, Himachal Pradesh, Sikkim, Punjab, Haryana, Rajasthan, Madhya Pradesh, Delhi, Uttar Pradesh and Bihar. At appearance of the disease, white lesions develops on leaves and sheath. Purplish or brown horizontal bands present on white lesions characterize the disease. Seed treatment with peat based formulation (*Pseudomonas fluorescence)* @ 16 g/kg of seed or as soil application @ 7 g/liter of water (soil drenching) or foliar spray of Sheethmar (Validamycin) @ 2.7 ml /liter water provides effective control of the disease. Stripping of 2 lower leaves along with leaf sheath also gives effective control of the disease.

Turcicum leaf blight (TLB): This disease is distributed in Jammu & Kashmir, Himachal Pradesh, Sikkim, West Bengal, Meghalaya, Tripura, Assam, Uttar Pradesh, Uttarakhand, Bihar, Madhya Pradesh, Gujarat, Karnataka and Tamil Nadu. At its appearance, it shows long, elliptical, grayish-green or tan lesions ranging from 2.5 to 15 cm in length on the leaves. For control of TLB, spray Zineb/Meneb @ 2.5-4.0 g/liter of water (2- 4 applications) at 8-10 days interval. The crop debris should be ploughed down. Also, the resistant cultivars should be grown.

Maydis leaf blight (MLB): MLB also occurs in wide range of maize growing states like Jammu & Kashmir, Himachal Pradesh, Sikkim, Meghalaya, Punjab, Haryana, Rajasthan, Delhi, Uttar Pradesh, Bihar, Madhya Pradesh, Gujarat, Maharashtra, Andhra Pradesh, Karnataka & Tamil Nadu. It shows the symptoms as lesions on the leaves elongated between the veins, tan, 2-6 × 3-22 mm with limited margins with buff to brown borders. For effective control of this disease, spray of Dithane Z-75 or Zineb @ 2.4 – 4.0 g/liter of water (2-4 applications) at 8-10 days interval after first appearance of symptoms of disease. In addition, the crop debris should be ploughed down. Also, the resistant cultivars should be grown.

Polysora rust: This disease appears mainly in peninsular India i.e. Andhra Pradesh, Karnataka and Tamil Nadu. The main symptoms of the disease shows appearance of circular to elongate light cinnamon brown, circular to oval 0.2-2.0 mm long densely scattered lesions on the upper leaf. The uredospores are yellowish to golden in colour. For effective control of polysora rust, three sprays of Dithane M-45 @ 2-2.5 g/liter beginning from first appearance of symptoms at 15 days interval are required. It is always advisable to use resistant varieties.

Post flowering stalk rot of maize (PFSR): The PFSR occurs mainly in Rajasthan, Uttar Pradesh, Bihar and Andhra Pradesh. Disease appears when the crop enters in senescence phase. The pathogen commonly affects the roots crown regions and lower internodes. When split open, the stalk shows pink-purple dis-colouration. For effective control of the disease, water stress at flowering should be avoided. Use balance dose of nutrients wherein potassium application helps in minimizing the disease. Use of bio-control agents (*Trichoderma* formulation) in furrows mixed with FYM @ 10 g/kg at 10 days prior to its use in the field. It is always advisable to practice crop rotation to minimize the disease incidence.

Production technology for quality protein maize specialty and other corn types

Other than grain, maize is also cultivated for various purposes like quality protein maize (QPM) and other special purposes known as 'Specialty corn' (Table 15). The various specialty corn types are QPM, baby corn, sweet corn, pop corn, waxy corn, high oil corn etc. In India, QPM, baby corn and sweet corn are being popularized and cultivated by the large number of farmers. The brief summary of different type of specialty maize is as follows.

***Quality protein maize*:** As more than 85% of the maize is used directly for food and feed, the quality has a great role for food and nutritional security in the country. In this respect, discovery of Opaque-2 (O2) and floury-2 (F2) mutant had opened tremendous possibilities for improvement of protein quality of maize which later led to the development of QPM. QPM which is nutritionally superior over the normal maize is the new dynamics to signify its importance not only for food and nutritional security but also for quality feed for poultry, piggery and animal sectors as well. QPM. has specific features of having balanced amount of amino acids with high content of lysine and tryptophan and low content of leucine & isoleucine. The balanced proportion of all these essential amino acid in QPM. enhances the biological value of protein. The biological value of protein in QPM. is just double than that of normal maize protein which is very close to the milk protein as the biological value of milk and QPM proteins are 90% and 80%, respectively. Whereas it is less than 50% in normal maize protein. There are 9 QPM. hybrids of different grain colours have been developed and released in India for their cultivation in different agro-climatic conditions across the country. The production technology of QPM is same as of normal grain maize except isolation as to maintain the purity of QPM, it should be grown in isolation with normal maize.

***Baby corn*:** Baby corn is a young finger like unfertilized cobs with one to three centimeter emerged silk preferably harvested within 1-3 days of silk emergence depending upon the growing season. It can be eaten raw as salad and in preparation of different recipes such as chutney, pakora, mix vegetables, pickles, candy, murabba, kheer, halwa, raita, etc. The desirable size of baby corn is 6 to 11 cm length and 1.0 to 1.5 cm diameter with regular row/ovule arrangement. The most preferred colour by the consumers/exporters is generally creamish to very light yellow. Baby corn is nutritive and its nutritional quality is at par or even superior to some of the seasonal vegetables. Besides proteins, vitamins and iron, it is one of the richest sources of phosphorus. It is a good source of fibrous protein and easy to digest. It is almost free from residual effects of pesticides. It can be cultivated round the year therefore, three to four crops of baby corn can be taken in a year. Cost of cultivation of baby corn in India is lowest in the world therefore; India can become one of the major baby corn producing country. It has great potential both for internal consumption and export.

In general, the cultivation practices of baby corn are similar to grain crop except (i) higher plant population (ii) higher dose of nitrogen application because of higher plant population (iii) preference for early maturing single cross hybrid and (iv) harvesting within 1-3 days of silk emergence.

Table 15. Specialty corn cultivars released in India (1969 to 2017)

Corn type	Cultivar	Nature of cultivar	Maturity (days)	Av. yield (t/ha)	Crop season
Quality Protein Maize (QPM)	Pusa HM-8 Improved (AQH-8)	SCH	90-95	6.3	K
	Pusa HM-9 Improved (AQH-9)	SCH	85-90	5.2	K
	Pusa HM-4 Improved (AQH-4)	SCH	90-95	6.4	K
	PusaVivek QPM-9 Improved (Andhra PradeshQH-9)	SCH	80-85	5.6	K
	Pratap QPM Hybrid1 (EHQ-16)	SCH	Medium	5.9	K
	HQPM-4	SCH	Late (95)	5.4	K
	HQPM-7	SCH	Late (97)	7.2	K
	Vivek QPM 9 (FQH 4567)	SCH	Extra-early (83)	5.0	K
	HQPM-5	SCH	Late (98)	5.8	K
	HQPM-1	SCH	Late (89)	7.5	K&R
	Shaktiman-3	SCH	Late (127)	9.5	K&R
	Shaktiman-4	SCH	Late (127)	12.0	K&R
	Sakthiman-2	SCH	Late	6.0	K
	Shakti 1	OPV	115	4.5	R
	Protina	OPV	Late (103)	4.3	K
	Rattan	OPV	Late (103)	6.3	K
	Shakti	OPV	103	6.3	K
	Shalimar QPMH-1 (KDQH -49)	N.A.	N.A.	N.A.	N.A.
Sweet corn	Hy-brix 53 (ADVSW-2)	SCH	90-100	13.7	K
	Hi-brix-39 (ADVSW-1)	SCH	90-100	13.8	K
	Central Maize VL Sweet Corn 1 (FSCH18)	SCH	90-95	10.8	K
	CANDY (KSCH333)	SCH	Early	11.9	K
	NSCH-12 (Misthi)	SCH	Medium	14944 (Green ear yield)	K
	HSC 1	SCH	Medium (81)	12	K
	Win Orange Sweet Corn	OPV	Medium	N.A.	K
	Madhuri (Sweet Corn)	OPV	Early (80)	1.6	K&R
Pop corn	DMRHP 1402	SCH	Early (75-77)	3.9	Kharif

	Shalimar Pop Corn-1 (KDPC-2)	OPV	95-100	3.9	Kharif
	Jawahar Pop Corn11	OPV	Early to Medium	2.3	K
	Pearl Popcorn	OPV	Medium	2.6	K
	VL Amber Pop Corn	OPV	N.A.	3.0	K
Baby corn	Vivek Hybrid 27 (Central Maize VL Baby Corn 2)	SCH	95-97	2.2	K
	HM-4	SCH	Medium (86)	8.4	K&R
	VL Baby Corn-1 (VL-78)	OPV	Extra-early (93)	1.2	K
	COBC-1	OPV	Early (103)	2.0	K
Starch hybrid	Hi-Starch	TWC	Late (103)	6.1	K
	Hybrid Maize Hi-starch	TWC	Late (103)	6.1	K
Fodder	Partap Makka Chari6 (EC 3135)	OPV	Medium (90)	4.3	K
	J-1006	OPV	Medium	4.0	K
	African Tall Composite	OPV	Late	3.0	K

Source: https://iimr.icar.gov.in/attachment/articles/52/Specialty% 20corn%20 Cultivars % 20 1969-17. pdf
SCH: Single cross hybrid, OPV: Open pollinated, TWC: Three way cross, N.A.: Not available
Extra-early maturity ≤ 75 days, Early maturity ≤ 85 days, Medium maturity ≤ 95 days, Late maturity ≥ 96 days

***Sweet corn*:** Sweet corn is one of the most popular vegetables in the U.S.A., Europe and other developed countries of the world. It is a very delicious and rich source of energy, vitamin C and A. It is eaten as raw, boiled or steamed green cobs/grain. It is also used in preparation of soup, salad and other recipes. It is becoming very popular in urban areas of country therefore; its cultivation is remunerative for peri-urban farmers. Besides green cobs the green fodder is also available to the farmers for their cattle. Generally, sweet corn is early in maturity. It is harvested in 70-75 days during *kharif* season. Green cobs are harvested after18-20 days of pollination during *kharif* but the duration may vary season to season. At the harvest time the moisture is generally 70% in the grain and sugar content varies from 11 to 20%. Sweet corn is generally dull yellow and white but dull yellow color is preferred.

Its picking should be done in the morning or evening time. Green cobs should be immediately transported to the cold storage at refrigerated trucks to avoid the conversion of sugar to starch. It loses flavor if kept in high temperature after picking. Sweet corn with high sugar content should not be planted when temperature is below 16 °C.

***Pop corn*:** Pop corn is one of the common snack items in many parts of the world, particularly in cities and is liked because of its light, porous and crunchy texture. The pop corn flour can also be used for preparing many traditional dishes.

It is consumed fresh, as it has to be protected against moisture absorption from the air. It is hard endosperm flint maize. Kernels of pop corn are very small and oval/round in shape. When heated at about 170 °C, the grains swell and burst, turning inside out. Quality of pop corn depends on popping volume and minimum number of non pop corn.

***Waxy corn*:** It is originated in China but largely used in U.S.A. Grain gives wax-like appearance and having 100% amylopectin starch. While in normal maize, the starch is nearly 30 per cent amylose and the remaining 70% is amylopectin. Waxy corn is mainly used for food and industrial purposes.

***High oil corn*:** Most of the normal maize lines have 3- 4% oil content. In general, lines with more than 6 % oil are considered high oil lines. 95% of the total oil is in the germ. When the oil percent increases the starch decreases. The wet milling industries are still in advantage with high oil content corn. In U.S.A. the high oil corn is cultivated on contractual basis and remunerative price is paid to the farmers. In India its cultivation is not economical because it is not sold on premium basis. Generally in normal maize crop, 15-20% population of high oil hybrids is used as pollen parent and there is detasseling of the normal corn plant. Due to xenia effect there is an increase of oil in normal maize and its cultivation is done in isolation. The corn oil has low content of saturated fatty acid and is considered to be one of the best quality cooking oil. In India more than 60,000 tonnes of corn oil is made available for various uses.

***Fodder maize*:** Maize fodder can be used at any crop growth stage. Its quality is adversely affected after anthesis. To maintain the fodder quality the detasseling is advised to the farmers for better digestibility and palatability. By grazing this fodder to the milch cattle, their milk is increased. The tall, leafy and longer duration cultivars are most preferred for maize fodder cultivation. The cultivation of maize for fodder can be done round the year. Very high seed rate is used. Generally the farmers grow composite varieties or advance generation of hybrid seed which is economical to the farmers.

Single cross hybrid maize seed production

***Isolation Distance*:** Seed production should be taken in fertile well drained, weed and disease free soil and preferably the fields where preceding crop was not maize to minimize rouging and maintain the genetic purity. At least 400-500 metre distance is required to avoid any contamination.

***Male: female ratio*:** The male: female ratio depends on (a) pollen shedding potential and duration of male parent; (b) male: female synchrony : for better seed setting flowering of female should be earlier than male or male pollen dehiscence should coincide with female silking and (c) season. In general the male: female ratio should be 1:2 or 1:3 or 1:4.

***Time of Sowing*:** To avoid flowering from heavy rains during *kharif* and low/ high temperature during winter season the optimum time of sowing is first week of July during *kharif* and first week of November during winter.

***Method of sowing and layout*:** It is desirable to plant the crop on ridges. Depending upon the plant type the row and plant spacing should be kept at 60-75 cm and 20 cm, respectively. Identification labels/tags should be put on the male and female lines to distinguish between them.

***Seed rate*:** The seed rate depends on size of seed/ test weight, plant type and male: female ratio. 15 kg/ha for female and 10 kg/ha for male is recommended.

***Removal of off-type plants and thinning*:** After 12-15 days of sowing: off-type and excess plants should be removed. Proper plant to plant distance should be maintained. At knee high stage all the dissimilar plants should be removed and at flowering remove dissimilar tassel bearing plant before anthesis from the male.

***Detasseling*:** Detasseling in female should be done before anthesis. It should be practiced row-wise. One person should follow to monitor the each row to check that no part of the tassel is left inside. The process of detasseling should continue for 8-10 days. While detasseling, leaf should not be removed which will reduce the yield. It has been observed that the removal of 1 to 3 leaves along with tassel reduces 5-15% yield. The removed tassel should not be thrown in the field but fed to the cattle as it is nutritive fodder.

***Harvesting*:** If possible male parent should be harvested after pollination. Optimum moisture content in grain at harvesting should be around 20%. The harvested cobs should spread evenly instead of making heap.

Stages of crop inspection

At the time of sowing: to monitor the land, isolation distance, planting ratio of male: female, proper sowing time, seed treatment.

During pre-flowering/vegetative stage: to verify the rouging and removal of off type plants.

During flowering stage: to check disease and pest infestation.

During post-flowering and pre-harvest stage: to remove the late and diseased plants differential type of tassel/silk plants.

Drying and sorting of seed parent cobs The drying of the cobs should not be done either on the kuccha or pucca flour, rather it should be dried on tarpoline sheets to avoid seed injury and during night the cobs should be kept covered. To maintain the purity, dissimilar, diseased and pest infested cobs should be removed before shelling. The female cobs should be dried up to 13-14% moisture content before shelling.

Shelling: Shelling of female parent should be done earlier than male to avoid mechanical mixture. Shelling can be done manually or by power operated maize Sheller.

Seed processing: All under size, broken, damaged etc seeds should be removed for maintaining the quality of seed.

Storage and marketing: Seed drying should be done till the moisture content of the seed is reduced to 8% and it should be kept in aerated jute bags. Seed should be stored at cool and dry place preferably in cold storage. Poor storage conditions will lead to loss of vigour and poor germination. Marketing should be done with specifications and standards.

Way forward for enhancing productivity

In addition to production push through speedy technology adoption, formation of farmer groups (FIGs/SHGs) and market linkages, creation of enabling infrastructures, innovative public private partnership (PPP) models across value chain and creating an enabling policy environment would be necessitated for overall sectoral growth.

Strategy 1: Enabling technology drivers to catapult maize production

Large scale adoption of single cross hybrid technology: In order to deal with productivity concern mentioned in earlier sections, adoption of single cross hybrids (SCH) at faster rate is essential. Only 30% area under maize is sown with SCH in India. Adoption of such hybrids is higher in South India, where seed production is concentrated and hence availability is not an issue. Private seed industry thus has to be incentivized to expand in other agro-ecologically suitable regions, where hybrid seed production is possible. Evolving around ICT enabled extension would play a major role in technology adoption and dissemination till last mile reach.

Assessing biotechnology as an enabler to increased maize productivity: Traits like drought tolerance, herbicide tolerance, multiple insect resistance, enhanced lysine, modified amylase and male-sterility are already available to maize growing farmers. Other important traits like high oil and nitrogen use efficiency, bio fortification, low phytate etc. are in advanced research and development pipeline of private sector. Private players have to be supported in both fiscal and non-fiscal terms to conduct multi location trials and related field trials. Regulatory and IPR regime needs to be strengthened to create an enabling environment for breeders and researchers.

Advancements in genomics and molecular breeding, next generation mutation techniques, cell wall engineering, adoption of RNAi technology for insect, virus and aflatoxin resistance etc. are other focus areas in maize research & development (R&D), which necessitate prioritization and time bound strategies for accomplishment.

Strengthening the pre-breeding activity to develop stress tolerant hybrids: The need of the hour is to develop climate resilient hybrids with enhanced water and nutrient use efficiency under changing climate. In order to achieve this, there is need to harness genetic diversity which was not fully unlocked till today. In this regard, identification of desirable alleles for traits of economic importance like resistance/resilience to biotic and abiotic stresses followed by their mobilization into new cultivars through pre-breeding approach will play an important role. Incorporating traits imparting tolerance to drought, water logging, high temperature etc. in high yielding backgrounds can play a crucial role in sustainable yield enhancement in long-run. Thus, simultaneous selections under combination of stresses should be the strategy to develop cultivars. In order to achieve this there is need to screen germplasm for combinations of stresses under controlled conditions suited for a target environment at managed screening sites by following standardized protocols.

Use of frontier technologies for enhancing genetic gains: Integration of the revolutionary tool of doubled haploidy (DH) with marker assisted selection (MAS) and marker assisted recurrent selection (MARS) coupled with high throughput precision phenotyping of traits of interest has potential to accelerate breeding cycle to breed new cultivars. Use of decision-support systems/tools for faster decision making can offer new opportunity for enhancing genetic gains and breeding efficiency. For example high throughput phenotyping coupled with high throughput genotyping opens up the opportunity to identify genomic regions contributing stress tolerance through genome wide association mapping (GWAS). Genomic selection (GS) opens-up new opportunities towards developing improved hybrids. Genetically modified (GM) maize is another frontier tool with immense potential. Herbicide tolerant GM maize has specific potential to reduce production cost particularly during *kharif* season. CRISPER CAS technique provides us new tools to engineer native genes selectively without much disturbance of native genome. These tools have opened-up new horizon of opportunities for maize improvement especially to develop new cultivars with resistant to biotic and abiotic stresses.

Resource conservation technologies: Conservation agriculture offers an opportunity to plant full season maturity cultivars, by reducing the time lag between two crops. It further cuts down the cost on fossil fuel, improves organic carbon, enhance nutrient use efficiency, and conserve moisture. It helps to reduce pollution due to burning of fossil fuel and improves environmental safety. However, the long term impact of conservation agriculture on dynamics of soil micro-flora, insect pests and diseases, etc. are needed to be studied under different cropping systems. In addition, the mapping of niche areas for adoption of conservation agriculture practices and cropping systems also need be established to promote farmers acceptance and technology targeting.

Strategy 2: Facilitating producer aggregation and market linkages

The private seed companies for maize are located majorly in Andhra Pradesh, Telangana, Karnataka and Maharashtra, where the farmers have turned into seed producers on contract basis and the entire stretch of villages has been converted into "Seed Production Hubs" and "Seed Production Villages". With high impetus on promotion of FPOs / FPCs in the Union Budget 2018, crop as well as seed production clusters in select pockets of the country , could emerge as formal institution like Maize Producers' companies (MPCs) / Maize Seed Producers' Companies (MSPCs). Once a formal institution is created, it could be facilitated by both public and private players in more effective manner. Successful business models around industry FPO linkages could be explored, implemented and scaled up.

Strategy 3: Supporting enabling dedicated infrastructure

Creation of dedicated seed cold storages for Maize, installing Maize dryers, creating modern storage structures like Maize silos, establishment of Maize value added units would be possible avenues in creating enabling infrastructure.

Creation of dedicated seed cold storages for maize: Government should incentivize seed storage infrastructure companies to take up initiative in seed cold storage infrastructure for certified as well as foundation seeds. To support seed cold storages, power tariff subsidy and warehouse receipt financing in seeds stores needs to be promoted. Such dedicated storage facilities are more needed in the seed consuming states, where because of lack of such infrastructure, reverse logistics cost is immense.

Installation of maize dryers for better price realization: Manual handling and poor storage infrastructure at farm level lead to increase in moisture content of maize up to 20%, which should ideally be up to 13%. Private players have to be incentivized to install maize dryers in the proximity of farm, which could be utilized for reducing moisture before storage. In addition to this, State governments may also plan to subsidize portable maize dryers to growers on its purchase.

Promoting maize silos for modern storage techniques **:** Modern silos are not only required to add the storage facility for meeting the storage requirement but also to modernize the existing storage system to reap the benefits of the scientific methods of storage. This will also reduce the intermediaries in supply chain and increase the overall efficiency of supply chain. State governments may devise and promote initiatives similar to the 'Maize Silo Scheme' of the government of Bihar.

Establishment of Maize value added units: Considering the high nutritional value of Maize, Maize based silage making units could be promoted around dairy farmers with end to end mechanization. The feed industry is growing at a CAGR of 8%, with poultry, cattle and aqua feed sectors emerging as major growth drivers. Similarly, scope of expanding product portfolio from Indian Maize starch, improving recovery percentage, and rising consumption from beverage and bio-fuel industry would certainly provide significant push to Maize processing and value addition. Although, industrial policies of state governments enable establishment of such units, additional fiscal and non-fiscal supports could be extended to provide necessary fillip.

Strategy 4: Forging innovative PPP models across value chain

Public Private Partnership (PPP) in ensuring availability of quality maize seeds, establishing maize based skill development centres (SDCs), promoting extension, running maize based farm machinery banks are several future considerations that would help in achieving Maize vision 2022.

Promoting PPP to ensure timely availability of seeds: On the lines of project golden days in Odisha and PPPIAD scheme of SFAC, both state governments and private players need to identify and collaborate effectively to ensure the timely availability of quality seeds to farmers.

Establishment of maize based skill development centers: Government of India (GoI) has given impetus on capacity building and skill enhancement of service providers and operators of precision farm equipment like maize combine harvester, maize planter, laser leveler etc. Few private sector initiatives like SDCs in MP are doing well in this direction. Basis this successful MP model, GoI may contemplate running such Skill Development centers (SDCs) for Maize in PPP mode.

Promoting PPP in running Maize based Farm machinery banks for small and marginal farmers: Instead of subsidizing implements, Government of Gujarat has subsidized usage fee or rental charge of equipment for small and marginal farmers through Agriculture Implement Resource Centre, established on PPP mode. Looking at the success of project, it is inferred that such initiatives could be replicated in states with high population of small and marginal farmers, engaged in Maize farming.

Promoting PPP in extension and marketing support: On the similar lines of Project Golden Rays in Rajasthan, extension support for the adoption of the recommended package of practices and market support for assured procurement of the produce could be formulated and implemented in PPP mode.

Strategy 5: Improved policy framework to unleash the potential

Need based adoption of Price Deficiency Payment (PDP) system, Scaling up of Pradhan Mantri Krishi Sinchayye Yojana (PMKSY), Scaling up of Pradhan Mantri Fasal Bima Yojana (PMFBY) and quality based customization of maize derivative contract post e- NAM integration would be possible avenues for industry and government to intervene and would provide lateral push to the sector.

Important policy decisions like Rashtriya Krishi Vikas Yojana (RKVY), National Food Security Mission (NFSM) and National Food Security Act, taken up by the GoI have significant impact on maize production. The National Food Security Act, 2013 aims at ensuring legal rights to food and nutritional security to every citizen of the country. This may boost up the demand for maize grain, particularly for QPM. Smooth implementation and grounding of such ambitious scheme will not only benefit the ecosystem at large, but will also bring in desired benefits to the Maize value chain.

Improved policy towards bio-fuels, investment promotion in Maize industries, better regulatory framework, IPR protection methods and ways to tackle the grey markets in seed supply chain have tremendous potential to support the maize sector. Devising practical ways to implement PDP (Price Deficiency Payment systems) and enhanced MSP (Minimum Support Prices) in maize will further corroborate the crop uptake amongst farmers. This also needs to be supported through future and derivative market development which would support the development of transparent price discovery mechanism and alternate commodity markets.

Conclusion

Maize is the most versatile crop which is used as food, fodder, feed and in recent past as source of bio-fuel. There are ample opportunities to enhance the yield levels of maize in India through development, popularization and adoption of new and improved technologies.

References

Anonymous. 2019. Pocket book of agricultural statistics 2019, government of India, ministry of agriculture & farmers welfare, department of agriculture, cooperation & farmers welfare, directorate of economics & statistics, New Delhi. (Website: http://eands.dacnet.nic.in)

Hiremath, S.M., Mohan Kumar R. and Gaddi, A.K. 2016. Influence of balanced nutrition on productivity, economics and nutrient uptake of hybrid maize (Zea mays)-chickpea (Cicer arietinum) cropping sequences under irrigated ecosystem. Indian J. of Agron. **61** (3): 292-296.

https://iimr.icar.gov.in/attachment/articles/52/Specialty%20corn%20Cultivars%201969-17.pdf

https://iimr.icar.gov.in/attachment/articles/6/Vision%202050.pdf

https://iimr.icar.gov.in/index.php?option=com_content&view=article&id=2&Itemid=114

Parihar, C.M., Jat, S.L., Singh, A.K., Kumar, R.S., Hooda, K.S., Chikkappa, G.K. and Singh, D.K. 2011. Maize production technologies in India. DMR technical bulletin. Directorate of Maize Research, Pusa Campus, New Delhi-110 012. Pp 30.

Ramachandiran K. and Pazhanivelan S. 2016. Influence of irrigation nitrogen level on growth, yield attributes and yield of maize (Zea mays). Indian J. of Agron. **61** (3): 360-365.

Singh, M. and Vashist, K.K. 2016. Enhancing productivity of spring maize (Zea mays) through planting methods, varieties and irrigations level in Punjab. Indian J. of Agron. **61** (3): 348-353.

Varma, V.S., Durga, K.K. and Neelima, P. 2014. Effect of sowing date on maize seed yield and quality: a review. Rev. of Plant Studies. **1**(2): 26-38.

4

Improved Technologies for Pearl Millet Cultivation

Rajesh C. Jeeterwal, Anju Nehra and Rupa Ram Jakhar*

Introduction

Pearl millet (*Pennisetum glaucum* L.) is the most widely grown type of millet and it is an important staple food grain crop for millions of people in arid and semi arid regions of the world. Pearl millet is very well adapted crop to growing areas characterized by drought, low soil fertility, and high temperature. It is also performs well in soils with high salinity or low pH. Because of its tolerance to difficult growing conditions, it can be grown in areas where other cereal crops, such as wheat or maize, would not survive. Physiologically, it is a C_4 plant which imparts it potential to grow under hot and dry climatic conditions. Among cereals, it is highly responsive to fertilizers and has highest water use efficiency. Pearl millet is being grown as dual purpose crop for both grain and fodder in dry areas like Rajasthan where animal production is complimentary with crop production. Pearl millet is also a good source of important essential minerals for human health like Fe and Zn hence it is useful to alleviating malnutrition.

Nutritive value

Pearl millet is a highly palatable cereal with one of the best nutritional profiles (rich in tryptophane and cystine amino acids). Pearl millet grain contains about 9.5 to 14% protein, 5.03 to 6.0% fat, 1.05 to 1.7% crude fiber and 65.5 to 70% carbohydrates. The grains are dehusked and ground into flour to make porridge, which is preferred over sorghum. Pearl millet grains are cooked like rice and chapatties are also prepared with its flour. As a food crop, pearl millet possesses the highest amount of calories per 100 g grain, which is mainly supplied through carbohydrates, fats and proteins. Its mineral content is also comparable with other cereals. Among micro-nutrients, it contains Fe ranging from 18 to 135 ppm and Zn 22 to 92 ppm.

**Corresponding author email: jeeterwal.raj.pbg@gmail.com*

Pearl millet fodder is also fairly rich in carbohydrates (67%), protein (11.6%), low lignin, high dry matter yield and digestibility. However, it possesses oxalic acid which is an anti-nutritional factor. Besides, the toxic component HCN is less in green fodder of pearl millet in comparison to sorghum.

Origin and distribution

Pearl millet is believed to have originated in Africa (Vavilov, 1950 and Murdock, 1959) from where it spread to India and other countries. According to Brunken *et al.* (1977) the Sahel zone of West Africa appears to be the original home of pearl millet from where it spread to other parts of the world.

Pearl millet is grown in semi-arid and arid climate of tropical and sub-tropical regions of world. Globally it is grown between 15° W and 90° E longitude and 5° S to 40° N latitude. Globally, among food grains it ranks sixth, following wheat, rice, maize, barley and sorghum in area and production. It is extensively cultivated over large areas in dry regions of Asia (mainly in India) and Sahel region of Africa (Nigeria, Niger, Mali, Chad and Tanzania). Globally, India, Nigeria, Niger, Mali, Burkina Faso, Sudan and Ethopia are the major pearl millet growing countries.

In India, it is grown in drier areas of central and western regions. The major pearl millet growing states are Rajasthan, Maharashtra, Gujarat, Uttar Pradesh and Haryana, covering nearly 90% of total acreage. In India, it is grown on 7.4 million ha with an average production of 9.13 million tonnes and productivity of 1237 kg/ha (Table 1) during 2017-18 (Directorate of millets development, 2019).

In India, Rajasthan covers the highest area accompanied with highest production in the country. In Rajasthan the acreage under this crop is governed by the onset of monsoon whereas, the production is governed by the amount and distribution of rainfall. The Rajasthan state occupies 42.40 lakh ha area, which is 57.38% of the national acreage. The total production of pearl millet in the Rajasthan state during 2017-18 is 42.40 lakh tonnes, which is 41.09% of the national production with an average productivity of 886 kg/ha (http://agricoop.nic.in).

Table 1. Area, production and productivity of pearl millet in major pearl millet growing states of India during 2017-18

State	Area ('000 ha)	Production ('000 tonnes)	Productivity (kg/ha)
Gujarat	400	920	2312
Haryana	450	720	1602
Karnataka	230	290	1241
Madhya Pradesh	310	760	2435
Maharashtra	680	610	903
Rajasthan	4240	3750	886
Uttar Pradesh	970	1770	1825
India	7380	9130	1237

Source: Directorate of Economic Survey & Statistics, DAC&FW

Botany and taxonomy

Pearl millet (*Pennisetum glaucum* L.) is a diploid species having 14 chromosomes and belongs to the family Poaceae (Gramineae). The genus *Pennisetum* is derived from two latin words *penna*, meaning 'feather' and *Seta,* meaning 'bristles'. *Pennisetum* is largest genera in the tribe Paniceae with five sections and approximately 140 species that are widely distributed in the tropics and subtropics.

It is placed close to the genus *Cenchrus* on the basis of spikelets that are arranged in groups surrounded by involucres. Inter-generic hybrids between pearl millet and *Cenchrus ciliaris* L. have also been reported. According to Brunken, *Pennisetum* includes two reproductively isolated species *viz. Pennisetum purpurium* Schumach which is a tetraploid (2n=28) perennial species which occurs throughout the wet tropics of the world; and *Pennisetum americanum* (syn. *P. glaucum*) (L) Leeke is a diploid annual species, native to the semi-arid tropics of Africa and India. *P. glaucum* (*P. americanum* according to Brunken) comprises three sub species: (a) Sub-species- *glaucum*- cultivated, involucres persistent at maturity, distinctly stalked (b) Plants wild or weedy, involucres readily deciduous at maturity, short stalked or sub-sessile. (i) Sub-species, *violaceum* (monodii) Involucres sub-sessile, stalks < 0.25 mm long; mature grains 0.6-1.0 mm deep. It is considered to be wild progenitor of cultivated pearl millets. (ii) Sub-species *stenostachyum*- Involucre short stalked; stalk > 0.25 mm long, mature grain 1.0-2.0 mm deep. *P. glaucum* is the only cultivated species for seed purpose, *P. purpurium* and its hybrids with bajra in both directions (bajra × napier and napier × bajra) are cultivated for fodder purposes in some areas of world and India.

The plant

Pearl millet is a tall, robust, erect and annual bunch grass. Its height varies depending upon the variety and environment, but plants are generally 6-15 feet in height. Seedling development occurs during the first two to four weeks and rapid stalk development occurs soon after. Pearl millet often tillers producing a profuse leaf growth. There are four different types of tillers: primary, basal, secondary and nodal. Tillering may be extensive in sparse stands, particularly if good soil moisture is available. Primary roots are also called as seminal roots, are basically elongation of the radicle. Over time, they deteriorate and finally die. Secondary roots start appearing 6-7 days after seedling emergence and occupy 5 to 15 cm area in the soil around the base of the stem. Adventitious roots are small, uniform and form a small portion of the root system. Another type of permanent adventitious roots develops from the second internode and above. These roots are branched laterally (about 1 m^2) interlacing the soil vertically. These roots mainly supply nutrients to the plant. Pearl millet roots will grow up to > 2 m in pursuit of water and soil nutrients.

Production technology

Climatic requirements

As it may grow under different day lengths, temperature and moisture stress, the crop has wide adaptability. It requires low annual rainfall ranging between 400-750 mm and dry weather. The crop may tolerate drought but cannot withstand high rainfall of 900 mm or above. The optimum temperature range for the growth of pearl millet is between 20 ^{0}C and 28 ^{0}C but the temperature optimum for vegetative growth is between 28-32 ^{0}C.

Soil management

Pearl millet is successfully grown in different kinds of soils but heavy soils are not suitable for pearl millet cultivation. Light soils with good drainage and mild salinity are best soil type for this crop. It does not tolerate soil acidity. The crop needs very fine tilth as the seeds are too small. For field preparation, 2-3 harrowing followed by cross cultivation is require for a fine tilth to facilitate the sowing and proper distribution of seed at appropriate depth.

Sowing

Method of sowing: There are three systems of pearl millet sowing are followed: (1) on a flat surface, or (2) using ridge and furrow system, or (3) on a broad-bed and furrow system. The seed should be sown at 2.5 cm – 3 cm depth.

Time of sowing: Sowing of *kharif* pearl millet should be done with the onset of monsoon i.e. first fortnight of July in north and central parts of the country. Gap filling should be done by transplanting seedlings after 2-3 weeks of sowing if scanty population exists. In Maharashtra, dry sowing prior to first monsoon rains is recommended. Summer pearl millet should be sown from last week of January to 1st week of February to obtain higher production of summer pearl millet in Zone B, which includes Gujarat, Maharashtra, Tamil Nadu and Telangana states. Maurya *et. al.* (2016) reported that the results revealed that sowing of pearl millet (var. Ganga kaveri-22) on 23rd July was recorded highest plant height (177.21 cm), plant dry weight (78.25 g), grain yield (3.579 t/ha) and stover yield (10.225 t/ha).

Seed treatment

1) Seed treatment with Thiram 75% dust @ 3 g/kg or biopesticides (*Trichoderma harzianum* @ 4 g/kg) seed will help against soil borne diseases.
2) Seed treatment with Vitavax @ 3 g/kg or 300-mesh sulfur powder @ 4 g/kg seeds management the smut disease.
3) For removing ergot affected seeds, they are soaked in 10% salt solution. Seed treatment with metalaxyl (Apron 35 SD) @ 6 g/kg is very effective after treating brine solution.
4) Seeds are treated with biofertiliser *Azospirillum* (600 g) and Phosphobacterium to enhance the availability of nitrogen and phosphorus, respectively.

Seed rate, spacing and plant population: The recommended seed rate for grain production is 4-5 kg/ha for drilling method but in case of dibbling method 2.5-3 kg/ha seed is sufficient. For green fodder crop 10 kg/ha seed is used. The seeds are planted at the spacing of 45 × 10-15 cm maintaining plant population of 1.75 to 2.0 lakhs/ha. For arid-western plain of Rajasthan, Haryana and Kutch of Gujarat, pearl millet should be planted in rows 60 cm apart, maintaining low plant population of 1.00 to 1.25 lakhs/ha. Seed rate for the crop should be taken @ 3 to 4 kg/ha for obtaining required plant stand.

Increased plant height and yield attributes with normal sown crop at 60 cm drip line spacing, thus, enhanced grain, straw and biological yield of pearl millet and highest grain (21.73 q/ha), straw (99.60 q/ha) and biological yield (121.33 q/ha) was recorded at normal sown crop at 60 cm drip line spacing. However, paired row sown crop at 120 cm drip line spacing also gave at par grain, straw and biological yield with normal sown crop at 60 cm drip line spacing (Bhunia *et al.*, 2015).

Table 2. Improved varieties and their characteristics

Varieties	Bred at	Year of release	Characteristics
HB1	Ludhiana	1965	80-85 days, medium tall, profuse tillering, succulent stem, green ligules, long ear heads.
Pusa Moti	New Delhi	1973	90-95 days, medium tall, bold grains, downy mildew tolerant.
WC-C75	ICRISAT	1982	82-85 days tall, thick and dense panicles, developed crossing seven full-sib progenies of World Composite.
RCB-2	Jaipur	1985	80-85 days, tall plants with 4-6 tillers, broad and glabrous leaves, long compact and cylindrical ear heads, yellowish grey grains, developed from 20 inbreds of diverse origin.
HHB 60	Hissar	1988	74-76 days, tall, high tillering, medium thin stems, compact candle ear head, obovate brown grains, resistant to drought and salt stress, good quality fodder.
Pusa Safed	Delhi	1989	83-87 days, medium tall, semi-compact ear heads, yellow anther, yellowish white grains of medium size.
HHB 67	Hissar	1990	60-62 days, medium tall, thin stem, medium narrow leaves, semi-compact spindle shape ear heads, yellow anthers, grey globular grains, escapes drought and tolerates salt stress, fits well in inter- and multiple-cropping.
Raj Bajara chari 2	Jobner	1990	85-90 days, tall, broad and succulent leaves, good quality green fodder, suitable for saline soils.
Raj 171	Jaipur	1992	82-85 days, tall, medium thick stem, long cylindrical semi compact to compact ear heads, obovate grey brown grains, resistance to downy mildew, bred from inter varietal composite.
CZP -1C 923	CAZRI, Jodhpur & ICRISAT	1997	72-80 days, tall, thick stem, long oblanceolate thick panicle, light yellow to brown anthers and light grey seed with yellow base.
RHB 30	Jaipur	1997	80-85 days, medium tall, non pubescent internodes, good tillering, ligules light red colored, yellow anthers, long bristles, creamy grey grains.
RHB-90	Jaipur	2000	85 days, medium tall, compact cylindrical ear heads, yellow anthers, purple bristles, obovate yellow brown grains.
CPZ 9802	CAZRI	2003	70-72 days, medium tall, good tillering, thin stem, narrow leaves, thin candleshaped ear heads, yellowish grains of medium size, drought tolerant, very high stover.
GHB-526	Jamnagar	2003	Summer cultivation areas across India. 85 days, medium tall, good tillering, narrow leaves with greenish white mid-rib, good exertion, yellow anthers, compact conical ear heads, obovate grey brown grains.

HHB67 Improved	Hissar	2005	Extra early maturity, highly resistant to moisture stresses, resistant to downy mildew. The first commercial cultivars developed using marker-assisted selection in India.
HHB 223	HAU Hisar	2010	Medium maturing, conical ear heads with long purple bristled, resistant to downy mildew, tolerant to drought
Raj 173	Jaipur	2011	Medium maturity, medium to tall plant height, compact cylindrical ear heads, resistant to downy mildew.
HHB-226	HAU Hisar	2011	Medium maturing, medium height, dark green leaves, candle shaped bristled ear heads, resistant to downy mildew.
Raj 177	Jaipur	2011	Early maturing, medium tall, cylindrical bristled ear heads, resistant to downy mildew, light yellow anthers.
Mandor Bajra Composite 2 (MBC 2)	Mandore, Jodhpur	2011	Early maturing, medium height, medium long semi compact cylindrical ear heads, obvate grey coloured seed.
CO 9	TNAU, Coimbatore	2012	Medium maturing, medium height, candle compact ear heads, grayish yellow seed colour.
HHB 234	HAU Hisar	2013	Early maturing, candle shaped ear heads with small bristles, medium seed size and tolerant to downy mildew.
MPMH 17	Mandor, Jodhpur	2013	Medium maturing, medium height, yellow anther colour, compact lanceolate ear heads with bristles, resistant to downy mildew, grey brown seeds.
RHB 233	Jaipur	2018	High iron and zinc, medium maturity dual purpose hybrid with high grain and dry fodder yield potential. Higher degree of resistance to downy mildew and other diseases.
RHB 234	Jaipur	2018	Bio-fortified (high iron and zinc) dual purpose hybrid with high grain and dry fodder yield potential. Higher degree of resistance to downy mildew and other diseases.

Nutrient management

Generally pearl millet crop requires lesser amounts of nutrients. But All India Co-ordinated Millet Improvement Project has proved that new plant types of pearl millet, especially hybrids respond to very high doses of fertilizers. Application of 40 kg N + 20 kg P_2O_5/ha for arid regions and 60 kg N/ha + 30 kg P_2O_5/ha for semi-arid regions is recommended for sole pearl millet as well as intercropping system. In light soils (sandy loams) the applied nitrogen may be lost due to leaching with heavy rains. So, only about half of the recommended nitrogen should be applied at seedbed preparation. The remaining half of nitrogen dose is side-dressed when the crop is 25 days old. On soils which do not leach easily like black soils, all of the nitrogen may be applied during seedbed preparation. Pearl millet seeds are sensitive to fertilizer burn. Do not apply fertilizer in the furrow with the seed or very near the seed in the row after sowing. It should be applied as side dressing. Use of biofertilizer (*Azospirillum* and PSB) can economize the N and P fertilizer application. Sivakumar and Salaam (1999) reported that fertilizer application (30 kg P_2O_5 and 45 kg N/ha) in pearl millet increased the water use up to 7-14 per cent. Increased yield due to the application of fertilizer was accompanied by an increase in the water use efficiency (WUE). The beneficial effect of fertilizer application could be attributed to the rapid early growth of leaves that could contribute to reduction of soil evaporative losses and increased WUE. To overcome the effects of sodium salt stress, the application of higher doses of N fertilizers up to 120 kg/ha was found effective to increase the dry matter yield, plant height, protein content and uptake of nutrients in pearl millet grown under sodic soils (Singh *et al.*, 2014).

In zinc deficient soils of the pearl millet growing area of the country, application of 10 kg $ZnSO_4$/ha is recommended. To correct the zinc deficiency in standing crop, spray of 0.2% $ZnSO_4$ at tillering to pre-flowering stage is recommended. Under prolonged dry spell, skip top dressing of N and spray 2% urea. Under excessive rain situation during vegetative phase, additional dose of nitrogen @ 20 kg/ha should be given.

Inter-culture and weed management

Two hoeings and weedings at 15 and 30 DAS are sufficient for management of weeds effectively which is comparable with the herbicidal weed management through pre-emergent application of atrazine @ 0.5 kg/ha or Propazine or Prometryne at 1.0 kg a.i./ha is recommended to superimposed with one hand weeding. Second weeding breaks the upper layer which act as dust mulching and helps to conserve soil moisture. Girase et al. (2017) and Singh et al. (2017) reported that the maximum grain yield was recorded with pre-emergence application of Atrazine 0.5 kg/ha + one hand weeding (HW) at 35 DAS, which was at par with two hand weeding and hoeing, and post-emergence application of Atrazine 0.4 kg/ha + one HW at 35 DAS and increased the yield by 62.14% over weedy check.

Intercropping in pearl millet

Intercropping refers to growing more than one crop in the same piece of land in rows of definite proportion and pattern. Adverse weather conditions like delay in the onset of rains and/or failure of rains for few days to weeks some time or other during the crop period is very common in the pearl millet growing areas. Thus, intercropping system should provide the necessary insurance against unpredictable weather. In case the year happens to be normal with respect to rainfall, the intercropping system, as a whole, should prove to be more profitable than growing either of the crops alone.

Pearl millet-based cropping systems in kharif

Rotation of cultivars also should be adopted to avoid the problems of downy mildew disease. Pearl millet hybrids and open-pollinated varieties should be used in alternate years/seasons. The same hybrid or open pollinated varieties does not grow continuously on the same piece of land. The suitable intercropping patterns for different states of the country are listed in table 3.

Table 3. Suitable intercropping patterns for different regions of the country.

State	Suitable intercrops
Rajasthan	Pearl millet + cluster bean/cowpea/moong bean/moth bean/sesame
Haryana	Pearl millet + moong bean/sesame/cluster bean/cowpea
Delhi	Pearl millet + pigeonpea/groundnut/castor
Punjab	Pearl millet + chickpea/fodder sorghum/wheat
Gujarat	Pearl millet + moong bean/sesame/cowpea
Madhya Pradesh	Pearl millet + urd bean/soybean/pigeonpea/cowpea
Uttar Pradesh	Pearl millet + moong bean/sesame/cowpea
Maharashtra	Pearl millet+ moth bean/pigeonpea/soybean/urd bean, moong bean/ cow-pea/sunflower
Karnataka	Pearl millet + pigeonpea/moong bean/sunflower/soybean
Tamil Nadu	Pearl millet + pigeonpea/moong bean/sunflower/soybean/cowpea
Andhra Pradesh	Pearl millet + pigeonpea/moong bean/sunflower/soybean/groundnut

Water management

Pearl millet is grown as rainfed crop and being drought resistant it requires about 250- 350 mm water. However, under prolonged dry spells it is observed that the yield may be significantly increased by irrigating the crop at critical growth stages like maximum tillering, flowering and grain filling stage. Therefore, need based light irrigations and efficient drainage is very essential for pearl millet production. In summer, pearl millet should be irrigated at regular intervals (0.75-1.0 IW/CPE with 40 mm) as per need of the crop.

Drip irrigation maintains moisture content at near about field capacity on one hand and eliminates water losses on the other hand drip system would be the most effective tool in increasing yield as well as water use efficiency in even close growing crop. Further, drip irrigation proved to be effective in improving yield and water use efficiency compared to level-basin irrigation method even under condition of deficit water. Bhunia *et al.* (2015) reported that drip irrigation levels from 40 to 80 per cent ETc saved water by 93.32 to 31.65 mm over surface irrigation which used 354.8 mm water. Hence, increased yield coupled with less water use in drip irrigation and recorded higher water use efficiency (WUE) of 4.92, 8.35 and 7.98 kg/ha/m mat 40, 60 and 80 per cent ETc, respectively, against 3.53 kg/ha/mm in surface irrigation. Kharrou *et al.* (2011) reported that drip irrigation applied to wheat was more efficient and saved 20 per cent of water in comparison to surface irrigation. Singh *et al.* (2009) also reported that water use efficiency was highest under drip irrigation as compared to surface irrigation. Lower water use efficiency in surface irrigation (absolute control) may be due to loss of irrigation water from sandy loam soil through deep percolation which resulted in higher water use but lowered grain yield.

Insect-pest and disease management

In India, insect pests are considered to be relatively less important in most of the pearl millet growing areas. The most important insect pests of pearl millet are white grub, red hairy caterpillar, shoot fly, grey weevil and termite.

White grub (*Holotrichia consunguinia*): It is a common and serious pest mainly in Gujarat and Rajasthan states. Its grubs attack on roots during July-August months causes complete plants damage which significantly reduces the plant population and ultimately the yield. Patchy gaps are formed due to death of plants which result in poor or uneven plant stand. The adults emerge from May to July with the pre- monsoon/monsoon showers and feed on leaves of different plants like neem and laid eggs in soil.

Management: For management of white grubs the following measures should be use.

1) Collect and destroy the adult beetles immediately after first showers when they feeding Neem/ Acacia trees and during mating.
2) Spray host trees with Carbaryl 0.2% or Chlorpyriphos 0.2% with onset of monsoon and the spraying within 2-3 days after receipt of first showers.
3) Mixing of Carbofuran 3 G @ 20-25 kg/ha with the seed and application in seed furrows at the time of sowing is very effective to manage.
4) Inter-cropping with Sunflower and Pigeon pea reduces the incidence of white grub.
5) Use Pheromone traps for adult beetles. Install pheromone traps at distance of 50 meter @ 4-5 traps per acre for each insect pest. Use specific lure for each insect pest species and change it after every 20 days. Trapped moths should be removed daily.

Shoot fly (*Atherigona approximata*): The larvae of shoot fly cut the growing point during the seedling stage (5 to 30 days after seedling emergence) will lead to the typical dead heart symptoms. The larva migrates to the upper side of the leaf, and moves along the leaf whorl until it reaches the growing point where the larvae cut the growing point. As a result the central leaf dries up forming a dead heart, which can be pulled out easily and produces a rotting smell. Late infestations may also damage the panicle in the formative stage, resulting in rotting or drying up of a portion of the panicle affected by shoot fly damage.

Management: For management of shoot fly the following measures should be use.

1) The crop should be sown with the onset of the monsoon or latest within 10-15 days of first shower of monsoon.
2) Staggered sowing to contain the buildup of shoot fly population.
3) Transplanting is suggested for late sown crop. In case direct seeding, a seed rate of 4 kg/ha is recommended and the affected seedlings are thinned within 15 days after sowing.
4) In case of heavy incidence of shoot fly in endemic areas, spray the crop with 0.07% Chlorpyriphos at 10 and 20 days after germination or dusting of Carbofuran 3G can be used.

Grass hoppers (*Hieroglyphus banian*): This insect laid eggs in the soil about 75-200 mm deep. The hoppers and adults of the insect feed on foliage thus causing severe defoliation of the plants. The adults are short winged and can fly short distances only.

Management: For management of grass hoppers the following measures should be use.

1) Follow weed free cultivation and deep summer ploughing after harvest of the crop is very effective to control it through expose egg pods in soil.
2) Scrapping of bunds and clean cultivation is also effective.
3) Dusting the crop with Methyl Parathion @ 20-25 kg/ha is very effective to control the damage.
4) Dusting the crop with Carbofuran 3G or Fenvalerate dust @20-25 kg/ha or spray the crop with 0.07% of Malathion.

Red hairy caterpillars: It attacks the crop sporadically. The larvae cause heavy defoliation of the plants. Pest is commonly found in Gujarat and semi-arid region of Rajasthan.

Management: Release of *Trichogramma chilonis* @ 75, 000 per ha/week and for chemical control dusting the crop with Carbofuran @ 20-25 kg/ha is effective control of red hairy caterpillar.

Termites: A social insect that live underground in colonies attacks young seedlings as well as grown up plants. Its attack is severe in dry areas. Infested plants wither and ultimately die.

Management: For proper management of termites the following measures should be use.

1) Deep ploughing after harvesting of previous crop followed by collection of stubbles/plant refuge and burning them.
2) Use well decomposed FYM.
3) Irrigate the crop timely and regularly in dry spells so incidence of termite is reduced.
4) Apply Chlorpyriphos 20 EC @ 1.25 litre along with irrigation water in standing crop is essential to save crop.

Major diseases: Although >50 diseases caused by different biotic factors have been reported in pearl millet in India but only few are important. These are downy mildew, smut, ergot, rust and blast. These diseases directly reduce grain yield by affecting grain formation. In addition ergot can also affect grain quality. Use of resistant cultivars is the most cost-effective method of the management of pearl millet diseases.

Downy mildew: It is major diseases of pearl millet and it caused by a fungus *Sclarospora granimicola.* The appearance of diseases is on ear as leafy structure on panicle so it is also knowns as "green ear disease". Downy mildew is widely distributed in all the pearl millet growing areas in the world. Systemic symptoms as chlorosis generally appear on the second leaf and all the subsequent leaves and panicles of infected plant show symptoms. Leaf symptoms begin as chlorosis at the base of the leaf lamina and successively higher leaves show a progression of greater leaf area coverage by the symptoms. Infected chlorotic area produces massive amount of asexual spores, generally on the lower surface of the leave a 'downy' appearance. Systemically infected plants remain stunted either do not produce panicle or produce malformed panicles. In many affected plants 'green ear' symptoms appear on the panicles due to the transformation of floral parts into leafy structure that may be total or partial and such plants do not produce seed or produce very few seeds. This disease can assuming alarming levels when a single genetically uniform pearl millet cultivar is repeatedly and extensively grown in a region. Yield losses within the region can reach 30-40%.

Management: The diseases of pearl millet can be best managed by integrating methods of chemical or biological management and cultural practices.

1) Rogue out infected plants and burn.
2) Rotate hybrids with variety alternately year to keep soil inoculums under management.

3) Seed treatment with Ridomil MZ-72 @ 8g/kg seed and a foliar spray of Ridomil MZ-72 2 g/l.
4) Seed treatment with Apron 35 SD @ 6 g/kg seed or *Bacillus pumulis* (INR7).
5) Use of resistant cultivars like HHB 67 Improved, Raj 171, Raj 173, Raj 177, MPMH 17, RHB 233 and RHB 234.
6) Foliar spray of Ridomil 25 WP (100 ppm) after 21 days of sowing, if infection exceeds 2- 5 % or Dithane M-45 or Dithane Z-78 @ 0.2%.

Ergot: Ergot is a fungal diseases and it is caused by fungus *Claviceps fusiformis*. The disease is easily identified as a honeydew substance of creamy to light pinkish ooze out of the infected florets which contains numerous conidia. Within two weeks these droplets dry out as hard dark black structures larger than seeds, protruding out from the florets in place of grain, which are called sclerotia. Here, the loss in grain yield is directly proportional to the percentage of infection as the infected seed is fully transformed into sclerotium. The disease occurrence and spread is highly influenced by weather conditions during the flowering time. It became more important due to cultivation of genetically uniform single-cross F_1 hybrids based on cytoplasmic male-sterility system in India. In India, two perennial grass weeds *Cenchrus ciliaris* and *Panicum antidotale* were found to harbor the pearl millet ergot fungus.

Management: For proper management of ergot disease the following measures should be use.

1) Mechanical removal of sclerotia from seed and washing of seed in 2% salt water.
2) Adjust sowing dates so that ear emergence does not ecocide with more rainy days.
3) Plough the field soon after harvest so that ergot is buried deep.
4) After appearing of symptoms of disease in the field it by three foliar application of Copper oxychloride @ 0.25% or Ziram @ 0.2% starting from 50% flowering.
5) Spraying the crop with Zineb 0.1%.

Smut: It is caused by *Tolyposporium penicillariae.* Smut disease is of greater importance in India especially with the adaptation of hybrids. The disease is more severe in CMS-based single-cross hybrids than in open-pollinated varieties. The smut infection is visible on ears as converting of grains as a black powder mass. The disease occurs during the month of September/ October. Early sown crop generally escapes from the smut infection.

Management: Use of tolerant cultivars *viz*., WC-C 75, CM 46 and MBH 110. For chemical management, spray with Captafol followed by Zineb on panicle at boot leaf stage which reduces infection. Removal of smutted ears from the field at initial level.

Blast or *Pyricularia* leaf spot: The disease is known as leaf spot of pearl millet caused by *Pyricularia grisea* (Cooke) has become a serious disease during the past few years. The disease affects both quality and production of forage and grain. The symptoms appear as distinct large, indefinite, water soaked, spindle shaped, grey centred and purple grey horizon with yellow margin, resulting in extensive chlorosis and premature drying of young leaves.

Management: Remove smutted ears from the field and spray with Captafol followed by Zineb on panicle at boot leaf stage which reduces infection.

Harvesting and storage: The best stage to harvest pearl millet is when the plants reach physiological maturity determined by the black spot at the bottom of the grain in the hilar region. When the crop matures, the leaves turn yellowish and present a nearly dried up appearance. The crop is harvested when grains become hard enough and contain moisture around 20 per cent. The usual practice of harvesting pearl millet is cutting the ear heads first and the stalks later. The stalks are cut after a week, allowed to dry and then stacked. Grain at or below 14% moisture is considered dry.

Irrigated crop yields 30-35 q/ha, while unirrigated crop yield about 15-20 q/ha. The fodder yield is about 100 to 150 q/ha. The separated grains must be cleaned and dried in sun to bring about 12% moisture. For long-term storage (>6 months), grain moisture content should be <12%. After proper drying the grains should be bagged and stored in a moisture proof room.

Conclusion

There are indications that pearl millet is becoming increasingly important as a nutritional food and forage crop in arid and semi-arid tropics. Further, it is also a fertilizer responsive food grain crop grown all around the world with high nutritional value for world's major nutritional deficient areas. However, these uses relatively shows importance of pearl millet in all the millets as food crops of the rural poor, primarily in the tropics. In sum, millet will remain largely associated with the food security of drought-prone human populations.

References

AICMIP. 2019. *Annual reports*. All-India Co-ordinated Millet Improvement Project (ICAR).

AICMIP. 2019. Varieties released. All-India Co-ordinated Millet Improvement Project (ICAR). (http://www.aicpmip.res.in/varietiesreleased.pdf.).

Bhunia, S.R., Mohd. A., Verma, I.M., Balai, K., Dhikhwal S. and Sharma, N.C. 2015. Effect of crop geometry and drip irrigation levels on pearl millet (*Pennisetum glaucum*). *Forage Res.* **41** (2): 118-121.

Brunken, J., de Wet , J.M.J. and Harlan, J.R. 1977. The morphology and domestication of pearl millet. **31**(2): 163–174.

Department of Agriculture, Cooperation & Farmers Welfare. 2019. http://agricoop.nic.in).

Directorate of millets development. 2019. Coordinators Review, AICRP on Pearl millet. http://www.millets.res.in/aicrp_pearl.

Girase, P.P., Suryawanshi, R.T., Pawar, P.P. and Wadile, S.C. 2017. Integrated weed management in pearl millet. *Indian J. of Weed Sci.* **49**(1): 41–43.

Kharrou, M.H., S. Er-Raki, Chehbouni, A., Duchemin, B., Simonneaux, V., Le Page, M., Ouzine, L. and Jarlan, L. 2011. Water use efficiency and yield of winter wheat under different irrigation regimes in a semi-arid region. *Agric. Sci.* **2**: 273-282.

Maurya, S.K., Nath, S., Patra, S.S. and Rout, S. 2016. Effect of different sowing dates on growth and yield of pearl millet (*Pennisetum glaucum* L.) varieties under Allahabad condition. *Int. J. Sci. & Nature*. **7**(1): 62-69.

Singh, N., Sharma, S.K., Kumar, R., Rajpaul and Singh, S. 2014. Effect of sodicity and nitrogen levels on dry matter yield, protein and nutrient uptake in pearl millet. *Forage Res.* **40**: 28-35.

Singh, R., Kumar, S., Nangare, D.D. and Meena, M.S. 2009. Drip irrigation and black polyethylene mulch influence on growth, yield and water use efficiency of tomato. *African J. Agric. Res.* **4**: 1427-1430.

Singh, R., Kumar, R., Kumar, A., Kumar, S., Kumar, J., Pal, L.K., Prajapati, M.K., Kumar P. and Sahu, Y. 2017. Effect of weed management on growth, yield and yield attributes of Pearl millet (*Pennisetum glaueum* L.) var. "Manupur" under rainfed condition. *J. of Pharma. and Phytochem.* **6**(6): 1133-1135.

Sivakumar, M.V.K., and Salaam, S.A. 1999. Effect of year and fertilizer on WUE of pearl millet (*Pennisetum glaucum*) in Niger. *J. Agric. Sci.* **132**: 139-148.

Vavilov, N.I. 1950. The origin variation immunity and breeding of cultivated plants. *Chronica Bot.* **13**(1): 366.

5

Advances in Pulses Production Technologies: A Holistic Approach for New Millennium

Narendra Kumar

Introduction

Pulses are the second most important group of crops after cereals in India. They are important food crops for nutritional security, soil health and sustainable agriculture. Consequently, they remained an internal component of Indian agriculture especially under rainfed since time immemorial. They are among the ancient food crops with evidence of their cultivation for over 8,000 years. India, China, Brazil, Canada, Myanmar and Australia are the major pulse producing countries with relative share of 25%, 10%, 5%, 5% and 4%, respectively. In 2017, the global pulses production was 95.98 million tonnes from an area of 95.17 million ha with an average yield of 1008 kg/ha. Dry beans contributed 32.7% to global total pulses production followed by chickpea (15.4%), dry peas (16.89%), lentil (7.91%) and pigeonpea (7.1%). About 70% of the global pigeonpea, 60% of chickpea and 17% of lentil area falls in India (FAO STAT, 2018). India is the largest producer and consumer of pulses in the world contributing around 25-28% of the total global production.

India can be proud of growing the largest number of pulse crops (grain legumes) in the world. Over a dozen pulse crops including chickpea (*Cicer arietinum*), pigeonpea (*Cajanus cajan*), mungbean (*Vigna radiata*), urdbean (*Vigna mungo*), cowpea (*V. unguiculata*), lentil (*Lens culinaris*), lathyrus (*Lathyrus sativus* L.), frenchbean (*Phaseolus vulgaris*), horsegram (*Macrotyloma uniflorum*), field pea (*Pisum sativum*), moth bean (*V. aconitifolium*), *etc.* are grown in one or the other part of the country throughout year. The latest data (2017-18) indicate that the present production of pulses is 25.23 million tonnes from an area of 29.99 million ha with productivity of 841 kg/ha (DAC, 2019) (Fig. 1). The stagnant growth of pulse production and continuous increasing human population in the country led to decline in per capita consumption of

**Corresponding author email: nkumar.icar@gmail.com*

pulses from 67 g/day/person during 1951 to 35 g/day/person during 2010 (Indian Council of Medical Research recommends 65 g/day/person).

The most important pulse crops grown are chickpea (48%), pigeonpea (15%), mungbean (7%), urdbean (7%), lentil (5%) and field pea (5%). To fulfill domestic demand of pulses in the country, India has to import 3-4 million tones of pulses every year. In order to ensure self-sufficiency, the pulse requirement in the country is projected to be about 50 million tonnes by 2050 which necessitates an annual growth rate of 4.0% (IIPR, 2013).

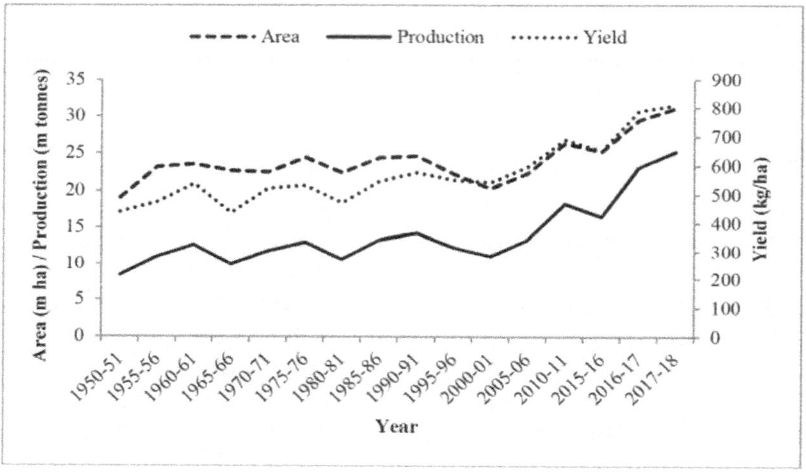

Fig. 1. Area, production and yield trend of total pulses in India

The major pulses producing states in the country are Madhya Pradesh (26.41%), Maharashtra (16.19%), Rajasthan (12.82%), Uttar Pradesh (8.89%), Andhra Pradesh (8.04%) and Karnataka (7.63%) which together share about 80% of total pulse production while remaining 20% is contributed mainly by Gujarat, Chhattisgarh, Bihar, Orissa, Jharkhand and Tamil Nadu 84.5%. In Indo-Gangetic Plains (IGP), pulses like chickpea, pigeonpea, lentil, urdbean, mungbean, peas, cowpea and lathyrus are grown since ancient long period. These crops find an important place in rainfed agriculture in state. Pulses are major constituent of diet of both, vegetarian and non-vegetarian population of IGP. In eastern IGP, lentil is major pulse, while chickpea and pigeonpea is major pulse crop of central IGP. Both, lentil and chickpea co-exist in Bundelkhand region

Constraints in pulses production

Pulses are considered as secondary crops by farmers and therefore grown mostly on marginal land under rainfed condition with minimum inputs and care. These resulted in severe yield losses not only due to edaphic, abiotic and socio-economic factors but also due to confounding effects of various biotic stresses. The greater effect of environmental factors and their interactions with genotype (G×E interaction) are the major yield limiting factors in pulses production. The major biotic and abiotic factors limiting pulses yields are discussed below.

Biotic stresses

Among the major biotic stresses, diseases, insect-pests and weeds are the most important. In case of chickpea, the important diseases affecting its production are, *fusarium* wilt caused by *Fusarium oxysporum* f.sp. *ciceri* and Ascochyta blight. Other diseases include botrytis grey mold (BGM) caused by *Botrytis cinerea*, leaf spot by *Alternaria* sp. black root rot by *Fusarium solani*, *Phytophthora* root rot by *Phytophthora megasperma* and *Pythium* damping-off by *Pythium ultimum* and rust by *Uromyces* and beet western yellow virus (BWYV) causing narrow leaf. In case of pigeonpea, *Fusarium* wilt, sterility mosaic and *Phytophthora* blight (*Phytophthora drechsleri*) are economically the most important diseases. In short duration pigeonpea (120-150 days maturity) *Phytopthora* blight is more common as compared to medium and long duration varieties of pigeonpea. However, in *Vigna* species, yellow mosaic disease cause by Mungbean Yellow Mosaic Virus (MYMV) and Mungbean Yellow Mosaic India Virus (MYMIV), powdery mildew, *Cerscospora* leaf spot, root disease caused by *Pythium* and *Fusarium* spp. cause significant losses. Lentil also suffers from many diseases, the most important being Rust (*Uromyces viciafabae*), *Fusarium* wilt (*Fusarium oxysporum* f.sp. *lentis*), *Ascochyta* blight (*Ascochyta lentis),* Stemphylium blight (*Stemphylium botryosum*) while pea suffers from Powdery mildew (*Erysiphe pisi),* Downy mildew (*Peronospora viciae),* Rust (*Uromyces viciafabae*), etc.

Among insect-pest in pulses (pigeonpea and chickpea) pod borer, *Helicoverpa armigera* (Hubner) is the most damaging pest worldwide and its frequent occurrence often results in complete crop failure. Besides *Helicoverpa*, other pests like maruca (*Maruca vitrata* Geyer), pod sucking bugs (*Clavigralla horrida* Germar) and podfly (*Melanagromyza chalcosoma* Spencer) pose a big threat to pigeonpea production. Mungbean and urdbean most suffer by spotted pod borer (*Maruca vitrata),* whitefly (*Bemisia tabaci*), aphids (*Aphis craccivora*) and thrips (*Caliothrips indicus, Megalurothrips distalis*).

Weeds are the other principal biotic constraints in pulses production. It is estimated that out of total annual losses of agricultural products from various pests, weeds alone account for about 37% loss which is higher than the losses caused by other pests taken together (Kumar *et al*., 2013). Besides direct reductions in crop yields inflicted due to the presence of weeds, there are many indirect ways by which the weeds may be troublesome in agriculture. For example in weedy fields, farm operation of fertilizers, insecticides and irrigation become cumbersome. Even when a crop is made despite the presence of weeds, it may be difficult to harvest eg. *Argemone maxicana* and *Convolvulus arvensis* in *rabi* pulses. Weeds are acting as an alternate host for many diseases and insect-pest. 31 to 110% yield improvement in *kharif* pulses like pigeonpea, mungbean and urdbean and 26-45% in *rabi* pulses (chickpea, field pea and lentil) was recorded due to effective weed management only (Chandra, 1982; Ali and Lal, 1989).

Abiotic stresses

The most common abiotic stresses affecting pulses production in country are drought accompanied by heat and cold. Other abiotic stresses specific to some regions are soil moisture stress, salinity, water logging, soil alkalinity and acidity, nutrient deficiency and toxicity. In *kharif* season water logging conditions in early stage of growth, especially in areas receiving good rains such as eastern India, highly affect the yield potential. Contrary to this, moisture stress is responsible for yield loss in low rainfall areas. In case of *Vigna* crops, during rainy season, the crops invariably witness rains at the time of pod maturity in the month of September, leading to deterioration of seed quality and pre-harvest sprouting (Singh *et al.,* 2011). This has a direct negative impact on both, productivity and marketability of the crop. During *rabi* season, low temperature in the month of December and January mostly in pigeonpea and lentil, and terminal drought cause considerable yield losses while in spring/summer grown pulses, terminal heat and drought stress are the major causes of concern. Soil salinity is mostly recorded in central and western parts of the country.

Production technologies of pulses

Sowing date

Sowing of pulses at optimum time is a key of success under both, rainfed and irrigated agro-ecosystem (Kumar *et al*., 2008). Sowing time of pulses is normally decided by onset and withdrawal of monsoon for rainy and winter pulses, respectively. Under normal monsoon season, long duration pigeonpea is sown in 1st fortnight of July, whereas mungbean and urdbean in 2nd fortnight of July in most parts of the country. Early pigeonpea is normally sown during 1st fortnight of June for proper yield and taking wheat under late sown condition. In late sown pigeonpea, crop establishment in major problem due to excessive soil moisture and wilting due to complex root diseases, however in early sown mungbean and urdbean harvesting coincides with withdrawal of monsoon which causes drastic reduction in yield and quality of produce due to pre-harvest sprouting of seeds inside the pod. Rainfed situation of Bundelkhand & central India, winter pulses like chickpea and lentil is sown in the month of October. Sowing time is generally decided by soil moisture condition. If the monsoon withdrawal is late then sowing will be delayed. Sowing of *rabi* pulses like chickpea, lentil and pea under irrigated condition can be done up to 15th November. Study showed that last week of October was the optimal period for sowing of chickpea which gave optimum yield (Table 1). Further delay resulted in poor plant establishment and growth due to dip in day and night temperature. Some of the early maturing chickpea varieties like "JG 14" can be sown even in 2nd fortnight of November. Rajmash in *rabi* season in Indo-Gangetic Plains (IGP) can be sown during 2nd fortnight of October to 1st week of November. Under rainfed rice fallows region of eastern and southern India, winter pulses like mungbean and urdbean are normally sown during December to 15 January depending on harvest of rice crop.

Table 1. Effect of sowing date on chickpea yield

Sowing date	Yield (kg/ha)
15 October	1920
25 October	2063
5 November	1772
CD (p=0.05)	140

Source: Chand *et al.* (2010)

Due to availability of extra-short duration (55-65 days) varieties, summer mungbean can be sown up to 15th April in northern plains and central India under irrigated ecosystems. Due to longer maturity period than mungbean, urdbean can be sown in spring season during 2nd fortnight of February to 1st fortnight of March.

Seed treatment

Pulses seeds must be treated with fungicides before sowing to check the damage caused by a number of soil-borne fungi. Some of the commonly available and cost effective fungicides are like Captan, Bavistin, Brassicol or Thiram, etc. which can be used at 2.5 g/kg of seed. At present, seed treatment with Trichoderma is also advocated for management of soil borne diseases in pulses through biological means. These treatments have proved beneficial in terms of limiting the incidence of soil borne diseases and enhancing seed yield.

Rhizobium inoculation

Pulses are known to fix nitrogen through symbiotic association with bacterium known as *Rhizobium*. There are seven genera of this bacterium namely, *Bradyrhizobium*, *Azorhizobium*, *Sinorhizobium*, *Allorhizobium*, *Mesorhizobium*, *Rhizobium* and *Methylobacterium*. A pulse crop having effective root nodules can not only meet its own nitrogen requirements but also enrich the soil nitrogen content and thus add to soil fertility. Studies reveal that generally 20-40 kg/ha of fixed nitrogen may be left unused in soil for next crop, although the amount varies with the crop, season and other factors. On an average yield advantage of 10% can be achieved in pulse crops with inoculation of *Rhizobium* culture. Estimates of nitrogen fixed by some of the pulse crops are presented in table 2. Owing to the important role of pulses in nutritional security and nitrogen economy the Government of India is laying emphasis on increasing pulse production in the country. Consequently, many microbiological laboratories in different agro-ecological zones of country have been funded to start production of *Rhizobium* culture on commercial scale. The rhizobial cultures are available in packets of 250 g with a suitable carrier, such as peat, lignite, etc. can be used for 10 kg of pulses seed. These bacteria can give best results under neutral pH. Thus, in acidic or alkaline soils pelleting of the inoculated seed with $CaCO_3$ or $CaSO_4$ is very helpful.

Table 2. Estimates of nitrogen fixed by some pulse crops

Crop	N fixed (kg/ha/yr)	Crop	N fixed (kg/ha/yr)
Chickpea	26-63	Mungbean	50-55
Cluster bean	37-196	Pigeonpea	68-200
Cowpea	53-85	Soybean	49-130
Groundnut	112-152	Peas	46
Lentil	35-100		

Source: Brahmaprakash *et al.* (2003)

Seed germination and sowing depth

There is no seed dormancy is observed in pulse crops. If enough moisture is available, the seedlings appear within 7-10 days after sowing in chickpea and pigeonpea, but in case of mungbean seedlings emerged within 3-4 days. However, certain percentage of hard seeds, especially in the *Vigna* group of crops is observed which do not imbibe water and hence fail to germinate. The formation of hard seed coat is a function of environmental conditions prevailing at the time of crop maturity. It is very easy to determine the percentage of hard seeds by simply soaking the seed in water for 6-8 hours. Seed lot containing high percentage of hard seeds should not be used or its seed rate should be calculated on basis of per cent hard seed in the lot.

Seed germination is also affected by seeding depth. Both, hypogeal and epigeal modes of germination are found in pulses. *Rabi* pulses including chickpea, pea, lentil and lathyrus follow hypogeal mode of germination wherein the cotyledons are left behind in the soil. Therefore, sowing of seeds of these crop deep down neither delays nor retards germination. However, epigeal mode of germination is observed in *kharif* pulses, except pigeonpea. Thus, such seeds do not germinate if sown deeper in soil. Uneven seeding depth is observed in manually sown crop which leads to non synchronous emergence of seedlings and growth in the field. Therefore, it is always advisable to sow the seeds at uniform depth (4-6 cm depth) through seed drill.

Conservation tillage

Excessive tillage of soils practiced in conventional agriculture results in short term increase in fertility, but degrades soil in long run. Structural degradation, loss of organic matter, erosion and falling microbial biodiversity are expected outcome of excessive tillage practices (Kumar *et al.*, 2018a). Soil degradation due to tillage has forced us to look for alternatives to reverse the process. Conservation tillage with suitable cropping systems is helpful to maintain soil health, increase water use efficiency and check erosion (Fuzisaka, 1990; So *et al.*, 2001; Mina *et al.*, 2008). In fact higher yield of pulses after wet season (rainy season) rice with reduced tillage was also reported by Pratibha *et al.* (1996) and Mahata *et al.* (1992) from the rainfed areas of eastern India. Minimum tillage with crop residue management is found to reduce soil water evaporation, soil sealing and crusting (Verma and Bhagat, 1992; Meelu *et al.*, 1994; Gangwar *et al.*, 2006) which prevent growth and development of pulse crops.

In a study on soil moisture conservation effect on rainfed chickpea after rice harvest at IIPR, Kanpur revealed that chickpea can be successfully grown on residual soil moisture after rice harvest under zero tillage + rice straw mulch. The improvement in chickpea yield was 23-28% due to zero tillage and mulching over conventional method. The highest relative water content at flowering stage was also recorded in zero tillage + dibbling sowing + mulching (72.4%) followed by zero tillage + no till drill sowing + mulching (69%) and lowest under conventional practice (61.2%) (Fig. 2). This was mainly due to more soil moisture under zero tillage + mulching at flowering stage which finally resulted in higher yield of chickpea under these practices. Similarly, in other set of studies, highest chickpea yield and system productivity was recorded in rice-chickpea-mungbean under conservation tillage (Kumar *et al.*, 2019).

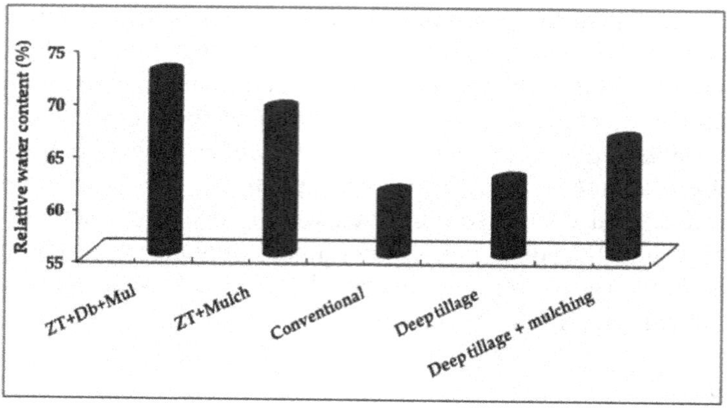

Fig. 2. Relative water content (%) of chickpea leaves at flowering stage *(ZT: Zero tillage; Db: Dibbling; Mul: Mulch)*

FIRB planting system

Furrow Irrigated Raised Bed (FIRB) system of planting is an agronomic intervention where crops are sown on raised beds of different size. The concept of raised bed planting is very advantageous in both water logged and limited water area. The system of planting crops on raised bed alters crop geometry and land configuration, imposes effective control over irrigation and drainage. Water logged situation is common features of rainy season pulses, however *rabi* pulses are normally grown under limited water condition or under rainfed. Further, 40-50% reduction in incidence of complex *Phytophthora* wilt is observed in pigeonpea under heavy rains. Furrows can be used to drain out the excess amount of water from water logged fields. In other hand, 40-50% saving in irrigation water was recorded when irrigation was applied through furrows. The problem of over irrigation or ponding at some parts in field can also be avoided. In a various studies at IIPR, Kanpur revealed that planting of 2 lines on raised bed size 75 cm enhances seed yield by 33.6% in urdbean, 15% in chickpea and 16% in lentil over conventional system of planting (Fig. 3). In addition, 40-45% saving of irrigation water and 25% saving of fertilizers and seeds were also recorded under FIRB planting (Kumar *et al.*, 2015a; Ali *et al.*, 1998).

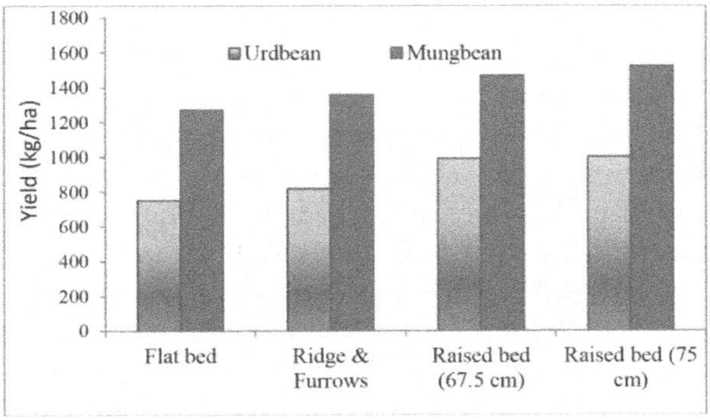

Fig. 3. Effect of sowing methods on urdbean and chickpea yield (mean of 2 years)

Pulses in conservation agriculture

The basic principles of conservation agriculture (CA) such as least disturbance of soil, retention of adequate crop residue on soil surface and sensible crop rotation for improving livelihood and ecological security are well met while bringing legumes in production systems. Pulses are endowed with unique gift of nature to trap atmospheric N_2 in their root nodules in association with *Rhizobium* bacteria besides adding huge amount of organic matter to soil and protect from erosion. Pulse crops fix 1.0-1.5 million tonnes N and thus help in cutting industrial production of green house gases (GHGs). Pulses have immense value in CA and fitted well in all three principles to achieve objectives of CA. Studies conducted at IIPR, Kanpur have shown better yield of pulses under zero-tillage and residue retention under both rainfed and irrigated ecosystems. Similarly, inclusions of pulses in cereal based cropping systems have sown overall improvement in productivity and soil health (Kumar *et al.*, 2019; Kumar *et al.*, 2018b; Nadarajan and Kumar, 2018). Therefore, inclusion of pulses in cereal based crop rotations under CA may be considered as one of the resource conservation technologies (RCTs) which will reverse the negative effect of cereal- cereal rotation systems in this region.

Crop diversification

Cropping system: Development of short duration and disease resistant varieties of different pulses led to the foundation of cropping system research in different agro-ecosystem. This also paved way for crop intensification in both irrigated as well as in rainfed conditions. Some of the examples are pigeonpea-wheat in middle and upper IGP, maize–*rabi* frenchbean in middle IGP, rice–wheat–mungbean in northern and central India and maize-potato/mustard- mungbean/ urdbean in central IGP. Many other cropping systems also evolved in different parts of the country.

Pulses in rice-wheat: After the green revolution, area under the rice-wheat system in India has increased considerably due to high productivity and profitability with less risk. Rice consumes > 40% of all the irrigation water resources available for agriculture. With time, many problems have been reported due to continuous following of rice-wheat system such as deterioration of soil structure and health, build-up of obnoxious weeds including resistance to herbicides, multi-nutrient deficiency (S, Zn, B, Mo, Fe), buildup of insect-pests and diseases, environmental pollution due to burning of crop residues and escape of N to sub-surface water (Gulati, 1999; Malik *et al*., 1998; Nayyar *et al*., 2001). Thus to overcome these problems diversification of rice-wheat cropping system with pulses is being advocated since long. In a long-term study at IIPR, Kanpur revealed that inclusion of pulses in the rice-wheat system increased the system productivity as well as yield of component crops. Results further revealed significantly higher yield under rice-wheat-mungbean followed by rice-chickpea and lowest under rice-wheat (Fig. 4). The rice-wheat system in which wheat crop was replaced by chickpea once in two years has also shown improvement in terms of system productivity and soil health as well as sustainability (Annual report, 2012-13). Similar results were also reported by Ghosh *et al*. (2012) and Ali and Kumar (2006).

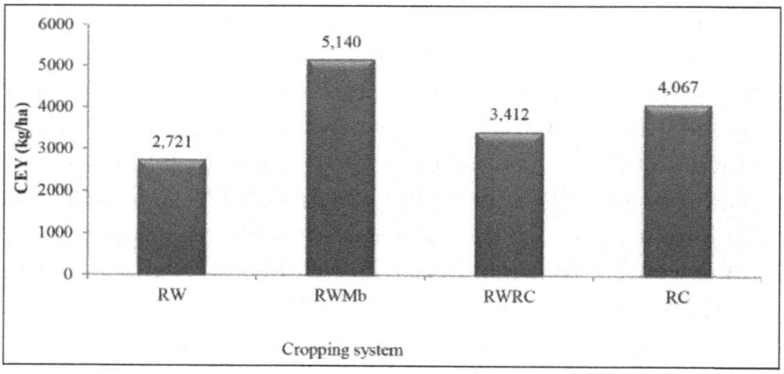

Fig. 4. Long-term effect of rice based cropping system on system productivity (CEY: Chickpea equivalent yield) (*Source:* Annual Report 2012-13)

Pulses in intercropping: The major considerations for intercropping are the contrast in maturities, growth rhythm, height and rooting pattern and variable insect pest and disease associated with component crops so that these complement each other rather than compete for the resources and guard against weather adversities. Growing of crops in intercropping systems is found more productive particularly under rainfed conditions. Pulses can be easily intercropped with oilseeds, cereals, coarse grains and commercial crops. Among pulses, late duration pigeonpea is planted in wider rows and its initial growth is slow which provides an opportunity for intercropping with crops like mungbean, urdbean, sorghum, etc. Being a deep-rooted crop it extracts nutrients and water from deeper soil layer and thereby minimizes the competition for these inputs with cereals. Pigeonpea intercropped with short duration pulses (mungbean and urdbean) is the most popular combination in Uttar Pradesh. The special feature of this system is that the productivity of the base crop i.e., pigeonpea remains unaffected and an additional 400-500 kg/ha of mungbean or urdbean or 600-800 kg/ha of sorghum can be obtained without any additional inputs. Intercropping of winter pulses like chickpea and lentil with oilseeds is common in rainfed areas. Literatures revealed that high productivity and monetary returns can be obtained from chickpea + mustard (Fig. 5), lentil + linseed and wheat + lentil intercropping systems (Kumar *et al.*, 2008; Kumar *et al.*, 2012). Intercropping of chickpea + mustard/linseed is commonly seen in Bundelkhand regions. Some of new intercropping systems like spring sugarcane + mungbean/urdbean and rajmash + potato have been advocated in different parts of country.

Studies on genotypic compatibility in intercropping system were initiated with development of new phenotypes in various pulse crops. Chickpea genotype 'KWR 108' was found more compatible than 'BG 256' and 'KPG 59' for intercropping with linseed cv. 'Neelam' and row ratio of 6:2. Similarly, lentil variety 'L 4076' was found more compatible than 'DPL 62' in lentil + linseed intercropping. Mungbean varieties 'PDM 11' and 'PDM 84-143' and urdbean variety 'DPU 88-31' were most compatible for intercropping with spring planted sugarcane (Ali, 1992 and IIPR, 2009). Recent released varieties of pulses can also fit well under intercropping. The erect growth behavior of pigeonpea variety 'IPA 203' is more suitable than 'Bahar' and 'Narendra Arhar 1'. Similarly, short duration variety of mungbean 'IPM 205-7' matured in 52-55 days is more suitable for intercrop than others.

Fig. 5. Intercropping of chickpea + mustard (6:2)

Residues management

Crop residues are good sources of plant nutrients and are important components for the stability of agricultural ecosystems. Green revolution during 1960s not only drastically enhanced the food grain production but also crop residue production. In areas where mechanical harvesting is practiced, a large quantity of crop residues is left in the field, which can be recycled through proper residue mechanism for nutrient supply in the system. Apart from organic carbon and minerals, about 25% of nitrogen and phosphorus, 50% of sulphur, and 75% of potassium uptake by cereal crops are retained in crop residues, making them valuable nutrient sources.

Incorporation of urdbean and mungbean residue was found beneficial to the succeeding mustard crop in terms of higher yield (6-7%) in a study at Kanpur. In rice-chickpea sequence, yield of chickpea was significantly influenced by rice-residue incorporation and highest seed yield was obtained with incorporation of chopped straw + irrigation, while lowest yield was obtained in rice residue removal treatment. Incorporation of chopped residue of mungbean + irrigation resulted in maximum wheat yield which was significantly higher (38%) than control (Fig. 6). In rice (upland)-lentil and rice-wheat–mungbean systems, incorporation of crop residues increased yield of all crops. Incorporation of both crop residues had shown an improvement of 17.6% in lentil yield over no residue in rice-lentil cropping system. Similarly, higher yields of all three crops in rice–wheat –mungbean were recorded due to incorporation of crop residues of either one crop or all crops in the system (Annual Report, 2010-11; Kumar *et al.*, 2012).

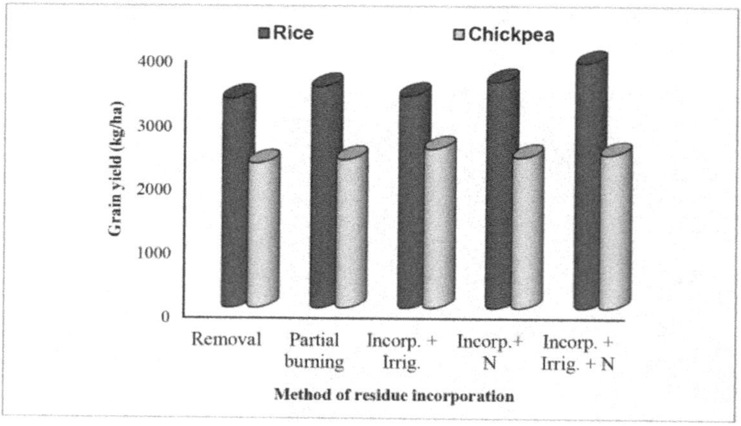

Fig. 6. Effect of residue incorporation on grain yield (kg/ha) of rice and chickpea (*Source:* IIPR, 2009)

Further, incorporation of urdbean and mungbean residue raised the organic carbon level by 35.48% over control. Residue incorporation also resulted in higher soil available N (24.6%), P (11.5%), and K (18.5%) over the initial fertility levels (Singh *et al.*, 2012). Aggarwal *et al.* (1997) also reported increase in soil organic carbon, nitrogen, phosphorus, phosphatase and dehydrogenase activity in the soil after incorporation of pulse crops residues. Soil physical parameters *viz.*, bulk density, particle density, percent pore space and water holding capacity (WHC) also improved under residue incorporation plots over residue removal plots (Table 3). In same set of study, periodic changes in soil microbial biomass carbon (SMBC) were also recorded. The results revealed that increase in SMBC up to 56 days after incorporation of urdbean and mungbean under chopping + incorporation + irrigation. Similar trend was also observed after harvest of wheat crop. The ratio of microbial carbon to soil organic carbon was also higher under chopping + incorporation + irrigation (Singh *et al.*, 2012). Similarly, other studies at IIPR revealed that incorporation of all crop residues in rice (upland)- lentil and rice – wheat – mungbean systems enhanced yields of all crops in the system, besides, improvement in soil physico-chemical properties including infiltration rate, nodulation and earthworm population were also observed.

Table 3. Effect of crop residue incorporation on soil physical properties

Residue management	Bulk Density (g/cc)	Particle Density (g/cc)	Pore space (%)	WHC (%)	SMBC (μg /100 g)	SOC (g/ kg)	Ratio of SMBC to SOC (%)
Mungbean[1]	1.38	2.42	45.5	37.3	262	3.9	6.71
Urdbean [1]	1.39	2.39	44.65	38.3	222	4.2	5.28
Mungbean[2]	1.38	2.38	46.80	38.3	322	3.9	8.25
Urdbean [2]	1.38	2.40	47.00	41.60	312	4.10	7.60
Mungbean[3]	1.34	2.38	47.32	42.50	327	3.6	9.08
Urdbean [3]	1.35	2.39	48.23	45.10	337	3.7	9.10
Mungbean [4]	1.32	2.36	49.63	46.40	320	3.5	9.14
Urdbean [4]	1.33	2.35	48.20	45.90	347	3.7	9.37
Control	1.44	2.50	38.15	33.40	132	3.10	4.25
CD (p=0.05)	0.05	0.10	3.51	3.8	39.8	0.29	NS

Note: 1- Incorporation; 2- Incorporation + irrigation; 3- Chopping + incorporation; 4 - Chopping + incorporation + irrigation; MBC: Microbial biomass carbon; SOC: Soil organic carbon; WHC: Water Holding Capacity (*Source*: Singh *et al.*, 2012)

Nutrient management

After green revolution, area under high yielding varieties of rice and wheat increased many folds in the country. Similarly, cropping intensity also increased which feeds heavily on soil nutrients. The use of concentrated fertilizer nutrients especially for N and P increased which has created imbalance in soil nutrients with time. Hence, nutrient deficiencies are inevitable unless steps are taken to restore soil fertility in this region. Pulses growing soils in different parts of country are now deficient in N, P and S. Same time at field level, application of N and P at recommended rate is not common. Further, it is generally thought that most Indian soils are adequate in K supply and its high cost, optimum K supply is never considered seriously especially in rainfed areas. In pulse growing areas S mining is greater than cereals. Thus, deficiency of S has cropped-up as serious obstacle to the sustainability of yields in pulses as S requirement of these crops is relatively high.

Micronutrient deficiencies are also emerging in different regions. The extent of deficiency of micronutrients in Indian soils is given in table 4. This also showed deficiency of micronutrients is increasing with time (Nadarajan *et al*., 2013). Among the micronutrients, zinc deficiency is widely encountered followed by B and Fe in that order. Field scale Zn and B deficiencies have increasingly reported by many researchers. Looking to the trend of nutrient deficiency, some more nutrients will be added to this list (Fig. 7). Thus, in order to realize the full potential of pulse crops/cultivars, it is essential to adopt an integrated and balanced fertilization which includes almost all nutrients deficient in soil.

Table 4. Micronutrients deficiencies in pulse growing states of India

State	Percentage deficiency of available micronutrients			
	Zn	Fe	Cu	Mn
Andhra Pradesh	51	2	1	2
Bihar	54	6	3	2
Gujrat	24	8	4	4
Haryana	61	20	2	4
Madhya Pradesh	63	3	1	3
Punjab	47	14	1	2
Tamil Nadu	53	15	3	8
Uttar Pradesh	45	6	1	3
Karnataka	78	39	5	19

Source: Ganeshamurthy *et al.* (2003)

On an average, for producing one tone of biomass, pulse crop remove 30-50 kg N, 2-7 kg P, 12-30 kg K, 3-10 kg Ca, 1-5 kg Mg, 1-3 kg S, 200-500 g Mn, 5 g B, 1 g Cu and 0.5 g Mo (Ahlawat and Ali, 1993). Since pulses have the capacity to fix nitrogen owing to their symbiotic relationship with *Rhizobium,* supply of high dose of nitrogen by application of chemical fertilizers inactivates the inherent plant system. This results in substitution of the nitrogen supply source rather than supplementation. However, initial application of a small starter dose of nitrogen @ 15-20 kg/ha is beneficial for proper initial plant growth till development of root nodules. Thus, starter dose of nitrogen helps the plant especially during the seedling stage.

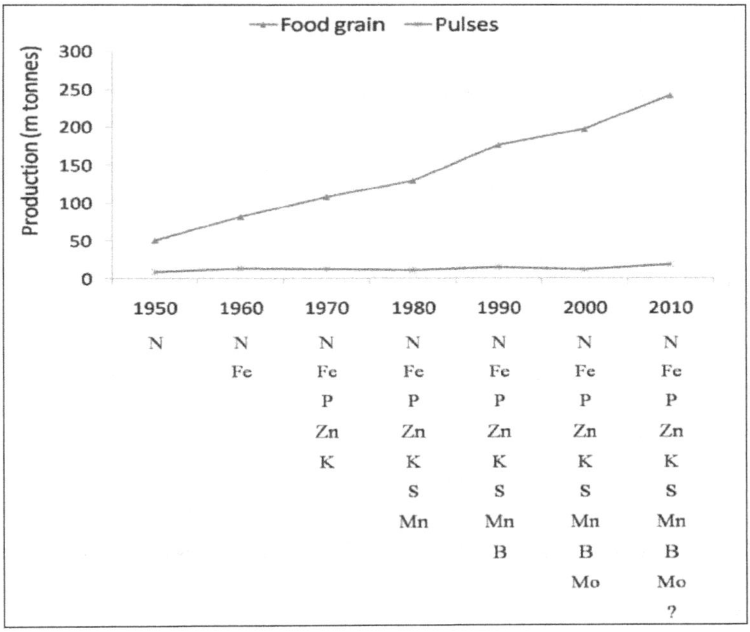

Fig. 7. Emerging nutrient deficiency due to increased production
(*Source:* Ganeshamurthy *et al.*, 2003)

The N-sparing and synergistic effects of pulses are well recognized. Pulses can fix 30-150 kg N/ha depending upon rhizobial population, host crop and varieties, management level and environmental conditions. In sequential crop involving pulses, the preceding pulse may contribute 18-70 kg N/ha to soil and thereby considerable amount of N can be saved in succeeding crops. In rice-wheat rotation growing of short duration mungbean in summer may brings nitrogen economy up to 30-60 kg N/ha in succeeding rice crop. Nitrogen economy due to inclusion of pulses in succeeding cereals under sequential cropping is given in table 5.

Table 5. Nitrogen economy due to inclusion of pulses in sequential cropping

Preceding pulse crop	Following cereal	Fertilizer N- equivalent (kg N/ha)
Chickpea	Maize	60
Chickpea	Rice	40
Pigeonpea	Wheat	40
Mungbean	Rice	40
Urdbean/mungbean	Wheat	30
Lentil	Maize	30
Fieldpea	Maize	25
Rajmash	Rice	10
Cowpea	Rice	40
Cowpea	Wheat	43

Source: Subbarao (1988)

For the growth and development of root nodules, phosphorus is absolutely essential and application of 40-60 kg P_2O_5/ha has been found to be optimum as per soil test values. The P_2O_5 should be applied in furrows below the seed before sowing. An application of 100 kg diammonium phosphate (DAP) per ha was found sufficient in many parts of the country to fulfill the requirements of nitrogen and phosphorus in almost all pulse crops. In acid soil, application of rock phosphate instead of super-phosphate reduces the fertilizer cost, as pulses have the capacity to mobilize phosphorus from the insoluble form. It has generally been observed that if pulse crops are grown after a cereal/vegetable crop, which receives substantial amount of fertilizer, the residual fertilizer is sufficient to meet requirement of pulse crops. This has been especially observed in summer mungbean which grown after potato cultivation in Kanauj and Fatehpur districts.

Table 6. Response of pulses to K under rainfed condition

Crop	No. of trials	Mean response of K (kg/ha)		
		N20P40K20 over N20P40K0	N30P60K30 over N30P60K0	N40P80K40 over N40P80K0
Chickpea	205	95	72	24
Pigeonpea	69	163	29	59
Mungbean	14	30	29	-
Urdbean	105	77	20	42
Lentil	90	112	85	73
Fieldpeas	15	148	87	81

Source: Yadav *et al.* (1993)

The response to foliar application of DAP has also been reported from different locations. Spraying with urea/DAP (2%) at flowering and pod-filling stages is also suggested under rainfed. The response to potassium is generally poor in pulse crops. But, an application of 20 kg K_2O/ha will be profitable (Table 6). The need of sulphur application in pulse crop was realized during early 1990s when multi-locations studies under AICRPIP showed good response to 20 kg S/ha (Ali and Mishra, 1996). Presently, application of 20 kg S/ha became part of fertilizer nutrient recommendation in pulses. With the development of agriculture and use of chemical fertilizers, deficiency of micronutrients observed in many parts of country. The deficiency of zinc, boron and molybdenum need to be corrected immediately by through suitable management practices. Molybdenum deficiency is most common in acid soils and can be corrected by liming. Low pH, low calcium, high aluminium and high manganese have also proved to be inhibitors of legume nodulation.

Water management

Water requirement of pulses is lower than cereals. Global water consumption by cereals is reported to be about 60% as against 4% in pulses. Pulses have ability to use water more efficiently than other crops due to their morphological and physiological features. Due to their deep root system, pulses are able to draw moisture from deeper layer of soil profile thereby having ability to thrive well under dryland situations. By consuming 1 ha-mm of water, chickpea could produce about 12.5 kg grain as against 7 kg in wheat and 2.5 kg in rice. The water requirement of rice crop is 900-2500 mm, wheat 400-450 mm and sugarcane 1400-2500 mm, however pulses need only 250-300 mm of water. In general under sub-tropical climate like in Indo-Gangetic plains (IGP), *rabi* pulses like chickpea and lentil need only one irrigation whereas wheat crop needs 5-6 irrigations. Therefore, it is recommended to replace one of the cereal crops by pulse crop to reverse the process of ground water depletion in rice-wheat regions of IGP.

Pulses are generally grown under rainfed conditions (84%). But, under moisture stress condition pulses respond well to irrigation. Some of the pulses like *rabi* frenchbean, early pigeonpea and summer mungbean/urdbean are cultivated under irrigated condition. Chickpea has maximum area (35%) under irrigation in the country. Various approaches such as crop growth stage, IW/CPE ratio and cumulative evaporation have been used in scheduling irrigations in different pulses. In field pea, 50% flowering stage was found most critical for irrigation. Similarly, in lentil and chickpea, one irrigation at the early pod filling stage was found most effective. Pigeonpea does also respond to irrigation. It uses about 20-25 cm water to produce about 1 tonne/ha grains. During soil moisture stress, irrigating pigeonpea crop at flower initiation and pod filling stages increased the yield by many folds (Table 7). Scheduling irrigation in pigeonpea can also be done through IW/CPE ration of 0.4 to 0.5. Whereas, scheduling of irrigation is recommended at IW/CPE ratio of 0.8 in *rabi* rajmash. Pre-plant irrigation is necessary to prepare land and ensure adequate moisture for germination of seed of summer mungbean/urdbean. Thereafter, the first irrigation should be given 20-25 days after sowing and subsequent irrigation at 7 days interval (Ali and Kumar, 2006). Under micro-irrigation on pulses, initial studies were conducted on sprinkler irrigation in closely planted short statured legumes like chickpea and mungbean.

Table 7. Effect of irrigation on grain yield of pigeonpea

Stage of irrigation	Seed yield (tonnes/ha)
Control (No irrigation)	1.28
Irrigation at flower initiation	1.30
Irrigation at peak flowering	1.22
Irrigation at pod filling	2.14
Irrigation at flower initiation and pod filling	2.43
Irrigation at 80 mm CPE	2.27
Irrigation at 120 mm CPE	2.36
C.D. (0.05)	0.49

Source: Praharaj *et al.* (2014)

Pulses are very sensitive to water logging even for short period. Localized ponding situation in field arises due to surface or conventional method of irrigation. Therefore, it is always advisable to apply water in furrows in case of raised bed system and sprinkler or drip under flat bed system of planting. Laser leveling can be done to minimize the local ponding situation. Three sprinkler irrigations of 60 mm each at sowing time, branching time and pod formation stages were found sufficient for chickpea crop with water saving of 44% (Chandegara and Yadavendra, 1998). Sprinkler irrigation in mungbean increased yield by 39.7% over surface irrigation and resulted in water saving of 49.8% (Velayutham and Chandrasekaran, 2002). The advantage of sprinkler irrigation on seed yield and water use efficiency (WUE) is also observed in french bean (Fig. 8). Drip irrigation has been found effective in pigeonpea due to wider row pacing or under paired row planting on raised bed.

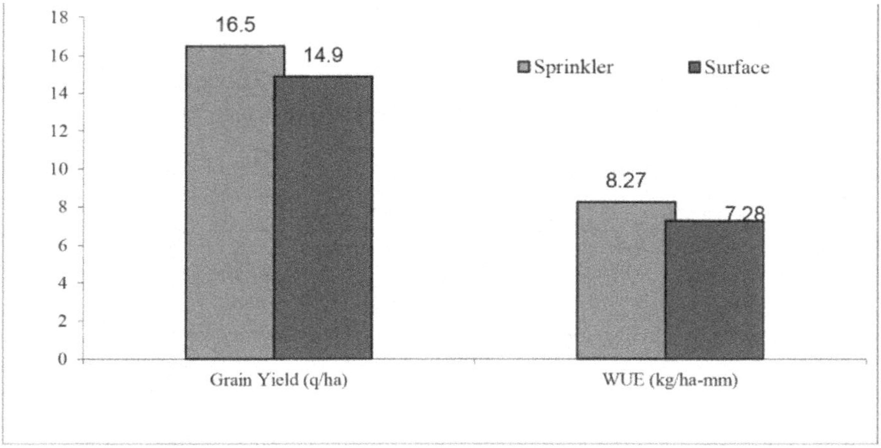

Fig. 8. Grain yield and WUE of frenchbean under different method of irrigation (Adopted from Praharaj *et al.*, 2014)

Optimum soil moisture at root zone can be maintained through drip irrigation which supports optimum plant growth (Fig. 9). At IIPR, Kanpur drip fertigation at branching or branching and pod development (with NK @ 10:10 kg/ha in 5-10 kg splits) was found most efficient (Praharaj and Kumar, 2011).

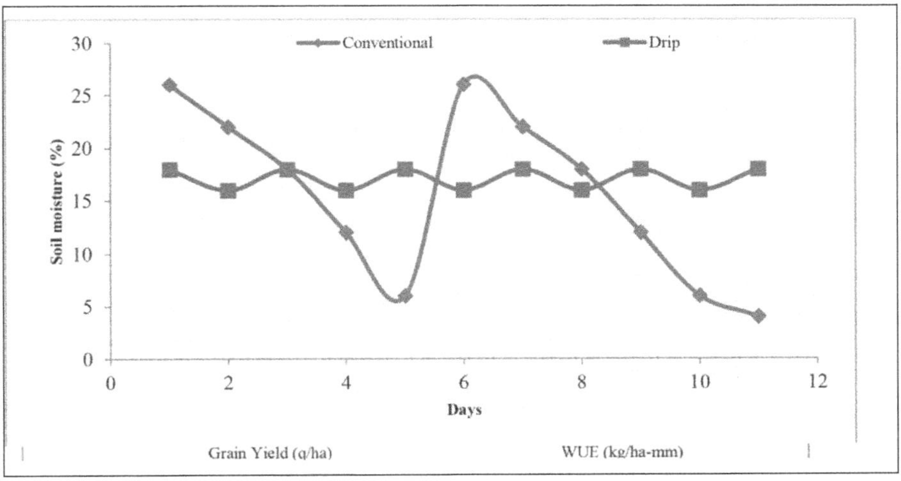

Fig. 9. Elimination of wide fluctuation of soil water content through drip irrigation

Weed management

Weeds are the principal biotic constraints to crop production. It is estimated that out of total annual losses of agricultural products from various pests, weeds alone account for about 45% loss. An estimate shows loss of agricultural products due to weeds is more than the losses caused by other pests taken together and account for about 37% loss. Pulses on account of their initial low vigour often suffer heavily due to weed infestation. Some of the important weeds commonly noticed in most of the pulses are given in table 8.

Table 8. Major weeds associated with pulses

Season	Type of weeds	Name of weeds
Kharif	Non -grasses	*Celosia argentea, Cleome viscosa, Commelina benghalensis, Cucumis trigonus, Digera arvensis, Eclipta alba, Euphorbia hirta, Phyllanthus niruri, Trianthema monogyna*
	Grasses	*Cynodon dactylon, Dactyloctenium aegypticum, Digitaria sanguinalis, Echinochloa colonum, Echinochloa crusgali, Eleusine indica, Eragrostis tenella, Fimbristylis* spp., *Panicum maxima, Setaria glauca*
	Sedge	*Cyperus difformis, Cyperus iria, Cyperus rotundus*
Rabi	Non -grasses	*Ageratum conizoides, Anagallis arvensis, Argimone mexicana, Asphodelus tenuifolius, Carthemus oxycantha, Chenopodium album, Convolvulus arvensis, Coronopus didymus, Fumaria parviflora, Gnaphalium indicum, Lathyrus aphaca, Launia nudicaulis, Medicago denticulata, Melilotus alba, Rumex dentatus, Solanum nigrum, Spergula arvensis, Vicia hirsuta, Vicia sativa*
	Grasses	*Avena ludoviciana, Phalaris minor, Lolium temulentum, Poa annua, Polypogon monspeliensis*
	Sedge	*Cyperus rotundus*
	Parasitic weeds	*Cuscuta spp., Orobanche* spp.
Zaid/ Summer	Non -grasses	*Amaranthus viridis, Chenopodium album, Eclipta alba, Physalis minima, Polygonum plebejum, Portulaca quadrifida, Solanum nigrum, Trianthema monogyna*
	Grasses	*Cynodon dactylon, Dactyloctenium aegypticum, Digitaria sanguinalis, Eleusine indica, Panicum maxicum, Setaria glauca*
	Sedge	*Cyperus rotundus*
Perennial weeds		*Convolvulus arvensis, Cynodon dactylon, Cyperus rotundus, Saccharum spontaneum, Sorghum halepense*

Source: Adopted from Kumar *et al.* (2016a)

The yield losses due to weeds have been estimated to range from 30-50% in chickpea and up to 90% in pigeonpea. 31 to 110% yield improvement in *kharif* pulses like pigeonpea, mungbean and urdbean (Table 9) and 26-45% in *rabi* pulses (chickpea, field pea and lentil) was recorded due to effective weed management only (Chandra, 1982; Ali and Lal, 1989). Besides direct reductions in crop yields inflicted due to the presence of weeds, there are many indirect ways by which the weeds may be troublesome in agriculture. For example in weedy fields, farm operation of fertilizers, insecticides and irrigation become cumbersome. Even when a crop is made despite the presence of weeds, it may be difficult to harvest.

Table 9. Yield reduction due to weeds in important pulses

Crop	No. of locations	No. of trials	Mean yield (kg/ha)		Losses (%)
			Weed free	Weed infestation	
Pigeonpea	16	21	1573	774	46.7
Urdbean	15	18	1001	474	55.4
Chickpea	10	13	1545	767	48.1
Lentil	6	8	1709	755	58.8
Field pea	9	13	2094	1054	47.1

Source: Yaduraju (2002)

The effective weed control in pulses is essential to maximize seed yield and quality and to reduce weed competition in following crops (Kumar *et al.*, 2016). Adequate weed control can be difficult to achieve because pulses have a relatively low growth habit and open canopy early in the growing season. Weed management takes away nearly one third of total cost of production of field crops. In India, the manual method of weed control is quite popular and effective. Of late labour has become non-available and costly, due to intensification/diversification of agriculture and urbanization. The alternative for this is the use of herbicides. The usage of herbicides is only 14% of the total agro-chemicals used in agriculture in India as against 48% used at the global level, and 55% used in USA. Additionally, few herbicides are registered for the use in pulses and that too for pre-emergence or pre plant application.

There are many ways by which weed population can be managed in pulses (Gopinath *et al.*, 2009; Kumar *et al.*, 2011). Prevention is one of prime important measures need to be followed to check establishment and spread of weeds in new niches. Among cultural methods, following intercropping fast growing crops like mungbean and cowpea in slow growing pigeonpea can suppress weed population and hence the weeding requirements have been found to be reduced substantially. Weed smothering efficiency (WSE) of important intercropping systems are given in table 10.

Table 10. Weed smothering efficiency of important cropping system

Intercropping systems	WSE (%)
Pigeonpea + urdbean	32.82
Pigeonpea + mungbean	31.01
Pigeonpea + cowpea	39.06
Pigeonpea + sesame	36.6
Pigeonpea + pearl millet	50.8
Maize + urdbean	17.3
Maize + pigeonpea	16.4

Selection of good crop rotation is must to prevent development of diverse weed population in any production system. Crop like mungbean and urdbean which grow fast and can compete with the weeds very well should be included in the cropping systems either as sole cropping or intercropping. In *rabi* crops like chickpea, *Phalaris minor* and *Chenopodium album* population can be minimized by following puddling in rice and rice straw mulching in chickpea. Puddling also helps in minimizing noxious weed population like *Cyperus rotundus*. The weed management using mechanical method is also a viable, eco-friendly and economical solution. The wider spacing between the rows and plants in pulses allow easy operation of mechanical weeders. Many hand operated tools and power weeders are available which can be used for weed management in pulse crops.

Among chemical weed management, pre-emergence application of pendimethalin @ 1.0-15 kg/ha is widely accepted in all pulse crops. Pendimethalin is a potential herbicide which controls some of the broad leaf weeds especially *Chenopodium* in *rabi* pulses during initial period of crop growth (up to 20-25 days of sowing). But, for effective control of weeds one hand weeding is recommended in all pulses at 25-45 days after sowing. In recent past, due to higher operational cost and limitation in availability of labours on time, and difficulties in carrying of the operation during rainy season, the option for use of post-emergence herbicides was explored. The work on post-emergence herbicides initiated during 2008 at IIPR, Kanpur and under AICRP showed that some of the post-emergence herbicides like quizalofop-ethyl and imazethapyr have great potential for management of weeds in pulse crops (Kumar, 2010; Kumar *et al*., 2015, 2016 and 2016a). The herbicides recommended in pulses are given in table 11 and 12.

Table 11. Herbicides recommended for mungbean, urdbean and pigeonpea.

Herbicide	Dose (g a.i./ha)	Product (g or ml/ ha)	Application time	Remarks
Alachlor	2000-2500	4000-5000	0-3 DAS	Controls many annual grasses and some BLWs
Fluchloralin	750-1000	1500-2000	Pre-planting	Incorporate into the surface soil immediately after application. Controls many annual grasses and some BLWs
Oxadiazon	250	1000	0-3 DAS	Control wide spectrum of weeds
Oxyfluorfen	100-125	400-500	0-3 DAS	Control wide spectrum of weeds
Pendimethalin	750-1000	2500-3000	0-3 DAS	Controls many annual grasses and some BLWs

Quizalofop-ethyl	100	2000	15-20 DAS	Excellent control of annual grasses
Imazethapyr	50-100	500-1000	20-25 DAS	Control wide spectrum of weeds
Pendimethalin (PI) + Imazethapyr (POE)	1250 + 100	4170 + 1000	0-3 (PI) and 20-25 (POE) DAS	Control wide spectrum of weeds

Source: Adopted from Kumar *et al.* (2016a)

Table 12. Herbicides recommended for chickpea, lentil and fieldpea

Herbicide	Dose (g a.i./ha)	Product (g or ml/ha)	Application time	Remarks
Fluchloralin	750-1000	1500-2000	Pre-planting	Incorporate into the surface soil immediately after application. Controls many annual grasses and some BLWs
Metolachlor	1000-1500	2000-3000	0-3 DAS	Controls many annual grasses and some BLWs
Metribuzin (in peas)	250	350	0-3 DAS or 15-20 DAS	Controls several annual grasses and some BLWs and some sedge
Oxyfluorfen (in peas)	100-125	400-500	0-3 DAS	Control wide spectrum of weeds
Pendimethalin	750-1000	2500-3000	0-3 DAS	Controls many annual grasses and some BLWs. Supplement with quizalofop as post-emergence for grassy weed control
Quizalofop-ethyl	50-100	1000-2000	15-20 DAS	Excellent control of annual grasses
Pendimethalin (PI) + Quizalofop-ethyl (POE)	1250 + 100	4170 + 2000	0-3 (PI) and 20-25 (POE) DAS	Control wide spectrum of weeds

Source: Adopted from Kumar *et al.* (2016a)

Pulses in new niches

Summer mungbean: The development of short duration (60-65 days maturity), disease resistant (mainly MYMV) and photo-insensitive mungbean varieties with synchronous maturity nature led to its cultivation in spring/season (mid March to mid June) as a catch crop in IGP. The possibilities of growing under intercropping with autumn planted sugarcane were also worked out by different institutions in this region (Rana *et al.*, 2006; Singh *et al.*, 2008). The work on development of agro-techniques like planting geometry, sowing time, weed management, irrigation scheduling, etc. were carried out extensively for this emerging crop which resulted in continuous increasing in area of summer mungbean in this region since last decade (Kumar *et al*., 2016b). The concerted effort led by IIPR, Kanpur resulted in success of summer mungbean in IGP.

The advantage of growing summer mungbean in cereal based cropping system is proven by many researchers. Growing of summer mungbean after harvest of wheat not only enhances total productivity and economics of system but also improves the soil health (Table 13). The success of growing summer mungbean by farmers in Fatehpur and other districts under supervision of IIPR, Kanpur is well documented. In Kanauj districts farmers are growing summer mungbean after harvest of potato on residual soil fertility.

Table 13. Long-term effect of summer mungbean on system productivity, economics and soil health

Cropping system	MBEY (q/ha)	Net return (Rs/ha)	Benefit: cost ratio	SOC (%)	Available nutrients (kg/ha)				Micronutrients (ppm)	
					N	P	K	S	Zn	Fe
Rice-wheat	24.3	69 936	1.47	0.28	192.3	18.0	157.5	16.4	0.83	1.07
Rice-wheat-mungbean	39.8	1 18 511	1.85	0.32	255.1	22.7	165.9	18.7	0.90	0.93
Maize-wheat	21.5	64 411	1.53	0.29	230.1	18.7	234.8	14.2	1.81	0.95
Maize-wheat-mungbean	36.3	1 02 356	1.75	0.33	238.3	22.5	295.1	20.1	1.91	0.99

Source: Venkatesh *et al.* (2015) (MBEY: mungbean equivalent yield; SOC: soil organic carbon)

Rabi rajmash: Rajmash is native to south and central America, probably Mexico. It is traditionally a crop of temperate region and mainly confined to northern hilly tracts of Jammu and Kashmir, Himachal Pradesh and Uttarkhand as a *kharif* crop and also as winter crop in some parts of Maharashtra, Andhra Pradesh, Western and Eastern ghats, where winters are mild and frost free. The introduction of rajmash in northern plains especially in Uttar Pradesh is possible with the development rajmash varieties 'PDR 14 (Uday)', 'HUR 15' and 'HUR 137' during late 1980s and early 1990s. During 1990s, agronomy of rajmash was worked out at IIPR, Kanpur for Uttar Pradesh condition. End of October has been found most ideal time for planting of *rabi* rajmash in this region (Table 14). Among pulse crops, poor nodulation is observed in rajmash. Thus, N recommendation for this crop is 120 kg/ha whereas in other pulses only 20 kg/ha. The recommendation for P, K and S for rajmash is 60, 40 and 20 kg/ha, respectively. Rajmash required 4-5 irrigation in northern India for cultivation under sole or intercropping with potato. Among different cropping system rice-rajmash-mungbean system was most suitable for this region. Intercropping of potato with rajmash in 2:2 rows ratio was also found promising (IIPR, 2009).

Table 14. Effect of sowing dates on productivity of rajmash

Sowing dates	Yield (kg/ha) -mean of 4 years (1985-89)
15 October	1364
30 October	1733
15 November	1666

Source: Ali and Lal (1991)

Early pigeonpea: The development of short duration varieties of pigeonpea during 1975-76 such as 'Pusa Ageti' and 'UPAS 120' led to introduction of pigeonpea in the irrigated area of northern plains under pigeonpea - wheat double cropping system. Many reports have been published during 1980s for successful cultivation of early pigeonpea under pigeonpea-wheat sequential cropping systems in irrigated areas of northern and central India including Uttar Pradesh. Its agronomic requirements were comprehensively worked out at IARI, New Delhi, GBPUAT, Pantnagar and HAU, Hisar under AICRPIP as well as in the institutional programmes of IIPR, Kanpur (Ali *et al.*, 2016). The optimum sowing time recommended for sowing of early pigeonpea is 20 May to 10 June. However, studies at CSUAT, Kanpur showed that a higher productivity from early pigeonpea could be obtained by advancing its planting to mid April at wider row spacing (75-100 cm) and intercropping of 2 rows mungbean (Panwar and Yadav, 1980). Early pigeonpea normally harvested in end of November or first week of December. Thus, wheat varieties suitable for late sown condition such as 'Unnat Halna' should be taken which fetches ₹ 200-300/ha higher price in market due to its bread making quality.

Intercropping with sugarcane: Sugarcane is cultivated in more than 2.0 million ha area in Uttar Pradesh. Due to wider row planting (90 cm), late emergence and slow initial growth, it offers good scope for intercropping with pulses. Two rows of lentil/chickpea in autumn planted sugarcane and mungbean/urdbean/ cowpea in spring planted sugarcane is successfully demonstrated in many parts of the state (Fig. 10). *Rabi* rajmash can also be taken as intercrop with autumn planted sugarcane. For growing intercropping of pulses with sugarcane, no additional inputs are required except in case of rajmash in which additional nitrogen is to be provided as per recommendation. Kumar *et al.* (2006) also reported advantage of pulse crops in spring planted sugarcane.

Fig. 10. Intercropping of mungbean in spring planted sugarcane

Pigeonpea transplanting: Pigeonpea is generally grown in *kharif* season. Due to the uncertain rainfall or variability in climatic condition, it often faces a temporal water-logging resulting in uneven plant population and poor yield. To mitigate the ill effect of climatic vagaries and to realize higher yield potential, transplanting of pigeonpea is a viable option as contingency measure in north-eastern plain zone. Transplanting of three weeks old seedlings in poly bag or paper bag was found effective as compared to tender or old seedling (Praharaj *et al.*, 2015). Similarly, crop geometry also played a vital role for higher yield in transplanted pigeonpea. Due to higher cost involved in transplanting of pigeonpea, it is advisable to use this technique as contingency measure for replacement of 15-20 per cent pigeonpea population (if died due to disease or water logging) up to 30 days of sowing.

Pulses in rice fallows: Out of 44 million ha of land under cultivation of rice across the country, about 11.7 million ha areas remains fallow during *rabi* season due a variety of biotic, abiotic and socio-economic constraints prevailing in these areas. Continuous cultivation of rice in the rice fallow areas causes disturbance in the ecology of the system and compactness of the soil. Incorporation of pulses improves the soil quality in terms of nutrient dynamics, soil organic carbon and biological activities thereby increasing nutritional security and improving sustainability of cropping system. Pulses are the ideal crops that can be grown in the areas vacated after rice, because of their property to establish with the surface seeding and suitability for relay cropping and resistance against soil moisture and temperature stress. However, a number of abiotic factors (water and soil related) limit their production in rice fallows. Low moisture content in the soil after rice harvest, faster decline in water table with advancement of *rabi* season, and risk of soil moisture stress towards flowering and pod filling stages are some of the water related constraints. The productivity of pulses under rainfed rice fallow area can be increased by following of improved management practices (Kumar *et al.*, 2016c; Kumar *et al.*, 2018).

Soil moisture conservation practices like retaining crop residue or stubble (20-30 cm height) can be useful in enhancing pulses yields under rainfed rice fallows (Bandyopadhyaya *et al.*, 2016). Using improved varieties will also reduce the problem of low productivity. Fast growing and early vigour pulses varieties which can establish even under surface seeding are most suitable for rice-fallows. The fast growing varieties can effectively utilize residual soil moisture for their growth and development. Early maturing varieties can minimize total crop water requirement and escape severe terminal drought. Pelleting of seeds with Super Phosphate and Rhizobial culture improves the establishment, nodulation and grain yield of pulses under rice fallows. Generally the soils of rice-fallow region are poor in N, P, Mo, Zn and native rhizobia. Application of Molybdenum through seed priming (at 0.5g sodium molybdate/kg seed) increased the nodulation by about 90% and grain yield upto 30%. Integrated management of diseases and pests including use of resistant cultivars (MYMV resistant urdbean/mungbean), use of healthy seeds, modification of cultural practices, and judicious use of chemical and bio-control agents may contribute substantially in stabilizing the yields. Further development of pulses plant ideotype that is amenable to mechanical harvesting may further improve the pulses productivity under rice-fallows. Weeds drastically affect pulse production and the problem complexes as the habitat changes. Generally under rice-fallows pulses are heavily infested with weeds of diverse nature. In order to get maximum yield benefits, the crops must be kept free from weeds. The commonly used chemical weed control measures like application of pendimethalin cannot be possible under rice-fallows relay cropping because seeding is done in standing rice crop before harvest. Post-emergence herbicides as quizalofop - ethyl and imazethapyr can be used for proper management of weeds.

Conclusion

Pulses are mostly grown under rainfed conditions with less or no fertilizers and other inputs. The farmers also consider these as secondary crops. Apart from soil moisture deficit stress, biotic stresses like diseases, insect-pests and weeds are the major limitation in the production of pulses in the state. The deficiency of major and micro-nutrients in the soils of pulses growing regions has further complicated the existing problems. Due to these reasons and safe return from cereal crops, pulses area in the state has shown declining trend since last 3 decades. The recent development in the pulses production technologies by different institutions as described in this chapter need to be promoted to overcome the problems associated with pulses production and to reverse the trend to increase the area and production of pulses in the country.

References

Aggarwal, R.K., Kumar, P. and Power, J.F. 1997. Use of crop residue and manure to conserve water and enhance nutrient availability and pearl millet yields in an arid tropical region. *Soil and Tillage Res.* **41**: 43-51.

Ahlawat, I.P.S. and Ali, M. 1993. Fertilizer Management in Pulses. In: Fertilizer Management in Food Crops (Ed. HLS Tandon). FDCO, New Delhi. pp. *114-138.*

Ali, M. 1992. Genotypic compatibility and spatial arrangement in chickpea (*Cicer arietinum*) and Indian mustard (*Brassica juncea*) intercropping in north-east plains. *Indian J. of Agril. Sci.* **62**: 249-53.

Ali, M. and Lal, S. 1989. Priority inputs in pulse production. *Fert. News.* **34**(11): 17-21.

Ali, M. and Lal, S. 1991. Rabi rajmash: Potentialities, limitations and Technology of Production. Technical Bulletin No. 3. Directorate of Pulses Research, Kanpur. Pp. 17.

Ali, M. and Mishra, J.P. 1996. Technology for late-sown gram. *Indian Farm.* **46**(9): 67-71.

Ali, M. and Kumar, S. 2006. Prospects of mungbean in rice-wheat cropping systems in Indo-Gangetic Plains of India. In: Improving income and nutrition by incorporating mungbean in cereal fallows in the Indo-Gangetic Plains of South Asia DFID Mungbean Project for 2002-2004. Proceedings of the final workshop and planning meeting, Punjab Agricultural University, Ludhiana, Punjab, India, 27-31 May 2004 (Ed. Shanmugasundaram, S.). pp. 246-254.

Ali, M., Kumar, N and Praharaj C.S. 2016. Agronomic research in pulses in India: Historical perspective, accomplishments and way forward. *Indian J. of Agron.* **61** (4th IAC Special Issue): S83-S92.

Ali, M., Mishra, J.P., Ahlawat, I.P.S., Kumar, R. and Chauhan, Y.S. 1998. Effective management of legumes for maximizing biological nitrogen fixation and other benefits. pp. 107-128. In: Residual Effects of Legumes in Rice and Wheat Cropping Systems in Indo-Gangetic Plain (Eds. JVDK Kumar Rao, C Johansen and TJ Rego), International Crop Research Institute for the Semi-Arid Tropics (ICRISAT), Patancheru, India.

Annual Reports. 2010-11, 2011-12 & 2012-13. Indian Institute of Pulses Research, Kanpur.

Bandyopadhyaya, P.K., Singh, K.C., Mondal, K., Nath, R., Ghosh, P.K., Kumar, N., Basu, P.S., Singh, S.S. 2016. Effects of stubble length of rice in mitigating soil moisture stress and on yield of lentil (*Lens culinaris* Medik) in rice-lentil relay crop. *Agril. Water Mangt.* **173**: 91–102.

Brahmaprakash, G.P., Girisha, H.C., Vithal, N. and Hegde, S.V. 2003. Biological nitrogen fixation in pulse crops. In: Pulses in New Perspective by Masood Ali, BB Singh, Shiv Kumar and VishwaDhar. Indian Society of Pulses Research and Development, Indian Institute of Pulses Research, Kanpur. pp. 271-286.

Chand, M., Singh, D., Roy, N., Kumar, V. and Singh, R.B. 2010. Effect of growing degree days on chickpea production in Bundelkhand region of Uttar Pradesh. *J. of Food Legumes.* **23**(1): 41-43.

Chandegara, V.K. and Yadavendra, J.P. 1998. Efficacy of sprinkler irrigation in chickpea. *Gujarat Agril. Uni. Res. J.* **24**(1):1-3.

Chandra, S. 1982. Increasing pulse production in India. In: Constraints and Opportunities, Report, All India Co-ordinated Pulse Improvement Project, Kanpur.

FAO STAT. 2018. "Online interactive database on agriculture", FAOSTAT. www.fao.org

Fuzisaka, S. 1990. Rainfed lowland rice: building research on farmer practice and technical knowledge. *Agri., Ecosys. & Env.* **33**(1): 57–74.

Ganeshamurthy, A.N., Subharao, A. and Rupa, T.R. 2003. Integrated nutrient management in pulse based cropping systems. In: Pulses in New Perspective by Masood Ali, BB Singh, Shiv Kumar and VishwaDhar. Indian Society of Pulses Research and Development, Indian Institute of Pulses Research, Kanpur. pp. 287-300.

Gangwar, K.S., Singh, K.K., Sharma, S.K. and Tomar, O.K. 2006. Alternative tillage and crop residue management in wheat after rice in sandy loam soils of Indo- Gangatic plains. *Soil and Tillage Res.* **88**(1–2): 242–252.

Ghosh, P.K., Venkatesh, M.S., Hazra, K.K. and Kumar, N. 2012.Long-term effect of pulses and nutrient management on soil organic carbon dynamics and sustainability on an inceptisol of Indo-gangetic plains of India. *Exp. Agric.* **48**(4): 473-487.

Gopinath, K.A., Kumar, N., Mina, B.L., Srivastva, A.K. and Gupta, H.S. 2009. Evaluation of mulching, stale seedbed, hand weeding and hoeing for weed control in organic garden pea (*Pisum sativum* Sub Sp. Hortens L.). *Arch. of Agron. and Soil Sci.* **55**(1): 115-123.

Gulati, A. 1999. Globalization of agriculture and its impact on rice–wheat system in the Indo-Gangetic Plains. In: Pingali, Prabhu, L. (Eds.), Sustaining Rice–Wheat Production Systems: Socio-economic and Policy Issues. Rice–Wheat Consortium Paper Series 5. Rice–Wheat Consortium for the Indo-Gangetic Plains, New Delhi, India, pp. 43–60.

IIPR. 2009. 25 years of pulses research at IIPR. Indian Institute of Pulses Research, Kanpur 208024, Uttar Pradesh, India. Pp 57-60.

IIPR. 2013. Vision 2050. Indian Institute of Pulses Research (ICAR), Kanpur -208 024. Pp. 28.

Kumar, N. and Yadav, A. 2018a. Role of Pulses in Improving Soil Quality and Enhancing Resource Use Efficiency. In Anup Das *et al.* (eds). Conservation Agriculture for Advancing Food Security in Changing Climate. *Today & Tomorrow's* Printers and Publishers, New Delhi. Vol. **2** (2018): 547-561.

Kumar, N., Yadav, A., Singh, S. and Yadav, S.L. 2018. Growing Pulses in Rice Fallow: Ensuring Nutritional Security in India. In Anup Das *et al.* (eds). Conservation Agriculture for Advancing Food Security in Changing Climate. Today & Tomorrow's Printers and Publishers, New Delhi. Vol. **1** (2018): 107-122.

Kumar, N., Nath, C.P., Hazra, K.K., Das, K., Venkatesh, M.S., Singh, M.K., Singh, S.S., Praharaj, C.S. and Singh, N.P. 2019. Impact of zero-till residue management and crop diversification with legumes on soil aggregation and carbon sequestration. *Soil & Tillage Res.* **189**: 158–167.

Kumar, N., Gopinath, K.A., Srivastva, A.K. and Mahajan, V. 2008. Performance of pigeonpea (*Cajanus cajan* L. Millsp.) at different sowing dates in mid-hills of Indian Himalaya. *Arch. of Agron. and Soil Sci.***54**(5): 507-514.

Kumar, N., Hazra, K.K., Singh, S. and Nadarajan, N. 2016c. Constraints and prospects of growing pulses in rice fallows of India. *Indian Farg.* **66**(6): 13-16.

Kumar, N., Mina, B.L., Chandra, S. and Srivastva, A.K. 2011. In-situ green manuring for enhancing productivity, profitability and sustainability of upland rice. *Nut. Cycl. in Agroecosys.* 90: 369-377.

Kumar, N., Hazra, K.K., Nath, C.P., Praharaj, C.S. and Singh, U. 2018b. Grain legumes for resource conservation and agricultural sustainability in South Asia. In: Meena R., Das A., Yadav G., Lal R. (eds) Legumes for Soil Health and Sustainable Management. Springer, Singapore. pp. 77-107.

Kumar, N. 2010. Imazethapyr: A potential post-emergence herbicide for *kharif* pulses. *Pulses News letter* **21**(3): 5.

Kumar, N., Hazra, K.K. and Nadarajan, N. 2016. Efficacy of post-emergence herbicide Imazethapyr in summer mungbean. *Legume Res.* **39**(1): 96-100.

Kumar, N., Hazra, K.K., Nath, C.P., Praharaj, C.S., Singh, U. and Singh, S.S. 2016b. Pulses in irrigated ecosystem: Problems and Prospects. *Indian J. of Agron.* **61** (4th IAC Special Issue): S199-S213.

Kumar, N., Hazra, K.K., Singh, M.K., Venkatesh, M.S., Kumar, L. Singh, J. and Nadarajan, N. 2013. Weed management Techniques in Pulse Crops. Indian Institute of Pulses Research, Kanpur. *Technical Bulletin* 10/2013. Pp. 47.

Kumar, N., Hazra, K.K., Yadav, S.L. and Singh, S.S. 2015. Weed dynamics and productivity of chickpea (*Cicer arietinum*) under pre- and post-emergence application of herbicides. *Indian J. of Agron.* **60**(4): 570-575.

Kumar, N., Nath, C.P., Hazra, K.K. and Sharma, A.R. 2016a. Efficient weed management in pulses for higher productivity and profitability. *Indian J. of Agron.* **61** (4th IAC Special Issue): S93-105.

Kumar, N., Prakash, V., Meena, B.L., Gopinath, K.A. and Srivastva, A.K. 2008. Evaluation of*toria* (*Brassica compestris*) and lentil (*Lens culinaris*) varieties in intercropping system with wheat (*Triticum aestivum*) under rainfed conditions. *Indian J. Agron.* **53**(1): 47-50.

Kumar, N., Singh, M.K., Ghosh, P.K., Hazra, K.K., Venkatesh, M.S. and Nadarajan, N. 2012. Resource Conservation Technology in Pulse based Cropping Systems. (*Technical Bulletin 1/2012*), Indian Institute of Pulses Research, Kanpur. Pp. 35.

Kumar, N., Singh, M.K., Praharaj, C.S., Singh, U. and Singh, S.S. 2015a. Performance of chickpea under different planting method, seed rate and irrigation level in Indo-Gangetic Plains of India. *J. of Food Legumes* **28**(1): 40-44.

Kumar, S., Rana, N.S., Singh, R. and Singh, A. 2006. Production potential of spring planted sugarcane as influenced by intercropping of dual purpose legumes under *tarai* conditions of Uttarakhand. *Indian J. of Agron.* **51**(4): 271-273.

Mahata, K.R., Sen, H.S. and Pradhan, S.K. 1992. Tillage effects on growth and yield of blackgram (*Phaseolus mungo*) and cowpea (*Vigna unguiculata*) after wet-season rice on an alluvial sandy clay-loam in eastern India. *Field Crops Res.* **29**(1): 55–65.

Malik, R.K., Gill, G. and Hobbs, P.R. 1998. Herbicide resistance—a major issue for sustaining wheat productivity in rice–wheat cropping system in the Indo-Gangetic Plains. *Rice–Wheat Consortium Paper Series.* **3**: 32p.

Meelu, O.P., Singh, V. and Singh, Y. 1994. Effect of green manuring and crop residue recycling on N economy, organic matter and physical properties of soil in rice-wheat cropping system. In: Proc. of symposium on sustainability of Rice-Wheat System in India, CCS, HAU, Regional Research Station, Karnal, Haryana, pp. 115–124.

Mina, B.L., Saha, S., Kumar, N., Srivastva, A.K. and Gupta, H.S. 2008. Changes in soil nutrient content and enzymatic activity under conventional and zero-tillage practices in an Indian sandy clay loam. *Nutrient Cycling in Agroecosys.* **82**: 273-281.

Nadarajan, N. and Kumar, N. 2018. Role of pulses in conservation agriculture. In: Singh V.K. and Gangwar B. (Eds.) System based conservation agriculture, Westville Publishing House, New Delhi. pp 134-154.

Nadarajan, N., Kumar, N. and Venkatesh, M.S. 2013. Fertilizer Best Management Practices in Pulses. *Indian J. of Fert.* **9**(4): 122-136.

Nayyar, V.K., Arora, C.L. and Kataki, P.K. 2001. Management of soil micronutrient deficiencies in the rice–wheat cropping system. In: Kataki, P.K. (Ed.), The Rice–Wheat Cropping Systems of South Asia: Efficient Production Management. Food Products Press, New York, USA, pp. 87–131.

Panwar, K.S. and Yadav, H.L. 1980. Response of short duration pigeonpea to early planting and phosphorus levels in different cropping systems. pp. 37-44. In: proceedings of the international Workshop on Pigeonpea, Vol. 1, 15-19 December 1980, International Crops Research Institute for Semi-Arid Tropics (ICRISAT), Patancheru, India.

Praharaj, C.S. 2014.Water management strategies for pulse based cropping systems. In-Ghosh PK, Narendra Kumar, Venkatesh MS, Hazra KK and Nadarajan N (Ed.) Resource Conservation Technology in Pulses. Scientific Publisher, Jodhpur. Pp. 199-212.

Praharaj, C.S. and Kumar, N. 2011. Drip fertigation in pigeonpea. *Pulses News letter* **22**(2): 5.

Praharaj, C.S., Kumar, N., Singh, U., Singh, S.S. and Singh, J. 2015. Transplanting in pigeonpea-A contingency measure for realizing higher productivity in Eastern Plains. *J. of Food Legumes*. **28**(1): 34-39.

Pratibha, G., Pillai, K.G., Satyanarayan, V. and Hussain, M.M. 1996. Tillage systems for production of black gram (*Vigna mungo*) succeeding rice crops. *Legume Res.* **19**(1): 23–28.

Rana, N.S., Kumar, S., Saini, S.K. and Panwar, G.S. 2006. Production potential and profitability of autumn sugarcane-based intercropping systems as influenced by intercrops and row spacing. *Indian J. of Agron.* **53**(2): 140-144.

Singh, A.K., Lal Menhi and Suman Archna. 2008. Effect of intercropping in sugarcane (*Saccharum*complex hybrid) on productivity of plant cane - ratoon system. *Indian J. of Agron.* **53**(2): 140-144.

Singh, B.B., Solanki, R.K., Chaubey, B.K. and Verma, P. 2011. Breeding for improvement of warm season food legumes. *In* A. Pratap and J. Kumar (eds.) Biology and Breeding of Food Legumes. CABI, Oxfordshire, UK. P. 63-80.

Singh, K.K., Srinivasarao, C.H., Swarnalaxmi, K., Ganeshamurthy, A.N. and Kumar, N. 2012. Influence of legume residues management and nitrogen doses on succeeding wheat yield and soil properties in Indo-Gangetic plains. *J. of Food Legumes*. **25**(2): 116-120.

So, H.B., Kirchhof, G., Bakker, R. and Smith, G.D. 2001. Low input tillage/cropping systems for limited resource areas. *Soil and Tillage Res.* **61**(1–2): 109–123.

Subbarao, N.S. 1988. Biological nitrogen fixation: Recent developments. Oxford and IBH, pub. pp. 1-19.

Velayutham, A. and Chandrasekaran, B. 2002. In: Proc. Training on Recent Advances in Irrigation management, Tamil Nadu Agric. Univ., Coimbatore. Pp: 164-167.

Venkatesh, M.S., Hazra, K.K., Singh, J. and Nadarajan, N. 2015. Introducing summer mungbean in cereal based cropping system. *Indian Famg.* **61**(1): 12-13.

Verma, T.S. and Bhagat, R.M. 1992. Impact of rice straw management practices on yield, nitrogen uptake and soil properties in a wheat-rice rotation in northern India. *Fert. Res.* **33**: 97–106.

Yaduraju, N.T. 2002. Importance of weeds in Agriculture. In teaching manual on Advances in Weed management. National research centre for Weed Science, Jabalpur. Pp. 4-5.

6

Advances in Oilseeds Production Technologies

Kartikeya Srivastava, Ayushi Srivastava and Akanksha*

Introduction

Oilseeds are the third most important group of crops after cereals and pulses in India. These are important food crops which also find their place in industrial use. Oilseeds have been an integral component of Indian agriculture since time immemorial. On the oilseeds map of the world, India occupies a prominent position, both in terms of acreage and production. India is the 4th largest edible oil economy in the world and contributes about 10% of the world oilseeds production, 6-7% of the global production of vegetable oil, and nearly 7% of protein meal. The oilseeds production statistics of the world for the year 2016-17 includes a production of 549.98 million tonnes from an area of 234.57 million ha with the productivity of 2.34 tonnes/ha. The top countries include U.S.A., China and India with U.S.A. ranking first in all area, production and productivity of 39.25 million ha, 126.94 million tonnes and 3.23 tonnes/ha, respectively. Next is China which ranks 2nd in production and productivity but 3rd in terms of area; the area, production and productivity being 22.72 million ha, 54.92 million tonnes and 2.42 tonnes/ha, respectively. India ranks 2nd in terms of area, i.e. 33.83 million ha but 3rd in both production and productivity being 36.32 million metric tonnes and 1.07 tonnes/ha respectively. (https://apps.fas.usda.gov/psonline/circulars/production.pdf).

Table 1. Area, production and yield of oilseed crops in India (2018-19)

Crop	Area (m ha)	Production (m tonnes)	Yield (kg/ha)
Rapeseed and Mustard	4.81	6.69	1393
Groundnut	6.23	9.34	1499
Soybean	11.33	13.79	1217

Source: DES, DAC & FW

**Corresponding author email: karstav7@gmail.com*

This situation therefore emphasizes that there is urgent need to improve the productivity of all oilseeds crops. Our first goal should be to upscale the oilseeds productivity up to at least the world average and then for the next higher levels.

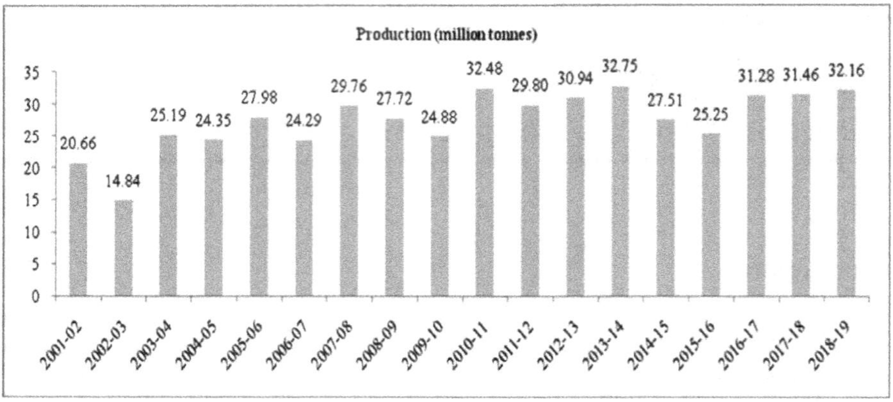

Fig. 1. Oilseed production in India

The combined area of *kharif* and *rabi* oilseeds during 2016-17 as per government of India estimates is 28.22 million ha and production of key nine oilseeds crops for the current year 2016-17 is estimated at 32.10 million tonnes (Fig. 1) compared with 20.41 million tonnes last year, showing 32% rise.

During 1951, India produced a meager 5 million tonnes of oilseeds. This number went up to 9 million tonnes in 1970 and there has been remarkable progress. Currently, oilseeds share 14% of the area under major crops.

Table 2. Demand and supply of vegetable oils in India over last 7 years

Year	Domestic demand (million tonnes)	Domestic availability (million tonnes)	Import (million tonnes)	Import (%)	Value of import (in crore)	Per capita consumption (kg/ year)
2012-13	19.82	9.23	10.81	54.54	53562	15.80
2013-14	21.06	10.08	10.98	52.14	44038	16.80
2014-15	21.71	8.95	12.71	58.54	64894	18.30
2015-16	24.04	9.19	14.85	61.77	68677	19.10
2016-17	24.75	10.75	14.00	56.57	73048	18.75
2017-18	25.74	10.38	15.36	58.43	74996	19.30
2018-19	25.38	10.50	14.88	58.63	69024	18.58

Source: http://www.sopa.org/india-oilseeds-area-production-and-productivity

The current demand for vegetable oil in India in 2018-19 was around 25.38 million tonnes. This demand is met from domestic sources and imports. The domestic sources of vegetable oil are of two types *viz.* primary and secondary.

- Primary sources includes: Groundnut, rapeseed and mustard, soybean, sunflower, sesamum, niger, safflower, castor and linseed.
- Secondary sources includes: Coconut, cottonseed, rice bran, Solvent extracted oils and oils from tree and forest origin.

From the table 2 it is clear that a huge demand-supply gap exists. The 60-65% import dependency worsens during the unfavorable monsoon years. The reason is that domestic demand for vegetable oils and fats has been rising rapidly at the rate of 6% per year but domestic output has been increasing at just about 2% per annum. In India, the average yields of most oilseeds are extremely low as compared to those in other countries of the world.

Oilseeds can be grown in diverse agro-climatic conditions ranging from north-eastern/ north-western hill to down south; under irrigated/rainfed, timely/late sown conditions, as a sole crop or under mixed or inter-cropping. India is blessed with varied agro-ecological environments ideally suited for growing a variety of oilseeds which include soybean (*Glycine max*), groundnut (*Arachis hypogea*), rapeseed and mustard (*Brassica* spp.), sunflower (*Helianthus annuus*), sesamum (*Sesamum indicum*), safflower (*Carthamus tinctorius*), castor (*Ricinus communis*), linseed (*Linum usitatissimum*) and niger (*Hyoscyamus niger*), two perennial oilseeds coconut (*Cocos nucifera*) and oil-palm (*Elaeis* spp.). Three main oilseeds namely, groundnut, soybean, and rapeseed-mustard accounts for >80 per cent of total oilseeds output in the country. Soybean is the most important oilseed crop, grown mainly in Madhya Pradesh, Maharashtra, and Rajasthan accounting for >90% of total production. The second most important oilseed crop is rapeseed-mustard mainly grown in Rajasthan, Madhya Pradesh, Haryana, Uttar Pradesh, West Bengal and Gujarat. Groundnut, which was the largest oilseed crop in the 1990s, lost its share and is now 3rd important oilseed grown in Gujarat, Andhra Pradesh, Tamil Nadu, Rajasthan, Karnataka and Maharashtra. In total oilseeds acreage, soybean (44%), rapeseed and mustard (22%) and groundnut (18%) share and in total production with 34%, 27% and 27% share, however, altogether contribute 88% of the total oilseed production in India, respectively (Table 3, Fig. 2 and 3).

Table 3. Per cent share in area and production of different oilseeds

Crop	% share in acreage	% share in production
Soybean	44.48	33.94
Rapeseed and mustard	22.02	26.92
Groundnut	17.62	26.67
Sesamum	7.48	3.37
Castor	4.07	6.94
Sunflower	1.87	1.17
Linseed	1.01	0.50
Niger	0.96	0.29
Safflower	0.49	0.21

Source:http://www.sopa.org/india-oilseeds-area-production-and-productivity

So far as their contribution to vegetable oil is concerned, rapeseed and mustard tops in the list (31%) followed by soybean (26%). Contribution of other four crops towards oilseed production and vegetable oil is 13% and 18%, respectively (Singh *et al.*, 2017).

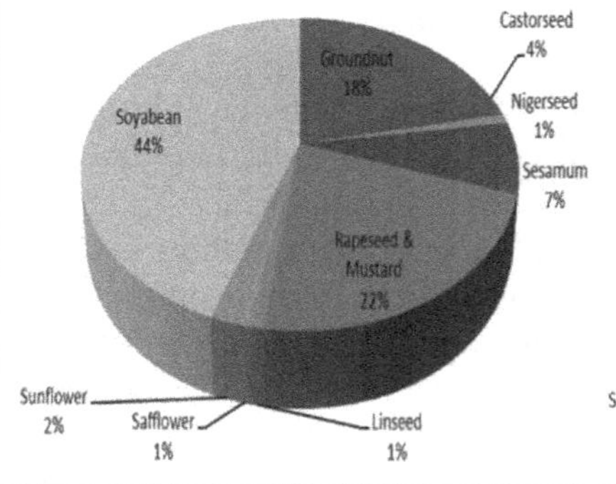

Fig. 2. Per cent contribution of different oilseeds in total acreage in India

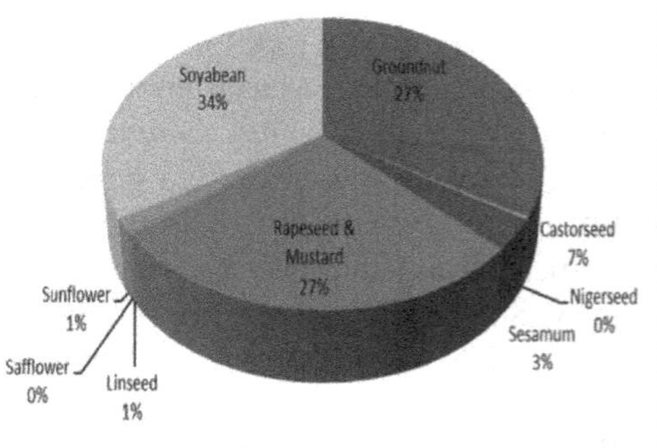

Fig. 3. Per cent contribution of different oilseeds in total production in India

The major oilseeds producing states in India are Madhya Pradesh (24.26%), Rajasthan (20.51%), Gujarat (16.85%), and Maharashtra (13.36%) while other states like Karnataka, Uttar Pradesh, Andhra Pradesh and Tamil Nadu etc. are also important in the production of various oilseeds in the country (Table 4).

Table 4. State-wise contribution of major oilseeds crops (Avg. 2015-18)

States	Area ('000 ha)	Production ('000 T)	Share (%)	Yield (kg/ha)
Madhya Pradesh	6988	7115	24.26	1018
Rajasthan	4523	6015	20.51	1330
Gujarat	2701	4943	16.85	1830
Maharashtra	4249	3919	13.36	992
Uttar Pradesh	1192	1020	3.48	856
West Bengal	844	990	3.38	1173
Haryana	536	972	3.32	1814
Andhra Pradesh	962	878	2.99	913
Tamil Nadu	370	858	2.93	2318
Karnataka	1226	842	2.87	687
Telangana	449	630	2.15	1403
Others	1552	1148	3.9	-
Total	25592	29330	100	1146

*Source:*DES, GOI; Values are average of triennium 2015-16 to 2017-18

In Uttar Pradesh, oilseeds such as soybean, rapeseed-mustard, groundnut, safflower, sunflower and linseed have been grown since ancient times. Like our country, almost 70% oilseeds production in Uttar Pradesh is also dependent on rainfed production system.

Constraints in oilseeds production

As already mentioned, the oilseed production is mainly rainfed based. One of the major constraints in oilseed production has been its low productivity as compared to other countries. In most of the regions of India, oilseeds are considered as secondary crops by farmers and therefore grown mostly on marginal land under rainfed conditions with minimum inputs and care. This accounts for the main reason behind the low productivity. Also, since the major part of crop improvement research is concerned with cereals and other economically important crops, less developments have been made in the field of oilseeds breeding. Superior varieties with respect to quality and yield have to developed and also resistant against various biotic and abiotic stresses. Poor soil health and lack of adequate quantities of soil macro and micro-nutrients is also an important factor.

In Uttar Pradesh, especially the *rabi* season oilseeds cultivation faces many difficulties such as biotic and abiotic stresses mainly because their sowing is delayed because of predominant wheat cultivation and also due to late harvest of paddy. Due to this terminal heat stress and aphid attack pose a serious problem which results in lower yields.

Production technology of oilseed crops

Soybean

Soybean is mostly grown as a *kharif* crop in states like Madhya Pradesh, Maharashtra, Rajasthan, Chhattisgarh, Andhra Pradesh and Karnataka. It is a short day plant and is highly sensitive to day length. It is a very important oilseed crop because in addition to 20-22% oil content, it also contains up to 40-45% protein which can be utilized as protein meal for animals. Several by-products of soybean such as soya nuggets, soy-milk, soy-flour, etc. have found its way to meet the protein requirements of human being.

Varietal advancement

In soybean, the varietal cafeteria approach may be followed where 3-4 soybean varieties which are high yielding and widely adaptable are cultivated simultaneously rather than cultivating a single variety. This approach facilitates timely sowing and harvesting, and provides insurance against biotic and abiotic stresses. Varieties such as Bragg, Clark (high yielding), Indira Soya-9, JS 80-21, PK-1024 (resistant to soybean rust), JS-335, PK-262, NRC-12 (resistant to stem fly), NRC 7, Pusa 16, Pusa 20 (defoliators), JS 71-05(resistant to girdle beetle and myrothecium leaf spot), PK-416, PK-472, PS-1042 (resistant to yellow mosaic) can be cultivated (Table 5).

Table 5. Improved varieties of soybean

Variety	Seed characteristics	Maturity duration (in days)	Yield (q/ha)	Special characteristics
PK-472	Yellow, round, bold seeded	120-125	30-35	Tolerant to yellow mosaic
JS715	Yellow, small seeded	100-105	25-28	Moderately tolerant to leaf miner
PS-564	Yellow, medium-seeded	115-120	25-30	Tolerant to yellow mosaic
PK-262	Yellow, round, bold seeded	120-125	28-30	Tolerant to yellow mosaic and bacterial wilt
JS-2	-	98-105	25-30	Tolerant to bacterial wilt
JS-935	-	102-108	25-30	Tolerant to root rot and leaf spot
JS-7244	-	105-110	20-28	Moderately tolerant
JS-7546	-	105-110	25-30	Tolerant to wilt
Pusa-20	-	110-115	30-32	Good germination percentage, defoliator
PK-416	Yellow, medium-seeded	115-120	30-35	Moderately tolerant to blight; tolerant to yellow mosaic virus and bacterial pustules.
PS-1024	Yellow, round seeded	115-120	30-35	Tolerant to yellow mosaic

Pusa-16	Yellow, Medium-seeded	110-115	25-35	Moderately tolerant to yellow mosaic
PS-1042	Yellow, Round, Bold seeded	120-125	30-35	Tolerant to yellow mosaic
JS-335	Yellow, Medium-seeded	100-110	30-35	Tolerant to wilt
MAUS-47	-	85-90	25-30	Tolerant to wilt
NRC-37	-	100-105	25-30	Tolerant to wilt

Source: Intensive farming techniques of *kharif* crops-2018, Department of Agriculture, Uttar Pradesh, India

Sowing time, seed rate and seed treatment

The optimum sowing time of soybean crop is from 20th June to 10th July. Timely sowing following a strict sowing schedule between 20th to 25th June ensures better yields. For rainfed ecosystem, sowing just after onset of rains always proves beneficial. Soybean seeds have a germination percentage of 75-80 and accordingly the seed rate can be adjusted to 75-80 kg/ha to get a good plant population of about 4.5 lakh plants/ha. The optimum seed rate is 75 kg/ha for small seeded varieties and 100 kg/ha for bold seeded varieties. Line sowing should be followed by maintaining a distance of 45 cm between lines. Plant to plant distance should be maintained at 3-5 cm. Care should be taken not to exceed the sowing depth beyond 3-4 cm to ensure proper germination.

Prior to sowing, seed treatment is done with 2 g Thiram mixed with 1 g Carbendazime into a 50% soluble mixture and 2 g Carbendazime/ Thiamethoxam/ Imidacloprid per kg of seed. Since, it is a leguminous crop, seed treatment also includes the use of biofertilizer like Rhizobium, *Bradyrhizobium japonicum* (400 g per 65-75 kg seed) and Phosphorus Solubilizing Bacteria (PSB). The culture should be mixed with jaggery applied thoroughly all over the seeds with the help of hands so that it forms a thin layer over the seed surface. Sowing should be done immediately after seed treatment.

A field experiment was conducted by Jaybhay *et al.* (2017) during the *kharif* season of 2013 and 2014 at Agharkar Research Institute, Pune, to assess the effect of microbial inoculants and inorganic fertilizers for sustaining the yield of soybean revealed that seeds inoculated with Rhizobium (*Bradyrhizobium japonicum*) and PSB (2480, kg/ha) gave significantly highest soybean yield over without inoculation (2191, kg/ha). Dual inoculation with Rhizobium and PSB recorded 13.19%, 12.11% and 3.76% increase in yield over control, inoculation with PSB and inoculation with Rhizobium, respectively.

Field preparation, manures and fertilizers

FYM decomposition supplies available plant nutrients directly to plants and creates favorable soil environment, ultimately increasing the nutrients and water-holding capacity of soil for longer time, which results in better growth, yield attributes and ultimately grain and haulm yields of soybean (Pattanashetti *et al.,* 2002; Bachhav *et al.*, 2012).

Soybean is predominantly cultivated in Vertisols. It is always important and recommended to carry out soil testing to determine the proper dose of fertilizers. If soil testing is not carried out, then fertilizers should be applied at the rate of 20 kg N: 80 kg P_2O_5: 20 kg K_2O per ha. Addition of gypsum at the rate of 200 kg/ha is recommended to meet the sulphur deficiency. Around 30-35 days after sowing, few plants should be uprooted to observe whether root nodule formation has started. If proper nodulation has not started, then 30 kg N/ha ha should be applied. Use of organic manure also ensures good nodule formation.

Water management

Soybean is a rainfed crop, but in case of moisture deficient soils, a pre-sowing irrigation prior to harrowing is must to ensure good germination. Irrigation is a must during flowering and pod formation stage. Adequate water availability should be ensured to the crop at critical stages of the crop growth i.e. germination and pod filling stages for better harvest. To prevent water stagnation, proper excess water drainage system should be made.

Water stress during vegetative growth and flowering affects soybean yield significantly. Water stress during August and early September reduces yield in Delhi region and supplemental irrigation in August, in the event of dry spells can minimize the yield losses (Jha *et al.*, 2018).

Cropping systems

The inter-cropping followed in different agro-climatic zones in India, is given below in table 6.

Table 6. Zone wise intercropping followed in soybean crop

Zone	Intercropping
Northern hill zone	Maize + Soybean
Northern plain zone	Maize + Soybean, Pigeonpea + Soybean, Rice + Soybean (on bunds)
Central zone	Cotton + Soybean, Pigeonpea + Soybean, Sorghum + Soybean
South zone	Pigeonpea + Soybean, Ragi + Soybean, Sorghum + Soybean, Cotton + Soybean, Sugarcane + Soybean, Maize + Soybean

Insect-pests, diseases and weed management

For proper management of insect-pests, deep summer ploughing should be done after harvesting of *rabi* crop. The major disease, pest and weeds of soybean are presented in table 7.

Table 7. Major insect-pests, diseases and weeds of soybean crop

Pests	Region
Insect-pests	
Stem fly (*Melanagromyza sojae*)	All soybean growing areas
Gridle beetle (*Obereopsis brevis*)	MP, Rajasthan, Uttaranchal
Tobacco cattter pillar (*Spodoptera litura*)	MP, Rajasthan, Maharashtra
Green semi-loopers (*Chrysodexis acuta* and *Diachrysia orichalcea*)	MP, Uttaranchal, Maharashtra
Blue beetle (*Cneorane spp.*)	MP, Maharashtra
Leaf miner *(Aproaerema modicella)*	Maharashtra, Karnataka
Bihar Hairy Caterpillar *(Spilosoma obliqua)*	Uttaranchal
White fly *(Bemisia tabaci)*	Delhi, Punjab, Uttaranchal
Cotton gray weevil *(Myllocerous* spp.)	Delhi,Punjab,Rajasthan, Karnataka
Pod borer *(Helicoverpa armigera, Cydia ptychora)*	All soybean growing areas
Diseases	
Soybean rust *(Phakopsora pachyrhizi)*	Parts of MP, Southern Maharashtra and Northern Karnataka
Yellow mosaic (Mungbean yellow mosaic virus)	Punjab, Uttaranchal, Pockets in western MP
Collarrot *(Sclerotium rolfsi)*	All soybean growing areas
Bacterial Pustule *(Xanthomonas campestris* pv. *glycines)*	All soybean growing areas
Leaf spot *(Myrothecium roridum)*	MP
Soybean mosaic (Soybean mosaic virus)	Uttaranchal, Punjab, Karnataka
Weeds	
I. Monocot *Echinochloa crusgalli, Diqitaria* spp., *Cynodon dactylon, Dinebra Arabica, Saccharum spontanium, Cyperus* spp.	All soybean growing areas
II. Dicot *Euphorbia* spp., Diqeria arvensis, *Celosia argentia, Eclipta alba, Commelina bengalensis /communis,* *Amaranthus* spp. & *Corchorus* spp., *Aclypha indica*	All soybean growing areas

Source: Status paper on oilseeds; DoD, GoI, Hyderabad

Regular field scouting and pest monitoring should be done to facilitate mechanical removal of plant parts/plants infested with girdle beetle, or gregarious phases of tobacco caterpillar or bihar hairy caterpillar or plants infected by yellow mosaic virus disease. This also helps in knowing the economic threshold level (ETL) of pests load. The details of ETL for starting management of pests through biological/mechanical/ chemical are given in table 8.

Table 8. ETL for insect-pests of soybean

Insect pests	Crop stage	ETL
Blue Beetle	Seedling	4 beetles/m row
Green semi-loopers	Flowering, Podding	4 larvae/m row 3 larvae/m row
Bihar Hairy Caterpillar	Pre-flowering	10 larvae/m row
Tobacco Caterpillar	Pre-flowering	10 larvae/m row
Pod borer	Podding	5 larvae/m row

Source: Status paper on oilseeds; DoD, GoI, Hyderabad

Based on the ETL, one foliar spray of any one of the insecticides Triazophos 40 EC @ 0.8 l/ha, Chlorpyrifos 20 EC @ 1.5 l/ha, Quinalphos 25 EC @ 1.5 l/ha, Ethion 50 EC @ 1.5 l/ha, Methomyl 12.5 L @ 2.0 l/ha, Ethofenprox 10 EC @ 1.0 l/ha is to be given preferably at the time of flowering. For the management of seed and seedling diseases (collar rot): seed treatment with Thirum + Carbendazim (2:1 ratio) i.e. 2 g + 1 g/kg seed should be done. Alternatively, seed treatment with *Trichoderma viride* (@ 3-4 g/kg seed) can also be done, where charcoal-rot and collar-rot are perpetual problems. In yellow mosaic virus prone areas: seed treatment with Thiomethoxam 70 WS @ 3 g/kg seed is recommended.

To prevent early seedling mortality due to stem fly and blue beetle infestation, soil application of phorate 10 G @ 10 kg/ha, should be done.

One spray of any of the microbial pesticides: (Dipel *(Bacillus thuringiensis*) @ 1 l/ha or Biobit (*Bacillus thuringiensis*) @ 1 kg/ha) or Dispel (*Beauveria bassiana*) @ 1 l/ha should be given 15 days after the spray of chemical insecticide for the control of defoliators. In rust prone areas, prophylactic sprays of Hexaconazol, Propiconazol, Triadimefon @ 0.8 kg/ha is recommended.

For the management of foliar diseases *viz. Myrothecium Cercospora* and *Altemaria* leaf spot diseases and *Rhizoctonia* aerial blight, two sprays of Carbendazim or Thiophenate methyl @ 0.5 kg/ha at 35 and 50 days after sowing may be given. For the control of bacterial pustule disease, crop may be sprayed with mixture of copper oxychloride (2 kg) + streptocyclin (200 g/ha) at the time of appearance of disease. For the control of yellow mosaic disease, spray of methyl dematon 25EC @ 0.8 l/ha or Thiomethoxam 25WG @ 100 g/ha is recommended for the control of vectors. In order to build up and conserve naturally occurring biocontrol agents/ fauna *viz.* Coccinellid beetles,

Chrysoperla etc., biodiversity should be created in the form of intercropping. In rainfed areas, intercropping soybean with maize, sorghum or short duration pigeonpea (in 4:2 row ratio) is beneficial.

Weeds also affect the yield of soybean and therefore, to minimize the losses, the chemicals which are effective in management, are shown in below table 9.

Table 9. Doses of weedicides for soybean

Chemical name	Mode of application	Dose (kg ai/ha)
Fluchloralin	PPI	1.00
Trifluralin	PPI	1.00
Alachlor	PE	2.00
Metolachlor	PE	1.00
Clomozone	PE	1.50
Pendimethalin	PE	1.00
Imazethapyr	PoE	0.10

PPI: pre-plant incorporation, PE: pre-emergence, PoE: post emergence

Source: Status paper on oilseeds; DoD, GoI, Hyderabad

Groundnut

Groundnut is most important oil seeds of India accounting for half of the major oilseeds produced in the country. Groundnut is predominantly a *kharif* crop but is also sown as a *rabi* crop. Out of total area 90-95% area is devoted to *kharif* crop. It is a legume which thrives best in tropical climate and requires 20 °C to 30 °C temperature; 50-75 cm rainfall. The crop is highly susceptible to frost, drought, continuous rain and stagnant water. It needs dry winter at the time of ripening. Well drained light sandy loams, red, yellow and black soils are well suited for its cultivation.

In India, groundnut is one of the most important oilseed crops growing in states like Gujarat, Tamil Nadu, Andhra Pradesh, Maharashtra, Karnataka and Rajasthan. It is also known as peanut, monkey nut or moongfali. Botanical name of groundnut is *Arachis hypogaea.* Greek word *Arachis* means legume and *hypogaea* means below ground, referring to formation of pods in the soil. Groundnut kernels are also used for the preparation of food products like chikkis, groundnut milk, butter, curd including different bakery products. Groundnut cake obtained after extraction of oil is used as valuable organic manure and feeding material for livestock. It consists of 7.3% N, 1.5% P_2O_5 and 1.3% K_2O. The peanut haulms contain crude protein 8–5 %, lipids 1–3% and minerals 9–10%. These are also used as cattle feed both in fresh and in dried stage and preparing hay or silage. The peanut shells or pod walls which constitute nearly about 25% of total pod weight are used as mulching material during summer season to reduce the evaporative losses. Shell material is also used as filler material for making mixed fertilizers and as insulation material for buildings or as fuel in boilers.

Erect or bunch type: include *Arachis hypogea* sub species *fastigiata* short duration (95-105), early maturing, and high yielding and almost free from dormancy, high germination percentage (90-95).

Spreading or trailing type: include *Arachis hypogea* sub species *procumbens. Lal.* Long duration (110-120 days), late maturity, high yielding ability and have dormancy (60-75%), low germination per cent (85-90).

Varietal advancement

In groundnut, the major reason for farmer's preference for old varieties is timely availability of seeds in abundance and non-availability of adequate seeds of improved varieties. Varieties like J-11, GG-2, GG-4, GG-5, GG-6, GG-20, GAUG-1, are recommended for summer cultivation. JL- 24 is sensitive to heat resulting in poor yield. Varieties such as ALR 1, ALR2, GIRNAR1 are resistant to early leaf spot, late leaf spot, and rust. J11 and OG-53-1(resistant to collar rot and stem rot), ICGV 86590 (*Spodoptera*), and NCAC 17149 (Alternaria leaf spot) are other improved varieties.

Sowing time, seed rate and seed treatment

In India, groundnut is grown in four seasons, namely, *kharif* (85%), *rabi* (10%), summer (4%) and spring (< 1%). *Kharif* groundnut is sown from June to November. The groundnut is sown in flat beds. Some of the improved methods to get higher yield over the conventional method are criss-cross sowing, broad-bed and furrow method (BBF) and ridge and furrow method. Early sowing escapes the damage caused by leaf miner and white grubs.

The normal seed requirement for bunch type groundnut varieties is 100-110 kg/ha, with spacing of 30 × 10 cm to get plant stand of 3.33 lakh/ha and spacing of 45 × 10 cm or 30 × 10 cm to achieve plant stand of 2.22 lakh/ha. Criss-cross sowing in groundnut gives 18% higher yield. A seed rate of 95-100 kg/ha is adequate for spreading and semi-spreading varieties.

Prior to sowing, seed treatment with commercial formulation of *Trichoderma harzianum* or *T. viride* or *Pseudomonas fluorescens* @ 10 g/kg seed or Thiram or Carbendazim or Captan or Mancozeb @ 3-4 g/kg seed or Tebuconazole (Raxil 2 % DS) @ 1.25 g/kg is desirable for controlling different groundnut diseases.

Field preparation, manures and fertilizers

Field is conventionally prepared by one ploughing followed by two harrowing and leveling but nowadays use of rotavator and reduced tillage based technology is under practice for groundnut. Sowing of groundnut following Broad Bed and Furrow (BBF) and ridge and furrow systems gives higher yield besides, helping in soil and water conservation. Deep ploughing and deep burial of surface organic matter and crop debris helps to control disease and insects pests.

For every one tonnes of pod yield and two tonnes of haulm yield, groundnut crop removes 63 kg N, 11 kg P_2O_5, 46 kg K_2O, 27 kg CaO and 14 kg MgO from the soil. To obtain higher yield well decomposed farm yard manure @ 10 t/ha should be applied at least 21 days before sowing of crop. Further, micronutrient deficiency like B, Cu, Mn, Zn, Mo, and Fe can be controlled by applying Borax (5-10 kg/ha), copper sulphate (5-10 kg/ha), manganese and zinc sulphate (10-50 kg/ha), sodium or ammonium molybdate (0.5-1 kg/ha), ferrous sulphate (10 kg/ha), respectively.

The effect of phosphatic fertilizers, gypsum and sulphur on growth contributing characters of groundnut reveals that the growth contributing characters like plant height, plant spread, number of branches/plant, dry matter/plant and number of nodules/plant increases by the application of recommended dose of fertilizers in addition with gypsum @ 500 kg/ha (250 kg gypsum/ha at the time of sowing and 250 kg gypsum/ha at the time of peg formation) + 5 t FYM/ha (Salke *et al.*, 2010).

Water management

Flowering (20-40 DAS), pod formation, development (40-70 DAS) and pod filling (70-100 DAS) are most sensitive to soil moisture stress with the water requirement of 450-650 mm. If irrigation water is not limiting, then a total of 8 irrigations are adequate for optimal yield i.e. pre–sowing irrigation followed by an irrigation at 25 DAS, 4 irrigations at 10 days interval and final two irrigations at 15 days interval.

The crop is usually irrigated by check basin method. Border strip is more suitable than other methods. Sprinkler irrigation is ideal for the crop on sandy soils. Recently drip irrigation is becoming popular among groundnut growers as it increase crop yield by 25-40% besides improving seed quality and saves up to 40-50% irrigation water compared to flood irrigation.

Drip irrigation with fertigation at 100% recommended dose of fertilizers signifies high nitrogen, phosphorus and potassium uptake by groundnut crop at harvest 128.19, 25.16 and 98.06 kg/ha, respectively. Therefore, drip irrigation with fertigation is very effective, that resulted in higher nutrient uptake that leads to a higher yield of groundnut followed by micro sprinkler fertigation and least under surface irrigation (Soni *et al.*, 2017) to a higher yield of groundnut followed by micro sprinkler fertigation and least under surface irrigation.

Cropping System

Maize (summer)- groundnut- chickpea or mustard

Groundnut – onion or garlic

Mungbean - groundnut – wheat

Maize - groundnut – pea

Lobia - groundnut –wheat

Groundnut crop can be grown in rotation with wheat, chickpea, pea, barley etc. It is grown as mixed crop with maize, pearl millet, castor, sorghum and cotton. Groundnut can also be followed by safflower where early varieties are used and moisture remains in the soil at the time of harvest.

Intercropping of soybean is done with soybean (for leaf miner control), castor (for Spodoptera control), cowpea (for hairy caterpillars, aphid and leaf-hopper control) and pearl millet (for thrips).

Weeds, diseases and insects-pests

Important weed flora in the groundnut crop are: *Amaranthus viridis* (jangli chaulai), *Boerhaavia diffusa* (vishakhapra), *Cyperus rotundus* (motha), *Cyperus esculentus* (yellow nut sedge), *Cynodon dactylon (Doob grass), Digera arvensis* (laksha), *Convolvulus arvensis* (hiran khuri), *Argemone maxicana* (satyanashi), *Anagallis arvensis* (krishna neel), *Desmodium trifolium* (tinpatia), *Commelina benghalensis* (kankawa), *Celosia argentea* (white cock's comb) and *Portulaca oleracea* (pig weed).

Weed control through weedicide, fluchloralin @ 0.9 kg/ha, pendimethalin @ 1.0 kg/ha or oxyfluorfen @ 0.18 kg/ha or oxadiazon @ 1.0 kg/ha as pre-emergence should be applied or fluozifop-p-butyle @ 0.25 kg/ha at 20-25 days after sowing should be sprayed along with inter-culturing at 30 days of sowing.

Table 10. Major insect-pests and diseases of groundnut

Name of insect-pest & diseases	Scientific name	Possible yield loss (%)	Period of occurrence
Insect-pests			
Leaf miner	*Aproaerema modicella* (Deventer)	16-92	Mar-Oct
Tobacco caterpillar	*Spodoptera litura* (Fab.)	15-30	Mar-Oct
Hairy caterpillars	*Amsacta albistriga* (Walker), *A. moorei* (Butler) and *Spilosoma obliqua* (Walker)	26-100	June-Oct
Thrips	*Caliothrips indicus* Bagnell, *Frankliniella schultzei* Trybom	15-28	Mar-Oct
Aphids	*Aphis craccivora Koch*	Up to 40	July-Sept
Leaf hoppers/ Jassids	*Empoasca kerri* Pruthi, *Balclutha hortensis* Lindb.	9-22	Mar-Oct
White grub	*Holotrichia consanguinea* Blanch and *H. serrata* (Fab.)	20-100	Aug-Oct
Termites	*Odontotermes obesus* (Rambur) and *Microtermes obesi* (Holgren)	5-46	Sept-Oct
Disease			
Collar rot	*Aspergillus niger* van *Tieghem.*	28-47	-
Aflaroot	*Aspergillus flavus* (Link) Fries.	-	-
Stem rot	*Sclerotium rolfsii* Sacc. Teleomorph: *Athelia rolfsii* (Curzi) Tu & Kimbrough.	27	-
Early leaf spot	*Cercospora arachidicola* S. Hori. Teleomorph: *Mycospharella arachidis* Deighton)	Up to 60	-
Late leaf spot	*Phaeoisariopsis personata* (Berk. & M.A. Curtis) Arx. Cercosporadium personatum	-	-
Rust	*Puccinia arachidis* Speg.	10-52	-
Alternaria leaf blight & leaf spot	*Alternaria alternata, A. tenuissima* and *A. Arachis*	up-to 22	Summer
Peanut bud Necrosis disease	Peanut bud necrosis Virus (Tospovirus)	30-90	-
Root knot	*Meloidogyne arenaria, M. hapla* and *M. javanica*	21.6	-

Source: Status paper on oilseeds; DoD, GoI, Hyderabad

The major insect-pests and diseases causing yield losses in groundnut are aphids, jassid, thrips, whiteflies, leaf miner, white grub, army warm and heliothis etc. (Table 10). The major sucking pests like aphids, jassids, thrips and white flies can effectively control by spraying of phosphamidon 0.03% or dimethoate 0.03% or methyle-o-demeton 0.025% at an interval of 10 days. Light trap attracts moths of leaf miner, which are collected and then destroyed.

These pests also controlled by spraying of dichlorvos 0.05% or monocrotophos 0.04% or quinalphos 0.05% or carbaryl 0.2% dust at an interval of 15 days. In areas where white grub problem is very severe, the soil may be drilled with phorate 10% granules @ 25 to 30 kg/ha in the furrows about 10 cm deep before sowing. Other measures include:

- Crop rotation with poor or immune host crops like cereals.
- Deep summer ploughing.
- Soil solarization by a transparent polythene sheet (25-50 μm) for 15 days during summer also helps to control nematodes. Soil amendments such as neem cake or castor cakes @ 1 tonnes/ha preferably seven days prior to sowing has been found to reduce nematode population. Their combination with seed treatment, with carbosulfan (25 DAS) @ 3% a.i. (w/w) further improves efficacy in reducing the nematode population and enhancing yield significantly.
- Avoidance of deep sowing and injury to the seedling.
- Crop rotation with wheat and gram mixed cropping with mothbean.
- Spray neem oil @ 5 ml/l water alongwith suitable surfactant like soap powder @ 1 g/l or NSKE 5% as it acts as oviposition deterrent.
- Erect bird perches @ 10-12/ha.
- Conserve the natural enemies like, coccinellids, spiders, hymenopteran and dipteran parasitoids.
- Release *Trichogramma chilonis* @ 50000/ha, two times at 7-10 days interval followed by release of *Bracon hebetor* @ 5000/ha two times at 7-10 days against leaf miner and defoliators.

Rapeseed-Mustard

Rapeseed and mustard is the second most important oilseed crop of India and plays a significant role in Indian economy by contributing about 27.8% of the total oilseed production. It includes Toria (*Brassica campestris/rapa* L. var. toria), Brown sarson (*Brassica campestris/rapa* L. var. brown sarson), Yellow sarson (*Brassica campestris/rapa* L. var. yellow sarson), Indian sarson (*Brassica juncea* (L.) Czernj & Cosson), Black mustard (*Brassica nigra)* and Taramira (*Eruca Sativa/vesicaria* Mill.), which have been grown since about 3,500 BC along with non-traditional species like Gobhi mustard (*Brassica napus* L.) and Ethiopian mustard or Karan rai (*Brassica carinata* A. Braun). Raya (*Brassica juncea*) is a common oilseed of this group and is cultivated in low hills both as a pure crop and in mixture with wheat. The crop can be raised well both under rainfed and irrigated conditions. Being more responsive to fertilizers, it gives better returns under irrigated conditions.

The oil content of the mustard seeds ranges from 38-46% and used as spices as whole seeds, ground or in powdered form, prepared pastes, sauces and oil are all used in cooking. Mustard oil possesses one of the best fatty acids profile, (low saturated fatty acids (8%), high mono unsaturated fatty acids (70%) and alpha linolenic acid (10%).

Varietal improvement

Varieties of mustard recommended for specific conditions/possessing particular trait includes:

Basanti, JM 1, JM 2 (white rust resistant), Narendra Swarna Rai 8, NRCDR 02, Rohini (as high oil content) RH 781, RH 819, RGN 48, Shivani, TM 2, TM 4, Vaibhav, RB 50 (as rainfed varieties) CS 52, CS 54, Narendra Rai (NDR 8501) (as salinity tolerant) RGN 13, RH 819, Swaranjyoti, RH 781, RGN 48 (as frost tolerant) etc. Genotypes in *B. juncea* such as T 6342, Glossy B 85, RH 7846, RH 7847 and *B. alba* have been reported to show less infestation against mustard aphid.

Sowing time, seed rate and seed treatment

Sowing time for mustard & raya is last week of September to end of October. Sowing time is 25th September to 15th October for Indian mustard, 30th September to 15th October for raya and taramira is sown throughout October. However, when sown in mixture other crops, the sowing time will depend on sowing date of main crop. For pure crop, seed should be sown in rows 30 cm apart by drill or pora. The seed should be placed at 2 to 3 cm depth. In case the moisture in the soil is insufficient, the seed may be mixed with moist soil and kept overnight for soaking. When sown pure, the seed rate of 6 kg/ha may be used for all rapeseed and mustard crops. In white rust and downy mildew endemic areas, seed treatment with apron SD 35 @ 6 g/kg of seed is advised. For other seedling diseases, seed treatment with carbendazim, thiram or captan @ 2 g/kg of seed is recommended.

Field preparation, manures and fertilizers

The rapeseed and mustard thrive well on light to heavy loam soils. Mustard grows well in light loam while raya can be grown in drier regions too. Raya, however, does well in medium and high rainfall areas. In general, one deep ploughing and 3-4 harrowing are recommended to prepare good seed beds for oilseed Brassica.

The recommended fertilizer doses for one hectare field of rapeseed mustard are 80-100 kg N, 30-40 kg P_2O_5, 20-30 kg K_2O depending upon the soil-site and cropping conditions. However, additional application of 20-40 kg S/ha,

5 kg Zn/ha and 1 kg B/ha in deficient soils improve the vigour and yield. Full amount of phosphorus and potash and half amount of nitrogen should be applied at the time of sowing, while the remaining dose of nitrogen should be top dressed at the pre-flowering stage if adequate moisture is present. For toria crop, the whole of nitrogen should be applied at sowing time otherwise the maturity would be delayed.

To study the economic feasibility and sustainability of Indian mustard (*Brassica juncea*) productivity through organics under semi - arid region of Rajasthan, a long term replicated experiment was initiated by ICAR-DRMR (Directorate of Rapeseed-Mustard Research), Bharatpur, in 2004-05, keeping conventional practices (CP), *Sesbania* green manuring (SGM) and 2.5 t/ha mustard straw recycle + SGM (MSGM) in main plot and eight combinations of NPK fertilizers in subplot. Sesbania green manuring (SGM) significantly increase a mustard seed yield by 32.6% over control. Mustard Straw Incorporation in addition to *Sesbania* green manuring (MSI + SGM) further augments the seed yield by 9.4% over SGM alone and by 45.0% over control. (ICAR-DRMR Annual report 2014-15).

Oilseed brassica (rapeseed mustard) is a high nutrient demanding crop but mostly cultivated by small and marginal farmers in nutrient deficient soil of semi-arid regions. Continuous cultivation of rapeseed mustard in such regions has resulted in stagnation of yield, increased cost of production and decline in factor productivity. This trend could be reverted through adoption of soil site specific fertilizer and organic input management strategies. With this hypothesis a field experiment was initiated in 2005-06, by ICAR-DRMR to evaluate the resilience capacity of four oilseed *Brassica* (rapeseed mustard) production systems. In general, addition of organic source increased the seed yield by 43.6% over absolute control (no fertilizer) and remained at par with conventional control (100% recommended dose of fertilizers). Supplementing organic source with fertilizers further augmented the seed yield of rapeseed mustard by 41.4% due to MFM (moderate– 50% RDF and SFM: subsistence-no fertilizer) and 72.7% due to CFM (conventional) strategies. (ICAR-DRMR Annual report 2014-15).

Water management

Rapeseed mustard is usually grown as a rainfed crop but generally, two irrigations at pre bloom and pod filling stages are recommended for higher seed yield.

Experimentations conducted at research stations under ICAR-DRMR (Directorate of Rapeseed-Mustard Research), Bharatpur reported an increase of 14.3 % seed yield under furrow irrigated raised beds over conventional tillage during five year experimentation on resource conservation technologies. (ICAR-DRMR Annual report 2014-15).

Cropping System

Pure crop: Cultivation of rapeseed-mustard as a pure line-sown crop gives higher monetary returns than mixed cropping with wheat, gram chickpea or barley.

Intercropping:

- Mustard + chickpea: Under rainfed conditions, chickpea inter-cropped with mustard (line sowing) results in higher monetary returns.
- Mustard + potato: Potato intercropped with mustard is more remunerative than potato alone. States like Assam, West Bengal, Orissa, Bihar, Uttar Pradesh, Madhya Pradesh, Haryana, Rajasthan and Gujarat, where potato and mustard crops are grown simultaneously, can follow this practice. Production practices are the same as recommended for the pure crop of potato. Three ridges are planted with potato and on every fourth ridge mustard is sown.

For unusual weather conditions: States like Madhya Pradesh, Uttar Pradesh, Punjab, Bihar and West Bengal where the monsoon is delayed and rainy season (*kharif*) crops like maize and pearlmillet suffer severely, toria may be planted under irrigated conditions to be followed by wheat. When a dry weather is anticipated, the traditionally grown rapeseed and mustard crops may be replaced by taramira in the states of Rajasthan, Haryana, Punjab, parts of Uttar Pradesh. Taramira can be planted upto the end of November in these regions.

To evaluate the effect of different crop establishment methods on crop productivity, soil properties and economics of mustard under various cropping systems, a field experiment has been initiated (by ICAR-DRMR) with five mustard-based cropping systems, *viz.* fallow-mustard, green manure-mustard, brown manure-mustard, cluster bean-mustard, and pearl millet-mustard, grown under conventional tillage (CT), reduced tillage (RT), zero tillage (ZT) and permanent furrow irrigated raised beds (FIRB) in a split-plot design replicated thrice. After five years of the experimentation, during 2013-14, the highest mustard seed yield was obtained under FIRB (2,456 kg/ha, 14.3% higher over CT), followed by ZT (2,362 kg/ha, 9.9% higher over CT) because of more dry matter accumulation, higher translocation efficiency, and greater sink/source potential at the seed filling stage. (ICAR-DRMR Annual report 2014-15).

Weed, diseases and insects-pests management

Weeds compete for all the resources required for plant growth and a drastic reduction in yield. The weed survey and surveillance in rapeseed-mustard is a continuous activity at all the centers of ICAR-DRMR to monitor the status of the present and upcoming weeds. A survey was conducted by ICAR-DRMR in the 25 districts across all the zones in India in 2015-16. The most common weeds found were *Chenopodium album/murale, Cynodon dactylon, Anagalis arvensis, Aphodelus tenuifolius, Cyperus rotundus, Melilotus alba/indica* and *Convolvulus arvensis*. The most serious upcoming weed was found to be *Orobanche aegyptiaca* which was reported in 7 districts, whereas in the previous year, it was reported in 3 districts only. (ICAR-DRMR Annual Report-2016).

Weed control includes two hands weeding at 40 and 70 days after sowing give effective control. Weeds can also be controlled by pre-emergence application of pendimethalin 30 EC @ 1.5 kg ai/ha 30 EC or isoproturon 1.0 kg/ha of the 30-35 days of sowing in 700-800 litres of water/ha.

Table 11. Major insect-pests and diseases of rapeseed-mustard

Pests of rapeseed and mustard	Crop stage attacked	Period of activity
Insect-pests		
Mustard aphid (*Lipaphis erysimi*)	Vegetative / flowering and pod formation	December-March
Painted bug (*Bagrada hilaris*)	Leaves	August - October
Tobacco caterpillar (*Spodoptera litura*)	i. Seedling	i. October-November
Mustard sawfly (*Athalia proxima*)	ii. Maturity stage	ii. March-April
Leaf miner (Chromatomyia	Vegetative	October-December
horticola)	Reproductive	February-March
Diseases		
White rust (*Albugo candida*)	i. Vegetative	i. November
Alternaria leaf spot (*Alternaria*	ii. Reproductive	ii. February-March
brassicae)	Throughout crop growth	February-March
Powdery Mildew (*Erysiphe*	Reproductive	October-November,
cruciferarum)	i. Vegetative	February-March
Sclerotina rot (*Sclerotina sclerotiarum*)	ii. Reproductive	i. October-November
i. Vegetative		ii. February-March
ii. Reproductive		

Source: Status paper on oilseeds; DoD, GoI, Hyderabad

Seeds are treated with carbedazim 0.1% or thiophanate methyl against seedling diseases and imidacloprid @ 5 g/kg of seeds. If the white rust means disease severity is > 3%, apply ridomil MZ 72 WP @ 3 g/l. If the alternaria blight means disease severity is more than 3%, spraying of mancozeb 50 WP @ 2 g/l needs to be taken up at 50 and 70 days after sowing. Disease affected plants should be uprooted and destroyed by burning at a distance from the crop field. If powdery mildew disease is observed at a later stage of the crop. However, if the disease appears before or at flowering, dusting of sulphur @ 1.5 kg/ha or spraying of sulfex 2 g/l may be done.

Sunflower

Sunflower (*Helianthus annus* L.) is an important edible oilseed crop which has high yield potential and oil quality. The cultivated sunflower is native of southern United States and Mexico and the crop was introduced to India during later part of 20th century. The sunflower seed contains oil content varying from 35-43%.

High concentration of unsaturated fatty acids such as oleic (42-57%) and linolenic acid (33-48%) in sunflower oil make it as healthy oil. High oleic sunflower seeds are used for confectionary purposes and oil has good cooking and keeping quality. Sunflower oil is used as a cooking medium; hulls are used in animal feeds as a source of roughage. It also finds industrial use as a fuel to generate steam or electricity and in production of furfural and ethyl alcohol.

Since, it is both photo- and thermo- insensitive crop, it can be grown in a wide range of climatic zones, both as a *kharif* and/or *rabi* crop. The major sunflower growing states of India are Karnataka, Andhra Pradesh, Maharashtra and Tamil Nadu, which contribute > 90% area and 80% production of the country. This crop has exhibited high potential in non-traditional areas of Punjab, Haryana, Bihar, West Bengal, Uttar Pradesh, Orissa and Chhatisgarh in spring (*zaid*) and *rabi* season.

Varietal advancement

The recommended hybrids for different states are as follows (Table 12).

Table 12. Sunflower hybrids recommended for various states in India

State	Hybrids	Varieties
All India	BSH-1, KBSH-1, Jwalamukhi, Sungene-85, PAC-36, PAC-1091, KBSH-44, Pro.Sun.09, DRSH-1	Morden, TNAUSUF-7, DRSF-108, DRSF-113
Andhra Pradesh	APSH-66, NDSH-1	
Karnataka	KBSH-41, KBSH 53, RSFH-1, RSFH-130	
Maharashtra	LSFH-35, MLSFH-47	
Tamil Nadu	BSH-1, MSFH-8, KBSH-1, MSFH-17, Jwalamukhi, Sungene-85,	Morden, TNAUSUF-7, CO-1, CO-2, DRSF-108, DRSF-113, COSFV-5
Punjab	BSH-1, KBSH-1, PSFH-67, Jwalamukhi, Sungene-85, PAC-36,	Morden, DRSF-108, DRSF-113
Haryana	PSFH-118, KBSH-44, DRSH-1, PSFH-118, PSFH 569	
Gujarat	BSH-1, KBSH-1, Jwalamukhi, Sungene-85, PAC-36, PAC-1091,MLSFH-47, KBSH-44, SH-41, DRSH-1.	GAUSUF-15, Morden, TNAUSUF-7, DRSF-108, DRSF-113

Source: Status paper on oilseeds; DoD, GoI, Hyderabad

Soil and land preparation

The soil for sunflower cultivation should be well drained and fertile; low lying soils with marginal fertility are not preferred. Sunflower performs well on a wide range of soils such as sandy loams, black soils and alluviums. As a rainfed crop during *rabi* season, sunflower can profitably be cultivated in moisture retentive soils like Vertisols. The soil should preferably be neutral with pH of around 6.5-8.0 because sunflower can tolerate slight alkalinity but not acidic soils.

For better germination, establishment and growth, a well prepared seed bed is required. In lightly textured soils, 1-2 ploughings followed by planking and harrowing is necessary. For medium to heavy textured soils, 1-2 harrowing should be done immediately after rains to obtain a fine tilth.

Sowing times, seed rate and seed treatment

Being a thermo and photo-insensitive crop, sunflower can be grown in all seasons but keeping in mind that the sowing time should not coincide with a continuous drizzle, cloudy period or temperature more than 38 ^{0}C during the flowering period. In traditional areas, during *kharif* season, sunflower can be planted from 2nd fortnight of June to middle of July in light soils and up to 2nd fortnight of August in heavy soils. During *rabi* season, sowing can be done from September to till the end of November. In non-traditional areas, it can be sown from January to February end in the spring (*zaid*) season.

Seed rate of sunflower varies as per genotype and test weight. Optimum seed rate for sunflower depends on the quality of the seed used and the seed bed condition. Normally, a seed rate of 5 kg/ha should be adequate to achieve the required plant stand. The suggested seed rates for rainfed is 6-7 kg/ha and for irrigated situation it is 5-6 kg/ha. To protect the seeds/seedlings at initial level of crop the use of recommended chemicals for seed treatment (Table 13).

Table13. Recommended seed treatment for sunflower

Seed Treatment	Utility
Pre-soaking in fresh water (1:1 w/v) for about 10 hours and shade dry.	For quick germination, better plant establishment.
Thiram/captan @ 2-3 g/kg of seeds.	To protect from seed borne diseases.
Metalaxyl @ 6 g/kg seeds.	To protect against downy mildew disease.
Imidacloprid @ 5 g/kg seeds.	To protect against insect vectors for necrosis management.

Source: Status paper on oilseeds; DoD, GoI, Hyderabad

A field trial was conducted to evaluate the performance of potential *Trichoderma* isolates capable of imparting salinity tolerance in sunflower with Chitosan through seed priming in natural saline soils (EC = 6.0 dS/m) at ARS Gangavathi,UAS, Raichur. The soil type was black cotton belonging to Vertisol soil order with clay texture and poor drainage. Seed priming with Th-4d had significantly improved the sunflower plant stand (18.3 plants/row) under salinity of 6.0 dS/m against control (13.8 plants/row). Seed treatment with *T. asperellum* (N13) and *T. harzianum* (Th-4d) resulted in better grain filling and head diameter compared to control. (ICAR-IIOR, Annual Report, 2017-18).

Manures and fertilizers

Nutrient management especially organic manures and bio-fertilizers based on the soil test results. If the dose of nitrogenous fertilizers is too high, the crop becomes too succulent and therefore, susceptible to insects and diseases. If the dose is too low, the crop growth is retarded. So, the farmers should apply an appropriate amount of nitrogen for best results. The phosphate fertilizers may not be required each and every season as the residual phosphate of the previous season may be available for the current season also (AESA, Based IPM Package: Sunflower; Department of Agriculture & Cooperation, Ministry of Agriculture & Farmers Welfare, GoI).

Incorporation of 7 tonnes/ha of well decomposed FYM or compost, 2-3 weeks prior to sowing is required for adequate and balanced nutrient supply. Soil testing is always beneficial to meet the nutrient supply and for soil health. In general, 60:60:30 :: N:P:K is recommended for rainfed crop and 60:90:30 :: N:P:K for irrigated crop. The fertilizers application is to be managed such that 50% of N + 100% of P and K is to be applied as basal dose and remaining 50% of N is applied in two splits at 30 and 55 DAS for irrigated crop. When SSP is used as a source of P, it also meets the sulphur requirement. In sulphur deficit soils, gypsum should be applied at the rate of 25 kg/ha. An essential micro-nutrient for sunflower is boron, which increases seed filling and yield. For boron, borax @ 2 g/litre should be applied as directed spray to capitulum at the star bud stage. Borax is first dissolved in hot water and then the final volume is made up.

In sunflower-sorghum cropping systems, in Alfisol indicated that application of 150% of recommended dose of fertilizers (RDF) to both crops recorded highest system gross and net returns. Long term integrated fertilizer management treatments showed highest sustainable yield index (SYI) for sorghum with 150% RDF application to both crops with mean seed yield of 4,300 kg/ha followed by 100% RDF+5 t/ha FYM and 100% RDF+B applied to sunflower in the system. Highest SYI of sunflower was recorded with RDF+S and RDF+S+B+Zn treatments. Nutrient inadequacy or imbalance treatments recorded lowest economics and sustainability (ICAR-IIOR, Annual Report, 2017-18).

Water management

Depending upon soils and season, apply irrigation water at an interval of 20-25 days in case of black soils and 8-10 days in case of red soils as per the schedule indicated. The details of season wise number of irrigations required in different soils are given below in table 14.

Table14. Irrigation management in sunflower

Season	Number of irrigations required		
	Light soils	Medium soils	Heavy soils
Kharif	3-4	2-3	1-2
Rabi	4-6	3-4	2-3
Summer	5-6	4-5	3-4

Source: Status paper on oilseeds; DoD, GoI, Hyderabad

Too frequent and excess irrigations should be avoided as such practice predisposes the crop to attack of wilt and root rots. When ever irrigation water is a constraint, provide irrigation at the three critical crop growth stages mentioned below (Table 15) for realizing optimum response. Avoid moisture stress at these sensitive crop growth stages, as it adversely affects seed set, filling and consequently the yield.

Table 15. Critical stages of irrigation in sunflower

Stage	Days after planting	
	Short duration varieties	Long duration varieties
Bud initiation	30-35	35-40
Flower opening	45-50	55-65
Seed filling	55-80	65-90

Source: Status paper on oilseeds; DoD, GoI, Hyderabad

Assessment of Pusa hydrogel in sunflower for summer season (2018) was carried out at ICAR-Indian Institute of Oilseeds Research (ICAR-IIOR) for irrigation water use and productivity. Data on soil moisture content by gravimetric method, total dry matter, leaf area index, harvesting parameters of sunflower and oil content were recorded. Physiological parameters like relative water content, membrane stability, chlorophyll content and carotenoids at different stages were also recorded. Irrigation was given at 60% depletion of soil moisture from field capacity. There were no perceptible changes in soil moisture content across treatments. The maximum variation was for one or two days. There were no significant differences in crop growth, yield and soil moisture and plant physiological parameters. (ICAR-IIOR, Annual Report, 2017-18).

Cropping system

The availability of wide range of early and medium duration varieties and hybrids responsive to inputs and management and its relatively less thermo- and photo-insensitivity renders sunflower an ideal crop for all seasons. Due to its wider adaptability, the crop is ideally suited for intercropping system. With major crops and varieties having 120 to 140 days duration, sunflower with 75 to 100 days duration forms a perfect match to provide best time complimentarily to achieve higher system productivity with minimum competition under intercropping situations. Efficient intercropping systems identified in different agro-climatic zones are presented in table 16.

Table 16. Cropping systems for sunflower

Efficient intercropping	States
Groundnut+sunflower	Karnataka, Maharashtra, Andhra Pradesh, Tamil Nadu and Gujarat
Pigeonpea+sunflower	Karnataka, Maharashtra, Andhra Pradesh
Fingermillet+sunflower	Karnataka
Soybean+sunflower	Maharashtra
Castor+sunflower	Andhra Pradesh, Gujarat, Tamil Nadu
Urdbean/mungbean+sunflower	Non-traditional areas

Source: Status paper on oilseeds; DOD, GoI, Hyderabad

A fixed plot field experiment was initiated during *kharif* 1999 to assess the need and response of major, secondary and micro-nutrients on a long-term basis for sustainable sunflower production in sorghum (*kharif*)–sunflower (*rabi*) cropping systems in Alfisols. Growth and yield of sunflower (DRSH-1) succeeding sorghum in *rabi* (2017) differed significantly due to long term nutrient management treatments. Seed yield was significantly highest with new organic treatments (1,560 to 1,630 kg/ha) compared to the highest of NPK+ crop residue (2,210 kg/ha).

Overall sunflower yield ranged from 553 to 1,629 kg/ha. Lowest growth parameters *viz*., plant height, stem girth, head diameter, seed filling, was recorded in N alone or no manure treatments. Soil fertility after *kharif* sorghum in 2017 varied significantly due to long term effects of fertilizer management treatments. With recommended NPK and balanced nutrition, the seed yield was maintained stable across *kharif* seasons of varied weather and productivity levels indicating the non-exploitative mining of soil nutrients and conservation.

The sunflower yields showed increasing trend over the years in treatments of adequate and balanced nutrition with secondary and micronutrients (ICAR-ICAR-IIOR, Annual Report, 2017-18).

Crop protection

Sunflower crop is affected by a number of pests, which normally reduces the yield of sunflower. The major insect pests and diseases and their management are given below (Table 17 and 18).

Table 17. Insect-pests of sunflower and their management

Insect-pests	Management practices
Seedling pests	
Cut worm (*Agrotis* sp.)	Sow the seeds on slopes of ridges (6-8 cm height) Apply chlorpyriphos (20 EC) @ 3.75 l/ha to soil with irrigation water.
Grass hoppers (*Attractomorpha crenulata)*	Follow clean cultivation by keeping bunds and fields weed free. Apply methyl parathion 2% dust @ 25 kg/ha.
Sucking pests	
Leaf hopper (*Amrasca biguttula biguttula*)	Seed treatment with imidacloprid 70 WS @ 5 g/kg of seed.
White fly (*Bemesia tabaci)*	Apply imidacloprid 200 SL @ 0.1 ml/lit of water at 15-20 days interval.
Thrips (*Scirtothrips dorsalis* and *Thrips* sp.)	Spray with phosphamidon (0.03%) or Dimethoate (0.03%) or monocrotophos (0.05%).
Foliage pests	
Tobacco caterpillar (*Spodoptera litura*)	Collect and destroy egg masses of early stage larvae of *S.litura* and *S.obliqua* on damaged leaves.
Bihar hairy caterpillar (*Spilosoma obliqua*) and green semi-looper (*Thysanoplusia orichalcea* and *Trichoplusia ni*)	Spray neem seed kernel extract (NSKE) 5% or endosulfan (0.07%) or dichlorvos (0.05%) or fenitrothion (0.05%) in 500-700 liter of spray solution/ha or dust methyl parathion (2%) @ 25 kg/ha.
Capitulum borer (*Helicoverpa armigera)*	Spray *Bacillus thuringiensis* @ 2 l/ha or *Helicoverpa* NPV @ 250 LE/ha or endosulfan(0.07%) or monocrotophos (0.05%) or fenvalerate (0.005%) or profenophos @ 0.05% in 500-700 ltr. of spray solution/ha.
Mealy bugs	Spray dichlorvos 76 EC (0.05%).

Source: Status paper on oilseeds; DoD, GoI, Hyderabad

Table18. Diseases of sunflower and their management

Disease/causal organism	Management practice
Alternaria blight and Leaf spot (*Alternaria helianthi*)	Treat the seed with captan/thiram @ 2.5 g or carbendazim 1.0 g/kg seed. Early planting (*kharif*) escapes the disease. Spray the crop with mancozeb (0.3%), 3-4 times at 15 days interval or rovral (0.05%) 2 sprays at 15 days interval.
Rust (*Puccinia helianthi*)	Removal and destruction of crop residues, volunteer sunflower plants reduce the disease severity. Foliar spray with mancozeb/zineb 0.2% or calixin 0.1% at 30 days interval.
Downy mildew (*Plasmopara halstedii*)	In endemic areas avoid continuous sunflower growing, follow 3-4 yearly crop rotations. Early sowing, shallow planting escapes from the disease. Clear cultivation, rouging of infected plants reduces the disease incidence. Treat the seed with metalaxyl 35 SD @ 6 g/kg of seed and followed by foliar spray of metalaxyl/ridomyl. In disease prone areas use resistant hybrids such as LDMRSH-1 and LDMRSH-3.
Sclerotium wilt (*Sclerotium rolfisii*)	Seed dressing with captaf/carboxin 3-6 g/kg of seed. Adding of soil amendments and antagonistic fungi such as *Trichoderma harizanum* incorporated into soil reduces the disease incidence. Crop rotation for 3-4 years to be adopted. Avoid moisture stress/water logging conditions in the field.
Charcoal rot (*Macrophomina phaseolina*)	Seed treatment with thiram 3 to 4 g/seed. Avoid moisture stress during high summer. Follow deep ploughing in summer and crop rotation.
Head rot (*Rhizopus arrhizus*)	Spray with copper oxychloride @ 0.4% or mancozeb 0.3% combined with endosulfan (0.05%) at 50% flowering stage.
Sunflower necrosis disease	Follow clean cultivation and remove weeds specially parthenium, commelina etc. both from inside and neighboring fields. Seed treatment with imidacloprid @ 5 g/kg of seed against insect vectors. Give prophylactic spray 2-4 times at 15-30 days interval with imidacloprid (0.01%) for vectors control.

Source: Status paper on oilseeds; DoD, GoI, Hyderabad

Harvesting and threshing

Sunflower can be harvested at physiological maturity when the back of the head turns to lemon yellow colour and the bottom leaves start drying and withering. For convenience of safe drying and handling, the crop can be harvested at harvest maturity when all leaves dry. Further, delay in harvesting causes reduction in yield due to lodging, breakage and termite attack. After separation of the heads, dry them for 2-3 days to facilitate easy separation of seed. Thresh the harvested heads either by beating with sticks or rubbing or through manual or power operated threshers. Dry the seed before storage so as to bring the moisture content to around 9-10%. The details of improved implements developed by SAUs are given below (Table 19).

Table19. Improved farm implements in Sunflower cultivation

Implements name	Developed by
APAU sunflower thresher	ANGRAU
Hand operated sunflower decorticator	CIAE
PAU axial flow sunflower thresher	PAU
Phule sunflowerthresher	MPKV
Sunflower seed sheller	TNAU
Sunflower thresher	PAU

Source: Status paper on oilseeds; DoD, GoI, Hyderabad

Safflower

Safflower (*Carthamus tinctorius*) (kusum, kusumbha, kardi) has been under cultivation in India for its brilliantly coloured florets and the orange red dye (carthamin) extracted from them and seed. The seed contains 24-36% oil. The cold pressed oil is golden yellow and is largely used for cooking purposes. The oil is as good as sunflower oil having enough amount of linolic acid (78%), which is very useful for reducing blood cholesterol content. The unsaturated fatty acids of safflower lower the serum cholesterol.

Safflower was believed to be originated in an area bounded by the eastern Mediterranean and Persian Gulf, encompassing southern parts of former USSR, Western Iran, Iraq, Syria, Southern Turkey, Jordan and Israel. In India, Maharashtra and Karnataka states is the major safflower growing states and these two states contribute about 87% of India's area and production of safflower.

Varietal improvement

The most commonly grown hybrids and varieties of safflower are mentioned below in table 20.

Table 20. Safflower hybrids/varieties cultivated in India

	Hybrids/varieties
Hybrids	DSH-129, MKH-11, NARI-NH-1, NARI-H-15, MRSA- 521,
Varieties	JSF-1, JSI-7, NARI-6, JSI-73, Parbhani Kusum, Malviya Kusum, Type-65, Phule Kusum.

Source: Status paper on oilseeds; DoD, GoI, Hyderabad

Sowing time, seed rate and seed treatment

The seed rate varies from 7-20 kg/ha depending on situation. Healthy seeds of improved varieties should be selected for sowing. For better stands, follow line sowing using improved seed drills or fertilizer-cum-seed drills available in different regions. Seeds can be sown behind the plough also. Safflower has branching ability, and the optimum population ranges between 1.0-1.1 lakh/ha.

Depending upon the availability of conserved/residual moisture/late *kharif* rains crop could be sown from late September to mid of November. In case soil moisture conditions are extremely favourable as in seasons with extended monsoon or late rains or where facilities exist for pre-sowing irrigations, plantings of rainfed safflower can safely be extended by 1-2 weeks beyond the suggested optimum period for different regions. Normal spacing in safflower is 45×20 cm.

However, based on research carried out in the Nimbkar Agricultural Research Institute (NARI) by Nandini, Nimbkar a trial with spacing 45×10 cm gave an increase of 10% in flower yield and 11% in seed yield over the recommended spacing of 45×20 cm. Seeds should be treated with thiram, captan or carbendizim @ 3 g/kg seed before sowing to prevent losses from seed and soil borne diseases.

Soils and field preparation

Safflower requires fairly deep, moisture retentive and well drained soils. The crop is fairly tolerant to saline condition and grows well under residual moisture in paddy cropping systems. Heavy soils with poor drainage must be avoided for growing safflower especially under irrigated conditions.

Safflower requires cold-free seed bed with firm subsoiland adequate moisture for good germination and stand establishment. In mono-cropped black soils of *rabi* areas, harrowing 3 to 4 times during the monsoon is as effective as deep ploughing or subsoiling to keep fields weed free. If the fields are infested with obnoxious weeds, resort to deep ploughing during summer as soon as the *rabi* crop is harvested.

Manures and fertilizers

In sandy soils, apply 15-20 tonnes/ha of compost or FYM at the time of last ploughing /harrowing. In areas, where irrigations are possible, 40:40:20 kg NPK/ha N:P_2O_5:K_2O should be applied at the time of sowing. The fertilizers should be applied in furrows 8-10 cm deep and 4-5 cm away from the seed at the time of planting. 50% N and 100% P_2O_5 at the time of sowing and remaining 50% N after 30 days of the sowing have to be given. Safflower also responds to sulphur fertilization in soils low to medium in available sulphur.

To save 25% P_2O_5 it is recommended to treat seed with PSB @ 200 g/10 kg seed. For getting higher seed yield and monetory return seed of safflower should be treated with *Azotobacter* and *Azospirillum* 20 g/kg seed along with 12.5 kg N/ha are recommended. An application of 80 kg N/ha, 80 kg phosphorus /ha and 60 kg S/ha significantly increases the seed yield with concomitant increase in yield attributes (Singh and Singh, 2013). Application of 100% RDF+*Azospirillum*+PSB improves the yield and significantly increases the uptake of NPK by safflower (Shillode *et al.*, 2016).

Cropping systems

Although sole crop of safflower is more profitable, it will be advantageous to intercrop safflower in a number of traditional crops. Some of the suggested intercrop combinations, which are more feasible, productive and profitable for different regions under rainfed conditions, are listed in table 21.

Table 21. Cropping systems involving safflower in India

Suggested intercropping system	State/region
Chickpea + safflower	Maharashtra, Karnataka, Andhra Pradesh, Chattisgarh, Eastern Uttar Pradesh
Wheat + safflower	Karnataka, Andhra Pradesh,
Coriander + safflower	Karnataka, Andhra Pradesh,
Linseed + safflower	Andhra Pradesh, Eastern U.P.
Mustard + safflower	Madhya Pradesh, Chattisgarh
Toria + safflower	Madhya Pradesh, Chattisgarh, East U.P.
Barley + safflower	Uttar Pradesh

Source: Status paper on oilseeds; DoD, GoI, Hyderabad

Irrigation management

Although the crop is grown without irrigation, but higher yields are obtained with irrigation. The seasonal consumptive use varies from 250-300 mm. Three to five irrigations should be given to safflower in medium to lighter type of soils for higher yields. Safflower should be irrigated twice i.e. at 35 and 55 DAS in medium to heavy soils for higher yields. If, only one irrigation is available, it should be given at 55 DAS. The oil yield of sunflower is highest under IW/CPE ratio 0.8 (795 kg/ha) irrigation schedule as compared to 0.4, 0.6 IW/CPE ratio and critical stages (Raghupati *et al.*, 2010). Number of capitula/plant, number of seeds/capsule and 1,000-seed weight of safflower increases under irrigation scheduled at IW/CPE 0.9 compared with two irrigations given at 40 DAS and 80 DAS (Paikara *et al.*, 2011)

Crop protection

For controlling Alternaria leaf spot, seed treatment with thirum or captan 3 g/kg seed is recommended and in field condition spraying with dithane M-45, 25 g in 10 l of water is recommended. For controlling root rot and wilt, seed treatment with thirum or captan 3 g/kg seed along with seed treatment of biological fungicide *Tricoderma* 4 g/kg seed is recommended.

Early sowing (i.e. during 25th September to 10th October) of safflower is recommended to escape from aphid infestation. It is recommended to adopt plant protection measures against safflower aphid when ETL of aphid colonies on 30% plants is reached. For management of safflower aphids spraying of fenthion 50 EC (10 ml) or quinalphos 25 EC (20 ml) or thiometon 25 EC (12 ml), or dimethoate 30 EC (10 ml), or acephate 75% WSP 4 g or malathion 50 EC (20 ml) or carbaryl 50% WSP 20 g in 10 lit of water or dusting of quinalphos 1.5 % dust or methyl parathion 2% dust or phosalone 4 % dust @ 20 kg/ha is recommended.

For effective management of gujhia weevil, application of phorate 10 G @ 10 kg/ha at sowing + foliar spray of chlorphriphos 20 EC @ 25 ml or lymbda chalothrin 2.5 EC @ 10 ml/10 l of water at 10 days after emergence and need based second spraying at 10 days after first application is recommended.

Harvesting and threshing

The crop is ready for harvest when the leaves and most of the bracteoles become dry and brown. Hand gloves may be used to protect legs and hands against spines. Effect of spines could also be minimized by harvesting of crop before rising of sun. Multi-crop threshers and combine harvester could be used for harvesting and threshing.

Linseed

Linseed (*Linum usitatissimum* L.) is one of the oldest crops cultivated by man for its seeds and fibres. All parts of the plant have-extensive and varied uses. The two products of seed are linseed-oil and linseed-meal. The oil content of the seed generally varies from 33 to 45 per cent. Linseed-oil has been used for centuries as a drying oil. Most parts of the linseed plant are used in the paints and varnish industry. It is also used for making linoleum, oil cloth, printer's ink, soap, patent leather and other products. Antibiotic linatine-found in the seeds of linseed-could cure diseases in man and animals against which there is no known medical treatment.

Varietal development

Suyog, Jawahar linseed-41, Azad Alsi-1(for irrigated conditions), Indira Alsi-32, Sharda (for rainfed conditions), Shekhar (resistant to PM, rust, wilt and moderately resistant to Alternaria blight), RL-914 (resistant to rust and wilt), Sheela, Banwa (resistant to rust), Biner (resistant to rust), RLC 76, etc are some of the improved varieties of linseed is given in table 22.

Table 22. State-wise recommended varieties for linseed

State	Name of varieties
Madhya Pradesh	Indira Alsi-32, Kartika, Suyog, Azad Alsi-1
Chattisgarh	RLC 92, Deepika, Kartika, Indira Alsi-32, Sharda
Uttar Pradesh	Sharda, Azad Alsi-1, Ruchi
Bihar	Shival, Ruchi, Azad Alsi-1
Jharkhand	Shival, Ruchi, Azad Alsi-1
Madhya Pradesh	JLS 9, Padmini, Parvati
Chattisgarh	J 552, Padmini
Uttar Pradesh	Shekhar, Padmini, Parvati, Garima, Shikha
Rajasthan	Meera
Bihar	Shekhar, Parvati, Shikha
Jharkhand	Shekhar, T-397, Padmini, Sweta, Shubhra

Source: Status paper on oilseeds; DoD, GoI, Hyderabad.

Sowing time, seed rate and seed treatment

Usually linseed is sown in month of October once rainy season is over. Sowing time depends on soil moisture and water for irrigation. The optimum time for rain fed Linseed is last week of October and irrigated linseed first fortnight of November. Sowing should be done with seed drill spaced at 30 cm. Care should be taken at the sowing time that seed should be placed in moist zone.

The prevalence of high soil temperature in the latter half of September adversely affects the germination of the crop. The optimum period for sowing linseed is the first fortnight of October, whereas mid-October sowing gives the maximum yield. Earlier or later sowings than this peak period gives significantly lower yields.

The seed-rate usually varies from 10 to 15 kg/ha when taken as a pure crop. When linseed is sown in a prepared seed-bed, 50 kg of seed per hectare is sufficient for variety 'K 2' (Anonymous, 1975). Seed should be treated with thirum @ 3 g/kg of seed or Bavistin 1.5 g/kg of seed to protect the crop from seed borne diseases and to some extent soil borne diseases also. Seed rate of linseed is 25 kg/ha.

Field preparation, manure and fertilizer

For rain fed linseed 25 kg N+25 kg P/ha should be given at the time of sowing whereas for irrigated linseed 60 kg N+30 kg P/ha should be given. 30 kg N+30 kg P/ha should be given at the time of sowing and remaining 30 kg N/ha should be given at the time of flowering. 5 kg PSB/ha is also given at the time of sowing. Application of 30 kg N/ha dose gave the optimum returns while 60 kg/ha gives maximum yield (Bhan and Singh, 1973). For achieving maximum production of linseed, the application of a mixture containing 9 kg each of N, P and K is necessarily followed by N and P. These two treatments doubled the linseed yields, indicating simultaneously the importance, of applying phosphorus (Kanwar and Joshi, 1964).

Water management

Two irrigations are essential for optimum yield. First irrigation should be given at flowering stage i.e. 40-45 days and 2nd irrigation should be given at capsule development stage i.e. at 60-65 days. Prasad and Sharma (1975) recorded a favourable influence of two irrigations on the yield and yield attributes of linseed grown on a sandy-clay *loam* soil in Ranchi. One and two irrigations registered an increase in grain yield of 19.67 and 42.20 per cent, respectively, over no irrigation.

Cropping system: Linseed+chickpea (4:2) intercropping system is recommended.

Weed, diseases and insects-pests

It is necessary to keep the crop free from weeds for the 35 days after sowing. Isoproturon 75 WP @ 1.0 kg/ha either with or without 2,4–D (Sodium salt) @ 0.50 kg/ha as post emergence at 35 days after sowing can control weeds effectively.

Pests and their control: Flax seed gall fly and flax seed caterpillar.

Control: Mixed cropping, light traps, killing of flies and caterpillars use.

Insects and their control:

Rust: Use resiatant varieties, application of diethane M-45 @ 1250 g 500 l water/ha.

Wilt: use of carbendazim @ 1.5 g/kg seed before sowing.

Powdery mildew: sulfex spray @ 3 kg/ha in 1000 litre water.

Seasemum

Sesamum indicum L. is the oldest oil plant known and cultivated by man and native of India. 70-75% of production used for oil extraction. Sesame contains 50% oil, 25% protein and 15% carbohydrates.

- Integral part of rituals, religion and culture. Oil contains about 40% oleic and 40% linoleic acid.
- Oil has long self-life due to presence of antioxidant called sesamol.
- Oil is used in manufacture of soaps, paints, perfumes, pharmaceuticals and insecticides.
- Sesame meal is an excellent high quality protein (40%) feed for poultry and livestock.
- The oil with 85% unsaturated fatty acids is highly stable and has reducing effect on cholesterol and prevents coronary heart diseases.
- The oil is used as base for Ayurvedic preparations and known as the queen of oils. Lignans found in sesame seed have remarkable antioxidant effect on human body.
- Sesame grown throughout the year in one or the other part of the country and in more than one season in a region.

India ranks first in world with respect to area as well as production with 19.47 lakh ha area and 8.66 lakh tonnes production. The average yield of sesame (413 kg/ha) in India is low as compared with other countries in the world (535 kg/ha). The main reasons for low productivity of sesame are its rainfed cultivation in marginal and sub-marginal lands under poor management and input starved conditions. However, improved varieties and production technologies capable of increasing the productivity levels of sesame are now developed for different agro ecological situations in the country. A well-managed crop of sesame can yield 1,200-1,500 kg/ha under irrigated and 800-1000 kg/ha under rainfed conditions.

Varietal advancement

For commercial sesame varieties, the optimum temperature ranges from 25 ^{0}C to 27 ^{0}C, while it requires 90–120 frost-free days to achieve optimal yields in cold regions. Sesame showed a considerable growth reduction below 20 ^{0}C, while seed germination and growth were completely inhibited below 10 ^{0}C (Oplinger *et al.*, 1990). Low temperature not only affects plant growth but also deteriorates oil quality by reducing the lignin content of seeds (Beroza and Kinman,1955). State wise varieties of sesame are listed in table 23.

Table 23. State wise varieties of sesame

State	Varieties	Reasons for preference by farmers
Andhra Pradesh	Swetha til	White seed, good for export, medium maturity, higher market price.
	Chandana	Brown seed, medium maturity, higher market price.
Kerala	Thilathara	Blackish brown seed, medium maturity.
	Thilarani	Dark brown seed with higher oil content, medium maturity, higher market price.
Karnataka	DS-1	Dark brown seed with higher oil content, higher market price.
	DSS-9	White bold seed, good for export, higher price.
Punjab	Punjab Til-1	White seed with higher oil content, medium maturity, higher market price.
	TC-289	White seed with higher oil content, higher market price.
Bihar	Krishna	Black seed.
Haryana	Haryana Til-1	White seed, medium maturity.
	Haryana Til-2	White seed, high seed yield.
Himachal Pradesh	Brijeshwari	White bold seed with higher oil content, good for export purpose, higher market price.

Source: Status paper on oilseeds; DoD, GoI, Hyderabad

According to IIOR, annual report, 2017-18, a set of 83 sesame lines were identified based on their maximum yield (>7g/plant) and oil content (>45%) in sesame core set under stress conditions. IC-203987 recorded maximum linoleic acid (62.31%) and minimum oleic acid content (25.55%). IC-26304 showed maximum oleic acid (52.68%) and minimum linoleic acid content (32.66%).

Sowing time, seed rate and seed treatment

A seed rate of 5 kg/ha is adequate to achieve the required plant population. Wherever, seed drill is used, the seed rate may be reduced to 2.5-3 kg/ha. For easy intercultural activities and to realize higher yield, adopt line sowing. Indian agro climate is so variable that sesame is sown and harvested throughout the year in all the months in one or the other part of the country. In general, sesame is sown during *kharif* in arid and semi-arid tropics and *rabi*/summer in cooler areas.

Maximum plant height and minimum number of branches/plant, dry matter production/plant, seed yield/plant, number of capsules/plant and number of seeds/capsule can be achieved with spacing of 30 cm ×10 cm. While, maximum seed yield, straw yield, oil yield, nutrient uptake, net realization and BCR is recorded with spacing of 45 cm×10 cm. For the prevention of seed borne diseases, seed should be treated with carbendazim 2 g/kg seed. Wherever bacterial leaf spot disease is a problem, soak the seed for 30 minutes in 0.025% solution of Agrimycin-100 prior to seeding (Patel *et al.*, 2018).

Soil and field preparation

Sesame can be grown on a wide range of soils but well drained light to medium textured soils are preferred. The optimum pH range is 5.5 to 8.0, acidic or alkaline soils are not suitable. One or two ploughing followed by two-three harrowing are recommended for pulverization and fine tilth required for good germination and plant stand. Keep the field weed free and perfectly leveled to avoid water logging to which sesame is highly sensitive. Prepare the soil into a fine tilth by ploughing 2-4 times and breaking the clods. Broadcast seeds evenly. To facilitate easy seeding and even distribution seed is mixed with either sand or dry soil or well sieved farm yard manure in 1:20 ratio. Work with harrow, followed by pressing with wooden plank so as to cover the seed in the soil.

Water management

- Usually the crop is grown under rainfed conditions. When facilities are available, the crop may be irrigated to field capacity after thinning operation and thereafter at 15-20 days interval. Irrigation should be stopped just before the pods begin to mature. Surface irrigation at 3 cm depth during the critical stages, *viz.*, 4-5 leaves; branching, flowering and pod formation will increase the yield by 35-52 per cent.
- Two irrigations of 3 cm depth each in the vegetative phase (4-5 leaf stage or branching) and in reproductive phase (at flowering or pod formation) are the best, registering maximum yield and water use efficiency.
- Consumptive use and water use efficiency increased with the increase in irrigation levels from two to three irrigations, one each applied at branching, flowering and capsule development stages reported by Dutta *et al.*, (2000).
- Garai and Datta (2002) reported that irrigation at branching, peak flowering, and capsule developments stages gave higher seed yield and consumptive use.

- In the case of single irrigation, it can be best given in the reproductive phase.
- Sesame crops experience a reduction in growth and yield after 2–3 days of water logging, which frequently occurs when they are grown on soils that are poorly drained. This results in immediate senescence and a decline in crop conditions. Crops may experience substantial yield losses under excessive irrigation (Ucan *et al.*, 2007). At various stages of growth, water logging considerably reduces plant growth, leaf axils per plant, biomass, seed yield, and net photosynthesis (Sun *et al.*, 2009).

Manures and fertilizers

For improving soil physical conditions and to obtain good yield, apply about 5 tonnes/ha of well decomposed farm yard manure before the last ploughing and incorporate it thoroughly into the soil. Sesame responds well to inorganic fertilizers (Table 24). The fertilizer dose would however, vary depending on the variety, season, soil fertility status, previous crop grown, rain fall and soil moisture.

Table 24. State wise recommended dose of fertilizers

State	Recommended dose of N:P:K (kg/ha)	Specific recommendation
Andhra Pradesh	40-40-20	-
Gujarat	30-25-0	Apply sulphur 20-40 kg/ha.
Madhya Pradesh /Chhattisgarh	40-30-20	Apply 25 kg/ha zinc sulphate once in three years in zinc deficient soils.
Maharashtra	50-0-0	Half N at 3 weeks after sowing and remaining half after 6 weeks
Rajasthan	20-20-0	For areas with less than 350 mm rainfall.
Orissa	30-20-30	-
Tamil Nadu	25-15-15	Apply full dose of N, P_2O_5, K_2O as basal. Seed may be treated with *Azospirillum*.
Uttar Pradesh	20-10-0	-
Haryana	30-0-0	-
Bihar	40-40-0	-
West Bengal	25-13-13	No fertilizer if sown after potato.
Assam	30-30-20	Entire as basal.

Source: Status paper on oilseeds; DoD, GoI, Hyderabad

The increasing K_2O level up to 30 kg/ha results in significantly higher seed yield (Rajiv *et al.,* 2012). Integrated nutrient management involving the use of 50% recommended dose through chemical fertilizer+50% N through FYM or vermicompost along with *Azospirillum* is found most effective for achieving higher growth and yield attributes, seed, oil and protein yield and higher gross and net return of sesame (Ghosh *et al.*, 2013).

***Cropping system*:** The inter-cropping systems followed in various states are given below in table 25.

Table 25. State wise inter cropping systems involving groundnut

Intercropping system	States
Sesame+Groundnut	Gujarat, Karnataka, Orissa, Tamil Nadu, Gujarat
Sesame+Pearlmillet	Gujarat, Maharashtra, Rajasthan
Sesame+Cotton	Gujarat
Sesame+Mungbean	Madhya Pradesh, Orissa, Tamil Nadu, Uttar Pradesh
Sesame+Soybean	Madhya Pradesh,
Sesame+Urdbean	Orissa, Tamil Nadu,
Sesame+Pigeonpea	Uttar Pradesh, Tamil Nadu,
Sesame+Urdbean	Tamil Nadu, Orissa, Madhya Pradesh

Source: Status paper on oilseeds; DoD, GoI, Hyderabad.

Crop protection

For control of leaf and pod caterpillar, remove affected leaves and shoots and dust with carbaryl 10 per cent. Azadirachtin 0.03 per cent at 5 ml per l spray at 7th and 20th DAS and thereafter need based application can manage the incidence of leaf and pod caterpillar, pod borer infestation and phyllody incidence. For control of gall fly, give preventive spray with 0.2 per cent carbaryl. For control of leaf curl disease, remove and destroy disease affected sesame plants as well as the diseased collateral hosts like chilli, tomato and zinnia. Remove plants affected with phyllody and destroy them. Do not use seeds from affected plants for sowing.

Harvesting and threshing

Harvest the crop, when the leaves turn yellow and start drooping and the bottom capsules are lemon yellow by pulling out the plants. Harvest during the morning hours. Cut the root portion and stack the plants in bundles for 3-4 days when the leaves will fall off. Spread in the sun and beat with sticks to break open the capsules. Repeat this for 3 days. Preserve seeds collected during the first day for seed purposes. Clean and dry in sun for about 7 days before storing.

Issues in oilseed production

The policy impetus to oilseed production in India came for the first time in 1986 when the government launched Technology Mission on Oilseed. This was a golden period for oilseed production in India when productivity jumped from 670 kg/ha in the 1980s to 835 kg/ha in the 1990s. However, after that there has been a slow pace of growth. Today, the major problem in oil seeds production is low productivity. India is way behind the developed countries and neighboring countries like China in its productivity of oil seeds per hectare. Since the increase in production could not keep pace with increased demand, India became more and more dependent on edible oil imports. One of the biggest constraints to raising oilseed output has been that production is largely in rain-fed areas. Only one fourth of the oilseed producing area in the country remains under the irrigation.

Measures to improve oilseed production: Key measures to improve oilseeds production include:

- Bringing additional oilseed areas under irrigation
- There is a need of concerted research on oilseeds, technology diffusion & dissemination and institutional intervention to re-energize the oil sector.
- Increase public research spending in oilseed crops for development of high yielding varieties coupled with biotic and abiotic stress tolerance as well as better oil quality (particularly in rapeseed mustard towards the development of '00'& '000'varieties) .
- Strengthen the oilseed crop seed chain to match the variety specific demand for higher yield.
- Provide incentives to private sector participation in processing and value addition in oilseed crops. Also, constraints for low capacity utilization should be addressed.
- Assured availability of key physical (seeds, machineries, fertilizers, weedicide & pesticides), financial (credit facilities, crop insurance) and technical inputs (extension services) in major crop ecological zones for oilseed crops.
- Contract farming and public-private partnership in production and processing
- Ensure a competitive market for oilseeds and edible oil along with adequate protective measures to avoid unfair competition from the international market.

References

AESA Based IPM Package: Sunflower; Department of Agriculture & Cooperation, Ministry of Agriculture & Farmers Welfare, GoI.

Anonymous. 1982. National Seminar on *"Production, problems and prospects of rapeseed-mustard in India" and Annual Rabi Oilseeds Workshop on rapeseed, mustard, safflower and linseed.* IARI, New Delhi.

Bachhav, S.D., Patel, S.H. and Suryawanshi, P.K., 2012. Influence of organic and inorganic fertilizers on growth, yield, and quality of Soybean (*Glycine max* (L.) *Merril.*) *An Asian J. of soil sci.* 7(2): 336-338.

Bhan, S. and Singh, A. 1973. Optimum fertilizer rates for linseed in the Central' tracts of Uttar Pradesh. Bhartiya Krishi Anusandhan Patrika **1**: 51-54.

Beroza, M., Kinman, M.L., 1955. Sesamin, sesamolin, and sesamol content of the oil of sesame seed as affected by strain, location grown, ageing, and frost damage. *J. Am. Oil Chem. Soc.* **32**(6): 348–350.

Dutta, D., Jana, P.K., Bandyopadhyay, P. and Maity, D. 2000. Response of summer sesame (*Sesamum indicum* L.) to irrigation. *Indian J. of Agron.* **45**(3): 613-616.

Directorate of Oilseed Development, GoI-Hyderabad, "Status paper on oilseed crops" 2012.

Garai, A.K. and Datta, J.K. 2002. Effect of different moisture regimes and growth retardant on consumptive use and water use efficiency in summer sesame (*Sesamum indicum* L). *Agril. Sci. Digest* **22**(2): 96-98.

Ghosh, A.K., Ghosh, D.C. and Duary Buddhadeb. 2013. Nutrient management in summer sesame (*Sesamum indicum* L.) and its residual effect on black gram (*Vigna mungo* L.). *Int. J. of Bio-resource and Stress Mangt.* **4**(4): 541-546.

https://www.gktoday.in/gk/current-data-on-oil-seeds-production-in-india/

ICAR-Indian Institute of Oilseeds Research, Hyderabad. Annual Report, 2017-18.

Jaybhay, S.A., Taware, S.P. and Varghese, P. 2017. Microbial inoculation of rhizobium and phosphate solubilizing bacteria along with inorganic fertilizers for sustainable yield of Soybean (*Glycine max* (L.) *Merril.*). *J. of Plant Nutrition.* **40**(15): 2209-2216.

Jha, P.K., Kumar, S.N., Ines and Amor, V.M. 2018. Responses of Soybean to water stress and supplemental irrigation in upper Indo-Gangetic plain. *Field Crops Res.*, **219**: 76-86.

Kanwar, J.S. and Joshi, M.D. 1964. Fertilizer response of major crops on hill soils of the Punjab. *Res. Bult.*, Punjab Agricultural University, Ludhiana.

Nimbkar Nandini. 2008. Proceedings of 7th International Safflower Conference, Wagga Wagga, Australia.

Oplinger, E.S., Putnam, D.H., Kaminski, A.R., Hanson, C.V., Oelke, E.A., Schulte, E.E. Doll, J.D., 1990. Sesame: alternative field crops manual. University of Wisconsin Extension, Madison, WI, USA, University of Minnesota Extension, St. Paul, USA. http://www.hort.purdue.edu/newcrop/afcm/sesame.html

Paikara, K.K., Jangre, A., Choubey, N.K. and Lakpale, R. 2011. Growth, yield and quality of safflower varieties as influenced by irrigation schedules and fertility levels in vertisols of Chhattisgarh plains. *Envt. and Eco.* **29**(2A): 774-777.

Patel, S.G., Leva, R.L., Patel, H.R. and Chaudhari, N.N. 2018. Effect of spacing and nutrient management on summer sesame (*Sesamum indicum*) under south Gujarat conditions. *Indian J. of Agril. Sci.* **88** (4): 647-650.

Paul, V. 2014. "Problems and prospects of oilseeds production in India", Report for centre for management in agriculture (CMA), Indian Institute of Management (IIM), Ahmedabad.

Prasad, B.N. and Sharma, N.N. 1975. Note on the optimum seeding date and irrigation Level for linseed. *Indian J. agric. Res.* **9**: 159-61.

Raghupati, D., Kamble, A.S., Channagoudar, R.F., Janwade, A.D. and Bhat, S.N. 2010. Effect of irrigation schedules and mulches on yield, soil temperature, water use and economic of sunflower (*Helianthus annuus* L.). *Int. J. of Agril. Sci.* **5**(2): 459-462.

Rajiv, Singh, D.P. and Prakash, H.G. 2012. Response of sesame (*Sesamum indicum* L.) varieties to sulphur and potassium application under rainfed condition. *Int. J. of Agric. Sci.* (8): 476-478.

Richharia, R. H. 1950. *Wealth from Waste. The Indian Linseed Plant from Fibre Point of View (A"new source of rextile/ibre).* United Press Ltd., Bhagalpur, India.

Salke, S.R., Shaik, A.A., Dalavi, N.D. 2010. Influence of phosphatic fertilizers, gypsum and sulfur on growth contributing characters of Ground-nut (*Arachis hypogea* L.). *Adv. Res. J. of Crop Improv.* **1**(2): 106-110.

Soni, J.K., Raja, N.A., Kumar, V. and Kumar, A. 2017. Influence of pressurized irrigation with fertigation on nutrient uptake, yield and quality parameters of groundnut. *Indian J. of Eco.*, **44**: 115-119.

Shillode, G.U., Patil, D.S., Patil, S.D. and Joshi, S.R. 2016. Soil properties and yield of safflower as influenced by different fertilizers. *Res. J. Agric. and Forestry Sci.* **4**(2): 13-16.

Singh, G.K. 1987. "Linseed" Publications and Information Division, ICAR, Pusa- New Delhi-110012.

Singh, R.K. and Singh, A.K. 2013. Effect of nitrogen, phosphorus and sulphur fertilization on productivity, nutrient-use efficiency and economics of safflower under late-sown condition. *Indian J. of Agron.* **58** (4): 583-587.

Sun, J., Zhang, X.R., Zhang, Y.X., Wang, L.H. and Huang, B., 2009. Effects of waterlogging on leaf protective enzyme activities and seed yield of sesame at different growth stages. *Chin. J. Appl. Environ. Biol.* **15**: 790–795.

Ucan, K., Killi, F., Gencoglan, C. and Merdun, H., 2007. Effect of irrigation frequency and amount on water use efficiency and yield of sesame (*Sesamum indicum* L.) under field conditions. *Field Crop Res.* **101**(3): 249–258.

7

Advance Production Technologies of Sugarcane: A Step Towards Higher Productivity

A.K. Mall, Varucha Misra, A.D. Pathak, B.D. Singh and Rajan Bhatt*

Introduction

Sugarcane is a vital commercial crop of our country that occupies around 5.11 million ha of land with an annual cane production of around 400.16 million tonnes and yield of 78.25 t/ha. Sugarcane occupies 2.67% of the cultivated land area and about 7.5% to the agricultural production in the country (DAC, 2020). This crop is a source of earning and livelihood for about 35 million farmers and approximately equal number of labourers. After textile industry, the sugar industry holds its place for the largest agro-based industry in country. Around 40-50% of the cane is utilised in 435 sugar industries present in India for the manufacture of about 15 million tonnes of sugar.

This crop is also a source of raw material for two important small scale industries, *i.e.*, *gur* and *khandsari*. These industries utilises about 50-55% of the sugarcane produced for production of about 10 million tonnes of these products altogether. Besides, sugarcane crop is also the producer of molasses, chief by product, which is the main source for production of alcohol. In the year 2018-19, about 13.79 million tonnes of molasses is produced by our country. Apart from this, the left over fibrous material of sugarcane termed as sugarcane bagasse is also a source of production of electricity in the sugar mills. The left over extra bagasse is now-a-days utilised for even production of paper in paper industries. Also, co-generation of power by this product of sugarcane as fuel is considered to be best practical thing in most of the sugar mills. Using this as a raw material, about 3500 MW power may be generated each year without the use of any extra fuel and money being less than normally utilised in case of fuel production from thermal powers. Press mud is also one

**Corresponding author email: ashutoshkumarmall@gmail.com*

of the by-products obtained from sugarcane which is considered to be rich source of organic matter and micro and macro nutrients. Even the green tops of sugarcane are being used as cattle fodder. Besides, the cool sugarcane juice full of energy is in great demands in urban area to quench the thirst of people, especially in hot summer seasons (Anonymous, 2008). The trend of sugarcane production over the years is given in Fig. 1.

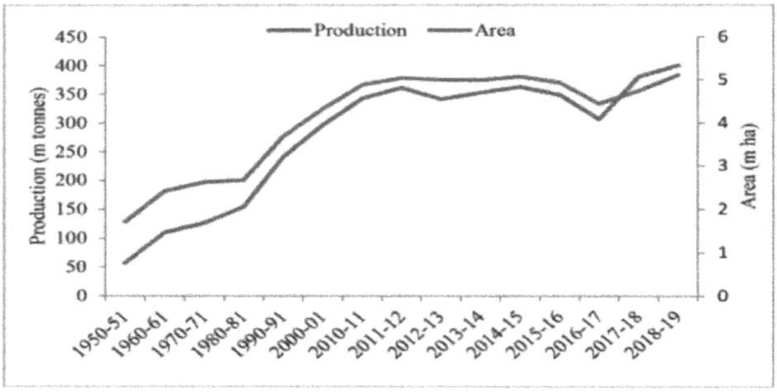

Fig. 1. Sugarcane area and production trend over the year in the country

In India, sugarcane is broadly grown in two distinct agro-climatic regions, *viz*., tropical and sub-tropical (Table 1). The former consist of the southern region below the Tropic of Cancer and the latter comprises of the northern states (26°N-32°N latitude). However, there are five agro-climatic zones, chiefly for the purpose of varietal development. These are North Western Zone, North Central Zone, North Eastern Zone, Peninsular Zone and Coastal Zone. About 45 and 55 per cent of the total sugarcane area and production in the country is shared by the tropical region having an average productivity of 78.77 t/ha (2016-17) while about 55 and 45 per cent of total area and production of sugarcane is shared by the sub-tropical region having an average productivity of about 66.9 t/ha during 2016-17 (Table 2).

Table 1. Distribution of sugarcane producing states on basis of different zones in India

Zones of India	States
Sub-tropical	Uttar Pradesh, Bihar, Punjab, Haryana
Tropical	Maharashtra, Gujarat, Tamil Nadu, Andhra Pradesh, Karnataka, Madhya Pradesh, Goa, Pondicherry, Kerala and Gujarat

Table 2. Zone wise productivity (t/ha) of sugarcane in India

Region		2011-12	2012-13	2013-14	2014-15	2015-16	2016-17
Sub-tropical	P	63.9	63.9	64.7	66.6	67.3	66.9
	+/-	(-6.7)	(-6.3)	(-6.5)	(-6.5)	(-0.4)	(+0.2)
Tropical	P	79.7	76.5	81.5	79.6	79.8	78.8
	+/-	(-25.3)	(-21.7)	(-25.7)	(-21.8)	(-16.1)	(-14.9)
Others	P	56.9	57.9	58.7	61.4	60.1	59.7
	+/-	(4.8)	(3.4)	(3.15)	(1.3)	(11.4)	(12.2)

All India	P +/-	66.8 (-10.8)	66.2 (-9.4)	68.3 (-11.4)	69.2 (-10.1)	69.1 (-3.1)	68.4 (-1.5)

Source: Cooperative sugar (2018)

Scenario of area, production and productivity

Being one of the most important cash crops, sugarcane occupies a pivotal role in agricultural economy of India. It is well known that 3/4th of the total sugar produced in the world is obtained from this crop while the rest 1/3rd is from sugarbeet. More than 115 countries in the world cultivate this crop and total sugar production of 177 million tonnes is obtained from it, of this about 172 million tonnes of sugar is being intake all through the world. Being the second largest producer of sugarcane after Brazil, India's share in world sugar production was 18.9 per cent in 2014-15 and 21.3 per cent in 2015-16. Uttar Pradesh has contributed 25.0 in 2014 and 24.6 per cent in 2015 out of the India's share for production of sugar. In respect to its contribution in the world, 4.7 in 2014 and 5.2 per cent in 2015 for sugar production (Table 3). Out of the total global sugar output, developing countries contribute approximately 70 per cent of the total production.

Against the national average productivity of ~71.0 t/ha of sugarcane, there is variation in productivity in both parts of India, *i.e.*, tropical as well as sub-tropical parts wherein the former had higher productivity ranging from 80.0 to 105.0 t/ha while the latter had around 65 t/ha. The reason behind this variation is basically due to shorter period favorable environmental conditions of 3-4 months for its growth and shorter crop cycle of 10-12 months in sub-tropical zone than in tropics as the crop growth is of 12-14 months. From past many years, the sugar recovery is floating around 10.5 per cent with highest recovery of 12.4 per cent in southern Maharashtra.

Table 3. Cane sugar production and per cent contribution of Sugarcane

Year	2009	2010	2011	2012	2013	2014	2015
World (m tonnes)	110.99	115.67	126.14	135.35	133.60	137.10	135.53
Uttar Pradesh (m tonnes)	4.06	5.18	5.89	6.97	7.49	650	7.10
India (m tonnes)	14.59	19.18	25.85	26.86	22.97	26.03	28.87
India (%)	13.14	16.58	20.49	19.84	17.19	18.98	21.30
UP (%) (wrt India)	27.86	27.01	22.77	25.97	32.58	24.95	24.60
UP (%) (wrt World)	3.66	4.48	4.67	5.15	5.60	4.74	5.24

Source: ISMA (*http://www.indiansugar.com /PDFS/STOCKS_OF_SUGAR.pdf. Accessed 30.08.2018*)

In spite of its great importance in national economy, the productivity of sugarcane in India and also in tropical and sub-tropical region, especially Maharashtra and Uttar Pradesh, respectively, is far below than its potential productivity. Present scenario of sugarcane area, production and yield per unit is shown in table 4. The trend of cane area, production and productivity of India and Uttar Pradesh revealed the yield of sugarcane has decreased although the cane production in India has increased from 2012-2015 but in 2016 it was decreased due to the climate change, however, the area remains almost same from past 10 years but decreased in 2016 due to the low production. In respect to Uttar Pradesh, the highest cane producer of sub-tropical zone, the yield of sugarcane has increased in 2015-16 but in 2016-17, the yield is low, although the cane production was almost same from past many years but in the last year, it was increased due to western disturbances, however, the area remains almost same.

Table 4. Current scenario of area, production and yield in India and Uttar Pradesh

Parameters	Uttar Pradesh			India		
	2015-16	2016-17	2017-18	2015-16	2016-17	2017-18
Sugarcane area (Lakh ha)	20.52	20.54	22.99	49.27	44.36	47.74
Sugarcane production (Lakh tonnes)	1364.12	1486.57	1820.75	3484.48	3060.7	3566.9
Yield (t/ha)	66.47	72.38	79.19	70.7	69.0	74.4
Sugarcane crushed	645.66	827.16	1111.90	2364.98	1934.34	2962.62
Sugar recovery (%)	10.62	10.61	10.84	10.62	10.48	10.70
Number of factory	117	116	122	526	489	-
For sugar (Lakh tonnes)	68.55	87.73	120.50	251.25	202.62	317.0
Production of molasses (Lakh tonnes)	30.43	38.89	53.72	108.37	90.26	-

Source: Cooperative sugar (2017), Indian sugar (2017), Cooperative sugar (2018)

Factors limiting sugarcane production and productivity

Being a long duration crop, it requires 10 to 15 and even 18 months, in case of *Adsali* planting, to become a full bloom crop, depending upon the geographical conditions. For sugarcane to give highest yield and production, hot and humid climate with the optimum temperature ranging between 20-32 °C is required for its overall growth but for its germination optimum temperature is about 26.8 °C. Cane production and productivity is controlled by a multifarious internal and external factors. The significance of various factors differs according to the climatic variability of different places. Following factors limit the production and productivity of sugarcane:

Light energy and growth: Light energy is favorable for sugarcane growth. The optimum solar energy required for favorable cane growth lies between 18-36 MJ/m^2. Sugarcane, being a C_4 plant, has higher rates of photosynthesis and shows a high light saturation. Light intensity as well as its duration influence tillering. The amount of dry matter produced by a crop per unit time depends on the amount of availability of light energy, at the time when the water is surplus for its growth. It also depends on the proportion of the light being intercepted and used up by the leaves. During the first 32 weeks of growth, the crop is directly related to the mean of daily temperatures as well as the amount of radiation received during that period. There is ample evidence that sugarcane is a sun-loving plant. In full sunlight, the stalks are thicker but shorter while the leaves are broader and greener. Under such a condition, tillering is promoted. In respect to plants grown under deficient light intensity, they have longer and slender stalks with thinner and narrower leaves and the latter show a distinct yellow coloration. Moreover, the percentage of dry matter is decreased which renders the plant more succulent. In addition to, day length also plays an important role. Plants under full sunlight produces more dry matter and have low moisture content than plants that receive direct sunlight from sunrise to noon or from noon to sunset. Apart from its direct effect day length exerts an indirect influence on cane growth for day length is largely responsible for flowering that checks the vegetative growth of the stem. Exposure to long light and high intensity encourages tillering while the reverse happens during short cloudy days. A day light duration of 10-14 hrs increased stalk elongation (Fernanado *et al.*, 2015) and leaf area index (Ramanujam and Venkataramana, 1999). The average radiation received by the sugarcane crop in its life cycle amounts to around 6350 MJ/m^2 and 3.0-3.5 is optimum leaf area index for effective usage of radiant energy (http://www.sugarcanecrops.com/climate/, 27.07.2015).

Weather factors: Sugarcane production and productivity are majorly dependent on the weather factors. As per the present scenario, the way climate is changing; it is causing huge variation in production and productivity of sugarcane. Temperature is the first limiting factor for cane production and productivity. Different temperatures are required for different stages of sugarcane growth (Table 5).

Table 5. Optimal temperatures for growth/processes/activities of sugarcane

Activity/Process	Optimal temperature/range	Reference
Germination of seed	30 – 35°C	Singh (1983)
Germination of sets	22 – 36°C	Anon. (Coimbatore) (1987)
Sprouting	26 – 33°C	Blume (1985)
Root growth	35°C (in soil)	Mathur and Haider (1940)
Shoot growth	36°C (in soil)	
Tillering	33.3 – 34.4°C	
Stalk elongation	30°C	Brandes (1958)
Carbon assimilation	30°C	Singh and Lal (1935)
Sugar synthesis	30°C	Hartt (1940)
Sugar transport	30 – 33°C	Julien *et al.* (1989)
Sugar storage	17°C	Rozeff (2002)

Initiation and development of floral primordia	Min. 27.2°C; 21.1°C* Max. 30.0°C; 23.9°C*	Singh (1985)
Ripening	12-14°C	Fageria *et al.* (2010)

*In excised leaf blades; 0 – for early flowering varieties; * - for late flowering varieties*
Source: Shrivastava and Srivastava (2006 & 2012), Shrivastava *et al.* (2016)

Temperature prevailing at germination stage is not a problem in tropical India. Under sub-tropical conditions very narrow time span is available during both autumn (October and November) and spring (February- March) seasons for germination to take place at both the seasons are followed by extremes of low and high temperatures. Low temperature is a major restricting factor for the growth of sugarcane. This causes enhancement in time taken by crop to reach to its full potential. Low temperature (< 10 °C) at ripening stage impairs sucrose accumulation. This results in low sugarcane and sugar productivity (Misra, 2004). In sub-tropical India, Mathur and Haider (1940) observed that when a crop harvested in the months of December/January, the time when temperatures are generally low, under such condition the subterranean buds do not sprout till the lower temperature prevails and growth of the sprouts, if any, was also affected. Thus, in sub-tropical India, February to middle April is the optimum period for harvesting to have a good ratoon crop (Dixit and Bharadwaj, 1974). Shrivastava and Srivastava (2006) had showed that harvesting cane between 15th January and 25th February at Lucknow was advantageous for achieving a high shoot population and cane yield. Sugar recovery and sugar production are also being affected by fall in temperatures. The ripening and maturity phase of the crop in the peak season of the crop in the year 2002-03, experienced prolonged winter-chill in some parts of sub-tropical India, where minimum and maximum temperatures both remained low in comparison to the long term average and the trends of the preceding years (Shrivastava and Srivastava, 2006). A study conducted in 15 districts of Uttar Pradesh revealed that maximum temperature (30-35 °C); humidity (50-60%) and bright sunshine hours were related with cane yield (Samui *et al.*, 2003). Also, crop-weather interactions showed effect on quality and sugar recovery. Sharma *et al.* (2003) studied this effect during 1990-91 to 2000-2001 in Uttar Pradesh. In eastern Uttar Pradesh, the relative temperature disparity (RTD) during July to September showed a positive correlation. In U.P. (India), a study on the effect of climate change on its productivity showed increasing trend of yield, *viz.*, 30 t/ha and 50 t/ha in east and west U.P., respectively. One of the possible reason for major decline in cane yield of east U.P. in comparison to west U.P. was the increase in temperatures (higher than the optimum, *viz.*, > 36 °C) during germination to elongation phase, whereas, in west U.P., better cane yield was observed because the temperature lied between 30-36 °C (Samui *et al.*, 2003). Due to variation in temperature, rate of photorespiration increased which decreased the amount of sugar accumulated (Binbol *et al.*, 2006; Gawander, 2007). Meteorological and yield data for the crop growing seasons from 1980/81 to 1988/89 and from 1994/95 to 1996/97, in Lucknow, India, indicated that the maximum and minimum temperatures above 31.7 °C and 14.7 °C, respectively, at the germination and tillering stages were beneficial for cane yield (Samui *et al.*, 2001). High temperatures along with scarcity of rainfall due to climate change are causing drought conditions to occur which is also causing huge effect on sugar production and post-harvest losses in sugarcane (Misra *et al.*, 2016).

Rainfall is another important factor for superior growth of sugarcane crop. For a crop to achieve high yield and production requires proper amount of water. Sugarcane is known to be a water loving crop and it requires different optimum water for different phases of its growth (Table 6). The table clearly depicted that at the time of grand growth phase of sugarcane, higher water is needed by the plant. On comparing the two climatic zones of sugarcane in India, high amount of water requirement is more in tropical zone in comparison to sub-tropical one in grand growth phase whereas it is reversal in case of other phases of sugarcane growth. It is necessary during the vegetative growth of sugarcane crop as it enhances elongation of cane and formation of internodes. Heavy rainfall is favorable during tillering (Samui *et al.*, 2003). During rainy and winter seasons, increase in temperatures leads to improve cane growth. It has been reported that any scarcity in water will affect cane yield to a great extent as it may cause alteration in physiological activities of the plant such as photosynthesis, stomata conductance, respiration, *etc*. (Gardener *et al.*, 1984; De Silva *et al.*, 2007). Matthieson (2007) had reported that during rainfall deficient periods and harvesting period, declination in precipitation rate may possibly increases efficacy of harvesting and seed quality.

Table 6. Optimum water requirements for different growth phases of sugarcane

Growth Phase	Water Requirement (%)	
	Sub-tropical	Tropical
Germination	17	12
Tillering	24	22
Grand Growth	37	40
Maturity	22	26

Variety: Genetic potential of a variety plays a vital role in cane yield determination. While some of the varieties have a potential of high yielding but it lack so in others. In sugarcane, in general, an opposite relationship exists between the sucrose content and cane yield. This implies that high tonnage varieties have lower sugar potential while the low tonnage ones, higher sugar potential. There may be a difference in growth rate as well as time of vegetation in different varieties. On basis of the maturity levels, varieties have been divided into three groups, *viz.*, early varieties, mid late varieties and late ripening varieties. The former ones are characterized by the fact that they achieve their grand growth phase (16 and 85 per cent, sucrose and purity, respectively) within a short time at 12 months while the mid-late varieties are the ones that achieve at 14 months and the latter are the ones that achieve it at 16 months after planting. For conditions of Uttar Pradesh, the favorable cane varieties are mentioned in the table 7. Even there are some varieties that are particularly suited for one year cropping system while others for cropping systems.

Table 7. Favorable cane varieties for the conditions of Uttar Pradesh

Zone	Division	Districts	Early ripening	Mid-late ripening
West U.P.	Meerut, Saharanpur	Meerut, Ghaziabad, Hapur, Bulandshahr, Bagpat, Saharanpur, Muzzafarnagar, Shamali	CoJ 64, CoS 03251, CoLk 9709, Co 0237, Co 0239, Co 05009, CoPk 05191	CoS 94257, CoS 96269, UP 39, CoPant 84212. CoS 07250, CoH 119, CoPant 97222, CoJ 20193, Co 0124 CoH 0128
Central U.P.	Lucknow, Bareilly, Moradabad	Lucknow, Lakhimpur, Sitapur, Hardoi, Raibairelly, Kanpur, Kanpur-Dehat, Unnao, Farukhabad, Bareilly, Pilibhit, Shahjahanpur, Badau, Aligarh, Eta, Mathura, Moradabad, Sambal, Amroha, Rampur, Bijnor	CoJ 64, CoSe 01235, CoLk 9709, Co 0237, Co 0239, Co 05009, CoPk 05191	CoS 94257, CoS 96269, UP 39, CoPant 84212, CoH 119, CoPant 97222, CoJ 20193, Co 0124, CoH 128
East U.P.	Devariya, Gorakhpur, Devipatan, Faizabad	Devariya, Kushinagar, Azamagarh, Mau, Ballia, Gorakhpur, Maharajganj, Basti, Sidharthnagar, Santkabirnagar, Gonda, Balrampur, Shrivasti, Bahraich, Faizabad, Varanasi, Badohi, Jaunpur, Barabanki, Ambedkarnagar, Sultanpur, Amethi, Allahabad, Mirzapur, *etc.*	CoSe 01421, CoLk 94184, Co 87263, CoSe 01235, Co 87268, Co 89029	CoSe 96436, Co 0233, CoSe 08452
For water logging areas	-	-	-	UP 950 and CoSe 96436

Source: Shukla *et al.* (2017)

With respect to changing climate and occurrence of frequent drought, water logging and salinity, sugarcane varieties also play an effective role in determining cane yield and production. Use of tolerant varieties helps in combating the effect of drought on sugarcane. Some of the approved varieties for drought areas are mentioned in the table 9.

Table 9. Abiotic stress tolerant varieties developed by the AICRP (Sugarcane)

North central zone		North western zone	
Co 87263	WL, RF, LIC	CoH 92201, CoS 95255	LPC
Co 87268	DR,WL, High pH	CoPant 90223	DR, WL, LT
CoSe 96234	Stress conditions	CoPant 93227	LIC, Sub-optimal environments
BO 128	Saline-Sodic soils	CoH 119	DR
CoLk 94184	DR, WL	Co 98014, Co 0118, Co 0239	DR, WL
Co 0232	WL	CoPant 97222	DR, WL, salinity
BO 146	DR,WL	CoJ 20193	Late planting

DR- Drought; WL- Waterlogging; LT- Low temperature; LPC- Late planted crop; LIC- Low input conditions; RF- Rainfed conditions (*Source:* Shrivastava *et al.*, 2016)

Abiotic/biotic stresses

***Drought*:** Drought is one of the major disasters that adversely affects the production and productivity of sugarcane (Table 9). Strike of drought in IGP is not a common phenomenon. In 2002, 15 districts of U.P. had been declared as drought affected areas after then 50 districts were declared severely drought affected in 2015-16. The reason behind was the nil or scanty rainfall (http://nidm.gov.in/pdf/dp/Uttar.pdf, 09.02.2017). As per the reports of Indian MET department (2015), Uttar Pradesh had received 53.5 per cent of rainfall between months of June and September (Kaur and Purohit, 2015).

Table 9. Sugarcane scenario during drought years in Uttar Pradesh

Parameters	2008-09	2011-12	2012-13	2013-14	2014-15	2015-16
Area (Lakh ha)	20.84	22.51	24.24	23.6	21.32	21.69
Production (Lakh ton)	11,090.48	1,335.72	1,493.98	1,480.93	1,389.02	1,453.85
Average yield (Tonnes/ha)	52.3	59.35	61.63	62.74	66.15	67.0
Sugar production (Lakh tonnes)	40.64	69.74	74.85	64.95	70.9	68.41
Sugar recovery (%)	8.94	9.07	9.18	9.26	9.55	10.61

Source: Cooperative sugar (2017), Indian sugar (2017)

Out of the 49 districts declared drought affected, 33 had received rainfall between 40-60 per cent while the remaining 16 districts had faced rainfall < 40 per cent (Mathur, 2015). It also affects the sucrose content of sugarcane and with the increase in time period after harvest cane grown under such condition deteriorates faster leading to low sugar recovery with more production of polysaccharides (Misra *et al.*, 2016). There is decline in ground water table apart from scarcity of rainfall in Uttar Pradesh due to which a large number of blocks are being affected (Table 10).

Table 10. Ground water levels in Uttar Pradesh

Decline in ground water level (cm/yr)	No. of affected blocks
1-10	296
10-20	102
20-30	37
30-40	11
40-50	4
>50	11
Total	469

Source: Bhattacharya (2010)

Water logging: Water logging and drought are the common phenomenon occurring in India (Table 11) and the former abiotic stress frequently occurs in the state of Uttar Pradesh. The year 2015 is considered to be exclusive year as half of the country faced the former problem while the rest the latter one (Shrivastava *et al.*, 2016). The criterion for the severity of such stress varies from agency to agency as well as states. National Commission on Agricultural (1976) and Ministry of Water Resources (1991) gave general criteria for the severity of such stress (Table 12). Sugarcane crop is adversely affected by this abiotic stress. Increase in water level during the active stages of the cane growth negatively affects stalk weight and plant population which on other hand results in loss in yield @ about 1 t/acre from one inch rise in surplus water although it is well known that cane is tolerant to high water levels as well as flooding (Deren *et al.*, 1991a; Deren *et al.*, 1993). Such conditions may deprive the cane root system with oxygen along with enhancement of production of toxic compounds. This condition even slows down the uptake of nutrients (Glover *et al.*, 2008).

Table 11. Frequency of drought and flood in India

Decade	Deficient year	Excess year
1901-10	3	0
1911-20	4	3
1921-30	1	0
1931-40	1	1
1941-50	1	1
1951-60	1	3
1961-70	2	1

1971-80	3	1
1981-90	2	2
1991-2K	0	1
2001-10	3	1
Mean	< 2.5	1.27

Source: Shrivastava *et al.* (2016)

Table 12. Criteria adopted by different agencies for defining waterlogging

Status of Waterlogging	National Commission on Agriculture (1976)	Ministry of Water Resources, Govt. of India (1991)
Waterlogged / Critical	Water table <1.5 m	Water table <2m
Potentially waterlogged	-	Water table <2-3m
Safe area	-	Water table >3m

Source: Country level Report, India (2009) http://www.indiawaterportal.org/sites/indiawaterportal. org/files/introduction.pdf, 14.03.2017

Salinity: Salinity is another abiotic stress which is often correlated with water logging and is even seen in places where problem of water logging prevails. Salinity in simple terms means excess of salts in soil profile and these accumulations of salts are toxic to plants. This in turn causes poor and spotty crop stands with irregular and stunted growth. This results in poor crop yields. Depending on the different pH levels, the salinity is grouped into three types, *i.e.*, saline (< 8.5), saline sodic (> 8.5), sodic/alkaline (> 8.5). The difference in pH also varies from severity level of salinity as well as alkalinity. Depending on the severity of soil salinity the effect on crop is observed. When the salinity is mild then in such a case plants often have blue/green tinge. Presence of white crusts on the soil surface and appearance of barren spots on crop is regarded as an indicator of such condition and it depends mainly on the extent of severity of such stress. Sugarcane is a glycophyte which is well known to be moderately sensitive to salinity (Rozeff, 1995). Sprouting and early growth stages are the two stages that are considerably more sensitive to such stress. At 13.3 dS/m and 9.5 dS/m, sprouting of sugarcane was affected by reduction of 50% (Rozeff, 1995). The excess salts present in soil affect the sugar production in two ways, *viz*., growth rate and yield and sucrose content (Lingle and Weigand, 1997).

Incidence of pest and diseases: Sugarcane crop grown in sub-tropical region is subjected to biotic stresses, *viz*., red rot, smut, wilt, grassy shoot, ratoon stunting disease and leaf scald diseases (Fig. 2a). The main insect pests are root borer, top borer, pyrilla and scale insect, *etc*. (Fig. 2b). During 2014-15 crop season, the incidence of *Pokkah Boeng* was observed not only in sub-tropical states like Haryana, Punjab, U.P. but also in the tropical states, like Maharashtra and Tamil Nadu (Rajula *et al.*, 2016).

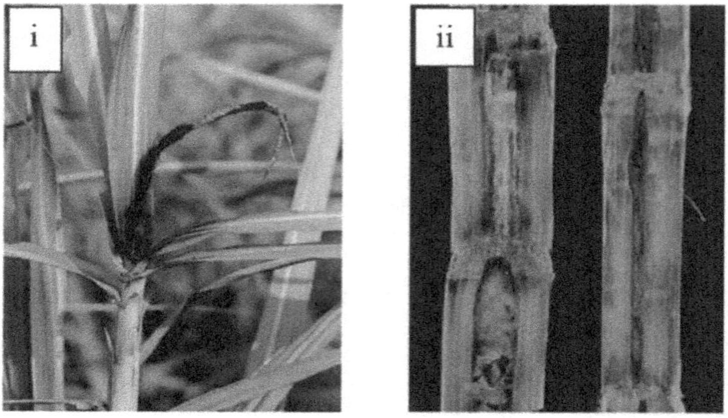

(i) Smut infected sugarcane (ii) Red rot infected sugarcane stalk

Fig. 2a. Common biotic stresses in sugarcane

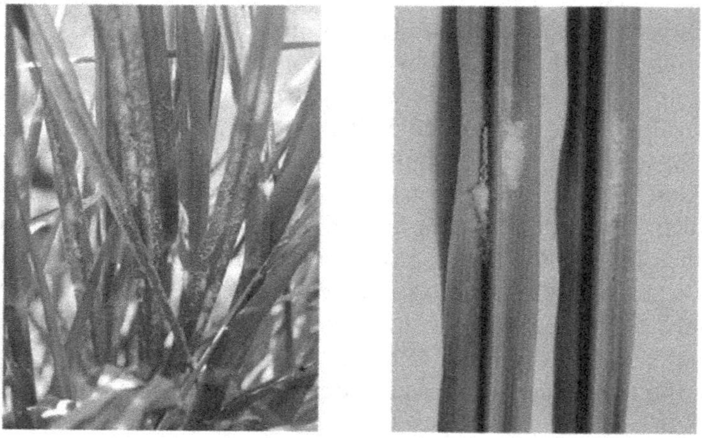

Fig. 2b. Common insect-pests of sugarcane

Soil related problems: In India, soil degradation is one of the major soil problems that lead to about 4.5-7.9 per cent loss in cane yield. Cultivation of cane year after year in the same soil will result to nutritional imbalance of soil. Excess nitrogenization (unbalanced application of N alone) is another concerned problem. Micronutrient deficiency especially that of Fe, Zn, Mo were reported from Northern Karnataka and parts of Tamil Nadu. In peninsular zone, sugarcane response to sulphur nutrition was observed. The eastern U.P. has a sizeable area of problem soils such as Bhat (calcareous) soil and waterlogged and sodic soils in Eastern Plain zone. These particular soil environments of U.P. have an adverse effect both on growth and maturity of cane. Under such anaerobic conditions during the monsoon period of July, August and September, the plant roots stop functioning resulting in less water and nutrient intake and also the photosynthetic activity and transpiration. Also the photosynthetic activity and transpiration rate are retarded. It has been estimated that there is 0.5 units less sugar recovery in such areas. The soils in the tropical regions are laterite, black heavy textured and alluvial soils along with coastal area. In the Indo-gangetic plains of North India mostly alluvial and alkaline soils are dominant. These soils need reclamation/specific management (Tripathi and Kumar, 2010). Quality of irrigation water is deteriorating and water table depth is also going down. Water quality will influence soil and crop health. Therefore, wherever poor quality irrigation water is available, its suitability for sugarcane, particularly for drip irrigation must be assessed prior to cultivation of sugarcane.

Production technologies of sugarcane

As per the estimates of the National Commission on Agriculture (1976) and by various agencies, the population of the country is expected to enlarge to 1.65 billion by the upcoming 50 years. It is estimated that by that time, the per capita consumption of sweetener will probably enhance up to 35 kg (28 kg white sugar and 7 kg *gur*). Keeping in mind, this rate of consumption and predicted increase in population, nearly 51 million tonnes of white sugar may be required by the country in the near 50 years (IISR Vision, 2050). As eating trends of *gur* and *khandsari* are decreasing rapidly the rate of consumption of sugar would be increasing in near future. In order to meet the growing demand of sugar and energy in upcoming fifty years in India, around 630 million tonnes of sugarcane will be required along with the expected recovery of 11.5 per cent (IISR Vision, 2050). Of the total sugarcane that will be produced, 70 per cent would be crushed for production of white sugar, 20 per cent for *gur* and *khandsari* and remaining 10 per cent for seed and other purposes. On this basis, the average cane productivity required will be 105 t/ha. It is been estimated that area under cane cultivation may not exceed 6.0 million ha. In respect to increase in cane area, it may be achieved through intercropping of the crops. Intercropping of sugarbeet with sugarcane enhances the sugar production along with a potential raw material for even bio-ethanol production.

Since the possibility of increasing the cane area is very less in view of emerging low input high value crops, the estimated aim can only be achieved by enhancing the cane productivity and improving its sugar recovery simultaneously. In this view, the development of package of agro-technologies is the only master key to unlock the potential of sugarcane yield. However, the combination of ecological sustainability with technological development is essential to maintain the high level of cane productivity without disturbing the soil environment. India has been bestowed with certain key environments, *i.e.*, bright sunshine all the year, diversified soil types, vast array of climatic changeability and a large collection of sugarcane varieties adapted for particular agro-ecological niches (Anonymous, 2016). Following are the agro-technologies for sustainable sugarcane based cropping systems suggested for better yield and production of sugarcane:

- *Improving germination to obtain early bulking of the crop.*
- *Modification in planting geometry accommodating higher plant population to obtain desired number of millable canes.*

Several improved planting techniques like nursery-cum-transplanting system, deep furrow-cum-trash vein system, *etc.*, were developed by ICAR-Indian Institute of Sugarcane Research, Lucknow for enhancement in cane production. Some of them are mentioned below:

i. For summer planting during May after wheat harvest, transplanting single-bud setts raised in polythene bags gave a better establishment of the settlings in the field, thereby leading to better yield.

ii. Three bud setts are currently planted end-to-end horizontally in 12-15 cm deep furrows at 30:60 cm spacing. This method gave a higher yield than conventional method (Fig. 3).

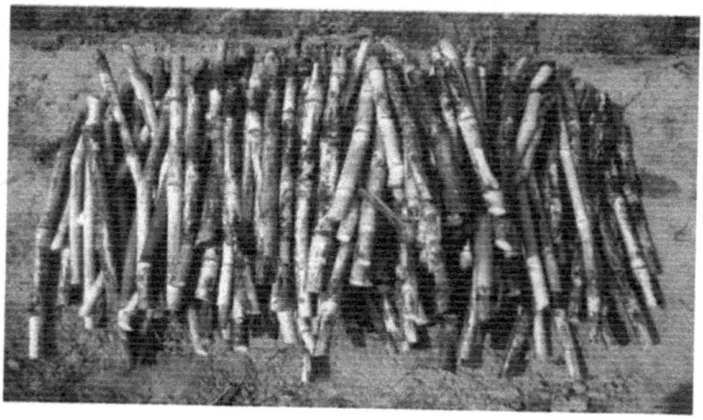

Fig. 3. Three bud sett planting method of sugarcane

iii. *Spaced transplanting technique* (STP):The technique gave a relatively higher population of millable canes and a uniform crop stand. Moreover, the ratio of seed cane and its output was improved to 1:40 compared to that of 1:10 in case of conventional planting. A better yield was also reported in the ratoon crop and this method was found to be suitable for late-planted conditions (Fig. 4).

Fig. 4. Spaced transplanting method of sugarcane

iv. *Cane node priming technique*: Another improved method, which is being recently developed, is the cane node priming technique (Fig. 5). It uses primed cane nodes with organic slurry (cattle dung: cattle urine: water in the ratio 1:2:5), which thereby saves 50 per cent of seed cane and even reduces the period of germination from conventional method of 30-45 days to 20 days. Within 5-6 days of priming, the buds start sprouting. This technique also maintains a uniform crop stand. With favorable growth conditions, this technology guarantees 25 per cent higher sugarcane production over the conventional method.

Fig. 5. Cane node priming technique of sugarcane

v. *Ring pit method of planting*: In this method, tillers are eliminated through altered crop geometry and soil manipulation (Fig. 6). Pits (90 cm width and 45 cm depth) are made to accommodate higher seed rate and setts are placed in circular horizontal manner. With the normal agronomic practices, up to 100 per cent increase in cane yield has been reported with 0.1 per cent unit higher pol percent compared to conventional flat planting method in sub-tropics. This method of planting is particularly suited for drought-prone areas, undulating topography and areas where saline-sodic soils are found. It also overcomes the problem of cane lodging. About 9,000 pits can be dug in one ha area. With this method of planting, cane yield up to 125 t/ha has been reported with an increase by 1.5-2.0 times and a benefit-cost ratio of 1.8. This method also helps in saving the irrigated water up to 20-40 per cent (Singh *et al.*, 1984). Yadav (2004) has illustrated that optimum dosage of nitrogen is less in ring pit method than in conventional method. It has also been found that for production of 1 tonnes of cane yield, 8.481 kg N and 2.3421 kg N has been required in conventional and ring pit method.

(i) Three sett cane sowing in ring pit

(ii) Arrangement of cane setts in ring pit method

(iii) Cane germination in ring pit method

Fig. 6. Ring pit method of planting of sugarcane

vi. *Furrow irrigated raised bed (FIRB) method of planting*: For timely planting of sugarcane in wheat-sugarcane cropping system, FIRB method of planting was developed (Fig. 7). This method averts yield losses due to delayed planting of sugarcane after wheat harvest prevalent in cane belt of western Uttar Pradesh. 2-3 lines of wheat are grown in the ridges and sugarcane is planted in the furrows in spring with saving in land and irrigation water. The method ensures optimum time-span for tillering in sugarcane. Compared to sequential planting of sugarcane with wheat, this enhances sugarcane yield by 30 per cent apart from saving in irrigation water. This is suitable for late planting situations particularly for western Uttar Pradesh, where late planting of sugarcane after wheat harvest is a common practice.

Fig. 7. Furrow irrigated raised bed (FIRB) method of planting of sugarcane

1. Special planting techniques for rapid cane multiplication with reduced seed rate, like, STP (saves seed cane to the tune of 4 t/ha, *i.e.*, approx. 70 per cent), moist hot air treated seed (15 to 20 per cent of higher cane yield, especially in the ratoon crop), polythene bag transplanting technique and tissue culture technique, both, help in quick propagation of seed culture.
2. Integrated nutrient management is another way to improve cane yield and production. This includes organic, inorganic and legume sources vis-à-vis waste recycling. Fertiliser recommendations vary from State to State. Restoration and sustaining the soil fertility as well as crop productivity is the main highlighting aspect of this management. It also plays a role in checking the emerging deficiencies of nutrients other than N, P and K. The physical, chemical and biological environment of the soil is favourably affected by this management technique (Yadav, 2007; Tandon, 1992). Integrated nutrient management optimizes all the aspects related to nutrient cycling, supply and even its uptake as well as minimize environmental pollution. Phonde *et al.* (2005) had showed that significantly higher cane yield, sugar recovery and relative economic benefit in site specific nutrient management based fertilization than blanket recommendation and farmer's practice. Mixture of 50%

recommended N, P_2O_5, K_2O, S and Zn with pressmud (20 t) and rice mill ash (10 t/ha) revealed that the highest total economic profit and better evident nutrient balance in soil (Paul *et al.*, 2005). Farm yard manure applied @ 15 t/ha accompanied with chemical fertilizers (N, P_2O_5, K_2O, S and Zn @ 178, 53, 54, 26 and 2.6 kg/ha) produced 108.4, 96.8, and 73.5 t/ha cane yield in plant, first, and second ratoon crops, respectively. Significant higher sugarcane yield of 170.33 t/ha was obtained when nutrients were applied with 50 per cent N through press mud and 50 per cent NPK through fertilizers + bio-fertilizers (Keshaviah *et al.*, 2013). By the application of 100 per cent NPK + 25 per cent N through FYM + bio-fertilizers (*Azotobacter* + PSB) in plant cane following 100 per cent NPK + trash incorporation with cellulolytic culture + bio-fertilizers in ratoon, the yields of both plant and ratoon cane were enhanced by 27.7 and 16.2 per cent, respectively (Kumar and Chand, 2012). Venkatakrishnan and Ravichandran (2012) showed that basal application of seasoned pressmud @ 25 t/ha and application of 100 per cent RDF + lignite flyash @ 25 t/ha + humic acid 50 kg/ha was the finest INM combination for constant sugarcane productivity as well as soil fertility on sandy loam soil.

3. Efficient irrigation management and water saving techniques to enhance water use efficiency. In spite of the fact that many of the rivers run through the eastern U.P., 40 per cent of total agricultural area is under rainfed conditions. Fully irrigated area is just 18 per cent and 42 per cent remaining area is under partial irrigation. The lack of water at the proper time particularly due to roaster in canals and non-availability of power supply (includes both electricity as well as diesel) constrains the crop production in Uttar Pradesh (Tripathi and Kumar, 2010). Ideally a crop of 12 months duration requires 35-40 irrigations in tropics and 6-7 in sub-tropics for its proper growth and development. For economic use of irrigation following approaches are being used-

(i) *Trench method of planting*: In sub-tropical India, trench planting is adopted to economise on the water supply (Fig. 8). After land has been ploughed and properly prepared for sugarcane cultivation trenches are made in October-November by keeping 90-100 cm distance from the centre of one trench to another. Each single trench is approximately 20 cm deep and 40 cm wide with separation by ridge of 50-60 cm wide from the next trench. These ridges are formed by taking out soil from the trenches. Cane setts are placed with last bud overlapping at the bottom in the centre of the furrow and then it is covered with the soil. By this method, the crop gets a good basal anchorage which helps in preventing lodging at an early stage. A saving of about 15 per cent in irrigation water and a yield increase of 5-10 t/ha can be anticipated over flat planting (Lal and Singh, 2007).

Fig. 8. Trench method of planting

(ii) *Furrow irrigation/Surface irrigation*: It is most commonly used method of irrigation. However, the efficiency of irrigation is low being 30-40 per cent due to loss of water through many processes, such as, evaporation, surface run-off, deep percolation and seepage. As per efficiency aspect, furrow irrigation is more efficient than surface irrigation method as water comes in direct contact with only 1/3rd to half land surface, thereby reducing evaporation losses.

(iii) *Skip furrow irrigation/Alternate furrow irrigation*: This sort of irrigation has more water economy control as water is applied in furrows that did not receive irrigation at the preceding irrigation (Prasad *et al.*, 1980). Cane yield and water economy in sugarcane under different methods of irrigations is mentioned in table 13.

Table 13. Cane yield and water economy in sugarcane under different methods of irrigations

Irrigation methods	Lucknow		Seorahi	
	Cane yield (t/ha)	Water economy (%)	Cane yield (t/ha)	Water economy (%)
Every furrow	63.2	16.7	97.3	24.4
Skip furrow	77.1	37.3	94.0	33.2
Check basis	74.3		89.3	
CD (5%)	9.0		5.4	

Source: Prasad *et al.* (1980), Singh *et al.* (1994)

(iv) *Drip irrigation/Trickle irrigation*: It is an advanced method of supplying required quantity of water to the root zone through network of turbing and drippers/emitters/nozzles placed along with water delivery (Fig. 9). This method showed the precise application of water with great water economy. There is slow release of water in the root zone, almost at the rate at which plant can absorb the same. About 35-40 per cent of irrigated water is saved by this method along with the increase in yield by 30-50 per cent and less growth of weeds. By the use of this method, there is reduction in consumption of electricity as well as cane cultivation cost. In respect to sugarcane crop, improvement of water utilization efficiency from 50-60 per cent in flooding method to 90-95 per cent in micro-irrigation method was seen. Due to this water saving technique, 40 to 60 per cent more land can be irrigated with the same amount of irrigation water. Application of this method even reduces the consumption of fertilizers and chemicals up to 30 per cent as these are applied through water directly at the root zone area of the plant in a homogeneous and effective way. Due to this, there is 50 per cent lesser germination of weeds between plants and plant rows. However, there are certain drivers and challenges for the application of this method (Table 14). Cane yield and water economy in furrow and drip methods of irrigation is mentioned in the table15.

i. Initial stage of crop ii. mature crop

Fig. 13. Drip irrigation system

Table 14. Water consumption for cane production under different methods of irrigation

Methods	Conventional crop	Drip crop
Number of irrigations	30	270
Quantity irrigated per ha	500000	62500
Total quantity irrigated (litres)	15000000	16875000
Cane yield (ha)	112.5	137.5
Water required to produce one ton of cane	133000	122700

Source: http://www.agsri.com/images/documents/symposium_1/Pdf%20files%20of%20PPTs/Technical%20Session%20I/Wider%20Row%20Spacing%20and%20Drip%20Irrigation%20in%20Sugarcane.pdf, 07.03.2017

Table 15. Drivers and challenges of micro-irrigation

Drivers	Challenges
Scarcity of water; Increasing the demand; Emphasis on yield increase; Timely requirement of water; Monsoon disparity; Efficiency in Fertilizer use; Emphasis on cultural practices; Reduce weed growth and Water distribution in field	High initial cost; Maintenance is high and Interfere in harvesting.

(v) Effective weed management strategies at reduced cost. About 15-20% of cane yield is reduced due to the growth of weeds in sugarcane crop. This amount of reduction may increase up to 40% if not controlled. An effective integrated method of weed management which involves cultural and chemical methods has been developed (Table 16). It comprises of one hoeing after 1stirrigation and application of atrazine @ 2.0 kg ai/ha after 2nd irrigation. It is effective in checking weed growth and increases cane yield and saves 50% cost as compared to manual hoeing. Metribuzin 1.0 kg ai/ha or ametryn 2.0 kg/ai as pre-emergence atrazine 2.0 kg/ai as pre-emergence application followed by 2,4 D @ 1.0 kg ai/ha at 60 days after planting (DAP) and one hoeing at 90 DAP has been recommended for effective and economic weed management in sugarcane (Shukla *et al*., 2017). Recent studies have showed that application of sulfentrazone pre-em @ 1200 g a.i./ha and sulfentrazone pre-em @ 1050 g a.i./ha was found statistically similar to thrice the manual weeding (Anonymous, 2015-16).

(vi) Effective pest management strategies for better yield and production- Some of the effective management techniques are as follows-

i. Use of temperature/insecticide tolerant strains of *Trichogramma* sp. need to be adopted on area wide mode against important borer pests.

ii. Phermone technology: Mass collection of beetles using light devices along with anisole (methoxy benzoate) is being adopted in several parts of sub-tropical India (Viswanthan *et al*., 2017).

iii. Cultural practices: Early planting of sugarcane as per normal routine may be possible to avoid the coincidence of susceptible stage of the crop with the insect pest incidence. Autumn cane planting in sub-tropical India will avoid the coincidence of destructive third brood of top borer and reduce the incidence level. Adoption of wide spacing in rows tends to facilitate processes such as de-trashing (process for management of pests like borers, mealy bugs and scale insects, *etc.*).

iv. Chemical application for control of insect/pest: Application of chlorantraniliprole 18.5 SC 375 ml/ha is effective against shoot borer.

De-trashing followed by application of manocrotrophos @ 0.75 kg a.i./ha twice in two months, *viz*., August and September at monthly intervals is effective against stalk borer. Application of chlorpyriphos 20EC 5 l/ha is recommended for root borer (Fig. 10).

Fig. 10. Application of chemical for control of insect-pests

Table 16. Weed dynamics, millable cane and cane yield as influenced by integrated weed management in spring season

Treatment	Dose (kg/ha)	Weed density (No/m^2) @ 60 DAP					Total weed dry biomass (g/m^2)	Millable cane ('000/ ha)	Pol %	Can yield (t/ha)
		Echi-nochloa colona	*Dactylocte-nium* sp	*Panicum repens*	*Cyperus rotundus*	Total weed density (g/m^2)				
Atrazine 50% WP (Ramicide sample)	2.0	6.3	6	4.3	23.3	39.9	155.2	82.5	19.0	50.13
Atrazine 50% WP (Market Sample)	2.0	5.3	3.3	3.7	24.7	37	97.8	98.4	19.14	53.50
Sulfentrazone	0.8	6.7	7.3	11.0	24	49	184.2	120.3	18.43	52.57
Metribuzin fb 2,4-D	1.0+1.0	6	6.7	5.0	24.7	42.4	143.8	120.9	18.63	53.26
Sulfentrazone fb 2,4-D	0.8+1.0	4.7	2.0	10.3	19.7	31.7	111.0	123.7	18.26	54.03
Atrazine fb hoeing + LaybyAtrazine fb 2,4-D	2.0+1.5+1.0	1.3	1.3	5.3	10.7	23.6	99.7	152.1	18.08	56.97
Atrazine fb 2,4 D hoeing	2.0+1.0	3.3	4.3	2.3	22.0	31.9	93.3	153.1	18.32	58.07
Atrazine + trash mulching (10t/ha) fb, 2,4-D fb hoeing	2.0 fb 1.0	2.3	2.3	2.3	17.3	24.2	49.5	149.4	18.35	58.73
Sulfentrazone + trash mulching (10t/ha) fb, 2,4-D fb hoeing	2.0 fb 1.0	2.3	1.7	3.0	20	27	74.1	150.1	17.86	57.47
Hand Boeing	-	4.7	2.0	1.3	20	28	35.9	153.9	18.63	59.63
Weedy check (Control)	-	15.0	18.3	15.7	42.7	91.7	301.9	56.4	17.72	35.07
LSD (P=0.05)		2.7	3.6	3.6	11.9	12.0	40.3	19.8	0.96	7.93

Source: Annual Report (2015-16)

(vii) Bed configuration and soil management to enhance productivity as well as cane quality.

(viii) Intercropping for better land utilisation, crop diversification and effecting input economy: Intercropping in sugarcane (a long duration crop) with short duration, high value and income generating crops facilitates economic security and household nutrition, especially for small and marginal cane growers. This also helps in improving productivity of winter-initiated ratoon and also promotes autumn planting.

Fig. 11. Intercropping systems with sugarcane (i) Sugarcane +Maize (ii) Sugarcane + Rajmash (iii) Sugarcane + Ramdana (iv) Sugarcane + Sarso (v) Sugarcane + Potato

Several intercrops have been advocated in autumn planted and spring planted sugarcane and also in winter- initiated sugarcane ratoon (Table 17). Sugarcane-based intercropping system with potato, wheat, mustard and short duration pulse crops have gained acceptance among the farmers, thereby adding to their economic and nutritional security (Fig. 11). Besides an additional income of up to one lakh rupees can be anticipated through intercropping with these crops.

Table 17. Important sugarcane based intercropping systems with their agronomic

Intercropping systems	Plant geometry/ row ratio	Fertilizer schedule for component crops (NPK kg/ha)	Additional Area of adoption
		Autumn planted cane	
Sugarcane + rajmash	1:2	65:40:30	Uttaranchal
Sugarcane + lentil	1:2	20:40:30	
Sugarcane + mustard	1:2	60:30:30	Maharashtra
Sugarcane + linseed	1:2	45:30:30	
Sugarcane + maize	1:2	120:60:60	Uttaranchal
Sugarcane + potato	1:2	120:60:80	
Sugarcane + onion	1:3	80:40:40	Maharashtra
		Spring planted crop	
Sugarcane + cowpea	1:2	20:40:20	Uttaranchal, Maharashtra
Sugarcane+mungbean	1:2	20:40:20	
Sugarcane+ urdbean	1:2	20:40:20	
Winter initiated ratoon			
Sugarcane ratoon+ber-seem	Broadcast	20:60:40	Uttaranchal, Punjab
Sugarcane ratoon+ shaftal	Broadcast	20:60:40	
Sugarcane ratoon+ lucerne	1:3	20:60:40	
Sugarcane ratoon+ senji	1:3	20:45:40	

Source: Solomon *et al.* (2014)

(ix) Agro-technologies for sugarcane production under abiotic stresses (drought conditions, water logged areas, saline sodic soils, *etc.*) double row planting system was developed for increasing population of millable canes under stress situations.

(x) Ratoon management to reduce production cost and phase out cane supply to maintain sugar recovery in mills. For good ratoon yield, at initiation dismantling of ridges, stubble shaving and off-baring is preferred. Gap filing with slip setts/pre-germinated setts/polybag raised settlings is must, in case of gap exceed 1 per cent of normal crop stand. Paired row planting (120:30) reduces gaps and optimises plant population in subsequent ratoon. Thus, it reduces higher yields compared to sole planting at 90 cm. Trash mulching (10 cm thick) in alternate rows for conserving soil moisture, minimising weed infestation and maintaining soil organic carbon. Application of potassium with irrigation water in standing plant cane one month prior to harvesting improves bud sprouting NMC and yield of succeeding ratoon crop.

Prospects of sugarcane production

Sugarcane crop is majority being used for the production of sugar globally. Sugar is used as a sweetener, blender as well as good preservative for many food commodities all over the world. Some of the main industrial users of the product are the pharmaceutical industries, the food and beverages industries, bakeries, soft drinks bottling plants as well as biscuits and other confectionery manufacturers. In a common man's life, it is used in large amounts as a table sweetener. Although a number of other by-products are produced when sugarcane is processed like bagasse, molasses, *etc.*, its major product is sugar. It is one of the main product for which it is commercially cultivated all over the world. With the increasing population day by day, the demand for sugar and its products is predicted to increase but the amount of production seems to have languished for some time. Sugarcane cultivation is the most planned sector of agriculture which is directly associated to the sugar industry and plays a distinguished role in Indian economy. During the last 15 years, India's yield of sugarcane has been ranging between 59.4-71.3 t/ha. This amount is very low in comparison to the international standard of sugarcane production and productivity. Thus, there is an urgent need to augment the average cane yield of 100 t/ha. Even during the last 15 years, the recovery rate of sugar from sugarcane in India has been altering between 9.42–10.55 per cent which is also at a very low rate in comparison to the sugar recovery rate of other major sugarcane producing countries in the world. The augmentation in cane production is not only the solution but also enhancement in average sugar recovery rate, substantially so as to cope up with the increasing demands of sugar production and consumption (Das *et al.*, 2016).

Conclusion

Sugarcane occupies an important position in agrarian economy of India and supports one of the largest agro-processing industries of the country and > 6 million farmers are engaged in its cultivation. Besides, about half a million skilled and semi-skilled workers, mostly from rural areas are also engaged in the sugar industry. By 2030 AD, India will require nearly 33 million tonnes of white sugar for domestic consumption alone. Production of alcohol for partial replacement of fossil fuel and use of bagasse in cogeneration of electricity have great potential in future and thus requirement of cane will increase further. With an average sugar recovery of 10.75%, about 520 million tonnes of sugarcane will have to be produced and this will entail an increase in sugarcane productivity to the tune of 100 to 110 tonnes/ha, as area may stabilise around 5 million ha. It is apparent that, in future, the production target of sugarcane has to be met mainly by increasing the productivity and quality of the crop. Hence, the research and development focus is needed to develop and promote those technologies that raise farm income and ensure employment opportunities.

References

Binbol, N.L., Adebayo, A.A., Kwon-Ndung. 2006. Influence of climatic factors on the growth and yield of sugar cane at Numan, Nigeria. *Climate Res.* **32**: 247–252.

Blume, H. 1985. Geography of sugarcane, Verlag Dr. Albert Barlens, Berlin, Germany, pp. 371.

Brandes, E.W. and Artschwager, E. 1958. Sugarcane (*Saccharum officinarum* L.): Origin, classification, characteristics, and descriptions of representative clones. USDA Agric. Handbook. U. S. Gov. Print Office, Washington, DC., Pp. 122.

Cooperative Sugar. 2017. **48**(5): 38-52.

Cooperative Sugar. 2018. **48**(6): 46-62.

DAC. 2020. Pocket book of agricultural statistics 2019. Directorate of Economics & Statistics, Government of India. Pp. 122.

Das, K.K., Jhunjhunwala, J. and Dhancholia, S.K. 2016. Problems and prospects for sugarcane growers in India: An Empirical study. *Int. Edu. and Res. J.* **2**(7): 52-57.

Deren, C.W., Cherry, R.H. and Synder, G.H. 1993. Effects of flooding on selected sugarcane clones and soil insect-pests. *J. Am. Soc. Sugarcane Tech.* **13**: 22-27.

Deren, C.W., Snyder, G.H., Miller, J.D. and Porter, P.S. 1991a. Screening for and heritability of flood tolerance in the Florida (CP) sugarcane breeding population. *Euphytica* **56**: 155-160.

Deren, C.W., Snyder, G.H., Tai, P.Y.P., Turick, C.E. and Chynoweth, D.P. 1991b. Biomass production and Biochemical methane potential of seasonally flooded intergeneric and interspecific Saccharum hybrids. *Bioresource Tech.* **36**: 179-184.

Dixit, R.S. and Bhardwaj, K.M. 1974. Response of sugarcane varieties to different dates of planting and rationing. Proc. Annl. Conv. STAI, **40**: 29-34.

Fageria, N.K., Baligar, V.C. and Jones, C.A. 2010. Growth and mineral nutrition of field crops, 3rd Edn. Boca Raton, FL: CRC Press. pp. 586.

Fernando, S., AluízioBorém, Celso Caldas. 2015. Sugarcane: Agricultural, Production, Bioenergy and Ethanol. Academic Press, Elseveir, pp 465.

Gardener, F.P., Pearce, R. and Mitchell, R. 1984. Physiology of crop plants. 1st Edition. Iowa State University Press, Iowa. pp 328.

Gawander, J. 2007. Impact of climate change on sugarcane production in Fiji. WMO Bull., **56**(1): 34-39.

Glover, J., Johnson. H., Lizzio. J., Wesley, V., Hattersley, P. and Knight, C. 2008. Australia's crops and pastures in a changing climate – can biotechnology help? Australian Government Bureau of Rural Sciences, Canberra.

Hartt, C.E. 1940. The synthesis of sucrose by excised blades of sugarcane: time and temperature. Hawaiian Planter's Rec., **44: 89-116**.

http://nidm.gov.in/pdf/dp/Uttar.pdf, 09.02.2017

http://www.sugarcanecrops.com/climate/, 27.07.2015

Indian Sugar. 2017. (10): 44

IISR Vision 2050. http://iisr.nic.in/download/publications/IISRVision2050.pdf. Accessed 30.08.2018.

Julien, M.H.R. Irvine, J.E. and Benda, G.T.A. 1989. Sugarcane anatomy, morphology and physiology. In: Recaud. C.B.T. Egan, A.G. Gillaspie Jr. and C.G. Hughes (Eds.) Diseases of sugarcane: major diseases, Elsevier, Amsterdam, Netherlands, pp. 1-17.

Kaur, S. and Purohit, M.K. 2015. Rainfall statistics of India-2015. Indian MET department. Ministry of Earth Sciences, pp 103.

Keshavaiah, K.V., Palled, Y.B., Shankaraiah, C., Jagadeesh, K.S., Channal, H.T. and Nandihali, B.S. 2013. Studies on performance of sugarcane under different nutrient management practices. *Karnataka J. Agril. Sci.* **26**(4): 506-510.

Kumar, V. and Chand, M. 2012. Effect of integrated nutrients management on cane yield, juice quality and soil fertility under sugarcane based cropping system. *Sugar Technology* **15**(2): 214-218.

Lal, M. and Singh, I. 2007. Conventional methods of sugarcane planting and their Improvement. In: Yadav, R.L. and Yadav, D.V. (eds) Sugarcane planting Techniques and Crop Management. ICAR-Indian Institute of Sugarcane Research Lucknow, pp 1-5

Lingle, S.E. and Weigland, C.L. 1997. Soil salinity and sugarcane juice quality. *Field Crop Res.* **54**: 259-68.

Mathieson, L. 2007. Climate change and the Australian Sugar Industry: Impacts, adaptation and R & D opportunities. Sugar Research and Development Corporation. Australia.

Mathur, R.N. and Haider, I.M. 1940. A summary of five years of physiological research on sugarcane at the Sugarcane Research Station, Shahjahanpur. *Proc. International Soc. Sugar Cane Technol.* **9**: 11-26.

Mathur, S. 2015. Drought declared in 50 of UP's 75 districts Times of India (18th Nov.), 22.02.2017

Misra, S.R. 2004. Constrains of improving sugarcane productivity in physiological and molecular approaches for improving sugarcane productivity. ICAR-Indian Institute of Sugarcane Research, Lucknow, pp 9-13.

Nielsen, A.C. 2006. Business world marketing white book. Sugar production for the year 2007 has been estimated to be 27 million MT. ISMA

Paul, G.C., Rahman, M.H. and Rahman, A.B.M. 2005. Integrated nutrient management with organic and inorganic fertilizers on productivity of sugarcane ratoon in Bangladesh. *Sugar Tech.* **7**(2&3): 20-23.

Phonde, D.B., Nerkar, Y.S., Zende, N.A., Chavan, R.V. and Tiwari, K.N. 2005. Most profitable sugarcane production in Maharashtra. *Better Crops* **89**(3): 21-23.

Prasad, S.R. Srivastava, N.S.L. and Alam Masood. 1980. Irrigation efficiency in sugarcane. *Indian J. Agric. Sci.* **50**(3):252-260.

Ramanujam, T. and Venkataramana, S. 1999 Radiation interception and utilization at different growth stages of sugarcane and their influence on yield. Indian Journal of Plant Physiology 4:85-89.

Rostron, H. 1971. The growth and productivity of sugarcane. Proc. of the South African Technologists Association, pp 152-157.

Rozeff, N. 1995. Sugarcane and salinity-a review paper. *Sugarcane* **5**:8-19.

Rozeff, N. 2002. Important temperatures for sugarcane production. *Sugar J.* **65**(6): 8.

Samui, R.P., John Gracy, and Kulkarni, M.B. 2003. Impact of weather on yield of sugarcane at different growth stages. *J. Agric. Physics.* **3**(1&2):119-125.

Shahi, H.N., Lal, M., Sinha, O.K. and Srivastava, T.K. 2000. Sugarcane Production: Strategies and Technologies. IISR Tech. bull. No. 40, ICAR-Indian Institute of Sugarcane Research, Lucknow. Pp. 87.

Shanthy, R., Bakshi Ram, Alarmelu, S., Premachandran, M.N., Viswanathan, R., Palaniswamy, Bhaskaran, Geetha N., Malathi, P., Jayabose, C., Arumumganathan, T. and Sobhakumari, V.P. 2016. Sugarcane Technologies. Sugarcane Breeding Institute, Coimbatore, pp. 110.

Sharma, B.L., Singh, S. and Singh, R.R. 2003. Impact of climatic constraints on the sugar accumulation and recovery in the Eastern U.P. *Co-op. Sugar* **34**(5): 399-402.

Shrivastava, A.K. and Srivastava, S. 2012. Sugarcane: Physiological and molecular approaches for improving abiotic stress tolerance and sustaining crop productivity. In: Tuteja N., S. S. Gill, A.F. Tiburcio, and R. Tuteja (eds.) Improving Crop Resistance to Abiotic Stress Vol. 2, Wiley-Blackwell, Germany, pp. 885-922.

Shrivastava, A.K. and Srivastava, M.K. 2006. Abiotic Stresses Affecting Sugarcane: Sustaining Productivity, International Book Distributing Company, Lucknow, India. pp. 322.

Shrivastava, A.K., Srivastava, T.K., Srivastava, A.K., Misra, V., Srivastava, S., Singh, V.K. and Shukla, S.P. 2016. Climate change induced abiotic stresses affecting sugarcane and their mitigation, ICAR-Indian Institute of Sugarcane Research, Lucknow (ISBN:978-93-5265-983-8), pp108

Shukla, S.K. Sharma, L. Awasthi, A.K. and Pathak, A.D. 2017. Sugarcane in India (Package of practices for different agro-climatic zones). ICAR-All India Coordinated Research Project on Sugarcane. pp 51.

Silva, de A. M. Jifon J.L., da Silva, J.A.G. and Sharma, V. 2007. Use of physiological parameters as fast tools to screen for drought tolerance in sugarcane.Braz. *J. Plant Physiol.* **19**(3): 193-201.

Singh, B. N. and Lal, K.N. 1935. Limitations of blackman's law of limiting factors and harder's concept of relative minimum as applied to photosynthesis. *Pl. Physiol.* **10**: 245-268.

Singh, S. 1983. Germination of sugarcane seeds during storage in relation to temperature. Paper presented at the National Seminar on Research Perspective on Seed Storage. Seed Technology Department, TNAU, Coimbatore, April.

Singh, S. 1985. Natural inductive temperature for primoridia initiation and development of floral primordia in sugarcane at Coimbatore. *Phyton.* **45**(1): 85-91.

Singh, S.N. Singh, S.B. and Singh, N.P. 1994. Effect of moisture regimes and method of irrigation on growth yield, quality and water economy in sugarcane crop. Proc. of water management held at Jalgaon, pp 7-12.

Singh, K. Yadav, R.L. Singh, B. Hora, B.S. Singh, R.V. and Singh R.A. 1984. Ring method of planting sugarcane: A new technique. *Biological Memoirs*, **9**: 161–166.

Solomon, S., Jain, R., Chandra, A., Shukla, S.K., Lal, R.J., Venugopalan, V.K., Nithya, K., Holkar, S.K., Singh, M.R., Prakash, B. and Mohd. Ashfaque. 2014. Sugarcane: A voyage from sett to sweeteners. ICAR-Indian Institute of Sugarcane Research, Lucknow, pp 111.

Tandon, H.L.S. 1992. Fertilizers, organic manures, recyclable wastes and biofertilizers: Fertilizer Development and Consultation Organization. pp. 148.

Tripathi, H.P. and Kumar, A. 2010. Compendium of cropping system in three decades. All India Coordinated Research Project on Cropping System. N.D. University of Agriculture and Technology, p 66.

Viswanthan, R. Jayanthi and Sankaranarayanan, C. 2017. Integrated disease and pest management. *Indian Farmg.* **67**(02): 28-32.

Yadav, D.V. 2007. Nutrient Management in Sugarcane in relation to planting techniques. In: Yadav, R.L. and Yadav, D.V. (eds) Sugarcane planting techniques and crop management. ICAR-Indian Institute of Sugarcane Research, Lucknow, pp. 63-67.

Yadav, R.L. 2004. Enhancing efficiency of fertilizer N use in sugarcane by ring-pit method of planting. *Sugar Tech.*, **6**:169.

8

Advances in Vegetable Production Technologies

Hari Har Ram

Introduction

The general recommendation for intake of fruits and vegetables is at least 400 g/person as recommended by World Health Organization (WHO), which amounts to five servings of 80 g/day or about 146 g/person/annum. For vegetables, the recommended intake requirement is at least 300 g per day per individual. This 300 g requirement must include 50 g leafy vegetables, 50 g roots and tubers and 200 g other vegetables (Kapur, 2016). Current annual vegetable production including potato, based on recent three years (2015-16, 2016-17, 2017-18) average is around 177 million tonnes from 10 million ha giving a productivity of 17.7 tonnes/ha. With this average and stagnating production and assuming Indian population at 133 crores (1.33 billion) and 25% post-harvest losses, per capita per day availability of vegetables comes to 273 g leaving a gap of 27 g between the recommended dose and the availability of vegetables/head/day in India. By 2050, the projected population of India is 1.5 billion and to meet the requirement of this population, vegetable production will have be 220 million tonnes, of course making allowance of 25% post-harvest losses. Considering all kind of constraints on the available land area of vegetables, the vegetable area will be around 10 million ha and consequently, major gain of production will have to come from productivity increase from current level of about 18 to 22 tonnes/ha and this seems to be a gigantic task as during last several years, the productivity of vegetables in India has been between 17 and 18 tonnes/ha, an indication that perhaps productivity has reached to a plateau. This situation will call fall massive interventions on technological fronts in terms of innovative products with focus on better hybrids with higher yields, and resistance to biotic and abiotic stresses coupled with horizontal scaling up of modern production technologies. This chapter looks into the modern production technologies as applicable to vegetable crops in India to boost the productivity further from current 18 tonnes/ha to at least 22 tonnes/ha.

**Corresponding author email: hhram13@yahoo.co.in*

The infrastructure available on vegetable research is quite adequate. All India Coordinated Research Project on vegetables (AICRPV) started in 1971 with headquarter at IARI, New Delhi. This was upgraded to Project Directorate level in 1986 and shifted to new location, now ICAR-Indian Institute of Vegetable Research (IIVR), Varanasi which has been upgraded to level of institute since, 1992. By now a total of 461 cultivars has been released in 23 vegetable crops (brinjal, tomato, chilli, capsicum, pea, okra, onion, garlic, cauliflower, cabbage, carrot, cowpea, frenchbean, dolichos bean, muskmelon, watermelon, ash gourd, bitter gourd, bottle gourd, cucumber, pumpkin, ridge gourd, smooth gourd). The break-up of these cultivars has been as open-pollinated types (273), hybrids (141) and resistant types (47). The breeding activities are expanding rapidly due to involvement of about 300 seed companies in India which are selling seeds of improved cultivars and hybrids. Based on the three activities of a seed company like research and development, production of proprietary seeds, and sale and marketing of seed in their own brand, 20% of the companies have all the three activities and intensively do research and produce proprietary seeds. Around 60% of the companies produce and market the seed without any research component. These companies either produce open-pollinates of public or private bred or licensed hybrids or do in-house production. Remaining 20% are either importing seeds from other companies are sourcing produced material from the market and packing and selling then seed. In vegetable seed sector, there are large seed companies with turnover of more than ₹ 100 crores, medium companies with turnover between ₹ 50 crores to less than ₹ 100 crores, small companies with which are from ₹ 20 crores to less than ₹ 50 crores and still smaller companies which are less than ₹ 20 crores. It is well established that in order to raise productivity, breeding and seed production of superior cultivars and hybrids will continue to play a dominant role along with improvement in production technologies. The progress on seed front especially in vegetables is quite comfortable and public and private sector together are fully equipped to produce quality seed as per demand provided regulatory regime does not become obstructive.

Global seed market is worth 60 billion USD of which Indian seed market accounts for 4 billion USD (₹ 28,000 crores). From 4 billion Indian seed market, vegetable seed market is worth 560 million USD (₹ 4,000 crores). USA tops the list of world seed market (12 billion USD) followed by China (10 billion USD), France (4 billion USD), India (4 billion USD), Brazil (2.6 billion USD), Japan (1.8 billion USD), Germany (1.7 billion USD), Italy (1 billion USD), Argentina (1 billion USD), Canada (0.8 billion USD) and Russian Federation (0.7 billion USD). It is clarified that these figures relate only to the traded seed in value terms and do not include the value of seeds saved and used by the farmers (Kapur, 2019).

Further, the usage of new breeding technologies (NBT) like gene insertion, genome editing, etc. and also seed technologies like seed coating, pelleting, priming and enhancement of seed performance are fuelling the growth of agriculture and horticulture sectors and in turn driving the demand for performing seed. However, in parallel, there are production innovations and these are briefly described.

Nursery raising and management

In case of several vegetable crops, first seedlings are raised in the nursery bed and then they are transplanted. These in general include tomato, eggplant, chilli, sweet pepper, cauliflower, cabbage, knol-khol, broccoli, Brussels sprouts, kale, endive, chicory, celery, lettuce, parsley, onion, etc. However, many of these are now direct seeded in the field using precision planting machines in USA, Japan, the Netherlands, Italy and several other countries. Many other vegetables like okra, peas, beans, radish, carrot, beetroot, turnip, and all cucurbits are conventionally planted directly in the field.

Advantages of raising seedlings in nursery

- Uniform, vigorous, healthy planting material.
- Economy of seed rate, useful particularly for hybrid seeds which are costly.
- Helps in raising off-season vegetables.
- Ensuring favourable growing conditions in limited nursery area.
- Efficient time management for main field preparation.
- Opportunity for increasing cropping intensity.
- Facilitates inter-cropping and relay cropping.
- Easy adjustment of date of planting to manage incidence of diseases and pests and to coincide with market demand to get better price.
- Ensuring optimum plant population in the field.
- Efficient management of pests and diseases in limited area initially.

Nursery soil and growing media

The nursery soil should be light to medium in texture with high porosity, good fertility, adequate water holding capacity and good drainage. The optimum pH is 6.5 to 7.5. Well drained sandy loam soil is the best for raising seedlings. In case of heavy nursery soil, sand @ 2 kg per m^2 should be added. In case of light soils, farm yard manure @ 2.5 kg per m^2 should be added. Well decomposed and powdered FYM or compost is used as organic mulch to improve water holding capacity of soil.

Seedling growing media are important factors in raising nursery as growth and development of nursery plants and roots proliferation depend on it. Common growing media used are soil, sand, coco-peat, compost, vermin-compost, peat-moss, perlite, pumice, saw dust, ash, etc. Under open field conditions, generally soil, sand and well rotten FYM are used. Growing media should be easily available, and have the properties like neutral in reaction, sufficient water and air retentive capacity and allow proper drainage, free from weeds, nematodes, fungi and bacteria inoculums and capable of being sterilized. A few common media are vermin-compost (prepared by rearing of earthworms), saw-dust (by-product of wood and timber industry and generally free from harmful substances), peat (remains of aquatic, marsh or swamp vegetation formed as a result of partial decomposition under anaerobic or semi-anaerobic conditions), Sphagnum moss (commercial sphagnum moss) as dehydrated young residue or living portions of plants in the genus *Sphagnum*, such as *S. papillosum*, *S. capillaceum* and *S. palustre*, relatively sterile, light in weight, high water holding capacity, 10-20 times of its weight, shredded by hand or mechanically before use, pH 3.5 to 4.0), perlite (grey white aluminum silicate mineral from lava flows, rich in moisture and nutrients, free from disease causing inoculums and nematodes, low cation exchange capacity not influencing pH, highly porous), vermiculite (hydrated maganesium, aluminum/iron silicate, light in weight, 90-150 kg/m^3, neutral in reaction, greater water holding capacity with 80% porosity), pumice (mostly silicon dioxide and aluminium oxide with small amounts of iron, maganesium and sodium in oxide form, increases aeration and drainage in rooting mixture, to be used alone or in combination with peat-moss), rock-wool (man-made mineral fibre, highly porous, contains 40-50% easily available water and 40-45% air capacity by volume, acts as support medium for the root system and as a means for holding nutrient solution and air in close proximity to the roots) and coco-peat (biodegradable and eco-friendly growing medium prepared natural coir fibre, completely organic and soil-free potting medium).

Nursery soil treatment

Nursery soil must be treated by physical, chemical and biological means to kill soil borne pathogens particularly *Pythium*, *Rhizoctonia*, *Fusarium*, *Sclerotinia*, *Botrytis*, *Phytophthora*, *Phomopsis*, *Sclerotinum*, *Alternaria*, *Pseudomonas*, nematodes (*Meloidogyne* spp. and insect-pests). These organisms mostly cause damping off. The effective treatments include soil solarization (whole nursery area after irrigation to be covered with transparent polythene sheet of 400 gauge for 5-6 weeks during summer when air temperature may raise up to 40-48 °C and soil temperature inside polythene covering up to 48 to 56 °C sufficient to kill most of the pathogens), steam treatment (nursery area covered with polythene sheet in air-tight manner followed by passing steam inside for 4-6 hours), formalin treatment (formalin solution prepared by adding 250 ml commercial grade formaldehyde (40%) per 10 litres of water and drenching the soil @ 4.5 litres of solution/m^2 so as to saturate top soil surface up to a depth of 15-20 cm, 15-20 days before sowing and covering drenched area with 400 gauge polythene sheet so that fumes of formalin penetrate inside the soil to kill the pathogens, removal of polythene cover after 24 hours and keeping bed open for 15-20 days prior to seed sowing), soil treatment with fungicides (Captan or Thiram @ 5 g/m^2 for dry treatment or 0.25-0.30% solution @ 4-5 litres/m^2 as soil drenching), soil treatment with insecticides (granular insecticides like Furadon, Phorate 10 G, etc. to be applied to the soil @ 4-5 g/m^2 followed by mixing to a depth of 15-20 cm to kill insects/eggs/nematodes), soil treatment with bio-agents (*Trichoderma* spp, *Pseudomonas fluorescence* and *Aspergillus niger* to be added @ 110-25 g/m^2 and well mixed in the soil followed by seed sowing 2 days after this application), seed treatment (seed treatment with Captan/Thiram/Carbendazim @ 2-3 g/kg of seed, hot water treatment of cabbage and cauliflower seed at 50 °C for 30 minutes to control black rot) and seed coating treatment (seed coating/pelleting common in tomato, brinjal, chilli, sweet pepper where seed are often coated with pesticides).

Nursery bed preparation, seed sowing in nursery and pricking out of seedlings

Nursery bed should be 15-20 cm raised from the ground level with fine soil and optimum moisture. Length of the bed should be 3-5 m with width of 1 m to facilitate intercultural operations. The bed should preferably be in east-west direction with lines in north-south direction to capture maximum radiation energy especially during winter. After seed bed preparation, seed are either broadcast or sown in lines, about 1 cm deep and 5 cm apart in rows. Within rows, seed are spaced at 1 cm. After sowing seed-bed should be covered with 0.5 cm thick mixture of farm yard manure, sand and soil in ratio of 1:1:2 for better moisture conservation and germination. Spraying with fungicides at regular intervals is recommended (Captan/Thiram, 0.25%). Seedlings in the nursery should be protected against sucking insects particularly white fly through spraying Imidacloprid (0.3 ml/l of water). Mulching the seed beds with paddy straw or sugarcane trash in hot weather with intermittent sprinkling of water and polythene mulch during winter helps in quick and uniform germination of seeds. If temperature reaches above 30 °C, then bed should be covered by green or black colour 50-60% shade nets about 1 m above the ground with suitable support to protect the seedlings from scorching sunlight.

In pricking out of seedlings method, young seedlings with first true leaves are taken out and transplanted to another bed or seed box or grown individually in small plastic thumb pots of 5 cm diameter to have healthy and hardened seedlings.

Seedling raising under poly-house

It is now a common practice to sow seeds of different cucurbits in winter months (December-February) in small poly bags (15 × 10 cm size, 200-300 gauge) filled with soil, farm yard manure/compost, and sand mixture in a ratio of 1:1:1 to get early crop. Seeds having hard/leathery seed coat like bottle gourd, bitter gourd, pumpkin, watermelon should be soaked overnight in water before sowing in polythene bags. The poly bags are kept in poly-house where temperature is raised above 20 °C during day time in winter months. The poly-house could be based on improvised structure with bamboo poles/ iron pipes with a covering of polythene sheet of 200 gauge. Micro-sprinklers should preferably be used to irrigate poly-house grown seedlings. Under poly-house, seedlings become ready in 25-30 days in winter months while in open conditions the time required may be 50-60 days. The seedlings are transplanted in the field either by removing the polythene or by cutting lower portion of the polythene to allow root penetration into the soil. This method allows 30-45 days early crop in comparison to direct sowing in the field in winter months.

Seedling hardening

Seedling hardening is done to have better field stand. This is accomplished by reducing soil moisture by 20% by withholding watering to the seedlings 5-7 days before transplanting, lowering the temperature to retard the seedling growth, by application of 4,000 ppm NaCl with irrigation water, by spraying of 2,000 ppm cycocel + 0.25% zinc sulfate and by spraying of 0.2% muriate of potash.

Raising seedling in root plug trays

Root plug trays should be disinfected with 1% solution of chlorine bleach followed by thorough washing with clean tap water. Correct cell size trays should be used. Cucurbits require larger cell size than other vegetable crops. Growing media should be prepared by mixing coco-peat, perlite and vermiculite in a ratio of 3:1:1. A mixture of peat-moss, compost and coconut coir can also be used. pH of the mixture should be 5.5 to 6.5 and if needed lime can be added. Fill the plugs tightly with pre-moistened media and put one seed in the centre of the cell, 0.5 to 1 cm deep and cover the seed with same mixture. Water the trays thoroughly to moisten the plugs. Avoid watering late in afternoon.

Apply water soluble fertilizer with irrigation water at 50-75 g per litre of water at weekly intervals. Excessive use of fertilizer makes seedlings lanky. Put trays under a net-house/poly-house. Spray the seedlings with Mancozeb + Carbendazim mixture @ 2.5 g per litre of water if damping off occurs. Harden the seedlings by withholding water 4-5 days before transplanting. If seedlings are grown under a protected structure, they should be moved outside for several days to acclimatize them for transplanting. Drench the plugs with water one day before transplanting and transplant 4-6 week old seedlings preferably in afternoon hours.

Nutrient management

Nitrogen (N), Phosphorus (P) and Potash (K) constitute primary nutrients and Calcium (Ca), Magnesium (Mg) and Sulphur (S) constitutes secondary nutrients under macronutrients for vegetable crops. The nutrients which are essential but used in small quantities by the crops are micronutrients and these include Iron (Fe), Manganese (Mn), Copper (Cu), Zinc (Zn), Molybdenum (Mo), Boron (B) and Chorine (Cl). Nutrient requirement of vegetable crops vary according to vegetables/cultivars/environment. In general, vegetable crops exert huge pressure on the soil for nutrients due to their high productivity. A fair degree of idea of nutrient requirement of vegetables can be gauzed from the quantities of nutrients removed by a particular crop. This information is summarized in table 1.

Table 1. Removal of primary nutrients from the soil by vegetable crops

Crop	Nutrient removal (kg/100 kg yield)		
	Nitrogen	Phosphorus	Potassium
Tomato	0.33	0.11	0.44
Brinjal	0.29	0.08	0.50
Cabbage	0.53	0.16	0.50
Cauliflower	0.40	0.16	0.50
Knolkhol	0.50	0.42	0.85
Carrot	0.42	0.18	0.67
Radish	0.60	0.30	0.60
Beetroot	0.24	0.16	0.45
Onion	0.30	0.13	0.40
Leek	0.34	0.13	0.80
Cucumber	0.18	0.13	0.30
Pumpkin	0.18	0.14	0.32
Muskmelon	0.36	0.11	0.64
Bottle gourd	0.15	0.16	0.16
Pea	1.25	0.45	0.90
Beans	0.87	0.27	1.07
Okra	0.30	0.13	0.45

Celery	0.67	0.27	1.01
Lettuce	0.30	0.12	0.53
Spinach	0.50	0.15	0.30
Asparagus	2.40	1.20	3.00
Cassava	0.38	0.18	0.88
Sweet potato	0.48	0.19	0.85
Elephant yam foot	0.34	0.08	0.49
Yam	0.57	0.14	0.61

Source: Hazra *et al.* (2011)

In any natural ecosystem, the nutrients absorbed by the crops return to the soil after decomposition of the plant into the soil. However in case of commercial crop cultivation the nutrients absorbed by the harvested biomass never return to the soil necessitating application of nutrients from the external sources. Before, coming to the external supply of nutrients to the individual vegetable crops, let us briefly discuss the role and importance of macro-nutrients in relation to vegetable crops.

Nitrogen: Nitrogen is the most important nutrient in plant growth and vegetable crop production. It is highly mobile and easily lost from the soil. Crops can utilize only 50-60% of available nitrogen in the soil. When nitrogen sources are applied to the soil, they all are eventually converted into ammonium (NH_4^+) and nitrate (NO_3^-) forms which are taken up by plant roots. Nitrate if left unused is leached into the soil downward but the unused ammonium is held by the soil particles and remains un-leached. However, both forms may be consumed by micro-organisms or converted to gaseous forms and lost to atmosphere. Nitrogen use efficiency in vegetables can be increased as per following procedures.

- Split application of nitrogenous fertilizers at the time of peak requirements of the crop which decreases nitrate leaching and thereby increases nitrogen use efficiency. In general, half of the total required nitrogen is applied as basal dose during the final land preparation and the rest half in different splits. In case of leguminous vegetables, entire nitrogen is applied as basal dose. In fruit vegetables, split applications after periodical harvests have been found to be beneficial.
- Nitrogenous fertilizers should be applied 5-10 cm deep in the soil for full utilization and to avoid surface leaching.
- Optimum soil moisture increases nitrogen use efficiency.
- Nitrogenous fertilizer should not be applied in large quantity near the root zone to avoid salt damage to the roots.
- Most efficient use of nitrogenous fertilizer happens when P and K are applied in required quantity.

Phosphorus: Young plants assimilate phosphorus rapidly and hence need water soluble forms which can easily migrate to the initial root system. Plants generally need about 75% of their P requirement in the early stages of the growth and this needs application of water soluble phosphorus in the soil zone of the germinating seed. Most soluble P (HPO_4^- and $H_2PO_4^-$) become fixed before plants can absorb them and hardly 10-15% of the applied P is utilized by the plants. Thus, leaching loss of P is practically nil and the major loss is through surface erosion of the soil. P availability is higher at pH of 6.0-6.5. In acidic soils, availability of P is low due to formation of complex compounds of P with aluminum and iron. Phosphorus use efficiency can be increased by adopting the following procedures.

- Maintenance of soil pH at 6.5 to 7.0.
- Application of entire P as basal near the root zone.
- Band placement is better and too deep placement should be avoided.
- Late application of P is not effective for plant growth.
- Adequate moisture and combined application of N and P is better.
- Organic matter in soil increases P availability.
- In cold weather, more P should be applied as absorption of P is less in cold weather.
- Rock phosphate or basic slag should be applied 2-4 weeks before sowing.

Potassium: Potassium occurs as mobile ion K^+ rather than as an integral component of any specific compound. Its luxurious consumption is common in vegetable crops. Its rapid uptake occurs in the early stage of crop growth and occurs faster than N and P and therefore ample K should be made available to the seedlings. It is less mobile than N but more mobile than P. It is applied in full as basal dose. It is also leached substantially downward through soil and therefore, there is often K deficiency in high rainfall areas. Potassium use efficiency can be increased through following measures.

- Split application of K in loose soils and in high rainfall areas.
- Soil pH maintenance at 6.0 to 7.0 by liming if needed to reduce leaching loss.
- Returning of crop residue into the soil and application of K rich manures into the soil.

General fertilizer recommendation to the vegetable crops is listed in table 2.

Table 2. General fertilizer recommendation (nutrient in kg/ha) to vegetable crops in the soils with moderate fertility level

Crop	Open–pollinated and hybrids					
	Open-pollinated cultivars			Hybrids		
	N	P_2O_5	K_2O	N	P_2O_5	K_2O
Tomato	100	80	80	200	100	100
Brinjal	125	80	80	200	100	80
Chilli	100	60	60	150	80	80
Sweet pepper	120	80	60	200	150	100
Cauliflower	125	80	60	200	100	100
Cabbage	150	60	80	200	125	150
Onion	100	60	100	-	-	-
Garlic	100	65	100	-	-	-
Okra	125	75	60	200	100	100
Pumpkin	100	60	80	-	-	-
Bottle gourd	90	60	60	200	100	100
Bitter gourd	100	60	60	200	100	100
Cucumber	100	60	60	150	90	90
Pointed gourd	150	60	60	-	-	-
Ridge gourd	100	60	60	150	90	90
Bean	30	75	60	-	-	-
Cowpea	25	75	60	-	-	-
Radish, beet, carrot, turnip	100	60	80	-	-	-
Amaranth	90	50	50	-	-	-
Palak	100	60	40	-	-	-
Sweet potato	75	50	75	-	-	-
Elephant foot yam	150	100	150	-	-	-
Colocasia	120	60	80	-	-	-

Source: Hazra *et al.* (2011)

In recent years, there has been increasing emphasis on higher application of phosphorus and potassium than normally recommended as in table 2 and this is reflected in table 3 specifically applicable to U.S.A. situation but this gives a fair idea about the requirement of high doses of N, P and K along with stage of split applications wherever applicable. Private sector seed companies in India too are now recommending almost similar fertilizer application to have full potential yield of hybrids which produce higher biomass.

Table 3. General fertilizer suggestions for vegetable crops

Crop	Desirable pH	N (kg/ha)	P_2O_5 (kg/ha) at medium soil P level	K_2O (kg/ha) at medium soil potassium level	Nutrient timing and methods of application
Bean, snap	6-6.5	40-80	60	60	Total recommended.
		20-40	40	40	Broadcast and disc-in.
		20-40	20	20	Band place with planter.
Beet	6-6.5	75-100	100	100	Total recommended.
		50	100	100	Broadcast and disc-in.
		25-50	-	-	Side-dress 4-6 weeks after planting.
			Apply 2-3 kg boron per ha with broadcast fertilizer		
Broccoli	6-6.5	125-175	100	100	Total recommended.
		50-100	100	100	Broadcast and disc-in.
		50	-	-	Side-dress 2-3 weeks after transplanting.
		25	-	-	Side-dress every 2-3 weeks after initial side dressing.
		Apply 2-3 kg boron/ha with broadcast fertilizer			
Cabbage and cauliflower	6-6.5	100-175	100	100	Total recommended.
		50-75	100	100	Broadcast and disc-in.
		25-50	-	-	Side-dress 2-3 weeks after transplanting.
		25-50	-	-	Side-dress if needed, according to weather conditions.
		Apply 2-3 kg boron/ha and 0.5 kg sodium molybdate /ha with broadcast fertilizer.			
Carrot	6-6.5	50-80	100	100	Total recommended.
		50	100	100	Broadcast and disc-in.
		25-30	-	-	Side-dress if needed.
		Apply 1-2 kg boron/ha with broadcast fertilizer			
Corn, sweet	6-6.5	110-155	120	120	Total recommended.
		40-60	100	100	Broadcast before planting.
		20	20	20	Band-place with planter.
		50-75	-	-	Side-dress when corn is 30-45 cm tall.
		Apply 1-2 kg boron/ha with broadcast fertilizer. On very light sandy soils, side-dress 40 kg N/ha when corn is 15 cm tall and another 40 kg N/ha when corn is 30-45 cm tall.			

Cucumber	6-6.5	80-160	100	150	Total recommended
		40-100	75	125	Broadcast and disc-in.
		20-30	25	25	Band-place with planter 7-14 days after planting.
		20-30	-	-	Side-dress when vines begin to run or apply in irrigation water.
Eggplant	6-6.5	100-200	150	150	Total recommended.
		50-100	150	150	Broadcast and disc-in.
		25-50	-	-	Side-dress 3-4 weeks after planting.
		25-50	-	-	Side-dress 6-8 weeks after planting.
		Apply 1-2 kg boron/ha with broadcast fertilizer.			
Leek	6.6-5	75-125	150	150	Total recommended.
		50-75	-	-	Broadcast and disc-in.
		25-50	-	-	Side-dress 3-4 weeks after transplanting.
		Apply 1-2 kg boron/ha with broadcast fertilizer.			
Cantaloupes and mixed melons	6-6.5	75-115	100	150	Total recommended.
		25-50	75	125	Broadcast and disc-in.
		25	25	25	Band-place with planter.
		25-40	-	-	Side-dress when vines start to run.
		Apply 1-2 kg boron/ha with broadcast fertilizer.			
Okra	6-6.5	100-200	150	150	Total recommended.
		50-100	150	150	Broadcast and disc-in
		25-50	-	-	Side-dress 3-4 weeks after planting.
		25-50	-	-	Side-dress 6-8 weeks after planting.
		Apply 1-2 kg boron/ha with broadcast fertilizer.			
Onion	6-6.5	75-125	100	100	Total recommended.
		50-75	100	100	Broadcast and disc-in
		25-50	-	-	Side-dress 4-5 weeks after planting.
		Apply 1-2 kg boron/ha with broadcast fertilizer.			
Pea	5.8-6.5	40-60	80	80	Total recommended. Broadcast and disc-in before seeding.

Pepper	6-6.5	100-130	150	150	Total recommended.
		50	150	150	Broadcast and disc-in.
		25-50	-	-	Side-dress after first fruit set.
		25-30	-	-	Side-dress later in season, if needed.
Pumpkin and winter squash	6-6.5	80-90	100	150	Total recommended.
		40-50	100	150	Broadcast and disc-in.
		40-45	-	-	Side-dress when vines begin to run.
Radish	6-6.5	50	100	100	Total recommended. Broadcast and disc-in.
		Apply 1-2 kg boron/ha with broadcast fertilizer.			
Turnip	6-6.5	50-75	100	100	Total recommended.
		25-50	150	150	Broadcast and disc-in.
		25-50	-	-	Side-dress when plants are 10-15 cm tall.
		Apply 1-2 kg boron/ha with broadcast fertilizer.			
Spinach	6-6.5	75-125	150	150	Total recommended.
		50-75	159	150	Broadcast and disc-in.
		25-50	-	-	Side-dress or top-dress.
Summer squash	6-6.5	100-130	100	100	Total recommended.
		25-50	100	100	Broadcast and disc-in.
		50	-	-	Side-dress when vines start to run.
		25-30	-	-	Apply through irrigation system.
		Apply 1-2 kg boron/ha with broadcast fertilizer.			
Sweet potato	5.8-6.2	50-80	100	200	Total recommended.
		-	60	50	Broadcast and disc-in
		50-80	40	150	Side-dress 21-28 days after planting.
		Add 0.5 kg of actual boron/ha 40-80 days after transplant.			
Tomato	6-6.5	80-90	150	200	Total recommended.
		40-45	150	200	Broadcast and disc-in.
		40-45	-	-	Side-dress when first fruits are set, as needed.
		Apply 1-2 kg boron/ha with broadcast fertilizer.			

Watermelon	6-6.5	100-150	100	150	Total recommended.
		50	100	150	Broadcast and disc-in.
		25-50	-	-	Top-dress when vines start to run.
		25-50	-	-	Top-dress at first fruit set.
		Note: Excessive rates of N may increase the incidence of hollow heart in the seedless watermelon.			

In general, the vegetable crops whose above-ground vegetative parts are consumed for example cabbage, cauliflower, knol-khol, amaranth, palak, spinach, lettuce, celery, etc. require more nitrogen. The bulb, root and tuberous vegetables (carrot, radish, beet, turnip, onion, garlic, leek, shallot, potato, sweet potato, elephant foot yam, taro, etc.) require more potash and the fruit vegetables like tomato, eggplant, chilli, okra, pumpkin, bottle gourd, bitter gourd, ridge gourd, smooth gourd, summer squash, cucumber, gherkin, watermelon, muskmelon, etc. require more phosphorus. Peas and beans require high nitrogen but most of the required nitrogen is supplied through symbiotic nitrogen fixation by *Rhizobium* bacteria in the root nodules. In these crops, a low dose of nitrogen should be applied as basal to stimulate early plant growth.

Micronutrients: Boron is the most widely deficient micronutrient in vegetable crop soils. Its deficiencies are most likely to occur in asparagus, most bulb and root crops, cole crops and tomatoes. Excessive amount of boron can be toxic to plant growth. Boron application should not exceed the amount as given in Table 3. Manganese deficiency often occurs in plants that have been over-limed. In this case, broadcast 20-30 kg/ha or band 4-8 kg/ha of manganese sulfate to correct the deficiency. Do not apply lime or poultry manure to such soils until the pH has dropped below 6.5 and be careful not to over-lime again. Molybdenum deficiency of cauliflower (which causes whiptail) may develop when this crop is grown on soils more acidic than pH 5.5. An application of 0.5 to 1.0 kg sodium or ammonium molybdate per hectare will usually correct this. Liming acid soils to a pH of 6.0 to 6.5 will usually prevent the whiptail in cauliflower.

Common disorders of vegetables associated with imbalanced nutrition

There are several physiological disorders associated with imbalanced nutrition in vegetable crops. This is summarized in table 4.

Table 4. Common disorders of vegetables associated with imbalanced nutrition.

Disorder	Causal factor	Remedy
Blossom end rot of tomato, sweet pepper and watermelon, club root of cabbage and cauliflower, damping off of onion.	Low pH and calcium deficiency in tomato, sweet pepper and watermelon; in cauliflower and cabbage club root is incited by *Plasmodiophora brassicae* but aggravated by low pH and calcium deficiency.	Soil application of ground limestone 10-14 days before planting.
Black rot in cabbage and cauliflower.	Incited by the bacteria *Xanthomonas campestris* pv.*campestris*.	Application of P and K in proportionately higher quantity.
Browning of cabbage and broccoli; hollow stem and head of cabbage, broccoli and cauliflower; black heart of beet; internal browning of turnip; stem resetting of tomato; fruit cracking in tomato; cracked stem in celery.	Boron deficiency.	Soil application of borax with basal application of NPK fertilizer.
Root rot of cucumber.	Deficiency of zinc and manganese.	Soil or foliar application of sulfates of zinc and manganese.
Interveinal chlorosis and necrosis in older leaves in tomato, brinjal and chilli; interveinal chlorosis and puckering of older leaves in cabbage, interveinal yellow spots followed by chlorosis in leaves in okra; dieback of leaves in onion.	Magnesium deficiency.	Soil or foliar application of magnesium sulfate.
Root splitting in radish.	High ammonium in soil.	Application of nitrogen through non-ammonium fertilizer or organic matter.
Whiptail of cauliflower; upwardly rolled leaves with diffused marginal and interveinal yellow mottling in potato; purple young leaves in tomato; discoloured leaves and death of growing tip in radish; yellow or white veins in leaves of pea; pale green leaves with interveinal mottling and scorching in bean; deep blue leaves with conspicuous yellow and green mottling in onion.	Deficiency of molybdenum.	Application of ammonium molybdate.

Source: Hazra *et al.* (2011)

The summary of micronutrients to be applied is given in table 5.

Table 5. Soil status and guide for soil application of micronutrients in vegetable crops.

Micronutrient	Critical limit (mg/kg)	Recommended dose of chemical to be applied
Zinc	0.6-1.0	Zinc sulphate @ 25 kg/ha with organic manure.
Boron	0.5	Borax @ 10 kg/ha with basal NPK fertilizer.
Molybdenum	0.2	Ammonium molybdate @ 1 kg/ha with basal NPK fertilizer.
Iron	2.5-4.5	Ferrous sulphate @ 50 kg/ha with organic manure.
Manganese	1.0	Manganese sulphate @ 30 kg/ha with basal NPK fertilizer.
Copper	0.2	Copper sulphate @ 5 kg/ha with basal NPK fertilizer.
Sulphur	10-12	Elemental sulhur @ 25-45 kg/ha.
Application of micronutrient as foliar spray		
Iron	0.4 % ferrous sulphate + 0.2 % lime.	
Manganese	0.4-0.6 % manganese sulphate + 0.2 % lime.	
Zinc	0.2-0.6 % zinc sulphate + 0.1-0.3 % lime.	
Copper	0.1-0.2 % copper sulphate + 0.5 % lime	
Boron	0.5-0.6 % borax.	
Molybdenum	0.05 % sodium or ammonium molybdate	

Note: Lime is added to neutralize the solution, otherwise leaves may get scorched. Minimum 500 litres of water must be used per ha.

Source: Hazra *et al.* (2011)

Water management and irrigation

Water requirement, frequency of irrigation depend upon a number of factors like depth of root system, water use efficiency, crop growth stage, soil type, prevailing weather conditions and actual consumptive use of the crop. When plants are young, they need less water but highly moist soil for root growth. Vegetable crops need more water in the later stage of the growth with more luxuriant growth, and greater biomass in terms of more and bigger leaves. There are critical stages of plants for assured supply of irrigation water but few crops like cauliflower, radish, carrot, beet, celery, etc. require optimum soil moisture level constantly during the whole vegetation period for proper growth and yield. Based on root depth, the vegetables have been classified into 5 types. Very shallow rooted (15-30 cm) crops include onion, lettuce, and small radish. Shallow rooted (30-60 cm) vegetable crops include broccoli, brussels sprouts, cabbage, cauliflower, celery, Chinese cabbage, endive, garlic, leek, parsley, potato, palak, spinach, radish and cowpea. Moderately deep rooted (60-90 cm) crops include beet, carrot, cucumber, brinjal and muskmelon. Deep rooted (90-120 cm) vegetables include pea, chilli, summer squash and turnip while very deep rooted vegetables (120-180 cm) are artichoke, asparagus, lima bean, parsnip, pumpkin, winter squash, sweet potato and tomato. Deeper the root system, more is the water requirement, however the relationship is not always true as other factors also play a critical role in deciding the water/ irrigation requirement of vegetable crops.

Over-irrigation causes a few problems like bursting of heads in cabbage, branching of roots in carrot, fruit cracking in tomato, fruit rot in pointed gourd, etc. Abundant irrigation during ripening of fruits causes an increase in the water content of fruits and thus reduces the sweetness and storability. It is thus advised to stop irrigation before fruit ripening in watermelon, muskmelon, pumpkin, winter squash, etc. In potato, onion, elephant foot yam and sweet potato, etc., irrigation should be stopped 15 days before harvest to extend the storability of the produce. The critical stages of irrigation in vegetable crops as summarized by Hazra *et al.* (2011) are given in table 6.

Irrigation system and methods

Vegetable crops are irrigated by surface, sprinklers and drip systems. The surface irrigation is the most common and cheapest method for the Indian farmers who have small holdings. The commonly used methods under surface method are flooding the field or plot, furrow irrigation and ring and basin method. The flood irrigation is practiced in the crops grown in rows or by broadcast on well leveled land/field, such as onion, garlic, peas, beans, cowpea, palak, spinach, okra, coriander, fenugreek, brinjal and chillies. Many farmers grow radish and carrot also on flat beds. However, in this system large amount of water is required and there is considerable water loss due to evaporation. The vegetable crops planted on ridges or raised beds are irrigated by furrow system. Furrow irrigation is followed in tomato, brinjal, sweet pepper, chillies, radish, carrot, turnip, beet, okra, cowpea, beans, peas, potato, palak, lettuce, onion, garlic, cauliflower, cabbage, cucumber, muskmelon, watermelon and pumpkin. Several cucurbits like bitter gourd, bottle gourd, ridge gourd, smooth gourd, ash gourd, melons, watermelon, cucumber and others planted in rows or on raised beds are irrigated by ring-and-basin system. This method is followed in widely spaced crops and in the areas where water is in short supply, like arid and semi-arid regions.

Table 6. Irrigation needs, critical soil moisture (SMT) period and critical growth stages for irrigation in major vegetable crops

Crop	SMT (bars)	Available soil moisture (%)	Critical stage of growth for a must irrigation	Major defects caused by water deficit apart from low yield
Potato	-0.30	65	Stolon formation, tuberization and tuber enlargement.	Poor and mis-shaped tubers.
Tomato	-0.35	60	Flowering and fruit development and after each harvest.	Blossom end rot and fruit cracking.
Brinjal	-0.45	50	Flowering and fruit development and after each harvest.	Blossom end rot, misshapen fruit and poor fruit colour.
Chilli	-0.45	50	Tenth leaf to flowering and fruit development and after periodical harvests.	Low fruit set and shriveled fruits.
Sweet pepper	-0.35	60	Flowering and fruit development and after each harvest.	Poor fruit size and blossom end rot.
Cauliflower	-0.25	70	Curd formation and enlargement.	Ricey curd and buttoning.
Cabbage	-0.35	60	Head formation and enlargement.	Tough leaves.
Broccoli	-0.25	70	Head development.	Strong flavor and premature bolting.
Onion	-0.35	60	Bulb formation and bulb enlargement.	Small bulbs and early bolting.
Garlic	-0.35	60	Bulb formation and bulb enlargement.	Poor sized cloves.
Carrot, beet and turnip	-0.40	55	Root enlargement.	Growth cracks and misshapen roots.
Cucumber	-0.45	50	Flowering and fruit development.	Pointed and cracked fruits.
Pumpkin and all gourds	-0.50	55	Flower bud development and early fruit development.	Misshapen, small cracked fruits.
Watermelon	-0.60	40	Flowering and fruit development.	Blossom end rot.
Muskmelon	-0.60	40	Flowering and fruit development.	Poor fruit growth and quality.
Okra	-0.55	45	Flowering and pod development.	Toughness and fibre development in pods.
Peas	-0.60	40	Flowering and pod filling.	Poor pod filling and shriveled grains.

French bean and hyacinth bean	-0.45	50	Pod setting and pod enlargement.	Poor pod filling with stringy pods.
Cowpea	-0.45	50	Prior to flowering and after pod set.	Poor pod filling with fibrous pods.
Leafy vegetables	-0.30	70	Entire crop duration.	Toughness in leaves.
Sweet potato	-0.70	35	Tuber initiation, early bulking and late bulking.	Poor and misshapen tubers.

Source: Hazra *et al.* (2011)

In sprinkler irrigation, water under pressure is applied to soil surface in the form of a fine spray by a rotating head from above. Sprinklers are fitted evenly spaced along the main conduit pipe or hose which carries water from the main source. The sprinkler system can be stationary or portable. It is useful in case of sandy or sandy loam soils. This system has several advantages like high efficiency, uniform distribution of water, economy in water and labour, well controlled water application and suited to uneven fields and saline soils. It also helps in cooling the plants in summer and also saves plants from frost injury. Along with water, the soluble fertilizers, pesticides and weedicides can also be applied. However, high cost is still a limiting factor (Swarup, 2005).

There is another system quite common these days, known as drip irrigation. It is a low pressure, slow application direct to the root region of the plant and highly efficient system. The water carried through plastic (PVC) tubes is released through emitters by which the amount and rate of water applied to the plant can be regulated. The method is economical as it reduces water loss due to percolation and evaporation and its operational cost and water requirements are low. In this system, fertigation (application of fertilizers) can also be adopted. Both surface and sub-surface drip systems are used. It is well suited to arid regions and other areas with deficit water supply. It is expanding in Maharashtra, Karnataka, Gujarat and Tamil Nadu. The experimental results have shown that drip irrigation in tomato gave increased yield and reduced water use as compared to furrow irrigation. It has been found to better in saline soils also for tomato.

Weed management

Weeds cause considerable yield losses in all the crops and vegetable crops if not controlled. One-third of the total crop loss is estimated to be caused by weeds alone. Weeds are competitive and aggressive in nature. They are well adapted to adverse climatic conditions and thrive well even under drought and water logging. They have high reproductive capacity even under adverse conditions. They reproduce even asexually. They are persistent in nature due to high longevity, prolific and perennial nature, and seed dormancy. Herbicides are commonly used these days to manage weeds.

The first widely used herbicide was 2,4-dichlorophenoxyacetic acid (2,4-D) which was developed by a British team during World War II with first widespread use and production in late 1940s. It still remains one of the most commonly used herbicides in the world. However, it exhibits relatively poor selectivity and is also less effective against some broadleaf weeds. 1970s saw the introduction of atrazine, which has dubious distinction of being the herbicide of greatest concern for ground water contamination because it does not breakdown readily (within a few weeks) after being applied. Instead it is carried deep into the soil through rain water and causes ground water contamination. Glyphosate was introduced in late 1980s for non-selective weed control. It is now a major herbicide globally in selective weed control due to development of GM crops which are highly tolerant to this weedicides and only weeds are killed. Different herbicides have been classified as given in table 7.

Table 7. Classes of herbicides and their examples

Class of herbicide	Examples
Acetamides and analides	Alachlor, Acetochlor, Metalochlor, Propachlor, Propanil.
Carbamates and Thiocarbamates	Asulam, Terbucarb, Tiobencarb.
Chlorophenoxy herbicides	2,4-D, 2,4-DP, 2,4-DB, 2,4,5-T, MCPA, MCPB, MCPP, Dicamba.
Dipyridyls	Paraquat, Diquat.
Heavy metals	Lead arsenate, Arsenicals.
Nitrophenolic and Dinitrocresolic herbicides	Dinitrophenol, Dinitrocresol, Dinoseb, Dinosulfon.
Pentachlorophenol	Pentachlorophenol.
Phosphonates	Glyphosate, Glyfusinate, Fosamine ammonium.
Triazines	Atrazine, Simazine, Cyanazine, Propazine.
Urea derivatives	Diuron, Flumeturon, Linuron.

Source: Hazra *et al.* (2011)

Chemical weed management in vegetable crops is given in table 8.

Table 8. Chemical weed management in vegetable crops.

Crop	Weed control herbicide	Time of application
Tomato	Metribuzin @ 0.25-0.75 kg ai/ha	Post-emergence
	Butachlor or Alachlor @ 2.0 kg ai/ha	Post-emergence one week after transplanting
Brinjal	Pendimethalin 1.0 kg ai/ha or Fluchloralin 1.5 kg ai/ha or Butachlor 2.0 kg ai/ha + one hand weeding 30 days after transplanting	Pre-emergence
Chilli	Pendimethalin 1.25 kg ai/ha + one hoeing 40 days after transplanting	Pre-emergence
	Nitrofen @ 2.5 kg ai/ha or Oxidiazon @ 1.0 kg ai/ha	Pre-emergence
Cabbage and cauliflower	Pendimethalin @ 0. 56 kg ai/ha or Oxyflorfen @0.25 kg ai/ha or Fluchloralin @1.2 kg ai/ha or Nitrofen @ 2.0 kg ai/ha + one hand weeding 30 days aafter planting	One day before transplanting

Onion	Pendimethalin @0.75 kg ai/ha or Fluchloralin @ 1.5 kg ai/ha or Oxadiazon @ 1.0 kg ai/ha or Nitrofen @ 1.0 kg ai/ha + one handweeding 45 days after transplanting	Pre-planting
Carrot and radish	Tenoran @ 2.5 kg ai/ha or Alachlor @ 1.5 kg ai/ha	Pre-emergence, pre-planting One day after transplanting.
	Dalapon @ 1.0 kg ai/ha or Fluchloralin @ 1.5 kg ai/ha or Nitrofen @ 2.5 kg ai/ha	
Pea	Alachlor @ 0.75 kg ai/ha or Tribunil @ 1.5 kg ai/ha + one hand weeding 45 days after sowing	Pre-emergence
Cowpea	Alachlor @ 2.0 kg ai/ha	Pre-emergence
Okra	Pendimethalin @ 0.56-0.75 kg ai/ha or Alachlor @ 2.0 kg ai/ha or Fluchloralin @ 1.5 kg ai/ha + one hand weeding 45 days after sowing	Pre-emergence
Watermelon	Butachlor @ 2.0 kg ai/ha	Pre-sowing
Bottle gourd	Alachlor @ 2.5 kg ai/ha	Pre-sowing
Bitter gourd, pumpkin, ridge gourd	Alachlor @ 2.5 kg ai/ha	Pre-sowing

Source: Hazra *et al.* (2011)

Conclusion

The remarkable transformation from subsistence vegetable farming in sixties to the current modern commercial production has been possible through enactment of new seed policy in 1988 allowing vegetable seed import under open general license by a large number of private sector vegetable seed companies bringing seed of superior varieties and hybrids to the farmers for scientific cultivation of vegetables with lot of technological inputs by the stakeholders including modern production technologies. However, considering the demand to meet the requirement of 300 g vegetables/capita/day for our ever increasing population which at the moment is estimated to be 1.33 billion and projected to reach to 1.50 billion by 2050. We need to have 220 million tonnes of vegetable to serve at least 300 g vegetables/day/individual and this calls for raising vegetable productivity from current 18 tonnes/ha to 22 tonnes/ha and this change will call for increasing role of modern production technologies in cultivation of vegetable crops.

References

Hazra P., Chattopadhyay A., Karmakar A., Dutta S. 2011. Modern technology in vegetable production. New India Publishing Agency, New Delhi. pp413.

Hazra, P. and Som, M.G. 2005. Vegetable Science. Kalyani Publishers, Ludhiana. pp491.

Kapur A. 2016. Vegetable seed industry and its future prospects. *Seed Res.* **44** (2): 91-96.

Kapur, A. 2019. Vegetable seed industry creating avenue to enhance seed trade. Paper presented at Seed World 2019, World Seed and Technology Congress, 18-21 Sep. 2019, Bangalore, Indian Council of Food and Agriculture.

Kemble J.M. 2010. Vegetable Crop Handbook for Southern United States. GROWER, USA. pp 275.

Swarup V. 2005. Vegetable science and technology. Kalyani Publishers, Ludhiana. pp656.

9

Advances in Medicinal and Aromatic Crop Production Technologies

Neha Singh, Hemant Kumar Yadav and Sujit Kumar Yadav*

Introduction

The various kinds of plants present on earth have been used over the millennia for human welfare in the promotion of health and as drugs and fragrance materials. About 80% of the world's inhabitants rely mainly on herbal medicines for their primary health care, while medicinal plants continue to play an important role in the health care systems of the remaining 20%. The use of herbal medicine and medicinal plants in most developing countries, as a normative basis for the maintenance of good health, has been widely observed (UNESCO, 1996). According to WHO (1998), herbal medicines are finished, labelled medicinal products that contain as active ingredients, above ground or underground parts of plants or other plant materials, or combinations thereof, whether in the crude state or as plant preparations. The plant materials referred to above include juices, gums, resins, fatty oils, essential oils and any other substances of this nature. Herbal medicines may contain excipients in addition to the active ingredients (WHO, 1991). Remarkably, even today there is no real definition for this special group of plants that has been accompanying mankind throughout history. Most frequently, medicinal plants are defined as feral and/or cultivated plants that, based on tradition and literature records, can be directly or indirectly used for medical purposes. The basis of medicinal use is that these plants contain so called active ingredients (active principles or biologically active principles) that affect physiological (metabolic) processes of living organisms, including human beings.

The conception about aromatic plants is even less definite. The term aromatic indicates plants having an aroma; being fragrant or sweet-smelling, while the word aroma is supposed to imply also the taste of the material (*aromatic herbs)*. Since, large number of plants possess both medicinal and aroma properties

**Corresponding author email: h.yadav@nbri.res.in*

which leads to complexity and overlapping uses of active ingredients, making impossible to establish rigid categories or a practical classification for medicinal and aromatic plants. For example; Anise, dill, coriander, thyme, etc. are equally known as medicinal, spice and essential oil crops. Thus, frequently these plants are simply referred to as medicinal plants, disregarding their specific features. More recently, the term "Medicinal and Aromatic Plants" (MAPs) has been used in a slightly broader sense, distinguishing the fragrant (aromatic, ethereal) ingredients containing group of medicinal plants.

Classification of medicinal and aromatic plants

Medicinal and aromatic plants are generally classified on the basis of their growth habit. It may be tree, shrub, herb, annuals, biennial, tubers, rhizomes and climbers. The commonly used MAPs of different categories are listed below (Table 1).

Table 1. Commonly grown medicinal and aromatic plants and their uses

Common name	Botanical name	Parts used
Trees		
Babul	*Acacia niloticeDelite*	Pods, leaves, bark, gum
Bael	*Aegle marmelos L. Corr.*	Roots, leaves, fruit
Neerh	*Azaflirachta indica*	Bark leaves, flowers, seed, oil
Palas	*Butea monossperma* (Lam.)	Bark, leaves, flowers, seed, gum
Gugul	*Commiphora mukul Engh J*	Resinous gum
Olive	*Olea europeae*	Leaves, Oil
Arjun	*Terminalia arjuan* Roxb.	Bark
Behela	*Terminalia bellirica* Gaertu	Bark, fruit
Hirda	*Terminalia bellirica* Gaertu	Fruits
Nagakesar	*Mesua ferrea L.*	Blowers, oil
Markingnut	*Semecarpus& anacardium L.*	Fruits
Safed musli	*Aparagus adscendens* Roxb.	Tuberous roots
Belladonna	*Atropa belladonna*	Leaves and roots
Lavender	*Lavandula officinalis*	Flowers
Sarpagandha	*Rauvalfia serpentina L.*	Roots
Chitrak	*Plumbagezey lanica L.*	Leaves, roots
Herbs		
Brahmi	*Bacopa monnieri L.*	Whole plant
Am haldi	*Curcuma amada Roxb.*	Rhizomes
Haldi	*Curcuma domestica Valet*	Rhizomes
Datura	*Datura metel L.*	Leaves, flowers
Kalazira	*Nigella sativa L.*	Seed
Afim	*Papaver somniferum L.*	Latex, seed
Pipli	*Piper longum L.*	Fruits, roots
Babchi	*Psoralea corylifolia*	Seed, Fruit
Annuals		

Jangalimuli	*Blumeala cera*	Whole plant
Cockscomb	*Celosia cristala L.*	Inflorescence
Red poppy	*Papaver rhoeas*	Flowers
Bhui amla	*Phyllantius niruri*	Whole plant
Biennial		
Bankultthi	*Cassia abus L.*	Leaves, seeds
Caper spurge	*Euphorbia lathyrus*	Seed latex
Catchfly	*Melandrium firmum*	Whole plant
Tubers and rhizomes		
Satavar	*Asparagus adscendens Roxb*	Tubers
Safed musli	*Chlorophytum borivilianum*	Tubers
Puskarmul	*Inulara cemosa Hook*	Roots
Sakarkhand	*Manihot esculenta crantz*	Tubers
Biennial		
Chocloate vine	*Akebia quinata Deene*	Stem, fruit
Malkunki	*Celustrus paniculatus Wild*	Bark, leaves, seed
Hajodi	*Cissusquadr angularis L.*	Whole plant
Khira	*Cucumis sativus L.*	Fruit, seed
Gudmar	*Gymnema sylvestre Retzx*	Whole plant, leaves
Kali mirch	*Piper nigrum L.*	Fruit

Trends in area and production in India

India has been always a rich habitat of valuable medicinal and aromatic plant species. The Indian system of medicine uses over 1,100 medicinal plants and most of them are collected from forests regularly, and over 60 species among them are particularly in demand. More than 90% of medicinal plants used by the industries are collected from the wild. While about 800 species are used in production by industries, < 20 species of plants are produced by cultivation. The production of medicinal and aromatic plants in India has been more or less stagnant, fluctuating between 10 to 20 tonnes over the last three consecutive years (Table 1).The area under cultivation was also reduced with each progressive year but it does not affect the production. Total production of MAPs in 2012-2013 was 918.22 tonnes in 557.06 ha while in 2014-2015, the production has been raised to 925.81 tonnes with an area of 499.05 ha (Table 2). Madhya Pradesh, Tamil Nadu and Rajasthan are three major contributing states of total production in India. In the year of 2016-2017, the area and production of MAPs has been increased in substantial amount i.e. 630.02 ha and 988.00 million tonnes, respectively. These trends of increased area and production of MAPs is revealing their growing demand for medicinal purposes.

Table 2. State-wise area and production of medicinal and aromatic plants

(Area: 000' ha, Production: 000' mt)

States/ Union Territories	2012-2013		2013-2014		2014-2015		2016-2017	
	Area	Production	Area	Production	Area	Production	Area	Production
Andhra Pradesh	0.29	4.36	1.88	3.51	1.88	3.51	1.26	4.65
Arunachal Pradesh	5.15	109.18	5.15	109.18	5.15	109.18	0.46	0.99
Assam	0.00	0.00	4.35	0.16	4.39	0.16	4.49	0.17
Bihar	3.82	0.50	4.56	0.59	4.40	0.69	4.44	0.61
Chhattisgarh	10.27	60.28	8.44	50.25	8.99	53.53	8.56	60.40
Haryana	1.75	1.14	1.76	1.14	0.55	1.23	0.35	1.29
Himachal Pradesh	0.00	0.00	1.11	0.90	1.11	0.90	1.12	0.91
Karnataka	3.71	21.66	3.70	21.10	3.75	21.33	2.20	16.19
Madhya Pradesh	62.63	393.00	63.95	404.60	65.62	414.00	72.90	502.07
Mizoram	1.02	0.71	1.11	0.90	1.11	0.90	0.93	0.90
Odisha	1.92	0.64	1.92	0.64	1.92	0.64	1.92	0.61
Punjab	8.97	1.43	14.01	2.51	15.10	2.90	12.52	2.42
Rajasthan	308.68	164.53	231.24	124.30	234.53	125.11	370.00	190.00
Tamil Nadu	15.15	147.41	16.37	162.12	17.18	178.33	13.83	192.54
Uttar Pradesh	133.70	13.40	133.70	13.40	133.70	13.40	135.04	13.53
Total	557.06	918.22	493.25	895.30	499.40	925.81	630.02	988.00

Source: Horticulture statistics at a glance (Horticulture statistics division, Department of Agriculture, Cooperation and Farmer Welfare (DAC&FW), Ministry of Agriculture & Farmer Welfare, Government of India

Contribution of medicinal and aromatic plants in India

Plant based medicines played a pivotal role for health care in most developing countries since time immemorial (Bannerman, 1982). It has been estimated that the contribution of plant derived drugs is as much as 25% of the total drugs in developed countries, while this contribution is raised to 80% in developing countries. Thus, the economic importance of medicinal plants is much more in developing nation as compared to developed countries. Two third of the plants used in modern system of medicine are provided by these developing countries and the health care system of rural population depend on these indigenous systems of medicine.

Approximately 80,000 plants have medicinal properties out of 2,50,000 higher plant species on earth. India is one of the world's 12 biodiversity centres with the presence of over 45,000 different plant species. India's diversity is unique due to the presence of 16 different agro-climatic zones, 10 vegetation zones, 25 biotic provinces and 426 biomes (habitats of specific species) of these, about 15,000-20,000 plants have good medicinal value (Table 3). However, only 7,000-7,500 species are used for their medicinal values by traditional communities. In India, traditional systems of medicines such as *Unani* and *Ayurveda* are utilizing plant-based drugs since ancient times. These drugs are derived either from the whole plant or from different organs, like leaves, stem, bark, root, flower, seed etc. Some excretory plant product such as gum, resins and latex are also exploited to prepare drugs. Nowadays, several plant derived drugs form an important segment in Allopathic system of medicine. Some important chemical intermediates needed for manufacturing the modern drugs are also obtained from plants (e.g. diosgenin and solasodine etc).

Table 3. Number of plant species used medicinally worldwide

Country	Plant species	Medicinal plant species	%
China	26,092	4,941	18.9
India	15,000	3,000	20.0
Indonesia	22,500	1000	4.4
Malaysia	15,500	1,200	7.7
Nepal	6,973	700	10.0
Pakistan	4,950	300	6.1
Phillippines	8,931	850	9.5
Sri Lanka	3,314	550	16.6
Thailand	11,625	1,800	15.5
USA	21,641	2,564	11.8
Vietnam	10,500	1,800	15.5
Average	13,366	1,700	12.5
World	4,22,000	52.885	-

Source: Hamilton (2003)

MAPs derived speciality products

Medicinal and Aromatic plants are important source of different kinds of speciality products such as:

- Essential oils
- Pharmaceuticals
- Herbal health products
- Dyes and colorants
- Cosmetics, personal care products
- Plant protection products

Essential oils: Essential oils are concentrated, volatile, hydrophobic liquid (mixtures of chemicals) obtained from plants. The word 'essential' signifies the aromatic nature of the oils which captures the "essence" of the plant from which it was extracted. The oils usually consist of a complex mixture of tens to hundreds of low molecular weight terpenoids. Essential oils are generally extracted through steam distillation, while sometimes organic solvent extraction is also used. These oils possess characteristic flavour and fragrance properties along with other biological activities due to which they are used in many industries. The food industry uses them as flavourings (e.g. soft drinks, food and confectionary), the cosmetics industry uses them for their fragrance (e.g. perfumes, skin and hair care products), and the pharmaceutical industry uses them for functional properties (e.g. antimicrobial activity). Some essential oils are also used as insect repellents, and as detergents. About 90% of global essential oil production is consumed by the flavour and fragrance industries in the form of cosmetics, perfumes, soft drinks and food. The largest consumer of essential oils is the USA, followed by western European countries like France, Germany and the UK, and Japan (Holmes, 2005). Approximately 3,000 plants are used for their essential oils, with 300 of these being commonly traded on the global market (Centre for the promotion of imports from developing countries, 2009a).

Pharmaceuticals: In the conventional pharmaceutical industry, pharmaceutical companies produce medicines from compounds extracted from plant material, or use plant derived compounds as starting material to produce drugs (Houghton, 2001). For example anti-cancer alkaloid "paclitaxel" extracted from Pacific yew (*Taxus brevifolia*), vinblastine from the Madagascar periwinkle (*Cataranthus roseus*) and digoxin from the foxglove (*Digitalis lanata*). Plant derived compounds play an important role in the production of "single chemical entity" medicines. More than 25% of the pharmaceutical drugs used in the world today are derived from plant natural products (Farnsworth, 1979; Schmidt *et al.*, 2008).

Table 4. Lists of some important drugs obtained from plant extracts

Compound	Chemical class	Source	Therapeutic use
Morphine, codeine	Alkaloids	*Papaver somniferum*	Analgesic, antitussive
Digoxin	Steroidal	*Digitalis lanata*	Heart disorders
Atropine	Tropane alkaloids	*Atropa belladonna*, *Hyoscyamus* spp.	Antispasmodic, pupil dilator
Scopolamine	Tropane alkaloids	*Atropa belladonna*, *Hyoscyamus* spp.	Sedative, against motion sickness
Paclitaxel	Diterpene alkaloid	*Taxus* spp.	Anticancer (ovarian and others)
Quinine	Quinoline alkaloid	*Cinchona* spp.	Antimalaria
Vinblastine, vincristine	Bis-indole alkaloid	*Catharanthus roseus*	Anticancer (leukemia)
Camptothecin	Indole alkaloid	*Camptotheca acuminata*	Antineoplastic
Galanthamine	Isoquinoline alkaloid	Various members of Amaryllidaceae	Against mild Alzheimer's disease, Dementia
Artemisinin	Sesquiterpene lactone	*Artemisia annua*	Antimalaria
Podophyllotoxin.	Lignan	*Podophyllum* spp.	Antiviral, antineoplastic
Pilocarpine	Imidazole alkaloid	*Pilocarpus* spp.	Glaucoma, xerostomia

Herbal health products: The MAPs derived medicines in the form of extracts, teas, tinctures or capsules are referred to as phyto-pharmaceuticals, phyto-medicines or herbal medicines in Europe and botanical drugs in the United States. In many European countries there is not much difference between pharmaceutical and phyto-pharmaceutical companies. Medicines are produced to high standards and must undergo clinical evaluation of safety and efficacy (Schmidt *et al.*, 2008). Phyto-pharmaceutical medicines are standardized in terms of the active constituents. Examples of popular phyto-pharmaceuticals are *Gingko biloba* extract to improve cognitive function (O'Hara *et al.*, 1998), St. John's Wort for treatment of mild depression (Bilia *et al.*, 2002), ginseng as general tonic and cognitive enhancer (Attele *et al.*, 1999), ginger against nausea and vomiting (Ernst and Pittler, 2000) and saw palmetto for the treatment of symptomatic benign prostatic hyperplasia (Kaplan, 2005). Some more examples of popular herbal medicines are shown in table 5.

Table 5. Commonly used herbal health products and their uses

Plant	Active compounds	Source	Therapeutic use
Echinacea	Alkylamides and other compounds	*Echinacea purpurea*	Immune modulator
Ginseng	Ginsenosides, eleutherosides	*Panax ginseng*	Fatigue, stress
Saw palmetto	Phytosterols, fatty acids	*Serenoa repens*	Benign prostatic hyperplasia
Ginkgo	Ginkgolides (terpenetrilactones), flavonol glycosides	*Ginkgo biloba*	Mental fatigue, cognitive decline
St. John's wort	Hyperforins	*Hypericum perforatum*	Mild depression
Valerian	Iridoid glycosides, terpenoids, valerianic and isovaleric acid	*Valeriana officinalis*	Anxiolytic, sleep improvement
Garlic	Allicin	*Allium sativum*	High cholesterol, hypertension, respiratory infections
Feverfew parthenium	Sequiterpene lactones	*Tanacetum parthenium*	Migraines, inflammation

Nutraceuticals/functional foods: Nowadays, a plant-based health products, nutraceuticals or dietary supplements are gaining popularity. There are no universally accepted definitions for functional food and nutraceuticals, but generally they are described as foods (fortified with added ingredients or not) with health benefits beyond basic nutrition (Wildman and Kelley, 2007). Sometimes the plant materials used to prepare health products are similar to those in the phyto-pharmaceutical industry, but products are marketed as nutraceuticals to avoid the costly and time-consuming process of licensing a medicinal product (Kuipers, 1997). Often limited health claims are made about a food component when clear evidence for its role in reduced risk of a disease or health benefit is emerging, but not strong enough to meet standards set by regulatory authorities such as the Food and Drug Administration (Wildman and Kelley, 2007). Garlic and ginseng extracts for example have been developed as nutraceuticals. Other examples include the use of red grape constituents for its antioxidant properties, lycopene from tomatoes, and broccoli for its cancer preventative properties.

Dyes and colorants: Compounds extracted from plants can be used as natural colorants or dyes. Plants were traditionally sources of natural dyes used to colour textiles. This was replaced by synthetically produced dyes in the 19th century. Recently, an increasing public awareness of the detrimental environmental impact of synthetic dye production, together with a need for sustainable sources of dyes has led to natural products becoming popular again. Since, the 1990s much research into the re-introduction of natural dyes has been initiated (Bechtold *et al.*, 2003; Gilbert and Cooke, 2001). The production of natural dyes from waste materials of the timber, food and other agricultural industries is of particular interest (Bechtold *et al.*, 2007a,b). Natural dyes are used to colour natural fibres such as wool, cotton, linen and hemp. They are also increasingly used in other industries, for example in paints and varnishes, cosmetics, food, in the eco-building industry, and in painting restoration (Centre for the promotion of imports from developing countries, 2009c). Natural dyes of various chemical classes can be obtained from organs of specific plants (Table 6).

Table 6. Plants containing natural dye/colorant compounds.

Common name	Source plant	Dye/colorant chemical class	Colour
Marigold	*Tagetes patula*	Flavonoids	Yellow
Walnut	*Juglans regia*	Naphthoquinone	Brown
Henna	*Lawsonia inermis*	Naphthoquinone	Red
Woad	*Isatis tinctoria*	Alkaloid	Blue (indigo)
Dyer's knot weed	*Polygonum tinctorium*	Alkaloid	Blue (indigo)
Madder	*Rubia tinctorum*	Anthraquinone	Red-brown
Barberry	*Berberis vulgaris*	Alkaloid	Yellow-brow
Goldenrod	*Solidago* spp.	Flavonoid	Yellow-olive
Hollyhock	*Alcearosea*	Anthocyanin	Brown-green
Privet	*Ligustrum vulgare*	Anthocyanin	Blue-green
Ash tree	*Fraxinus excelsior*	Flavonoid	Beige-black

Cosmetics: A cosmetic product is defined as "a substance or preparation intended for application to any external surface of the human body, teeth or mouth for the purpose of cleaning, perfuming or protecting them, keeping them in good condition or changing their appearance" (Dweck, 1996). Cosmetic companies produce different kinds of products; beauty and personal care, hair care, perfume and fragrances, and the recently growing group of cosmaceuticals (products that contain one or more bioactive compound and are intended to enhance health and beauty) (Centre for the promotion of imports from developing countries, 2008b). Some important plant-derived ingredients used in the manufacture of cosmetics are oils, fats and waxes, essential oils and oleoresins, plant extracts and colorants. Vegetable fats and oils are composed of triglycerides, and are usually extracted from the seeds of oilseed plants. These substances are referred to as oils or fats depending on whether they are liquid or solid at room temperature. Solid fats with a brittle texture are also referred to as waxes (Centre for the promotion of imports from developing countries, 2008a). These ingredients have numerous roles in the final cosmetic products, such as a fragrance or colouring function, moisturizers, thickening agents and stabilizers. Many plant extracts are used in cosmetics for their functional properties, such as free-radical scavenging, sun protection, whiteners and anti-microbial effects (Aburjai and Natsheh, 2003).

Plant protection products: There is a long history of use of plants as sources of plant protection products against insects, microbes and weeds. Records of plant-derived insecticides date back more than 2000 years ago in China, India, Egypt and Greece (Isman, 2006). In the middle of the 20th century synthetic chemical pesticides became dominant. The potent and fast action of the synthetics led to great increases in yields of crops in many parts of the world. However, with these improvements came unwanted effects such as destruction of non-target organisms, disruption of pollination, groundwater contamination, chronic and acute human poisoning and resistance in pest populations. In recent years concerns over these problems have resulted in the banning or restriction of dangerous compounds, and a move towards the use of less harmful ones. Plants are an attractive source since their metabolism has adapted to deal with pests in their environment. Some compounds or extracts can be used directly, and others serve as precursors for the production of agents used as weed, insect or plant pathogen protectants (Copping and Duke, 2007; Dayan *et al*., 2009). Various products derived from plants are currently being used to control insect pests, microbial pathogens and weeds. The best-known ones are the pyrethrins extracted from *Tanacetum cinerariaefolium*, rotenone isolated from *Derris lanchocarpus* roots and *Azadirachtins* isolated from seeds of the Indian neem tree (*Azadirachta indica*) (Guleria and Tiku, 2009). The essential oils of some well-known aromatic plants have also been applied to plant protection, and can be found in commercial preparations (Isman, 2006).

Challenges in MAPs production

The continuous increase in human population is one of the causes for concern in meeting the daily requirements of food and medicine as the economy and livelihoods of human societies living in developing countries primarily depend on forest products. This phenomenon is leading to continuous erosion of forest and the forest products (Samal *et al.*, 2004), thus making challenge to meet the requirements as well as to conserve useful bio-resources. The market prices for medicinal plants and derived materials provide only a limited insight into the workings of the market, and not on the precise information of profits, supply and demand. The Major feature and challenges present in medicinal and aromatic plant production are as following:

Rising demand

As per the WHO estimation, the demand for medicinal plant-based raw materials is growing at the rate of 15 to 25% annually the demand for medicinal plants in terms of values it may reached about 5.0 trillion USD in 2050. According to an estimate of NMPB (National Medicinal Plant Board), the medicinal plant related trade is approximately 1,34,000 million tonnes (2014-2015) in India. The projected escalating demand of medicinal plants has led to the over-harvesting of many plants from wild, which subsequently results in the loss of their existing populations. For example, the large quantity of Himalayan yew (*Taxus baccata*) has been gathered from the wild. *Aconitum heterophyllum, Nardostachys grandiflora, Dactylorhiza hatagirea, Polygonatum verticillatum, Gloriosa superba, Arnebia benthamii* and *Megacarpoea polyandra* are other examples of north Indian medicinal plant species which have been overexploited for therapeutic uses and have subsequently been placed today in rare and endangered categories. Many medicinal plant species are used in curing more than one disease (Kala *et al.*, 2004, Kala *et al.*, 2005) and as a result, these species are under pressure due to over collection from wild. For example, *Hemidesmus indicus* is used to cure 34 types of diseases; *Aegle marmelos* 31, *Phyllanthus emblica* 29 and *Gloriosa superba* 28. Over-exploitation and continuous depletion of medicinal plants have not only affected their supply and loss of genetic diversity but have seriously affected the livelihoods of indigenous people living in the forest margins (Rao *et al.*, 2004). More than 95% of the 400 plant species used in preparing medicine by various industries are harvested from wild populations in India (Uniyal *et al.*, 2000).

Limitation factors

1. Harvesting medicinal plants for commercial use.
2. Destructive harvest of underground parts of slow reproducing
3. Slow growing and habitat-specific species, are the crucial factors in meeting the goal of sustainability (Kala, 2005, Ghimire *et al.*, 2005).
4. Harvesting shoots and leaves of medicinal plants may decline their photosynthetic capacity, and as well as the potential for survival and effective propagation.
5. Medicinal plants tolerance to harvest varies with climatic conditions as the temperate herbs become highly vulnerable to harvest of individuals (Ticktin, 2005).

Furthermore, rising demand with shrinking habitats may lead to the local extinction of many medicinal plant species.

Increasing rarity

The continuous exploitation of several medicinal plant species from the wild and substantial loss of their habitats during past 15 years have resulted in population decline of many high value medicinal plant species over the years. The primary threats to MAPs are those that affect any kind of biodiversity used by mankind (Rao *et al.*, 2004; Sundriyal and Sharma, 1995). The weakening of customary laws, which have regulated the use of natural resources, is among the causes of threatening the medicinal plant species (Ghimire *et al.*, 2005; Kala, 2005). These customary laws have often proved to be easily diluted by modern socio-economic forces (KIT, 2003). There are many other potential causes of rarity in medicinal plant species, such as habitat specificity, narrow range of distribution, land use disturbances, introduction of non-natives, habitat alteration, climatic changes, heavy livestock grazing, explosion of human population, fragmentation and degradation of population, population bottleneck, and genetic drift (Kala, 2000; Weekley and Race, 2001). Additionally, natural enemies (i.e., pathogens, herbivores, and seed predators) could substantially limit the abundance of rare medicinal plant species in any given area (Bevil *et al.*, 1999; Dhyani and Kala, 2004)

Cultivation of MAPs

Information on the propagation of medicinal plants is available for <10% and agro-technology is available only for about 1% of the total known plants globally (Lozoya, 1994; Khan and Khanum, 2000). This trend shows that developing agro-technology should be one of the thrust areas for research. Furthermore, in order to meet the escalating demand of medicinal plants, farming of these plant species is imperative. Apart from meeting the present demand, farming may conserve the wild genetic diversity of medicinal plants. Farming permits the production of uniform material, from which standardized products can be consistently obtained. Cultivation also permits better species identification, improved quality control, and increased prospects for genetic improvements. Selection of planting material for large-scale farming is also an important task. The planting material therefore should be of good quality, rich in active ingredients, pest- and disease-resistant and environmental tolerant. For the large-scale farming, one has to find out whether monoculture is the right way to cultivate all medicinal plants or one has to promote poly culture model for better production of medicinal plants.

Technologies for improving MAPs production

Most of the medicinal plants, even today, are collected from wild. The continued commercial exploitation of these plants has resulted in receding the population of many species in their natural habitat. Vacuum is likely to occur in the supply of raw plant materials that are used extensively by the pharmaceutical industry as well as the traditional practitioners. Consequently, cultivation of these plants is urgently needed to ensure their availability to the industry as well as to people associated with traditional system of medicine. In-situ conservation of these resources alone cannot meet the ever increasing demand of pharmaceutical industry. Its, therefore, inevitable to develop cultural practices and propagate these plants in suitable agro-climatic regions. Commercial cultivation will put a check on the continued exploitation from wild sources and serve as an effective means to conserve the rare floristic wealth and genetic diversity.

It is necessary to initiate systematic cultivation of medicinal plants in order to conserve biodiversity and protect endangered species. In the pharmaceutical industry, where the active medicinal principle cannot be synthesised economically, the product must beobtained from the cultivation of plants. Systematic conservation and largescale cultivation of the concerned medicinal plants are thus of great importance. Efforts are also required to suggest appropriate cropping patterns for the incorporation of these plants into the conventional agricultural and forestry cropping systems.

Location specific cultivation

Till today, > 80% of the plant material used in pharma and aromatic industries is presently collected from non-descriptive, uncharacterized land races from forest and other wild resources. More often than not, such collections are usually made using destructive harvests from the wild stands that pose a serious threat to country's biodiversity. This is neither sustainable nor acceptable in a globally competitive scenario. In order to arrest extinction and rapid decline in biodiversity of MAPs and to maintain a sustained supply of quality raw material for the production of herbal formulations, there is a pressing need for a change in the shift of our mind-set from "collection to cultivation" of MAPs. There are at least 35 major medicinal plants that can be cultivated in different agro-climatic zones of the country and there is an established global demand for their raw produce and high value bioactive principles. The cultivation of MAPs on marginal and degraded lands (saline/alkaline/waterlogged and ravines) is a highly promising approach for additional income generation and land quality improvement simultaneously. At present about a significant area in the country is either left barren or unculturable/ culturable waste land due to different aberrant like soil and water erosion and saline soils which can be used for cultivation of medicinal and aromatic plants (Table 7). Cultivation of MAPs in marginal and degraded lands can not only improve the economic condition of small and marginal farmers owing to higher price and/or higher volume of their main and by products but also conserve the most important natural resources and sustainable utilization of marginal lands.

Table 7. Prioritized medicinal plants and their suitability to different regions

Regions	Medicinal plants
Tropical	*Adhatoda vesica, Andrographis paniculata, Aqui laria agallocha, Centella asiatica, Dioscorea alata, Ocimum basilicum, Plantago erosa, Withania somnifera, Vitex negundo, Paedaria foetida, Cryptolepis buchanani, Acorus calamus.*
Subtropical	*Mucuna pruriens, Hydnocarpu skurzii, Lavendula vera, Litsea cubeba, Mucuna pruriens, Pogostemon calelin, Zanthoxylum armatum, Curcuma caesia, Leonotis nepetaefolia*
Temperate	*Coptis teeta, Geranium nepalensis, Panax pseudoginseng, Swerlia chirata, Picrorrhiza kurooa, Satyrium nepalensis, Rubia cordifolia, Taxus baccata, Orchis latifolia.*
Alpine	*Aconitum ferox, A. heterophy lum, Illicium griffithii, Berberis* spp., *Podophyllum hexandrum, Rheum emodi, Delphinium subulatum.*

Improved varieties

The quality of herbal products is dependent on quality of raw material used in the preparation. Several varieties rich in quality and yield have been developed in more than 35 MAPs in our country. The high yielding varieties of some MAPs released from CSIR-CIMAP are given in table 8.

Table 8. Improved varieties of MAPs developed by CSIR-CIMAP

Botanical name	Common name	Variety
Acorus calamus	Sweet flag	CIM-Balya
Aloe vera	Aloe	Sheetal
Andrographis paniculata	Kalmegh	Megha
Artemisia annua	Wormwood	CIM-Arogya, Jeevanraksha

Asparagus racemosus	Satavari	Shakti
Bacopa monnieri	Bramhi	Jagriti, Pragyashakti
Cassia senna angustifolia	Senna	CIM-Sona
Chamomila recutita	Chamomile	CIM-Sammohak, Prashant, Vallary
Chlorophytum borivilianum	Safed musli	CIM-Oj
Chrysnthemum cinerariaefolium	Pyrethrum	Avadh
Curcuma longa	Turmeric	CIM-Pitamber
Cymbopogon flexuosus	Lemon grass	Pragti, Krishna, Praman, Chirharit, Nima, Suwarna, Cauvery, T-1, GRL-1
Cymbopogon martini	Palmorosa	PRC-1, Trishna, Tripta, CIM-Harsh, Vaisnavi
Cymbopogon winterianus	Citronella	CIM-Jeeva, CIM-Bio13, Manjusha, Manjari, Mandagini, Jalpallavi
Foeniculum vulgare	Fennel	CIM-Sujal
Hyoscyamus niger	Black Henbene	Aela, Aekala
Lippia alba	Bushy matgrass	Kavach
Mentha arvensis	Menthol mint	Kranti, Kushal, Kalka, Kosi, Himalaya, Gomti, CIM Saryu, Saksham, Sambhav, Damroo, MAS-1
Mentha citrata	Bergot mint	Kiran
Mentha piperita	Peppermint	CIM-Indus, Madhuras, Pranjal, Tushar, Kukrail
Mentha spicata	Spearmint	Neerkalka, Neera, Arka, MSS-5
Mucuna pruriens	Velvet bean	CIM-Ajar
Ocimum africanum	Lemon-scented basil	CIM-Jyoti
Ocimum basilicum	French/Indian basil	Kushmohak, Vikarsudha, Saumya, Sharada
Ocimum tenuifiorum	Holy/sacred basil	CIM-Ayu, CIM-Kanchan, CIM-Angana
Papaver somniferum	Opium poppy	Ajay, Sampada, Rakshit, Sanchita, Sapna, Shyama, Shweta, Sujata, Shubhra, Vivek
Pelargonium graveolens	Geranium	CIM-BIO 171, CIM-Pawan
Phyllanthus amarus	Carry me seed	Jeevan, Navyakrit, Kayakirti
Plantago ovata	Psyllium	Mayuri, Nimisha, Niharika
Pogestemon patchouli	Patchouli	CIM-Samarth
Rauvolfia serpentina	Sarpagandha	RS-1, CIM-Sheel
Rosa damascena	Damask Rose	Noorjahan, Ranisahiba
Rosmarinus officinalis	Rosemary	CIM-Hariyali
Salvia sclarea	Clarysage	CIM-Chandni
Silybum marianum	Milk Thistle	CIM-Liv, CIMAP-Sil 9
Stevia rebaudiana	Sweet herb	CIM-Mithi, CIM-Madhu
Tagetesminuta	Little marigold	Vanphool

Vetiveria zizanioides	Vetiver/Khus	CIM- Samriddhi, KS-1, KS-2, Dharani, Gulabi, Kesari, CIM-vridhi, CIM-Khus15, CIM-Khus22
Withania somnifera	Ashwagandha	Poshita, Rakshita, NMITLI-118, Chetak, Pratap

Source: Improved varieties portal of MAP's: Director; CSIR-CIMAP's contribution

Alternative cropping system

Integration of MAPs in the existing cropping system is most important and promising intervention for increasing income of small and marginal farmers. South Asian countries including India have a tradition practicing of mixed farming system that may include medicinal and aromatic plants. MAPs have ease of their incorporation in the exiting cropping system due to availability of a large number of species and choice of plant types such as tree, shrubs, herbs, vines and their suitability to grow in different agro-physical conditions offer great opportunities and advantages (Table 9).

Table 9. Promising medicinal and aromatic plants (MAPs) suitable for integration in prevailing cropping system

Kharif season crop	*Rabi* season crop	*Zaid* season crop
Ocimum basilicum, O. sanctum, Andrographis paniculata, Cassia angusfifolia, Abelmoschus moschatus, Withania somnifera	*Withania somnifera, Papaver somniferum, Plantago ovata, Chamomilla recutita, Aloe vera, Silybum marianum, Lepedium sativum*	*Mentha arvensis, Pelargonium spp., Aloe vera, Artemisia annua*

MAPs under agro-forestry

Medicinal and aromatic plants are generally grown under forest cover and are shade tolerant; therefore agro-forestry plays an important rolein promoting their cultivation and conservation. Several approaches are feasible: integrating shade tolerant MAPs as lower strata species in multi strata systems; cultivating short cycle MAPs as intercrops in existing stands of plantation tree-crops and new forest plantations; growing medicinal trees as shade providers, boundary markers, and on soil conservation structures; inter planting MAPs with food crops; involving them in social forestry programs; and so on.

Hi-tech cultivation

Though economic importance of medicinal plants is well known, it is considered as a forestry sub-sector (non-timber forest products) in India. Recognizing and addressing the needs of each of the stakeholders involved requires a holistic approach for overall development of the medicinal and aromatic plants sector. Several studies have clearly brought out the economic potential of medicinal plants in different agro-climatic conditions. The potential return to the farmers from cultivation of medicinal plants is reported to be quite high in case of certain high altitude Himalayan herbs. The success story on cultivation of opium poppy in Rajasthan and Madhya Pradesh, isabgol cultivation in Gujarat, Rajasthan and Madhya Pradesh, senna cultivation in Rajasthan, mentha cultivation in Uttar Pradesh, palmarosa cultivation in central India, jasmine cultivation in south India and saffron cultivation in Kashmir are well known. However, more research is needed for proper planning for cultivation and utilization of medicinal plants with modem science and technological interventions keeping in view their ecological and aesthetic values. The following science led hi-technological interventions are suggested to make the medicinal and aromatic plants more productive and profitable enterprise.

Quality seeds and planting material

Planting material plays an important role in the production of any crop. Inadequate availability of quality planting material is one of the important deterring factors in development of a sound MAP industry. Farmers do not have access to certified disease free material as a result of which production, productivity and quality of the produce suffers. The seed producer should follow all required practices and parameters for quality seed production. Several varieties of these crops which can be multiplied through seed chain system and supplied to the farmers have been developed.

Nursery management

Nursery management for production of disease and pest free quality planting material of several vegetatively propagated crops like bhrami, mentha, lemon grass, palmarosa, jasmine etc are very important. There is also a ready market in nearby towns, cities, arogyamelas, other social fairs and religious gatherings for medicinal plants due to awareness on use of household herbal products. Hi-tech nurseries with controlled climate can be employed for the production and supply of high-quality disease-free planting material of medicinal and aromatic plants.

Harvesting and post-harvest management

The MAP are used as raw materials where different plant parts (root, root bark, stem, stem bark, leaf, flower, fruit, seed and whole plant and the combination of any of the parts) are used as drug and also the extracts and secondary metabolites are used in various formulations. Therefore, the quality of the produce depends on the harvest at particular physiological stage of the plant.

This optimum stage of harvest may differ with special to species to species and place to place depending on prevailing environment. This is the stage where the biological yield and chemical content are optimum and this decides the quality as well profitability of the crop. Therefore, it is essential to harvest the crop at optimum stage of harvest. Post-harvest losses occur at each and every step of the supply chain. The losses at the farm level occur due to improper harvesting methods, handling techniques and aggregation of the produce and then in transportation. At the wholesale level, the major losses occur due to rough handling and inappropriate storage. Value addition which in simple terms denotes to make things valuable or important in terms of economic gain, time and money saving in preparation, quantity and quality improvement or modification of raw ingredients for specific desirable characteristics is also assuming lot of importance. Though the production of MAP requires full care at each stage of production but the post-harvest management, monitoring the shelf-life and need based value addition is more crucial and requires full attention. Removal of excess moisture from the produce through proper methods of sun drying and shade drying is very essential. It is advisable to use the solar driers and other mechanical driers as per the requirement of the produce to ensure the required moisture and colour and to fetch premium prices.

Efficient extraction

Extraction, as the term is used pharmaceutically, involves the separation of medicinally active portions of plant or animal tissues from the inactive or inert components by using selective solvents in standard extraction procedures. The products so obtained from plants are relatively impure liquids, semi-solids or powders intended only for oral or external use. These include classes of preparations known as decoctions, infusions, fluid extracts, tinctures, pilular (semi-solid) extracts and powdered extracts. Such preparations popularly have been called galenicals, named after Galen, the second century Greek physician. The purposes of standardized extraction procedures for crude drugs are to attain the therapeutically desired portion and to eliminate the inert material by treatment with a selective solvent known as. The extract thus obtained may be ready for use as a medicinal agent in the form of tinctures and fluid extracts, it may be further processed to be incorporated in any dosage form such as tablets or capsules, or it may be fractionated to isolate individual chemical entities such as ajmalicine, hyoscine and vincristine, which are modem drugs. Thus, standardization of extraction procedures contributes significantly to the final quality of the herbal drug.

Production technology of some important MAPs

Mints (*Mentha spp.*): Mints are aromatic perennial herbs with quadrangular stem and bearing leaves with essential oil present in glands located in the subcuticular region. Among the various types of mints, only *Japanese mint* is cultivated in the tropics or subtropics with a cooler climate. It is generally cultivated as a primary source of menthol, which is widely used for flavouring toothpastes, candies, beverages, confectionery, chewing gums, pan parag, and mouth washes and for scenting shaving creams, tobacco, cigarettes, aerosols, polishes, hair lotions and lipsticks. It is employed as a soothing ingredient in cosmetic preparations, colognes, deodorants, aftershave lotions and perfume bases. It is also employed in a number of medicinal preparations like ointments, pain balms, cough syrups, cough lozenges and tablets.

Agro-techniques: Mints grow well over a wide range of climatic conditions. Japanese mint grows well under subtropical conditions while others prefer temperate climate. Adequate and regular rainfall during the growing period and good sunshine during harvesting are ideal for its cultivation. Medium deep soil rich in humus is best suited for the cultivation of Japanese mint. The soil should have a pH range of 6-7.5 with good water holding capacity but waterlogging is detrimental. Japanese mint is propagated through stolons. 'CIMAP/MAS-1' and 'Hybrid-77' are improved varieties. Seed rate is 400 kg/ha. A hectare of well-established mint provides enough planting materials for 10 ha. Stolons are planted either on flat land or ridges. In plains, they are planted in shallow furrows of 7-10 cm deep at a spacing of 45-60 cm after incorporating compost or farm yard manure at 10-12 tonnes/ha. Inorganic fertilizers upto 160 kg N and 50 kg each of P_2O_5 and K_2O/ha are applied; nitrogen being applied in 2-3 split doses. Irrigation enhances growth and improves the yield. The field should be kept weed free, particularly during the initial stages of growth till proper establishment and coverage of the ground area (Singh and Singh, 1989). Termite attack observed during the dry months can be controlled by the soil application of 3% heptachlor at 50 kg/ha. Hairy caterpillars cause rapid defoliation. Cut worms, semi-loopers and red pumpkin beetle also attack the crop. These insect pests can be controlled by 5% DDVP, 2% methyl parathion dust or any other suitable insecticide. Nematode attack has also been reported for which application of Fenamiphos at 10-12 kg/ha is effective. Mentha rust is caused by *Puccinia menthae* which results in severe leaf shedding. Powdery mildew caused by *Erysiphe cichoracearum* can be controlled by wet table sulphur application. *Macrophomina phaseoli* and *Thielavia basicola* cause stolon rot which is effectively controlled by the application of maneb or PCNB.

Post-harvest technology: The harvested herb may be wilted in shade for a few hours for draining off the excess moisture thereby reducing the bulk. Both fresh and dry herb are employed for distillation. Steam distillation is usually preferred and the duration of distillation is 1.5-2.0 hours generally. Fresh herb contains 0.4 to 0.6% oil. On an average, 100-150 kg oil/ha is obtained annually.

Ocimums (*Ocimum* spp.): *Ocimums* are an important group of aromatic and medicinal plants which yield many essential oils and aroma chemicals and find diverse uses in the perfumery and cosmetic industries as well as in indigenous systems of medicine. *Ocimum* species with oil rich in camphor, citral, geraniol, linalool, linalyl acetatemethyl chavicol, eugenol and thymol are important and can be harnessed for successful utilization by the industry. Among the various *Ocimum* species *Ocimum basilicum* L. is commercially and extensively cultivated for essential oil production. Its oil is employed for flavouring of food stuffs, confectionery, condiments and in toiletry products such as mouth washes and dental creams. It is also used in the flavouring of foods such as spiced meats, sausages, tomato pastes, various kinds of sauces, fancy vinegars, pickles, ketchups and beverages. In the perfumery industry, the oil is used for compounding certain popular perfumes notably jasmine blends. It is recognized as a febrifuge and antimalarial plant. The juice obtained from the leaves gives relief to irritation of throat ear ache and ringworm infections. Seeds are used internally for the treatment of constipation and piles.

Agro-techniques: The crop comes up well under tropical climate up to an altitude of 1,800 m. The growth is poor in areas which receive heavy and continuous rainfall. Frost is harmful to the plant and hence frost prone areas are to be avoided. Basil can be cultivated on a wide variety of soils, though moderately fertile well drained loamy or sandy loam soils are considered ideal for its cultivation. Basil is tolerant to higher concentration of copper and zinc but is susceptible to cobalt and nickel. The plant is propagated through seeds. Seedlings are first raised in the nursery and then transplanted in the field. The seed rate is about 125 g/ha for transplanting. Seeds start germinating 3 days after sowing and germination is over in 7-10 days. When 6-10 cm tall the seedlings are transplanted in the field at 40-60 cm spacing in rows. At the time of planting, 10-15 tonnes of compost or farm yard manure is to be applied. A medium fertilizer dose of 40:40:40 kg/ha of N, P_2O_5 and K_2O is recommended for economic yield though good response has been received up to 120:100:100 kg/ha. Irrigation is required once a week when it is raised as a summer crop. The field should be kept weed free for the first 20-25 days, till the crop canopy completely covers the ground. Weeding is usually carried out once or twice.

A number of diseases are reported in basil crop. *Corynespora cassicola* (Berk. and Curt.) Wie.causes leaf spot disease which appears as small water-soaked spots turning brown. *Elsinoearxii* sp. nov. causes scab disease. The symptoms are little defoliation with pluckering, clipping of the leaves and distortion of the tender twigs. Blight caused by *Alternaria species* and *Colletotrichum capsici* (Sy.) Butler & Bisby can be controlled by spraying 0.2% zineb or maneb. Wilt is caused by *Fusarium oxysporum* at all stages of growth. But the attack is more pronounced in rainy season. This is controlled by dipping the seedlings in a solution of organo mercurial fungicide. Basil is harvested when the plant is in full bloom and lower leaves start turning yellowish. The crop comes to full bloom 9-12 weeks after planting. For high quality oil, only the flowering tops are harvested. 4-5 crops are obtained per year. In some areas it is possible to get four floral harvests. The first harvest is done when the plants are in full bloom and the subsequent ones after every 15-20 days. The last harvest comprises the whole plant. Floral harvests yield 3-4 tonnes of flowers and the final harvest of the whole plant is 13-15 tonnes of herb per ha. While harvesting the whole herb, plants are cut not less than 15cm from the ground for enabling regeneration of the crop.

Agro-techniques: The essential oil in young inflorescence or the whole plant is extracted by hydro distillation or steam distillation. Corresponding to the part employed, two grades of oil are obtained, i.e. *flower oil and herb oil*. The flower oil has a superior note and is more expensive. Steam distillation is preferred for large plantations it takes less time and gives better recovery of oil, while hydro distillation carried on in a direct fire still is cheaper and handier for small plantations. Distillation is carried out for 1-1.5 hours. The young inflorescence contains 0.3-0.5% oil and the whole herb 0.10-0.25%. Generally, a yield of 30-40 kg of flower oil and 20-25 kg whole plant oil is obtained per hectare.

Patchouli (*Pogostemon patchouli*): Patchouli is an erect, branched, pubescent aromatic herb, the essential oil of which is one of the best fixatives for heavy perfumes which imparts strength, character, alluring notes and lasting qualities. In fact, it is a perfume by itself and is highly valued in perfumes, soaps, cosmetics and flavour industries. The oil is extensively used as a flavour ingredient in major food products, including alcoholic and non-alcoholic beverages, frozen dairy desserts, candy, packed foods, gelatin, meat and meat products. It blends well with the oils of sandal wood, geranium, vetiver, cedar wood, clove, lavender, bergamot and many others. The oil gives one of the finest attars when blended with sandal wood oil. Tenacity of odour is one of the great virtues of patchouli oil and is one of the reasons for its versatile use. The oil possesses antibacterial and insect repellent activity. In Chinese medicine, patchouli leaves are used as decoction with other drugs to treat nausea, diarrhoea, cold and headaches. The dried leaves are used for scenting wardrobes. The leaves and tops are added in bath water for their antirheumatic action. It is also used as a masking agent for alcoholic breath.

Agro-techniques: Patchouli prefers warm humid climate with a fairly heavy and evenly distributed rainfall of 2,500-3,000 mm/annum, a temperature of 24-28 °C and an average atmospheric humidity of 75%. It grows successfully up to an altitude of 1,000 m above MSL. The crop grows well under irrigation in less rainfall areas, and in partially shaded conditions. It is relatively a hardy plant and adopts itself to a wide range of soil conditions. A well drained deep loamy soil rich in humus and nutrients, with a loose friable structure and with no impervious hard layer at the bottom is ideal. A pH range of 5.5-6.2 is suitable. Patchouli is as hade loving plant. It is generally grown as an intercrop in orchard crops like fruit trees, coconut or arecanut.

The plant is propagated vegetatively by stem cuttings having 4-5 nodes and 15-20 cm length. Improved varieties commonly cultivated are 'Johore', 'Singapore' and 'Indonesia'. Cuttings are prepared from the apical region of healthy stocks. The basal 2-3 pairs of leaves are carefully removed and the cut ends are treated with IBA, IAA or NAA at 500, 1,000 or 1500 ppm respectively for better rooting. Cuttings are planted 3.5 cm apart in nursery beds, seed pans or polythene bags. It is important to provide aeration, partial shade and regular watering in order to get early and good rooting. Rooting occurs in 4-5 weeks and they are ready for transplanting in 8-10 weeks. Before transplanting, the field is prepared well and laid into beds of convenient size. These beds are incorporated with organic manure at 12-15 tonnes/ha and N, P_2O_5 and K_2O at 25:50:50 kg/ha and levelled. Rooted cuttings are transplanted at 40-60 cm spacing and irrigated if there is no rain. After 2 months, 25 kg N is applied. Subsequently, 50 kg N/ha is applied in two split doses; the first dose just after the harvest and the other about two months later. Totally 150 kg N/ha/year is applied to the crop. Constant watering, regular weeding and light cultivation after every harvest are essential for proper growth and yield of the crop. The crop is highly susceptible to root-knot nematode, *Meloidogyne incognita* (Kofoid & White) Chitwood. An integrated approach consisting of crop rotation, application of neem oil cake, carbofuran andsystemic nematicide proved effective. Phyto-sanitory measures are to be adopted from the nursery stage itself. This is done by passing steam through the seed pans or poly bags for about an hour. Nematicides like carbofuran and carbofenthion are also effective in controlling the nematode problem. Leaf blight caused by *Cercospora* sp. is controlled by spraying 0.5% solution of zineb. Yellow mosaic disease is transmitted by white fly, *Bemisia tabacci* (Gen.). Caterpillar and leaf webber attacks can be controlled by spraying methylparathion (Sarwar, 1969; Sarwar and Khan, 1972). The crop is harvested when the foliage becomes pale green to light brown and the stand emits a characteristic patchouli odour. The first harvest of the leaves is taken after about 5 months of planting. Subsequent harvests can be taken after every 3-4 months depending on the local conditions and management practices. Harvesting is done in the cool hours of the morning to avoid loss of essential oil. Young shoots of 25-50 cm length which contain at least 3 pairs of mature leaves are cut. In practice, a few shoots are always left unplucked to ensure better growth for next harvest. The crop stands for 3-4 years.

Post-harvest technology: The harvested herb is dried in shade allowing free air circulation for about 3 days. Proper drying is very important for the quality of oil. During drying, the material should be frequently turned over for promoting uniform drying and for preventing fermentation. Completely dried material can be pressed into bales and stored in a cool dry place for some time. The dried herbage is steam distilled for its oil. Interchange of high and low pressures (1.4 to 3.5 kg/cm^2) produces better yield as more cell walls rupture in this process. Duration of distillation is 6-8 hours. Prolonged distillation gives higher yield and better quality of oil. But if it is distilled for too long, the oil will have a disagreeable odour. The oil yield varies from 2.5 to 3.5% on shade dry basis. On an average, from one hectare we get 8,000 kg fresh leaves annually which on shade drying yield 1,600 kg and on distillation give 40 kg of oil. Patchouli resinoid is also prepared occasionally by extracting the leaves with volatile solvents such as benzene. Such extraction gives 4.5-5.8% of resinoid which contains 70-80% of alcohol soluble absolute.

Rosemary (*Rosmarinus officinalis*): Rosemary is a dense evergreen undershrub with lavender like leaves, and a characteristic aroma. Its essential oil is used almost wholly in the perfumery industry in the production of soaps, detergents, household sprays and other such products. It is an excellent fixative material. The oil contributes a strong, fresh odour, which blends well with various other oil odours and also serves to mask the unpleasant smells of certain other ingredients. It is used in shampoo, toilet soaps and medicine. Rosemary oil is known to have antimicrobial activity. It can kill 90-100% of mosquito and larvae of *Culex quinquefasciatus*. It is carminative and mildly irritant. It is used in formulations of compounded oils for flavouring meat, sauces, condiment and other food products, It is used as a culinary herb. Distilled water is obtained from flowers which can be used as an eye wash (Graf and Hoppe, 1964).

Agro-technology: The plant rosemary comes up well in Mediterranean climate. It is susceptible to frost injury. In cooler areas it can be cultivated in summer season. It requires light dry soil, preferably lying over chalk. Neutral to alkaline pH is suitable. The plant is propagated through seeds and vegetatively by cuttings, the latter being generally adopted. The cuttings should be 15 cm long and leaves removed from the basal half portion. The cuttings are put in nursery beds of sandy soil at a depth of about 10 cm. The main field is prepared well incorporating 10-15 tonnes/ha of organic manures. The rooted cuttings are transplanted in rows, 120 cm apart with a plant to plant spacing of 30-40 cm. Fertilizers are applied at 100:40:40 kg N, P_2O_5 and K_2O/ha, N being applied in 4-5 split doses during each year. Irrigation is needed when the soil is depleted of water during non-rainy period. The field of rosemary is kept weed free by regular weeding and hoeing. Inter-cultivation keeps the soil loose and clean from weeds and promotes proper plant growth and development. *Phytocoris rosmarini* sp. nov. and *Ortholylusribesi* sp. nov. are reported to infest rosemary crop. The shoots are cut for distillation when they have reached their maximum size but before they have become woody. The hard wood imparts an undesirable turpentine odour to the essential oil. Harvesting is usually done during May-June. Frequent cutting of the bushes after 2-3 years keeps them free from becoming leggy and promotes the formation of numerous young shoots.

Post-harvest technology: Freshly harvested twigs and leaves are steam distilled to obtain the essential oil of rosemary. Steam distillation at 2-3 times atmospheric pressure gives an oil yield of 1.0-1.5% of freshly harvested plants and 1.5-2.5 % of dried leaves.

Clarysage (*Salvia sclarea*): Clarysage is a perennial herbaceous plant with a widely branched deep root system; the flowering tops, leaves and derivatives of which are extensively used in the flavour industry for the formulation of liquors and soft beverages. The plant was used in middle ages for *clearing the vision* and for this reason it received its popular name *clarysage or clary-eye* (Guenther, 1949). The essential oil is used in perfumery because of its coriander like notes. It is used as a flavour in liquors and as a modifier in spice compounds. The oil is also used in preparations of non-alcoholic and alcoholic beverages, ice-creams, candy and baked goods.

Agro-technology: Clarysage is tolerant to cold and drought and adaptable to a wide variety of situations. Higher altitude with ample sunshine and few good showers in spring results in good yield of oil having superior quality. The plant is generally grown on poor soils. Slightly acidic soils of pH 4.0-5.5 are better. It is propagated through seeds. Few high yielding hybrid varieties have been developed in Bulgaria . 'Zarya' is a medium early variety with 0.24% oil and 78% linanyl acetate in the oil. 'Lazur" is cold resistant with 0.23% oil and 73% linalyl acetate. A new biennial variety with 0.31% oil and 81.5% linalyl acetate has also been developed (Illieva, 1979). The seeds can be directly sown in the field or transplanted either in November or March-April depending upon the weather conditions. Seed rate is 3-4 kg/ha for transplanting. Seedlings appear in 10-15 days and are transplanted when 30-35 days old, at 1 m row spacing after incorporating 10-12 tonnes of organic manure in the field for optimum growth of the plants. 100-120 kg N and 30 kg each of P_2O_5 and K_2O are recommended per hectare. N may be applied in 4 equal splits. One or two irrigations may be given in case of a drought situation. One or two weeding should be done during March-April. Pre-emergence application of fluometuron or diuron at 2 kg/ha and post-emergence application of preforan or introchlor at 3 kg/ha have been recommended for weed control. 2-3 hoeings should be done before the flowering season. An aphid *Acyrtosiphon salviae* has been found in colonies on clarysage which can be controlled by a mild insecticide. Rootknot nematode *Meloidogyne incognita* infests the plant heavily. The fungus *Rhizoctonia solani* causes rot disease. All the above ground portions have been found infected. Under humid and wet condition, the whole plant collapses within 2-3 days. Drenching with copper oxychloride or Bordeaux mixture is recommended for rot disease. The flowering tops and leaves are harvested twice a year during July and September. Excessive stalk growth is removed as it contains no significant amount of oil. After harvest, a hoeing is given. Plants remain productive for 5-6 years and there after yields decline and new plantation is started in a different location.

Post-harvest technology: The harvested herb is to be distilled immediately with a view to avoid evaporation loss of essential oil. Distillation is carried out for a period 2-3 hours using live steam from a separate steam boiler. An average, recovery of 0.15% is obtained on poor soils whereas 0.2-0.3% is achieved with improved varieties and good management when the yield of oil will be 40-50 kg/ha.

Future prospective of MAPS

India is a proud possessor of an impressive medical heritage which encompasses various systems of medicine, viz., Ayurveda, Siddha, Unani, folklore and grandma medicine. It has an invaluable treasure trove of various scriptures on diverse medical systems. The vast range of agro climatic conditions in India, varying from alpine/mild temperate to tropical regions with abundant rains and sunshine make it an ideal place for the luxuriant growth of flora. Despite comprising only 2% of the land mass, India is blessed with 25% of the biodiversity of the world. Over 7,000 species of plants found in different ecosystems are said to be used for medicine in our country. The Indian pharmacopoeia records about 100 medicinal plants available in India and their preparations. Out of these, quite a few are also recorded in the pharmacopoeias of other countries of the world and there is a growing demand for them in the international market. Therefore, India could become a potential supplier of

Phyto-pharmaceuticals, alkaloids and raw medicinal herbs for the emerging world market. WHO has also emphasized the need for better utilization of the indigenous system of medicine, based on the locally available medicinal plants in the developing countries. Owing to the realization of the toxicity associated with the use of antibiotics and synthetic drugs, Western countries are increasingly aware of the fact that drugs from natural sources are safer. Therefore, there is an upsurge in the use of plant derived products. In addition, these crops have many virtues like drought hardiness, capability to grow on marginal lands. They are relatively free from cattle damage and hence, can be profitably grown in areas where stray cattle or wild animals or pilferage is a major problem. As it is, medicinal plants are better earners than many of the field crops. Since they are new crops, there is an immense scope for further improvement in their productivity and adaptability, in order to obtain further increase in returns. They are suitable for incorporating into various systems of culture like intercropping, mixed cropping and multi-tier cropping.

Conclusion

Plants play an indispensable role as a source of medicine and other products having natural origin and lesser side effects. The traditional systems of various ancient medications such as Chinese, *Unani* and *Ayurvedic* are still popular worldwide despite the advances in modern western medicines. About 80% population of developing countries depends on traditional plant-based medicine for cure of many diseases. Growing consciousness about health and side effects of modern medicines has again set the stage for innovation and use of herbal medicines. The chemical constituents of medicinal and aromatic plants can be found either in the roots, leaves, stems, flowers or bark which can be separated using an appropriate extraction techniques. The rising international demand, many important medicinal plant species are becoming scarce and some are facing the prospect of extinction. Therefore, it is important to conserve the extensively traded medicinal plants in its natural environment or cultivate it in favourable environments.

References

Aburjai, T. and Natsheh, F.M. 2003. Plants used in cosmetics. *Phytother. Res.* **17**: 987–1000.

Attele, A.S., Wu, J.A. and Yuan, C.S. 1999. Ginseng pharmacology: multiple constituents and multiple actions. *Biochem. Pharmacol.* **58**: 1685–1693.

Bannerman, R.H. 1982. Traditional medicine in modern health care. *World Health Forum* Vol. **3**: pp 8-13.

Bechtold, T., Mahmud-Ali, A. and Mussak, R. 2007a. Anthocyanin dyes extracted from grape pomace for the purpose of textile dyeing. *J. Sci. Food Agric.* **87**: 2589– 2595.

Bechtold, T., Mahmud-Ali, A. and Mussak, R. 2007b. Reuse of ash-tree (*Fraxinus excelsior* L.) bark as natural dyes for textile dyeing: process conditions and process stability. *Color. Technol.* **123**: 271–279.

Bechtold, T., Turcanu, A., Ganglberger, E. and Geissler, S. 2003. Natural dyes in modern textile dye houses—how to combine experiences of two centuries to meet the demands of the future? *J. Clean. Prod.* **11**: 499–509.

Bevill, R.L., Louda, S.M. and Stanforth, L.M. 1999. Protection from natural enemies in managing rare plant species. *Cons. Bio.* **13:**1323-1331.

Bilia, A.R., Gallori, S. and Vincieri, F.F. 2002. St. John's wort and depression: efficacy, safety and tolerability-an update. *Life Sci.* **70**: 3077–3096.

CBI. 2008a. The market for natural ingredients for cosmetics in the EU. CBI

CBI. 2008b. The market for natural ingredients for pharmaceuticals in the EU. Centre for the Promotion of Imports from Developing Countries.

CBI. 2009a. Natural ingredients for cosmetics: The EU market for essential oils for cosmetics. CBI.

Copping, L.G. and Duke, S.O. 2007. Natural products that have been used commercially as crop protection agents. *Pest Manage. Sci.* **63**: 524–554.

Dayan, F.E., Cantrell, C.L. and Duke, S.O. 2009. Natural products in crop protection. *Bioorg. Med. Chem.* **17**: 4022–4034.

Dhyani, P.P. and Kala, C.P. 2005. Current research on medicinal plants: Five lesser known but valuable aspects. *Curr. Sci.* **88:**335

Ernst, E. and Pittler, M.H. 2000. Efficacy of ginger for nausea and vomiting: a systematic review of randomized clinical trials. *Br. J. Anaesth.* **84**: 367–371.

FAO. 2003. State of the world's forest**.** Rome: Food and Agricultural Organization.

Farnsworth, N.R. 1979. Present and future of pharmacognosy. *Am. J. Pharm. Educ.* **43**: 239–243.

Ghimire, S.K., McKey, D. and Aumeeruddy-Thomas, Y. 2005. Heterogeneity in ethnoecological knowledge and management of medicinal plants in the Himalayas of Nepal: Implication for conservation. Ecology and Society 9 (36 [http://www.ecologyandsociety.org/vol9/iss3/art6/].

Gilbert, K.G. and Cooke, D.T. 2001. Dyes from plants: past usage, present understanding and potential. *Plant Growth Regul.* **34**: 57–69.

Guleria, S. and Tiku, A.K. 2009. Botanicals in pest management: current status and future perspectives. In: Peshin, R., Dhawan, A.K. (Eds.), Integrated Pest Management:Innovation-Development Process, vol. 1. Springer, Berlin, p 317–330.

Hamilton, A. 2003. Medicinal plants and conservation: issues and approaches (online) UK.WWF, Portable Document Format. Available from internet.http://www.wwf.org.uk/filelibrary/pdf/medplantsandcons.pdf

Holmes, C.A. 2005. IENICA summary report for the European Union 2000-2005.Interactive European Network for Industrial Crops and their Applications**.**

Houghton, P.J. 2001. Old yet new—pharmaceuticals from plants. *J. Chem. Educ.* **78**:175–184.

Isman, M.B. 2006. Botanical insecticides, deterrents, and repellents in modern agriculture and an increasingly regulated world. *Annu. Rev. Entomol.* **51**: 45–66.

Joshi, K., Chavan, P., Warude, D. and Patwardhan, B. 2004. Molecular markers in herbal drug technology. *Curr. Sci.* **87**:159-165.

Kala, C.P., Farooquee, N.A. and Dhar, U. 2005. Traditional uses and conservation of timur (*Zantho xylumarmatum* DC.) through social institutions in Uttaranchal Himalaya, India. *Conservation and Society* **3**(1):224-230.

Kala, C.P. 2003. Commercial exploitation and conservation status of high value medicinal plants across the borderline of India and Nepal in Pithoragarh. *The Indian Forester* **129**:80-84

Kala, C.P. 2000. Status and conservation of rare and endangered medicinal plant in the Indian trans-Himalaya. *Bio. Cons.* **93**:371-379.

Kaplan, S.A. 2005. Updated meta-analysis of clinical trials of *Serenoarepens* extract in the treatment of symptomatic benign prostatic hyperplasia. *J. Urol.* **173**:516.

Khan, I.A., Khanum, A. 2000. Role of biotechnology in medicinal and aromatic plants. Hyderabad: Ukaaz Publications.

KIT. 2004. Searching Synergy. In *Bulletin 359* Amsterdam: Royal Tropical Institute.

Kuipers, S.E. 1997. Trade in medicinal plants. In: Bodeker, G., Bhat, K.K.S., Burley, J., Vantomme, P. (Eds.), Medicinal plants for forest conservation and health Care, vol. 11. Food and Agriculture Organization of the United Nations, Rome, pp. 45–59.

Lozoya, X. 1994. Ethnobotany and the search of new drugs**.** England: John Wiley and Sons;

O'Hara, M., Kiefer, D., Farrell, K. and Kemper, K. 1998. A review of 12 commonly used medicinal herbs. *Arch. Fam. Med.* **7**: 523–536.

Rao, M.R., Palada, M.C. and Becker, B.N. 2004. Medicinal and aromatic plants in agro-forestry systems. *Agroforestry systems* **61**:107-122.

Samal, P.K., Shah, A., Tiwari, S.C. and Agrawal, D.K. 2004. Indigenous health care practices and their linkages with bio-resource conservation and socio-economic development in central Himalayan region of India. *Indian J. of Traditional Knowledge,* **3**:12-26

Schmidt, B., Ribnicky, D.M., Poulev, A., Logendra, S., Cefalu, W.T. and Raskin, I. 2008. A natural history of botanical therapeutics. *Metab. Clin. Exp.* **57**: S3–S9.

Sharma, A.B. 2004. Global medicinal plants demand may touch $5 trillion By 2050. Indian Express.

Shrinivas, P.K. and Kudli, A.P. 2008. Market research data on Essential oils and absolutes used in fragrance and flavor industry [Online]http://www.goarticles.com/cgibin/showa.cgi?C=1175757

Sundriyal, R.C. and Sharma, E. 1995. Cultivation of medicinal plants and orchids in Sikkim Himalaya. Almora: G.B. Pant Institute of Himalayan Environment and Development.

Ticktin, T. 2004. The ecological implications of harvesting non-timber forest products. *J. of Applied Ecology* **41**:11-21.

UNESCO. 1996. Culture and Health. Orientation Texts– World Decade for Cultural Development 1988-1997. Paris, France.

Uniyal, R.C., Uniyal, M.R. and Jain, P. 2000. Cultivation of medicinal plants in India: A Reference Book. New Delhi: TRAFFIC India and *WWF* India.

Weekley, C.W. and Race, T. 2001. The breeding system of *Ziziphus celata*Judd and D.W. Hall (Rhamnaceae), a rare endemic plant of the Lake Wales Ridge, Florida, USA: implications for recovery. *Biological Conservation* **100**:207-213.

WHO. 1991. Guidelines for the assessment of herbal medicines.

WHO. 1998. General guidelines for methodologies on research and evaluation of traditional medicine. WHO/EDM/TRM/2000.1, Geneva, pp 128.

Wildman, R.E.C. and Kelley, M. 2007. Nutraceuticals and functional foods. In: Wildman, R.E.C. (Ed.), Handbook of Nutraceuticals and Functional Foods. Second Edition. CRC Press, Boca Raton, pp. 1–22.

10

Advances in Forage Crop Production Technologies

D. Vijay, N. Manjunatha and Sanjay Kumar*

Introduction

Livestock is an important component of Indian agriculture contributing 28% of agricultural gross value added (GVA) of the country. Agriculture, as such, contributes 16.4% of country's GVA (Economic survey, 2017-18). Livestock as a source of income has seen a steady increase from 4% to 13% in a decade time (2002-2012). India, with 2.29% of the world land area, is maintaining about 10.71% of the world's livestock population. The milk production in the country increased from 17 million tonnes in 1950-51 to 187.7 million tonnes in 2018-19. The feed and fodder play an essential role in the health and productivity of livestock. Even though India is the world largest milk producer with 21% of global production, the productivity (1700 kg/year) is far below the world average (2574 kg/year) (FAO STAT, 2018). One of the main responsible factors (50.2%) for this low productivity is the deficiency in feed and fodder. Although forage crops are the critical component of livestock-based farming system, the area under cultivated fodder is static to around 8.4 million ha (5.23%) since last two decades due to the competition of land use with other crops. Availability of the fodder determines the productivity and profitability of the livestock rearing because fodder and feeds constitute about 60% of the total cost of milk production, which can be lowered by enhancing the green fodder-based feeding system. The traditional grazing lands are gradually diminishing because of urbanization, expansion of the cultivable area, grazing pressure, industrialization etc. These factors coupled with stagnation of area under cultivated fodder resulted in a severe shortage of feed and fodder to the extent of 26% in dry crop residues, 35.6% in green fodder and 41% of concentrates (IGFRI, 2011). This gap in demand and supply may further rise due to consistent growth of livestock population at the rate of 1.23% in the coming years. The share of the in-milk population to total population has increased from 23% to 27% in

**Corresponding author email: vijaydunna@gmail.com*

the case of cattle and from 33.64 to 34.74% in buffaloes from 2012 to 2019. The year-on-year growth rate of milk is approximate 6%. Thus, to sustain this growth rate and for further expansion to meet the demands of the ever growing human population, livestock needs a sustainable supply of feed material. To reduce the demand and supply gap, the production and productivity of fodder crops need to be enhanced. The productivity of cultivated fodder crops is low, due to the least attention and allocation of minimal production resources and lack of dissemination of the production techniques to stakeholders involved in the forage resource development. The present chapter highlights the recent advances in the forage crop production technologies. The comprehensive compilation of information at one place provides the opportunity to get the holistic picture of forage production altogether.

Forage statistics

The area, production and productivity statistics of individual forage crops are not available similar to food crops. The lack of crop cutting data at base level in forage crops is the main reason for this gap. The absence of statistics paralyses the policy decisions and is one of the reasons for low resource allocation to this sector. The forage resources can be grouped into two main categories *viz*., cultivated fodder crops and grasses and legumes of grass and rangelands. The important cultivated fodder crops in India are cowpea, guar, maize, teosinte, sorghum, pearl millet, rice bean in *kharif* (rainy) season and berseem, oat, lucerne and gobhi sarson in *rabi* (winter) season. Approximately 75 cultivars of these 11 crops are under seed chain that too for many varieties the breeder seed indents are meagre. The area under cultivated fodder crops remains static in last >15 years (Fig. 1) with minor fluctuations here and there. The stagnation of area is mainly due to severe competition from food crops.

The overall increase in cultivated area under fodder crops will be difficult until and unless there are radical policy changes towards fodder cultivation supported by the development of peri-urban commercial dairies.

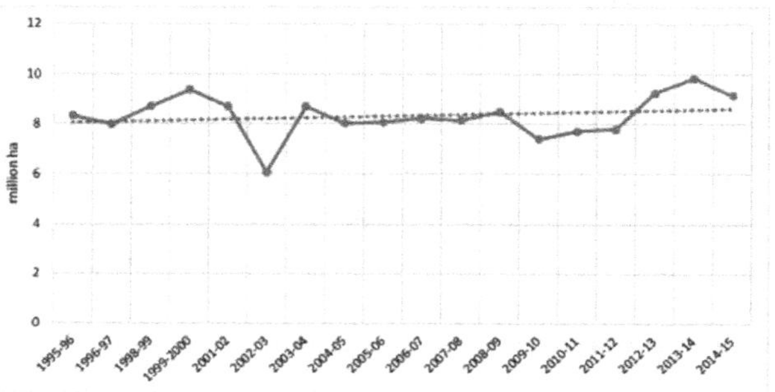

Fig. 1. Area under cultivated fodder crops in India over the years (*Source:* BAHS, 2019)

In India, the maximum area under fodder crops is in Rajasthan, which accounts for 54% of the total area under fodder crops (Fig. 2). The four states Rajasthan (4.853 million ha), Maharashtra (0.901 million ha), Gujarat (0.85 million ha), and Uttar Pradesh (0.8 million ha) combined have > 80% area under fodder crops. However, the livestock population of these four states combined is only 35% of the total livestock population of the country. This shows the clear disparity in the area of fodder crops in different states. Rajasthan with 11% of the total livestock population is having 53% of the total fodder crop area. The states of Andhra Pradesh (including Telangana), Bihar, West Bengal and Karnataka have far less area under fodder crops compared to their livestock population. Among the fodder crops, sorghum occupies maximum area (2.6 million ha) followed by berseem (1.9 million ha) (Table 1).

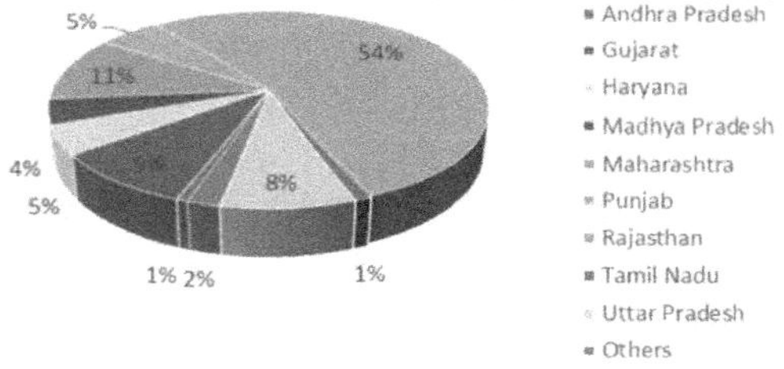

(Others: Jammu & Kashmir, Karnataka, Uttarakhand, Kerala, West Bengal, Dadra & N. Haveli, Himachal Pradesh, Assam, Bihar and Telangana)

Fig. 2. Statewise cultivated fodder crop area during 2014-15 (*Source:* BAHS, 2019)

Out of the total area, > 90% area is under five crops *viz.*, sorghum (33%), berseem (24%), lucerne (13%), maize (11%) and bajra (11%). Productivity wise the *rabi* legumes are having better fodder yield compared to other crops. Among cereals, sorghum and maize have better productivity (Table 1). In recent years, oat is gaining popularity and is spreading to various corners of the country. The seed supplied by ICAR-Indian Grassland and Fodder Research Institute, Jhansi in the last five years shows that oat crop was raised in >16 states. The oat cultivation has spread to southern states of Karnataka, Tamil Nadu, Andhra Pradesh and Telangana as well as to the north-eastern state of Assam apart from traditional northern and central India.

Oat

Guar

Cenchrus

Table 1. Crop wise area and green fodder yield of cultivated fodder crops in India

Common name	Botanical name	Area (million ha)	Green fodder yield (tonnes/ha)
Berseem (Egyptian clover)	*Trifolium alexandrinum*	1.900	60–110
Lucerne (Alfalfa)	*Medicago sativa*	1.000	60–130
Senji (Sweet clover)	*Melilotus indica*	0.005	20–30
Shaftal (Persian clover)	*Trifolium resupinatum*	0.005	50–75
Metha (Fenugreek)	*Trigonella foenumgraecum*	0.005	20–35
Lobia (Cowpea)	*Vigna unguiculata*	0.300	25–45
Guar (Cluster bean)	*Cyamopsis tetragonaloba*	0.200	15–30
Rice bean	*Vigna umbellata*	0.020	15–30
Jai (Oat)	*Avena sativa*	0.100	35–50
Jau (Barley)	*Hordeum vulgare*	0.010	25–40
Jowar/Chari (Sorghum)	*Sorghum bicolor*	2.600	35–70
Bajra (Pearl millet)	*Pennisetum glaucum*	0.900	20–35
Makka (Maize)	*Zea mays*	0.900	30–55
Makchari (Teosinte)	*Zea mexicana*	0.010	30–50
Chara sarson	*Brassica pekinensis*	0.010	15–35

Source: Handbook of Agriculture (2006)

Apart from cultivated fodder, the range species also provide fodder to livestock. The permanent pastures and grazing lands also provide green and dry fodder and are traditional resources in rural India. The area under grazing resources was decreased from 14 million ha to 10 million ha in the last five decades. Even the existing grasslands are in the denuded state due to severe grazing pressure. The optimum grazing pressure is one adult cattle unit (ACU) per ha of grazing land. In India, the grazing pressure calculated based on the total livestock population and potential grazing land comes to 2.95 ACU/ha.

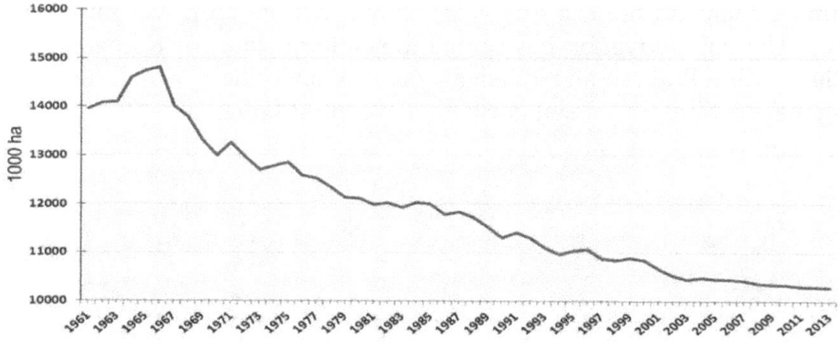

Fig. 3. Area under permanent pastures in India during the last five decades

The permanent pastures comprise only 10 million ha area. The potential grazing resources apart from permanent pastures comprises nearly 71 million ha (Table 2). Even though this area is not under pasture/meadows, it is either partially being used for grazing purpose or has the potential to develop into grazing land. The strong policies at the Union and state level are required to utilize these potential grazing resources. The main obstacles are the seed requirement and implementing agency. Since, these areas are under the control of different government departments, a unifying authority is required to coalesce the resources and inputs for an effective grassland establishment.

Table 2. Potential grazing resources in India

Potential grazing resource	Area (million ha)
Open forest area	30.449
Barren and unculturable land	16.996
Permanent pastures and other grazing lands	10.258
Culturable wasteland	12.469
Fallow other than current fallow	11.092
Total	**81.264**

Source: Land use statistics at a glance (2017)

The data on forage production and productivity is lacking. Based on the available resources, an estimate of demand and supply of forages was made by ICAR-Indian Grassland and Fodder Research Institute (Table 3). As per the estimate, by 2050 the demand for green fodder will reach 1,012 million tonnes and 631 million tonnes for dry fodder. With the current growth rate and availability, the per cent deficiency of dry fodder may rise from 10.95% to 13.2% while that of green fodder may decrease from 35.6% to 18.4% (IGFRI, 2011).

Table 3. Demand and supply estimates of dry and green fodder in India

Year	Demand		Supply		Deficit		Deficit as %	
	Dry	Green	Dry	Green	Dry	Green	Dry	Green
2010	508.9	816.8	453.2	525.5	55.72	291.3	10.95	35.66
2020	530.5	851.3	467.6	590.4	62.85	260.9	11.85	30.65
2030	568.1	911.6	500.0	687.4	68.07	224.2	11.98	24.59
2040	594.9	954.8	524.4	761.7	70.57	193.0	11.86	20.22
2050	631.0	1012.7	547.7	826.0	83.27	186.6	13.20	18.43

Source: IGFRI Vision 2050 (2011)

To meet out the deficit, the green forage supply has to grow at 16.9% annually (IGFRI, 2011). Thus, to realize this growth, there is a need to follow proper production technology in totality. The crop production technology for major forage crops i.e. both cultivated and range species, is discussed below.

Forage crop production technology

Climatic requirement

The *kharif* season fodder crops *viz.*, maize, sorghum, cowpea, NB (napier bajra) hybrid and guar (cluster bean) need warm weather for growth and these crops can not be grown in areas where the mean daily temperature is < 19 °C. Therefore, these monsoon season crops should be sown preferably during first June-July after the commencement of rains. Guar is a tropical plant, which requires warm growing season. Guar can give good fodder yield even in low rainfall areas also and it can tolerate temperature as high as 45-46 °C. NB hybrid grows well at high temperatures and can withstand the drought conditions for a long spell. It grows extremely well in areas with rainfall of over 1,000 mm but it can not tolerate the flooding/ waterlogging. The optimum temperature is 31 °C for its growth but it performs well in areas having temperature above 15 °C. It is a tropical grass, which can withstand moisture stress and high temperature for a short spell and regenerate with rains.

The *rabi* season crops *viz.,* berseem and oat are well adapted to cooler environment. Such conditions prevail during winter and spring seasons in north India, which is considered as a favourable and productive zone for these crops. The optimum temperature at the time of sowing of berseem and oat is 20 °C. For luxuriant vegetative growth, the temperature range of 25 °C to 27 °C is ideal. Due to the shorter winter period, berseem can not be cultivated in southern India. Lucerne is well adapted to relatively dry conditions and it may tolerate heat as well as cold.

The range species including both grasses and legumes *viz.*, Deenanath grass (*Pennisetum pedicellatum*), Guinea grass (*Panicum maximum*), Anjan grass (*Cenchrus ciliaris*), Dhaman grass (*Cenchrus setigerus*), Dhawlu grass (*Chrysopogon fulvus*), Saen grass (*Sehima nervosum*), Lampa grass (*Heteropogon contortus*) are suitable mainly for tropical conditions and can tolerate the severe summer and harsh winter of plains. Whereas, *Stylosanthes hamata* thrives well in the areas receiving annual rainfall from 500 to 1,270 mm with a pronounced dry season. Therefore, *Stylosanthes hamata* is well-adapted *kharif* legume crop in South India. The other species of *Stylosanthes* like *S. sebrana* can tolerate the tropical conditions better than *S. hamata.* Under severe cold conditions such as prevalent in Jammu and Kashmir and hill regions of northern and eastern India, the temperate species *viz.*, rye grass, *Dactylus glomeratus*, white clover, red clover etc. are suitable perennial forage species.

Soil and its preparation

The field should be weed-free, thoroughly prepared and levelled before sowing. Better crop stands and growth is possible only on well-prepared soils. One ploughing with soil turning plough followed by two harrowing (crosswise) and planking is sufficient to get a good seedbed. The suitable soil types for growth of different fodder and forage crops is provided in table 4.

Table 4. Suitable soil types for important fodder crops

Crops	Soil requirement
Maize	Sandy loam (pH 6.5 to 7.5)
Sorghum	Sandy loam (pH 6.5 to 7.5)
NB Hybrid	Clay loam
Guinea grass	All types of soil
Dinanath	All types of soil
Cow pea	Loam and sandy loam
Cluster bean	Alluvial soil
Stylosanthes	All types of soil
Berseem	Clay to clay loam soil
Oat	Loam to clay loam soil
Lucerne	All types of soil

Source: Sunil *et al.* (2012)

The forage species in general and range species, in particular, can grow even in problematic soils due to their higher adaptability (Table 5). Suitable species are to be selected along with proper soil management practices to obtain a good yield.

Field prepration: Harrowing

Table 5. Suitable grass and legume species for problematic soils

Problematic soil	Cultivated species	Range species
Saline soils (EC > 4 dS/m, exchangeable sodium <15%, pH <8.5)	Oat, Teosinite, Sorghum, Senji, Lucerne, NB Hybrid	*Cenchrus ciliaris, Pennisetum pedicellatum, Dichanthium annulatum, Stylosanthes hamata, Setaria* sp.
Sodic soils (EC < 4 dS/m, exchangeable sodium >15%, pH =8.5-10)	Berseem, Oat, Sorghum, Pearlmillet, Shaftal, NB Hybrid	*Leptochloa fusa, Cynodon dactylon, Brachiaria mutica, Chloris gayana*
Acid soils	Maize, NB Hybrid	*Dinanath grass, Guinea, Setaria*
Calcareous soils	Sorghum, Pearl millet, NB Hybrid, Teosinite, Oat, Lab lab bean	*Cenchrus ciliaris, Panicum antidotale, Dichanthium annulatum, Heteropogon contortus., Pennisetum pedicellatum, Stylosanthes hamata, Macroptilium atropurpurium, Clitoria ternatea, Cenchrus setigerus, Lasiurus scindicus, Saccharum munja and Sehima nervosum*
Waterlogged soils	*Lathyrus sativus*, chatari matari (*Vicia sativa*) (for temporary waterlogged conditions)	Almon grass (*Echinochloa polyptachya*), Para grass (*Brachiaria mutica*), *Coix sps., Iseilema laxum, Chloris gayana,* Signal grass (*B. decumbense*), Karnal grass (*Leptochloa fusca*), Khas (*Vitiveria zizanioides*) and Congo signal grass (*B. brizantha*)

Suitable varieties

Systematic forage crop breeding programmes at the research institutions under the Indian Council of Agricultural Research (ICAR) and the State Agricultural Universities (SAUs) have led to the development and release of a large number of improved varieties in different forage crops suitable for different agro-ecological zones. These varieties resulted in a substantial increase in the productivity and production of forages in the country. The improved varieties released/ notified during the past three decades are given in table 6.

Table 6. Some of the important varieties of fodder crops along with release year, recommended area and green fodder yield

S. N.	Crop/variety	Year of release	Institute developed the variety	Area for adaptation	Fodder yield (t/ha)
Sorghum (Single cut)					
1.	Pusa Chari 6	1979	IARI, New Delhi	All India	35-40
2.	Pusa Chari 23	1984	IARI, New Delhi	North India	35-40
3.	PCH 106 (Hybrid)	1985	PAU, Ludhiana	North India	35-30
4.	MP Chari	1985	JNKVV, Jabalpur	U.P., M.P., M.H & T.N.	35-45
5.	CO-27	1986	TNAU, Coimbatore	T.N.	45-65
6.	AS-16	1989	GAU, Anand	Gujarat	40-60
7.	Pant Chari 5	1999	GBPUAT, Pantnagar	All India	47
8.	Co (FS) 29	2001	TNAU, Coimbatore	T.N.	150
9.	PCH 109 (Hybrid)	2005	IARI, New Delhi	North India	40-50
10.	CSH 13-R	1991	NRC Sorghum	North-Western India	25
11.	COFS-31	2016	TNAU, Coimbatore	T.N., Kerala & Karnataka	183
Maize					
1.	African Tall	1981	MPKV, Rahuri	All India	55-80
2.	J-1006	1993	PAU, Ludhiana	Punjab	45-55
3.	APFM-8	1997	ANGRAU, Hyderabad	South India	35-40
4.	Pratap Makka Chari-6	2009	MPUAT, Udaipur	North-West India	45-50
Bajra					
1	Giant Bajra	1985	MPKV, Rahuri	All India	64
2	AVKB 19	2006	IGFRI, Jhansi	North-Western zone	38
3	BAIF Bajra-1	2010	BAIF, Pune	North-West & central zone	38-40
4	AFB-3	2011	AAU, Anand	North-West zone of India	50-52
5	FBC-16	2016	PAU, Ludhiana	All India	57

NB hybrid					
1.	Hybrid Napier-3 (Swetika)	1983	IGFRI, Jhansi	North & central zone	70-100
2.	NB-21	1987	IARI, New Delhi	All India	100-160
3.	Yashwant	1987	MPKV, Rahuri	Maharashtra	190-250
4.	CO-3	1996	TNAU, Coimbatore	South India	400-450
5.	APBN-1	2001	ANGRAU, Hyderabad	South India	260-295
Guinea grass					
1.	PGG 1	1982	PAU, Ludhiana	Hill & North/ North-West India	70-100
2.	PGG-14	1988	PAU, Ludhiana	Central India	95-140
3.	PGG-616	2001	PAU, Ludhiana	All India	40-45
4.	Bundel guinea-1	2001	IGFRI, Jhansi	All India	50-55
5.	Bundel guinea-2	2005	IGFRI, Jhansi	All India	40-45
Dinanath grass					
1.	Bundle Dinanath-1	1987	IGFRI, Jhansi	All India	35-40
2.	Bundle Dinanath-2	1990	IGFRI, Jhansi	All India	35-40
3.	TNDN-1	1996	TNAU, Coimbatore	South India	40-50
Stylosanthes					
1.	*Stylosanthes scabra*	1991	TNAU, Coimbatore	South India	-
2.	RS 95	2004	MPKV, Rahuri	Western zone	-
Cowpea					
1.	UPC-5286	1982	GBPUAT, Pantnagar	All India	30-35
2.	Bundel Lobia-1	1992	IGFRI, Jhansi	All India	30-35
3.	Bundel Lobia-2	1994	IGFRI, Jhansi	North-West India	35-40
4.	CL-367	2006	PAU, Ludhiana	Punjab	25-30
5.	UPC 625	2009	GBPUAT, Pantnagar	All India	30-35
6.	UPC 628	2010	GBPUAT, Pantnagar	North-East, North-West & Hill zone	35-40
Guar					
1.	Bundel Guar- 1	1993	IGFRI, Jhansi	Guar growing area in India	30-35
2.	Bundel Guar- 2	1995	IGFRI, Jhansi	Guar growing area in India	30-35
3.	Bundel Guar- 3	1999	IGFRI, Jhansi	Guar growing area in India	30-35
4.	Guar Kranti	2005	ARS, Durgapura (Raj.)	Rajasthan	30-35
Lucerne					
1.	Anand 2	1984	AAU, Anand	Maharashtra, Gujarat & Rajasthan	60-65
2.	Anand 3	1995	AAU, Anand	Cold, dry zone of Kinnaur, Lahul & Spiti valley	50
3.	RL-88	1996	MPKV, Rahuri	All India	70-75

4.	Anand Lucerne 3	2009	AAU, Anand	All India	70-75
5.	Anand Lucerne 4	2013	AAU, Anand	North-West zone of India	40-50
6.	RBB -07-1	2016	SKRAU, Bikaner	North-West zone of India	180
Oat					
1.	Kent	1978	PAU, Ludhiana	All oat growing areas	40-45
2.	JHO-822	1989	IGFRI, Jhansi	Central zone	30-35
3.	Bundel Jai 851	1998	IGFRI, Jhansi	All oat growing areas	40-50
4.	JHO-99-1	2005	IGFRI, Jhansi	All India	30-35
5.	JHO-99-2	2005	IGFRI, Jhansi	North -West & North- East zone	50-55
6.	JHO-2000-4	2006	IGFRI, Jhansi	All India	35-40
7.	Phule Harita (RO-19)	2007	MPKV, Rahuri	All oat growing areas	55-60
8.	Jawahar Oat 03-93	2010	JNKVV, Jabalpur	Central zone of oat growing area	46
9.	Jawahar Oat 03-91	2011	JNKVV, Jabalpur	Central zone of oat growing area	47
10.	NDO-2	2013	NDUAT, Faizabad	All oat growing areas	52-55
11.	NDO-10	2013	NDUAT, Faizabad	All oat growing areas	53-56
12.	OS 377	2013	CCSHAU, Hisar	Central zone of oat growing areas	53
13.	JHO 2009-1	2016	IGFRI, Jhansi	Central zone	53-62
14.	RO 11-1	2016	MPKV, Rahuri	North- East, North -West, Central & South zone	45-50
15.	OL1802-1	2017	PAU, Ludhiana	North-West zone	53
16.	OL 1769-1	2017	PAU, Ludhiana	Central zone	48.7
Berseem					
1.	Mescavi	1978	CCSHAU, Hisar	All India	60-62
2.	Wardan (S-99-1)	1982	IGFRI, Jhansi	All berseem growing area	60-65
3.	BL-22	1988	PAU, Ludhiana	North & Central zone	50-55
4.	BB-2 (JHB-146)	1997	IGFRI, Jhansi	North-West & Central zone	75-80
5.	BB-3	2001	IGFRI, Jhansi	Eastern & North-East zone	55-70
6.	Hisar berseem-1	2006	CCSHAU, Hisar	North-West & hill zone	65-70
7.	BL-180	2006	PAU, Ludhiana	North-West zone	55-60
8.	BL-42	2007	PAU, Ludhiana	Punjab, Haryana & H.P.	70-72

9.	JBSC-1	2018	IGFRI, Jhansi	single cut short-duration condition for Maharashtra, Rajasthan, Punjab, Haryana, U.P. & M.P.	38

Source: Database of forage crop varieties (2018)

Forage production systems

The forage production systems differ for irrigated and rainfed situations. In both the systems, proper management helps in availability of fodder either throughout the year or along with food crops or horticulture crops based on the requirement.

Intensive forage production system under the irrigated situation

This system is best suitable to meet the need of dairy farmers for the production of maximum green fodder throughout the year. The perennial NB hybrid intercropped with cowpea during summer and *kharif* and berseem during *rabi* provides the highest green fodder (270 tonnes/ha/year). In berseem non-growing areas, lucerne can be used with a slight reduction in yield. Similarly, in humid regions, the NB hybrid can be replaced with guinea grass and in temperate regions with *Setaria* grass.

Food-fodder intercropping system

The small and marginal farmers with livestock can adopt this system where either the fodder is intercropped with the food crop or fodder crop is grown as a sequence crop of food crop. The best examples are rice – berseem or cowpea - wheat sequences and cowpea intercropped with sorghum/ pearl millet. This system also improves soil fertility, as the fodder crop is a legume.

Round the year fodder production under the rainfed irrigated system

The availability of fodder round the year under rainfed conditions is a challenge. The hardy crops of *Pennisetum* species help in achieving this. The perennial tri-specific hybrid of *Pennisetum* species can tolerate harsh conditions and can give multiple cuts and supplies fodder until the main crop yields fodder. The double/multi-cut pearl millet (*Pennisetum glaucum*) acts as the main fodder crop. The total system productivity is around 85 tonnes green fodder/ha/year.

Several other combinations of perennial fodder, food crop and annual fodder crops are also there which have several advantages over mono-cropping. The suitable system may be adopted based on the requirement, irrigation facility and economic feasibility.

Sowing and seed rate

Sowing time is an important factor governing germination, seedling survival, number of cuts and herbage quantity in fodder crop production. Timely sowing extends the period of forage availability and thereby increases the total yield. The sowing of all crops should be done according to the recommended seed rate to get uniform field establishment (Table 7). In *kharif* season crops, sowing shall be taken up just after commencing the monsoon whereas, for *rabi* season crops October-November is the suitable period. Temperature sensitive crops like berseem should be sown when the temperature range is 25-27 °C. Line sowing results in higher output due to uniform plant population and seed drills are available even for very small sized seed like berseem. The range grasses and legumes are in general transplanted or broadcasted based on the field conditions and water availability.

Table 7. Seed rate and spacing of different fodder and forage crops

Crops	Seed rate (kg/ha)	Spacing
Maize	40-50	30 cm (row to row)
Sorghum	35-40	25 cm (row to row)
NB hybrid	35000 rooted slip	75×50 cm
Guinea grass	6-8	50×50 cm
Dinanath	5-6	25 cm (row to row)
Cowpea	35-40	25 cm (row to row)
Cluster bean	30-35	25 cm (row to row)
Stylosanthes	6-8	50×50 cm / Broadcasting
Berseem	20-25	35 cm (row to row)
Oat	80-100	20 cm (row to row)
Lucerne	20-25	25 cm (row to row)

Source: Modified from Sunil *et al.* (2012)

Water management

The rainy season (July) sown crop may require 1-2 irrigations depending upon the distribution of rains but for the summer-sown crop, 5-6 irrigations are required due to high evaporative demand. In southern region *rabi* season crops; require about four irrigations for better performance of the crop. Generally, fodder and forage crop production requires less water but in case of berseem huge quantities of water is required for production of green and succulent biomass. In berseem around 500 kilograms or more water is required for every kilogram of plant dry matter produced in a dry climate. Therefore, adequate and timely water supply is one of the basic inputs for obtaining potential crop yield which necessitates precise knowledge of irrigation techniques and approaches in the berseem crop. On an average berseem and lucerne crops require 16-18 irrigations in 10-12 days interval for good green fodder production. During seed production of berseem sufficient irrigations are necessary during the flowering season to get proper seed setting and seed yield. In the case of oat, timely irrigation improves the tillering remarkably, which contributes to higher forage yield and total 7-8 irrigations are required for multi cut varieties. Singh *et al.* (2014) reported that fodder crops like berseem, alfalfa, lucerne and cowpea require more water than sorghum and oat.

Weed management

Weed management is an important aspect of good crop production. In the case of fodder and forage, weed management should be taken in such a manner that residue of herbicides should not be reached up to cattle's feed. The cultivated fodder crops *viz.,* sorghum, maize, pearl millet and cowpea are infected with many seasonal weeds and crop weed competition is more during 30-35 days stage. Therefore, one hoeing through weeder cum mulcher at critical crop stages i.e. 3-4 weeks, is very effective to control the weeds and provide good aeration for crop growth. Pre-emergence application of atrazine @ 0.50 kg a.i. /ha in 600 litres of water or alachlor @ 1.0 kg a.i. /ha is also recommended for the management of weeds. The major associated weed of berseem crop is kasani /chicory (*Chicorium intybus*) which is an objectionable weed because of the difficulty to separate by ordinary methods. The major sources of chicory infestation are irrigation sources, soil seed bank, and seed sources. Chicory seed has a conical shape and is lighter in density compared to a berseem seed. Treatment with 10% NaCl solution separates chicory seeds to some extent. Chicory seed can also be separated during seed processing by using specific gravity seed separator and draper belt. The weeds in berseem and cowpea can be effectively controlled by the application of weedicides. Application of Imazethapyr @ 0.10 kg/ha at 20 days after sowing not only increased green fodder yield (18%) but also seed yield (38%) compared to the weedy check. Similarly, *Cuscuta* is an objectionable weed in lucerne and can be effectively controlled by the pre-emergence application of Pendimethalin @ 1-2 kg a.i./ ha or post-emergence application of Diquat @ 6-10 kg/ha (5-10 DAS). The winter season crop oat is heavily infested with grassy and broad-leaved weeds, similar to those mostly found in wheat crop. The dicot weeds can be effectively controlled by the application of 2,4-D @ 0.37 kg a.i./ha at six weeks crop stage. In most of the grasses during the early field establishment, regular hand weeding/hoeing ensures good aeration and weed control. The application of Imazethapyr @ 0.1 kg a.i. /ha as pre-emergence and one-wheel hoeing at 20 DAS, significantly controlled the weed species in cowpea with 75% reduction in weeds compared to weedy check and has at par green and dry fodder yield of a weed-free check. Similarly, in berseem, the practice of stale seedbed + Propaquizafop spray @ 0.1 kg a.i/ha as post-emergence at 20 DAS was successful in managing the weeds and has on par green and dry fodder yield of a weed-free check (IGFRI Annual Report, 2016-17).

Nutrient management

Nutrient management is an integral aspect of good crop production and timely application of different nutrients in sufficient quantity can boost crop growth and yield manifold. In case of cultivated fodder crops *viz*., sorghum, maize, pearl millet and NB hybrid before the sowing, field should be manured with FYM (10 tonnes/ha) to meet the requirement of secondary and micronutrients. The recommended dose of fertilizers should be applied to get good crop growth and yield. The recommended fertilizer should be applied in two split doses in which 50% of nitrogen and a full dose of phosphorus and potassium should be given at the time of sowing and remaining 50% nitrogen should be given as top dressing at 30-40 days after sowing (Table 8). In sorghum, maize, pearl millet, guinea grass and NB hybrids seed treatments of biofertilizer *viz., Azospirillum* and *Azotobacter* improves the yield and saves 15-20% lesser nitrogenous fertilizers. The cowpea, guar, berseem and lucerne are leguminous crops and have capacity to fix atmospheric nitrogen through *Rhizobium trifolii* bacteria which is present in the root nodules therefore; these crops required very less nitrogen from external sources for establishment of the plant before the formation of root nodules. In legume fodder crops, seed inoculation with *Rhizobium meliloti* is recommended, where legume crops are being cultivated for the first time. Similarly, application of 20 kg/ha each of S and Zn along with 2 kg/ha of Mo may enhance the effectiveness of biological nitrogen fixation in fodder legume crops.

Table 8. Nutrient requirement of different forage crops

Crops	Fertilizer dose		
	N	P_2O_5	K_2O
Maize	120	60	40
Sorghum	90	40	30
NB Hybrid	80	40	30
Guinea grass	100	50	40
Dinanath	60	30	30
Cowpea	20	60	20
Cluster bean	20	40	25
Stylosanthes	20	60	15
Berseem	50	60	40
Oat	80	40	30
Lucerne	20	60	40

Source: Sunil *et al.* (2012)

Pests and diseases of major forage crops

Forage crops are subjected to the depredation of pests and diseases in the same manner as in other agricultural produce. These pests and diseases hamper green fodder production, impair forage quality and seed yield. Besides this, they also cause indirect losses like nodule formation in legumes. Nematodes, aphids, leaf-hoppers and other sucking pests inflict direct losses to the crop, yet are also indirectly responsible for the transmission of a number of viral and other pathogens. Further, disease and pest interactions in the plant produce toxins, adversely affect animal health. Which sometimes can be fatal.

Although pests and diseases can cause spectacular damage, most losses go unnoticed because of its insidious nature along with lack of awareness among the farmers. In addition, sometimes damage tends to be patchy, and the plant stand is thinned or weakened rather than destroyed altogether. Many of the pests involved live either beneath the soil surface or hidden within plant tissues and are easily overlooked. Similarly, virus disease may reduce growth and vigour resulting in serious yield or herbage quality losses but producing no discernible symptoms on the plants. Some major diseases caused by the fungi in seedlings before and after emergence make the losses difficult to detect. Probably, however, the main reason why the damage passes unheeded is that there is usually no un-infested area with which to compare the infested ones. The value of many leguminous forages as a soil improver in the rotation further complicates assessing herbage yield losses at farm level. In addition, grasslands normally support and can tolerate a large number of pests and recent evidence suggests that there are some situations where damage is more likely to occur, especially during the establishing period of young seedlings. Moreover, grasslands act as a reservoir of pests which later can infect/infest cultivated food crops. Despite the above facts several researchers have been made to assess quantitative and qualitative losses caused by biotic stress factors in major forage crops (Fig. 4a & 4b).

Fig. 4a. Maize leaf blight

Fig. 4b. Berseem root rot

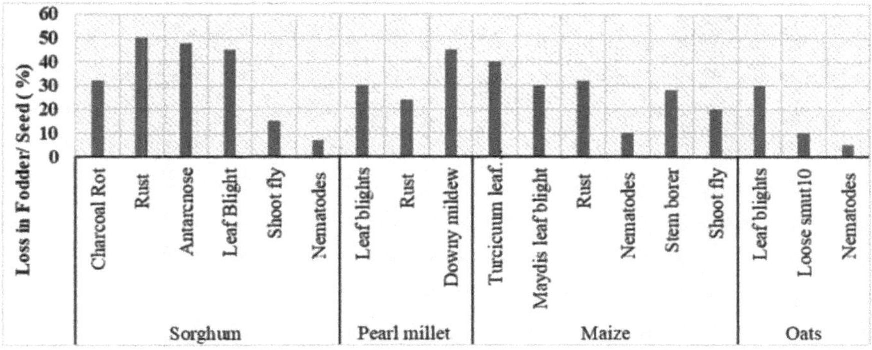

Fig 4a. Losses in cereal forage crops caused by various diseases and pests

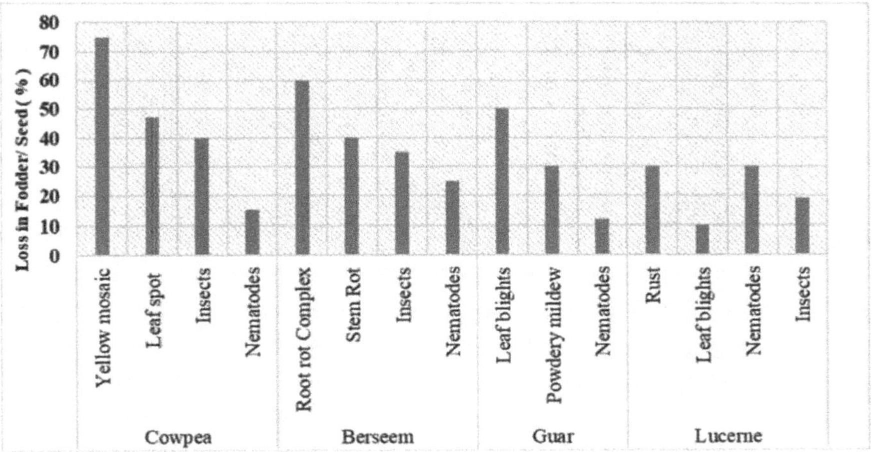

Fig 4b. Losses in legume forage crops caused by important diseases and pests

A large number of pests and diseases have been found associated with a variety of fodder crops grown in different parts of the country. The information relating to important fodder pests and diseases is given in table 9. Pests and diseases have been listed on the basis of their economic importance. Some specific insects and diseases are infecting particular forage crops.

Table 9. Important pests and diseases their nature of damage/ symptoms, distribution, favourable conditions of major fodder crops

Crops/pests & diseases	Insect/pathogen	Major symptoms	Distribution and Favourable conditions for disease occurrence
Berseem *(Trifolium alexandrinum)*			
Root rot complex	*Rhizoctonia solani, Fusarium semitactum, Tylenchorhynchus vulgaris*	The most prominent symptom is the sudden complete wilting of the plant. From the top to bottom, the entire plant is killed within a day. The affected plant can easily be pulled out. If the roots are looked into, it could be observed that except the taproot and few secondary roots all other roots are decayed and detached. The affected field gives a patchy appearance, which gives a patchy appearance, which is more or less circular areas.	Common in Gangetic and central plains. High moisture, dense plant population and poor drainage are some important factors responsible to increase the disease severity
Stem rot	*Sclerotinia trifoliorum*	The collar region becomes necrotic with depressed lesions. Stems are girdled at the collar region, and infected plants are easily pulled out with least resistance. Heavy mycelial growth can be seen in the infected crop field during low-temperature days.	Common in a temperate climate, the incidence also often Gangetic plains and northern parts during January and February when the temperature is quite low. The disease incidence is more in fields where the sowing is done by the broadcast method.
Pod borer	*Helicoverpa armigera*	The young caterpillars nibble the leaf surface, later as these grow they become voracious and destroys the whole crop. These cause more damage in crop left for seed.	It is a sporadic pest of berseem particularly in the raised for seed production
Cowpea (*Vigna unguiculata***)**			
Antracnose	*Colletotrichum lindmuthianum*	Circular, black sunken spots with dark centres and bright red-orange margins are formed on leaves and pods. In severe infections, the attacked parts wither off. Seedlings get blighted due to infection.	The disease favoured by warmer temperature and high relative humidity and periodic rainfall.

Root rot	*Macrophomina phaseolina*	Small circular brown spots are appearing on the cotyledon or the young leaves. The centre of the stem lesions assumes greyish colour and definite brownish margins with a black dot.	Dry weather following heavy rains and high soil temperature (35-37 °C).
Yellow mosaic Mosaic virus	*Begomovirus*	Initially, mild scattered yellow spots appear on young leaves. Spots gradually increase in size and ultimately some leaves turn completely yellow. Disease plants are stunted and mature late	Very common during warner months of the year when white fly population is more
Flea beetles	*Pagria signata*	The adult feeds on the cotyledons and leaves of young plants producing a shot hole effect with many small round holes all over the leaf surface.	This is the most damaging pest of cowpea during July and December
Leaf hoppers	*Empoasca kerri*	The edges of the leaf curled and turn yellow	Widely distributed in India and is one of the destructive pests of the north-western region.
Root-knot nematode	*Meloidogyne incognita*	The uprooted plant shows distinct galls, shaped like spindles, beads or club in the roots of the plant. The infected portion of the root also becomes curved and on the convex side of this curvature or above galls a large number of small thin roots	They can grow at a different temperature regime. However, the optimum soil temperature is 28-32 °C and Sandy type of soil.
Lucerne *(Medicago sativa)*			
Downy mildew	*Peronospora trifolii*	Small chlorotic spots on the upper surface of the leaves. These turn later to dark brown with a downy growth of mycelium on the lower surface.	The disease confined to north India and in central India during the cold months, i.e. December and January. Low temperature 21-30 °C, >90 % relative humidity coupled with drizzling.
Rust	*Uromyces striatus*	Small round or oblong dark brown pustules on leaves and petioles	Very common in north-west, western Himalayas, Gangetic plains and western plains of the country. Warm moist weather and high relative humidity
Common leaf spot	*Pseudopeziza medicagenis*	The leaves petiole and stalk of the plant have tea brown needle-like spots of 1-2 mm. Dark brown warty structures are produced in the central part of the lesion, becoming jelly like in wet conditions. The leaf turns yellowish brown and fall.	The disease is very serious in north-west, gangetic and central plains of the country. Disease is favoured by cool moist conditions.

Lucerne weevil	*Hypera postica*	Damage is caused by the larvae feeding within the plant tips on the leaves as they open and later on the lower foliage. In severe infestation they may consume all but main veins, causing the damaged field to appear whitish.	This is most damaging pest of Lucerne. When the weather suddenly warm and dry in spring and summer months favours the infestation
Aphids	*Acyrthosiphon pisum and Theriophis trifolii maculata*	These aphids suck sap from leaves and stems and are mostly confined to lower parts of the plant and ventral surface of leaves. Heavy infestation causes the plant to turn yellow, die and drop off. The seedling stage is more prone even under light infestations. The aphid secrets honey dew on which sooty mould grows which deteriorate the quality of hay.	Warm, dry conditions are most favourable, and peak activity is commonly observed during late spring, summer and autumn.
Guar *(Cyamopsis tetragonoloba)*			
Bacterial blight	*Xanthomonas campestris pv.cyamopsidis*	Small soaked lesions, which later becomes necrotic covering large areas of the leaves. These may appear on flowers and pods also.	Rainy season favours the disease development
Powdery mildew		White powdery patches appear on leaves and other green parts, which later becomes dull coloured. Severely affected plants get shrivelled and distorted.	High nitrogen, warm days with cool night and high humidity with cloudy days
Stylo (*Stylosanthes guianensis)*			
Anthracnose	*Colletotrichum gloesporioides*	Small, pinhead like necrotic spots on leaves and stem. These later become larger with dark margins and light brown centre. The tip of the branch may wilt, giving a blighted appearance to the plant.	The disease favoured by damp and hot climate.
Sorghum (*Sorghum bicolor*)			
Anthracnose	*Colletotrichum graminicola*	Appears first on the lamina of lower leaves as small circular spots later spreading to upper leaves. The colour of the spots may vary from red, purple to brown depending on the variety. These symptoms may be found on stems also.	The disease is favoured by low or unbalanced soil fertility.

Sooty stripe	*Ramulisporia sorghi*	On the leaves elongate, elliptical lesions with straw coloured centres and purplish to tan margins appear. These may coalesce to cover a larger area, which becomes necrotic.	The disease is prevalent in Haryana, Delhi, Uttar Pradesh, Andhra Pradesh and Tamil Nadu. In humid weather, the disease spreads rapidly. The optimum temperature for growth is 28 °C. The pathogen survives in the soil with fallen leaves and debris from season to season.
Zonate leaf spot	*Gloeocercospora sorghi*	On the leaf lamina, semi-circular to circular spots with alternating bands of dark purple or straw colour is formed. These may enlarge, become dark red and tend to elongate, parallel with leaf veins and form characteristic zonation's glumes may also be affected.	The disease is favoured by rains as the severity of the disease is high during the rainy season.
Downy mildew	*Scleropsora sorghi*	Abundant local rectangular chlorotic lesions appear which later coalesce to give a blighted appearance.	The incidence of this disease is higher in fodder sorghum. The fungus requires optimum temperature of 24-26 °C. High levels of systemic infection can occur at 11-32 °C with a wet period of 4 hours or longer.
Shoot fly	*Atherigona varia soccata*	Young seedlings of 5-20 days age are affected due to cutting of growing point by the larvae. This results in typical dead heart formation	Attacks the seedlings and late sown crops are attacked badly. The attack is severe from July to October. Cloudy weather favours multiplication of the insect. In rabi, early sown crop suffers more and hence sowing should be delayed possibly.
Stem borer	*Chilo partellus*	Young larvae feed on leaves producing scarification and shot holes. Third instar larvae bore at the base and produce dead heart and tunnels the stem.	The pest hibernates in the larval stage in stubbles. Seasonal occurrence: The pest is generally active from July to November. The infestation is more on rabi& summer crops.
Sorghum mildge	*Contarinia sorghicola*	The larva feeds on the developing seeds. The grain head is flattened with tiny shrunken heads.	

Sorghum cyst nematode	*Heterodera sorghi*	Affected crop gives unhealthy appearance with pale greenish yellow stunted plants. On uprooting the plants, elongated main root with excessive branching at the tip gives a twiggy appearance.	High soil moisture and sandy loam soil favours the multiplication of nematodes
Maize (*Zea mays*)			
Downy mildew	*Scleropthora rayssi* var. *zeae*	Narrow chlorotic or yellowish streaks which may be broken or continuous and 3-7 cm wide. The lesions extended in parallel fashion have well-defined margins and are delimited by veins which later become brown	The fungus requires an optimum temperature of 24-26 °C. High levels of systemic infection can occur at 11-32 °C with a wet period of 4 hours or longer.
Bacterial stalk rot	*Erwinia carotovora* var. *zeae*	The diseased plant shows rooting from base upward (basal rot) or the top downward (top rot). In basal rot, the leaves turn yellow and have brownish longitudinal discolouration at the base of the stalks. Top rot begins with and drying of the tips of the leaves. The decay spreads downwards and affects the stalk.	The disease is favoured by high temperatures and high relative humidity. It can be a problem in areas of heavy rainfall or where overhead irrigation is used and the water is pumped from a lake, pond, or slow-moving stream.
Stem borer	*Chilo partellus*	Young larvae feed on leaves producing scarification and shot holes. Third instar larvae bore at the base and produce dead heart and tunnels the stem.	The pest hibernates in the larval stage in stubbles. Seasonal occurrence: The pest is generally active from July to November. The infestation is more on rabi& summer crops.
Ear caterpillar	*Helicoverpa zea*	Caterpillars hatch and begin feeding on buds, new leaves, and flowers. Their distribution on plants changes as they grow and as plants develop. Their highest number usually occurs on leaves and the amount of leaf-feeding increases as caterpillars grow. Although not common, defoliation can become extensive in some fields, especially when large populations occur early before plants have set flowers.	Adult moths emerge from chambers from May through August, peaking in early July, and fly to suitable hosts including such as corn, sorghum, and many other wild and cultivated plants.

Maize cyst nematode	*Heterodera zeae*	Affected plants are stunted, pale in colour, with narrow leaves. In the field, stunting frequently occurs in irregular patches. The development of maize tassels may be noticeably delayed, and the maize plants bear smaller cobs with relatively fewer grains. The root system is poorly developed with a bushy appearance, and the presence of cysts on the root surface can be observed	Severe in coarse-textured soil, as well as under hot and dry conditions. Temperature plays an essential role in the biology of *H. zeae*, and favourable soil temperatures for most phases of the life cycle lay above 25 °C. At temperatures of 10-15 °C, only 10-20% of the juveniles emerge from cysts. The total life cycle from egg to reproducing adult is short and takes 15 to 17 days under favourable temperatures (about 27-39 °C). It is estimated that under these optimal conditions.
Pearl millet (*Pennisetum glaucum*)			
Downy mildew	*Sclerospora graminicola*	Chlorosis on the upper surface of the leaf and whitish sporangial growth on the under the surface. Downy growth is more clearly visible in the early hours. Sometimes the symptoms appear in the form of green ear a malformed earhead.	Under conditions of high humidity and moderate temperature, the infection could be very severe. Late sowing favour disease development. The atmospheric temperatures between 15-25 °C and relative humidity above 85 per cent. Light drizzling accompanied by cool weather is highly favourable
Ergot	*Claviceps fusiformis*	The florets produce a copious creamy, pink or red coloured sweet liquid called honeydew	Conditions favouring the disease are RH greater than 80% and 20 to 30 °C temperature
Smut	*Tolyposporium penicillariae*	The florets transform into large plum sacs containing black powder	Optimum environmental conditions for maximum infection include: temperatures of between 25 and 35 °C and slightly acidic soils favour the disease development.
Rust	*Puccinia peeniseti*	Lower leaves show typical erupt pustules containing a reddish-brown powder. Symptoms can occur lower and upper surface of the leaves.	A low temperature of 10 to 12 °C favours teliospore germination. A spell of rainy weather favours the onset of the disease.

Oats (*Avena sativa*)			
Leaf blotch	*Helminthossporium avenae*	The disease produces one or more narrow brown lesions on coleoptile. Long brick red blotches appear on young plants in the third and fourth leaf stage.	Disease development is favoured by soil temperature and moisture and slow growth of the host plant
Crown rust	*Puccinia coronata*	The disease is characterized by the appearance of reddish-brown to dark brown, oblong pustules on culms and leaf sheath. These pustules later turn black.	It is confined to the temperate climatic regions. The perpetuation and annual recurrence of the rust is aided by grasses belonging to the tribe *Avenae*, *Hordae* and *Phalaridae*
Rhizoctonia root rot	*Rhizoctonia solani*	The disease produces in patches with purplish look. These patches are defined with stunted plants, with a stiff erect habit of growth	Occurs primarily in early to mid-summer. High soil moisture, warm soil temperatures, soil types with high amounts of organic matter.
Aphid	*Rhapalosiphumm aidis*	The leaves, leaf sheath and inflorescence are covered with dark green aphid colonies with a slight white covering. Mottling and distortion of the leaf may occur. Honeydew production is quite prolific.	No rainfall during the season favours the peak population of aphids, and dense plant population helps to cause more damage in the standing crop.

Source: Pandey et al. (1995)

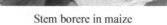

Alternaria and aphid damage in cenchrus

Stem borere in maize

Stem rot in berseen

Integrated pest and disease management in forage crops

Pests and diseases of forage crops have already been recognised as major thrust in agricultural production, especially in the developing countries. During past, information on pests and diseases of forage crops was generally restricted to record host disease list. Presently, changes are fast occurring in Indian agriculture like the use of intensive fodder production system is practised for producing maximum forage per unit time because of the availability of limited land for forage crops. Therefore, the same piece of land is intensively cultivated to maintain a continuous supply of green fodder round the year especially in milk shed areas. This provides more favourable conditions for the multiplication of pests. Therefore, it becomes important to assess the role of pests and diseases under varied conditions and utilize such information in developing variable and feasible pest and diseases management practices for the benefit of farmers.

Various cultural practices can reduce the infection to some extent; however, if the environmental condition favours the spread of diseases and pests, then the use of chemicals or resistant varieties is the only option of the management. However, the development of resistant varieties is the most desirable approach in the sustainable agricultural system. The All India Coordinated Research Project on Fodder Crops (AICRP-FC), with its 21 centres covering all the major agro-climatic situations of the country, is engaged in research on all the facets of technology development relating to forage crops. As a result, a wide range of germplasm material of various fodder crops has been collected and maintained at one or the other centres; such collections provide an excellent opportunity for screening the material for selection and breeding for resistance to major insects, diseases and nematodes. Some important plant protection technologies were developed by AICRP-FC centres for better yield and quality of forages (Mawar *et al*., 2015). An integrated pest and disease management (IPDM) module for lucerne and other management technology for berseem, cowpea, oat and white clover were also developed (Ahma *et al*., 1996). Moreover, it is important that the existing gap between fodder requirement and availability can be bridged to a considerable extent by avoiding losses caused by biotic stresses. The important integrated management approaches against pest and diseases are summarized below:

Pest and disease surveillance and forewarning

Crop health surveillance is an importance component of the IPDM system. Pest and disease surveillance involves the monitoring and severity, auxiliary crop information and other relevant data for decision making. In the present day context, surveillance focuses primarily on decision making for the proper timing of chemical pesticides usage thereby recommending the use of chemicals only if needed and also to reduce it to the bare minimum, thus avoiding unnecessary use in the agricultural system. For surveillance activities may help in identifying potentially, important diseases and pests early enough and indicate that the pest/ disease feared by farmers may not actually, warrant intensive crop protection. Thus, surveillance along with other logistic support can be an efficient tool of pest/disease forewarning.

Cultural methods

Although resistance varieties and chemical treatment are a very important method of pest and disease control, their efficacy can be improved, made more efficient, economical as well as environmentally acceptable by modification of cultural practices (Shah *et al.*, 2011). The judicious use of fertilizers, adjustment of sowing time, proper soil and water management, crop rotation, maturing, spacing, sanitation and proper weed management help in reducing the losses due to pests, thereby resulting in better yield. The cultural practices such as deep ploughing and burning results into decrease in soil borne diseases & pests of intensive fodder production system. Intercropping of mustard in wheat and barley was found to reduce the number of cyst/ plant and nematode infestation in barley (Rajvanishi *et al.*, 2002).

Sanitation

Sanitation is another effective method of controlling pests. Practices like removal of crop debris and regulating the irrigation water from fields with the diseased crop also helps in reducing the primary inoculums and spread of the disease. This is especially important in the case of soil borne diseases caused by species of *Fusarium*, *Verticillium*, *Rhizoctonia*, *Sclerotium* and *Sclerotinia* etc.

Seed health

The seed is one of the most important inputs in crop production. Apart from the replacement of old and obsolete varieties, many plant diseases and insect pests can be checked by the use of the healthy seed. Several crops suffer from diseases that are internally or externally seed borne. By properly identifying the disease and roguing those in the seed production plots will permit the production quality and healthy seed. Roguing and burning infected plants is a routine operation in healthy seed production.

Organic amendments

Organic amendments like the application of green manure, FYM, crop residues and various oil cakes are one of the important means of plant protection management (Mawar and Lodha, 2006). These instead of improving the plant health serve as very good substrates for the multiplication of soil micro-flora. Many of them have been found to act as a biocontrol agent to the plant disease causing organisms by hyper parasitism and competition. Incorporation of selected on-farm weeds after partial decomposition reduced the viability of the sclerotial population of *M. phaseolina* in the soil and the incidence of dry root rot in guar in the field under rainfed conditions.

Biological control

The natural mortality is the most desirable method in avoiding pest out breaks; hence, the use of parasite and predators is another method which can be useful in managing the diseases. In this method of diseases, control care has to be taken so that the organism used for biological control should not be pathogenic to the crop concerned or to other crops that are likely to be grown in the cropping sequence. Several examples of antagonism by micro-organisms to the plant pathogens have been studied. Root rot of berseem caused by *Rhizoctonia bataticola* are effectively controlled by the use of *Trichoderma harzianum* and *T. viride* applied as seed treatment (5 g/kg of seed). Beside diseases control these organisms have been found to enhance the plant growth and nodulation in roots there by significant increase in green fodder yield.

Chemical control

This is the most common method used for pest and disease management in the field and storage. The application of chemicals to plants for control pests and diseases has been practiced for many decades. Different groups of chemicals can be applied in several ways which are given below (Table 10).

Pre-emergence herbicides spraying using boom sprayer

Table 10. Seed treatment and foliar chemicals for important fodder crops

Crop	Chemical name	Fungicide/ insecticide	For controlling	Recommended doses
Maize	Thiram 75% WDP or Carbendazim or Vitavax	Fungicide	Seed borne diseases (for seed treatment)	2.5 g/ kg of seed
	Dithane M-45 (Mancozeb)	Fungicide	Leaf blight/Downy mildew diseases (foliar spray)	2-3 kg/ ha
Sorghum	Carbendazim /Thiram dust	Fungicide	Seed treatment to control seed borne diseases	2.5 g/ kg of seed
	Imidacloprid WS or Thiomethoxam 25 WSC	Insecticide	Shoot fly	Seed dressing @ 3g/ kg seed
	Imidacloprid	Insecticide	Shoot fly/ Army worm / Cut worm/ Sorghum midge / Stem borer	@ 0.5 ml/ litre of water
Pearl millet	Thiram 75% WDP or Carbendazim or Vitavax	Fungicide	Seed borne diseases (for seed treatment)	2.5 g/ kg of seed
Cowpea	*Tricoderma viridae*	Bio-fungicide	Root rot disease	5 g/ kg of seed
	Carbendazim/ Thiram dust	Fungicide	Root / Collar rot diseases (seed treatment)	2.5 g/ kg of seed
	Imidacloprid	Insecticide	Leaf eating insects/ aphids	0.5 - 1.0 ml/ litre of water as foliar spray
Lucerne and Berseem	*Tricoderma viridae*	Bio-fungicide	Root/stem rot disease (seed treatment)	5 g/ kg seed
	Carbendazim/ Thiram 75% dust	Fungicide	Seed borne disease (seed treatment)	2.5 g/ kg of seed
	Hexaconazole / Tebuconazol	Fungicide	Foliar fungal diseases such as rust, leaf spots and mosaic diseases	0.5 ml/litre of water
Oat and Barley	Vitavax or Carbendazim or Thiram	Fungicide	Seed borne fungal diseases (seed treatment)	2.5 g/ kg of seed

Source: Saxena et al. (2002)

Seed treatment

Chemicals applied to seeds before sowing will help in preventing their decay by pathogens carried by them or present in the soil. They are applied to seeds as dust or concentrated water suspensions followed by drying before sowing (Table 10). The chemical will diffuse into the surrounding soil, disinfect the root zone of seedlings, and help in the growth of healthy plants. The insecticides applied through seed treatment, which are mostly systemic in nature help in protecting the young seedlings from insect pests.

Harvesting management

The best harvesting time of green fodder varies with crop species (Table 11). However, in all the crop species during the flowering stage, the nutrient content will be maximum. Therefore, the ideal condition for green fodder harvesting is 50% flowering stage in most of the forage crops. In case of single cut varieties of sorghum and pearl millet, crop is to be harvested at 60-70 days after sowing (50% flowering) and multi cut varieties first cut should be taken at 40-45 days and subsequent cut at 30 days intervals. The maize crop is ready for harvest at the silk stage (60-75 DAS) for fodder purpose, which continues up to milk stage. The early harvesting produces good quality fodder but the yield is reduced, whereas the fodder quality is adversely affected due to late harvesting in maize. The proper stage of harvesting determines the herbage yield and quality of oat and harvesting of single cut oat varieties is done at 50% flowering stage. Whereas, in case of multi-cut varieties, first cut should be taken at 8-10 cm above the soil surface for good re-growth and recommended at 60 days stage while the second cut at 45 days after first cut and third cut at 50% flowering stage. The first cut in berseem will be at 55 DAS followed by 25-30 days interval based on irrigation facility available. Before leaving the crop for seed production purpose, maximum of three cuts for green fodder can be taken up under well-irrigated conditions. The date of the last cut is crucial in deciding the seed productivity of berseem (Yadav *et al.*, 2015). The seed harvesting of berseem during physiological maturity, when the head turns brown and contains yellowish mature seeds and the stalks are still green in colour, almost doubles the productivity by reducing the shredding losses, which is a major problem in berseem (Vijay *et al.*, 2016).

The first cut in case of grasses like NB hybrid, guinea is taken up at 60-65 days after planting and subsequent cuts at 30-35 days interval. In perennial grasses the seed harvesting is possible in two to three pickings due to the indeterminate nature of flowering. The morphological indices like spikelet colour, ease of separation etc. helps in identification of suitable harvesting stage. The harvesting stage is very crucial in grasses as the spikelet will start shedding attaining maturity. The severe shedding loss further reduces the harvest index in grasses.

Table 11. Harvesting and cutting management of various fodder crops

Crop	No. of cuttings	Harvesting stage
Berseem	Up to 6	First cut at 55 DAS later at every 25-30 DAS
Lucerne	Up to 5	First cut at 50-55 DAS later at every 30 DAS
Oat (multi-cut)	3	First cut at 60 DAS Second cut at 45 days after first cut Third cut at 50% flowering
Oat (single cut)	1	50% flowering
Sorghum (single cut)	1	60-75 DAS (50% flowering)
Sorghum (multi-cut)	2-3	First cut at 45 DAS and later at 30 days interval
Maize	1	60-75 DAS (silk stage)
Pearl millet	1	55-60 DAS (50% flowering)
Cowpea	1	50-60 DAS (50% flowering)
Guar	1	60-75 DAS (blooming to pod formation stage)
Perennial grasses like NB hybrid, guinea etc.	6-7	60-65 DAP later at 30 days except during extreme winter and summer months

Prospects of forage crop production

The demand for forage is to be realized with the enhancement of production and productivity both in cultivated and range species. The policy level interventions are needed to enhance the overall growth of this sector. The seed production of forage crops by the national and state seed corporations is the need of the hour. The development of the organized market will attract private players who help in the spread of forage crops at the commercial level. The research on climate resilient production technologies will help in ensuring the future. The advent of modern tools like marker-assisted breeding, development of stress-tolerant cultivars, enhancing the fodder value of crop residues, livestock-based farming systems for small and marginal farmers, round the year fodder production helps in augmenting the forage productivity *per se*. Some of these issues are already under study and the results will be available in a couple of years for the further enhancement of forage crop production to new levels.

Mechanization is another issue in forages. Even though in cultivated crops, it has penetrated to a large extent, the range species are still to be covered. Mechanization boosts the largescale production and reduces the labour cost. Along with crop production, there is a need to emphasize on forage seed production also as quality forage seed is one of the limiting factors in the spread of forage cultivation particularly in case of range species.

Conclusion

Forage crops are often neglected species among the cultivated lot resulting in the severe deficiency of green and dry fodder to the livestock. The stagnation in the area and yield levels of cultivated fodder and a gradual decline in permanent pastures made the productivity low in Indian livestock. With the increase in consumption of milk and meat products as well as the stability provided by livestock to the marginal farmers emphasises the importance of this sector in Indian farming. The increase in livestock population and its dependence on green fodder emphasizes the need for advancement in forage production technology under limited land availability. Starting from the selection of suitable crop species for the soil and climatic conditions to the infusion of various research outputs in the form of production techniques has been discussed in detail for effective implementation. The weed management is especially crucial in fodder crops as the weeds will decrease not only the yield but also the quality of the green fodder.

Similarly, the pest and diseases directly affect the yield and quality of main economic product i.e. green fodder apart from the seed yield. The integrated pest and disease management discussed in the chapter provides the best possible solution as the green matter is used as fodder for animals. The additional cutting management apart from the final harvest is vital for achieving the higher fodder output. Thus, the overall management is unique in fodder crops and various techniques were discussed under different categories for attaining the higher yields. The information of this chapter provides practical utility apart from fulfilling the academic interest and thus is useful to field-level workers as well as teachers and students.

References

Ahmad, S.T., Pandey, K.C. and Bhaskar, R.B. 1996. Integrated pest management for increased forage production. *Indian Farmg.* **45**: 34-37.

BAHS. 2019. Basic animal husbandry statistics-2019. Government of India. pp.132

Economic survey. 2017-18. Economics division, department of economic affairs, ministry of finance, Government of India, New Delhi.

FAO STAT. 2018. Food and Agriculture Organization of United Nations, Rome, Italy.

Handbook of Agriculture. 2006. Indian Council of Agricultural Research, New Delhi pp.1346.

IGFRI. 2011. Vision 2050. Indian Grassland and Fodder Research Institute, Jhansi.

IGFRI Annual Report. 2016-17. ICAR-Indian Grassland and Fodder Research Institute, Jhansi. pp 137.

Kumar, S., Agrawal, R.K., Dixit, A.K., Rai, A.K. and Rai, S.K. 2012. Forage crops and their management. Indian Grassland and Fodder Research Institute, Jhansi. pp 60.

Land use statistics at a glance. 2017. Directorate of economics and statistics, department of agriculture & cooperation, ministry of agriculture and farmers welfare, Government of India, New Delhi.

Mawar, R. and Lodha, S. 2006. Relative efficacy of on-farm weeds as soil-amendment for managing dry root rot of cluster bean in an arid environment. *Phytopathologia Mediterranea*. **45**: 215-224.

Mawar, R., Mall, A.K. and Kantwa, S.R. 2015. Eco-friendly plant protection technologies for forages. *BioEvolution*. **2** (1): 31-37.

Pandey, K.C., Hasan, N., Bhaskar, R.B. and Hazra, C.R. 1995. Pests and diseases of major forage crops. AICRP, IGFRI, Jhansi. pp 1-22.

Rajvanshi, I., Mathur, B.N. and Sharma, G.L. 2002. Effect of inter-cropping on incidence of *Heterodera avenae* in wheat and barley crops. *Annl. of Plant Protec. Sci.*. **10**: 406-407.

Roy, A.K., Agrawal, R.K., Shahid, A., Kumar, R.V., Mall, A.K., Bharadwaj, N.R., Mawar, R., Singh, D.N., Kantwa, S.R. and Faruqui, S.A. 2018. Database of forage crops: 2018. AICRP on forage crops & utilization. ICAR-IGFRI, Jhansi–284 003. pp. 330.

Saxena, P., Shah, N.K., Hasan, N., Pandey, K.C., Faruqui, S.A., Bhaskar, R.B., Padmavati, C.H., Roy, S. and Azmi, M.I. 2002. *Forage Plant Protec.* IGFRI, Jhansi. pp 38.

Shah, N.K., Sahay, G. and Tyagi, P.K. 2011. Field selection for resistance in cowpea (*Vigna unguiculate* (L) Walp.) to major insect pests. *Range Mangt. and Agroforst*. **32**(2): 138-140.

Singh, S., Mishra, A.K., Singh, J.B., Rai, S.K., Baig, M.J., Biradar, N. and Verma, O.P.S. 2014. Water requirement estimates of feed and fodder production for Indian livestock *vis* a *vis* livestock water productivity. *The Indian J. of Animal Sci*. **84**(10):1090-1094.

Vijay, D., Manjunatha, N., Maity, A., Kumar, S., Wasnik, V.K., Gupta, C.K., Yadav, V.K. and Ghosh, P.K. 2016. Berseem-intricacies of seed production in India. Indian Grassland and Fodder Research Institute, Jhansi. pp 47.

Yadav, P.S., Vijay, D. and Malaviya, D.R. 2015. Effect of cutting management on seed yield and quality attributes of tetraploid berseem. *Range Mangt. and Agroforst*. **36**(1): 47-51.

11

Restoration of Degraded Sodic Lands Through Agroforestry Practices

Y.P. Singh

Introduction

Worldwide, salt–affected areas are estimated to range from 340 million ha to 1.2 billion ha (FAO, 2007; Oo *et al.*, 2015; Ahmad *et al.*, 2016; Drake *et al.*, 2016). Millions of hectare of these salt affected soils are suitable for agricultural production but are unexploited because of salinity/sodicity and other soil and water related problems (Abrol *et al.*, 1988). Salt-affected soils are reported to comprise 42.3% of the land area of Australia, 21.0% of Asia, 7.6% of South America, 4.6% of Europe, 3.5% of Africa, 0.9% of North America and 0.7% of Central America (El-Mowellhey, 1998). In India, out of 329 million ha geographical land area of the country about 175 million ha suffers from different problems and is getting further degraded through natural or man-made processes. Majority of these lands is treated as wastelands as their productivity is low due to soil based constraints like water logging, salinity and sodicity. According to FAO, salinization of arable land will result in 30% to 50% land loss by the year 2050 if, remedial actions are not taken. Recent estimates indicate that 6.73 million ha (NRSA and Associates, 1996) land is suffering from salinity and sodicity problems in India. High salt deposits inherited by the soil from the original parent material during soil forming processes and poor drainage are important factors contributing to the development of such soils. These lands occur in different biogeographic zones and therefore consist of diverse morphological, physical, chemical and biological properties. These soils are universally low in fertility and due to the adverse edaphic environment; they are devoid of any vegetation because of excessive exchangeable Na^+ associated with high pH (>8.5) which impairs the physical condition of the soils and adversely affects water and air movement, nutritional and hydrological properties of the soils (Suarez *et al.*, 1984; Gupta and Abrol, 1990; Sumner, 1993; Garg, 1998).These lands are largely carbon-

Corresponding author email: ypsingh.agro@gmail.com

depleted but can be brought back to their native carbon-carrying capacity by reforestation through agroforestry systems. The presence of $CaCO_3$ concretions at various depths (caliche bed) causes physical impedance for root proliferation, therefore, making it difficult for tree establishment and restrict the choice of arable crops to be grown (Shukla *et al.,* 2011; Singh *et al.*, 2012a).

Unutilized sodic soils include village community lands, government lands reserved for specific purposes and the lands lying abandoned near roads, canals and railway tracks (Singh *et al*., 2011). In addition to this, a large part of salt affected soils belonging to the individual farmers is lying unproductive. However, such lands can effectively be utilized in productive land use systems applying suitable techniques. Since, no additional land resources are available for horizontal expansion of agriculture, we must find out viable technologies for utilization of existing land resources including the wastelands in order to meet future requirements of food, fodder and fuel. Therefore, it has become necessary to develop the salt affected marginal wastelands under suitable agroforestry systems. Agroforestry is one of the important land uses for restoration of these lands improving livelihood of resource poor farmers who cannot afford the reclamation expenditure of these lands (Pandey *et al.,* 2011).

There are two approaches for the rehabilitation of sodic lands. The first is by improving soil properties through suitable chemical amendments (Singh *et al*., 2016) and the second is to establish agroforestry plantations that tolerate sodicity (Singh *et al*., 2010; Singh *et al*., 2011; Singh *et al*., 2014; Singh *et al*., 2016). Judicious use of these lands can substantially contribute in meeting out the increasing demands of food, fodder, fuel and timber in the country and have tremendous potential for carbon (C) sequestration not only in above ground C biomass but also root C biomass in deeper soil depths.

Introduction of alternate land use planning with low capital intensive agroforestry systems provides an alternative to control further deterioration of degraded sodic soils and also maintain the health of these soils (Singh *et al*., 2008; Singh *et al*., 2012; Tripathi and Singh, 2005; Mishra *et al*., 2004). During last four decades the management of these soils has been done largely for crop production, and there were only a few attempts to afforest. Studies conducted at various locations in India have shown that sodic soils can be ameliorated by different agroforestry systems, which ameliorate the soil to various degrees through the addition of large amount of organic matter and nutrients from litter and fine roots and improve the physical and chemical properties as well as biological activity in the soil (Yadav, 1980; Singh, *et al*., 1998; Singh *et al*., 2010; Singh *et al*., 2014; Mishra *et al*., 2002; Singh and Dagar, 2005; Singh *et al*., 2011; Singh, 2009).

Extent of salt-affected soils in India

There are various reasons associated with the formation of salt-affected soils that are both natural and anthropogenic. The geological deposition of clay minerals comprises quartz, feldspars (orthoclase and plagioclase), muscovite, biotite, chloritised biotite, tourmaline, zircon and hornblende in their sand fractions (Bhargava *et al.*, 2011). Quartz and feldspars occur distinctly in the salt fraction. However, illite, mixed layer minerals, vermiculite, and chloride are common to both the silt and clay fractions. The mixed layer minerals vermiculites and smectite in these soils originate from biotite mica. Different workers have reported variable estimates of salt-affected soils in India. According to the latest estimation in India, salt-affected soils occupy about 6.73 million ha of land, which is 2.1% of the geographical area of the country (Sharma *et al.*, 2004). Out of 584 districts in the country, 194 have salt-affected soils (Fig. 1). Major part of salt-affected soils in India are commonly found in Indo-Gangetic plains (IGP) of Uttar Pradesh, Punjab, Haryana, Rajasthan, Bihar, and West Bengal which occupied 2.35 million ha salt affected soils (Mandal and Sharma, 2006) (Fig. 2).

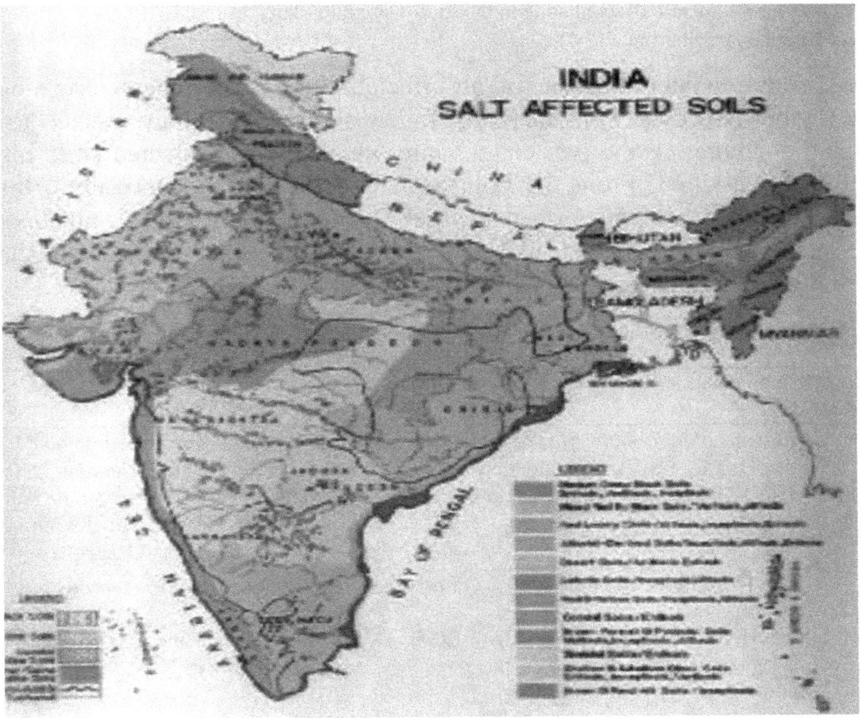

Fig. 1. State-wise extent of salt-affected soils in India (million ha) *Source*: Sharma *et al.* (2004)

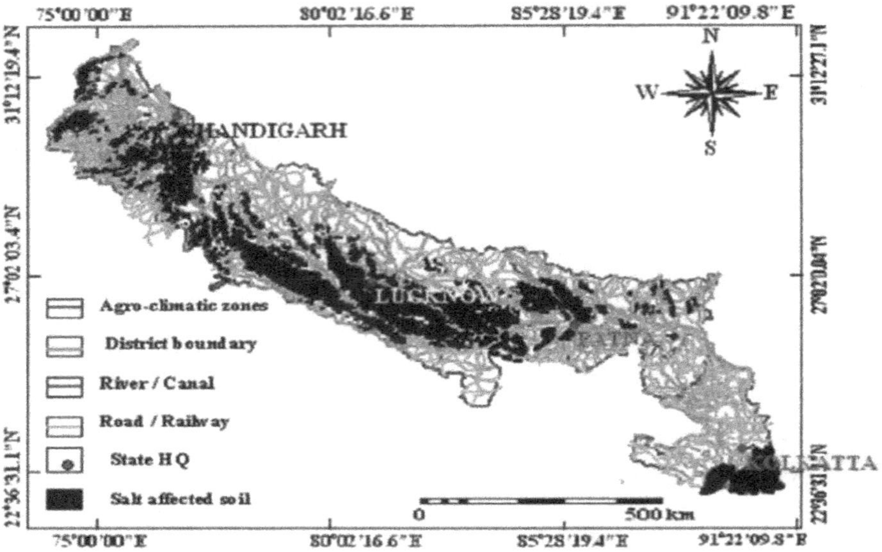

Fig. 2. Distribution of salt affected soils in the Indo-Gangetic Plains in India (*Source*: Mandal and Sharma, 2006)

There are three distinct categories of salt-affected soils from soil characterization point of view. However, from the point of view of clay mineralogy, micaceous/ illite and smectite are the two classes into which the salt-affected soils can broadly be classified. From the point of view of soil characterization, the salt-affected soils are categorized in three distinct categories, i.e., alkali or sodic, saline, and acid sulphate saline. The distribution of salt-affected soils in different geo-climatic regions of India is given in table 1.

Table 1. Distribution of salt-affected soils in different geo-climatic regions in India

S. No.	Main characteristics	Rainfall (mm/ annum)	Distribution
1	*Alkali soils of indo-Gangetic alluvial plain, developed on less calcareous alluvium*		
	High pH, EC, ESP and preponderance of sodium bicarbonate and carbonates	600 – 1,000	Parts of Punjab, Haryana, UP, South Bihar, Palwama and Badgam districts of Kashmir, Jammu region and Rajasthan.
2.	*Alkali soil of indo-Gangetic alluvial plain developed on fine, highly calcareous alluvium*		
	High pH, EC, ESP and preponderance of sodium bicarbonate & carbonates.	1,000 – 1,400	North Bihar and parts of Western UP

3.	*Inland saline soils of arid and semiarid regions*		
	Neutral to alkaline pH, high EC and preponderance of chlorides &sulphates.	< 500	Part of Punjab, Haryana, UP, Rajasthan, Gujarat, Leh district of Jammu & Kashmir.
4.	*Inland saline soils of sub-humid regions*		
	Neutral to alkaline pH, high EC and preponderance of chlorides &sulphates.	1,000 – 1,400	North Bihar
5.	*Inland salt-affected deep black soils (vertisols)*		
	Neutral to highly alkaline pH, variable EC and preponderance of chlorides and sulphate with miner amounts of sodium carbonate &smectitic mineralogy.	700 – 1,000	Parts of Madhya Pradesh, Maharashtra, Rajasthan, Andhra Pradesh, Gujarat, Karnataka and Tamil Nadu.
6.	*Medium to deep black soils of deltaic and costal semi-arid regions*		
	Neutral to highly alkaline pH, high EC, preponderance of chlorides and sulphates with or without sodium bicarbonate smectitic mineralogy.	700 – 900	Saurashtra region of Gujarat, and deltas of Godavari, Krishna and Cauveri river in Andhra Pradesh and Tamil Nadu.
7.	*Saline micaceous, deltaic alluvium of humid regions*		
	Neutral to slightly acid pH, high pH and preponderance of chlorides.	1,400 – 1,600	Sunderban delta in West Bengal and parts of Mahanadi delta in Orissa.
8.	*Saline humic and acid sulphate soils of humid tropical region*		
	Acid pH, high EC, presence of humus (Organic) horizon and preponderance of chlorides and sulphates, sulpher and pyritic material.	200 – 3,000	Malabar coast of Kerala and parts of Sunderban delta in West Bengal.
9.	*Saline marsh of the rann of kachh.*		
	Neutral to slightly alkaline pH, high EC and preponderance of chlorides and sulphates.	< 300	Rann of Kachh of Gujarat.

Source: Bhargava and Kumar (2011)

Characteristics of sodic soils

Sodic soils suffer with a varying level of degradation in structural, chemical, nutritional, hydrological and biological properties. These soils are compact and heavy with a high bulk density and silty clay loam texture (Typic Natrustalf). They also have a higher proportion of sodium in relation to other cations in soil solution and on the exchange complex. The sodic soils of the IGP are generally gypsum ($CaSO_4.2H_2O$) free but are calcareous, with $CaCO_3$ increasing with depth, which is present in an amorphous form, in a concretionary form, or even as an indurate bed at about 1m of depth. A high pH (>10) and high exchangeable sodium percent (ESP) of more than 60 imbalances the ionic equilibrium of the soil solution which leads to abnormal nutrient physiology. The high ratio Na

: Ca and low ratio of C : N cannot sustain the vegetation. The growth of most crops on sodic soils is adversely affected because of impairment of physical conditions, disorder in nutrient availability and suppression of biological activity due to high pH. Deficiency of some micronutrients (Zn, Fe, Cu, Mn) and toxicity of other elements (Na, B, Mo) further aggravate the situation for a stress growth of whatever plants exist on such land. Poor water permeability (hydraulic conductivity and infiltration rate) due to interlocked pore space as well as compactness impedes the root development of plants (Fig. 3). A wide range of microbial population and diversity do not exist in sodic soil due to hostile conditions, which retards the rate of litter decomposition and nutrient mineralization leading to poor nutrient availability in the growing plants. The soil properties of a typical sodic soil profile are given in table 2.

(A) (B)

Fig. 3. Typical sodic soil (A) during summer (B) after drying

Table 2. Physico-chemical properties of a typical sodic soil profile of IGPs

Soil parameters	Soil depth (cm)							
	0-15	15-30	30-45	45-60	60-75	75-90	90-105	105-120
Sand (%)	63.4	48.0	48.6	42.0	54.0	57.4	46.5	52.5
Silt (%)	18.8	25.2	38.0	23.0	28.7	25.0	22.4	25.0
Clay (%)	17.8	26.8	35.0	35.0	34.2	43.6	33.9	32.5
Textural class	l	sil	sil	cl	sicl	cl	cl	cl
Bulk density (g/cm^{-3})	1.64	1.57	1.55	1.51	1.50	1.48	1.48	1.46
pH (1:2)	10.5	10.6	10.6	10.4	10.2	9.8	9.8	9.7
EC(1:2) d/Sm	1.43	2.42	2.02	0.86	0.64	0.64	0.26	0.44
ESP	89.0	82.6	82.0	80.2	80.0	66.0	63.4	60.4
O.C. (g/kg)	0.8	0.8	0.6	0.6	0.8	0.8	0.6	0.6
$CaCO_3$ (g/kg)	14.1	12.6	23.2	23.2	37.7	89.4	116.9	124.6
Available N(kg/ha)	94.00	62.72	54.60	47.04	45.10	45.04	40.60	37.63
Available P(kg/ha)	19.5	18.6	17.5	17.7	18.2	17.0	16.1	16.6
Available K (kg/ha)	388	388	321	404	290	278	199	169
Ca^{++} + Mg^{++} (meq/l)	2.60	2.10	2.20	1.60	1.60	1.60	1.75	2.10
Na (meq/l)	242.0	119.0	110.0	52.7	21.4	9.00	7.5	5.00
K(meq/l)	0.22	0.12	0.06	0.04	0.04	0.03	0.02	0.01
CO_3 (meq/l)	188.0	84.00	54.5	26.00	8.2	2.00	2.0	1.00
HCO_3 (meq/l)	18.0	21.00	21.0	21.00	11.5	5.50	4.5	3.50
Cl (meq/l)	33.0	12.0	11.0	11.0	7.0	3.0	3.0	3.0

Source: Singh *et al.* (2011)
l= loam, sil= silty loam, sicl= silty clay loam, cl=clay loam; pH_2 and EC_2: Refers to soil: water suspension ratio of 1:2;
ESP: Exchangeable sodium percent

Species selection for sodic soils

Selection of species is the first criterion to undertake any plantation work on sodic lands. The selection of right species matching with site conditions may perform better than abrupt plantation of any species. The selection of suitable species for high biomass and bio energy production in sodic soils depends upon the tolerance of the species to sodicity, suitability to local agro-climate, and purpose of plantation. Several studies have been conducted to evaluate the performance of a large number of trees, grasses, hurbs and shrub species for sodic conditions in India. *P. juliflora, A. nilotica, Salvadora oleoides, S. persica, Capparis deciduas, C. sepiaria* and *Clerodendrum phlomidis* are among the prominent woody species found on high pH soils, while *Acacia leucophloea, A. eburnear, Mimosa hamata, Prosopis cineraria, Butea monosperma, Diospyro stomentosa, Balanites roxburghii* and *Maytenus emerginatus* are frequent on slightly low (up to 9.0) pH. Among bushes of *Capparis* and *Salvadora*, climbers such as *Asparagus racemosus, Cocculus pendulus, C. hirsutus, Cayratia trifolia, Momordica dioica, Mukia maderaspatana, Achyranthes aspera, Withania somnifera* and *Ichnocarpus prutescens* are quite common. *Calotropis procera, Datura metel, Adhatoda vasica* and *Ziziphus nummularia* form isolated patches. Among herbaceous species *Desmostachya bipinnata, Sporobolus marginatus, S. coromandelianus, S. diander, Chloris virgata, Cynodon dactylon, Dichanthium annulatum, Euphorbia hirta, E. thymifolia, Trianthema triquetra, Suaeda fruticosa, S. maritime, Pluchea lanceolata*, and *Kochia indica* are prominent. During rainy season *Cassia tora, C. occidentalis, Abutilon indicum, Croton bonplandianum, Eclipta prostrata, Phyllanthusf raternus, Amaranthus viridis, Corchorus spp., Chenopodium ambrosioides, Trianthemum portulacastrum* and many other herbs form association with many grasses and sedges particularly in protected areas. Grasses are in general more tolerant to the sodic conditions than most field crops. Some of the promising grasses identified for cultivation in highly sodic soils are: Karnal grass (*L. fusca*), Rhodes grass (*Chloris gayana*), Paragrass (*Brachiaria mutica*) and Bermuda grass (*Cynodon dactylon*) (Kumar and Abrol, 1986).

For fuel wood production *P. juliflora, A. nilotica* and *Tamarix articulata* were found most successful on highly sodic soil (pH 10 or more). These trees produced 51, 70 and 97 tonnes/ha biomass, respectively in 7 years of growth on these soils. These species not only produced economic yield but also improved soil conditions in terms of organic matter. *Beta vulgaris* and *Nypafruticans* have been identified as potential source of liquid fuels while species such as *Jatropha curcas*, *Pongamia pinnata* and *Euphorbia antisyphilitica* are among potential diesel-fuel plants and these can be grown successfully on degraded sodic lands (Singh, 2015; Singh *et al.*, 2016). The energy yield (in the form of biogas) from *L. fusca* has been estimated at 15×10^6 Kcal/ha (Jaradat, 2003).

Agroforestry systems for sodic soils

***Silviculture system*:** Field study to identify sodicity tolerant tree species was initiated in 1995 at ICAR-Central Soil Salinity Research Institute (CSSRI), Regional Research Station, Research farm, Shivri, Lucknow in north India (26^0 47 ' 45"-26^0 48' 13" N, 80^0 46' 7"-80^0 46' 32" E) on a virgin soil represents a typical alkali soil in subtropical region of central IGPs in India. The soil reaction was strongly alkaline in surface 0-15 cm (pH_2 10.5) and sub-surface horizons (pH_2 8.8-10 and $EC_2$1.43 d/Sm), ESP ranges from 85-92, organic carbon 0.8 g/kg, available N 94 kg/ha, available P 19.5 kg/ha and K 388 kg/ha. Six months old saplings of ten multipurpose tree species tree species viz. *Terminalia arjuna, Azadirechta indica, Prosopis juliflora, Pongamia pinnata, Casuarina equisetifolia, Prosopis alba, A.nilotica, Eucalyptus tereticornis, Pithecellobium dulce and C. siamea* were transplanted in randomized block design with 4 replications on 26th September 1995. These species were planted in auger holes of 45 cm diameter at the surface and 20 cm at the bottom and 120 cm deep which were filled with a uniform mixture of original soil + 4 kg gypsum + 10 kg FYM + 20 kg silt keeping row to row and plant to plant spacing of 4 m and 3 m, respectively. For proper establishment of the saplings, three irrigations were given at monthly interval during first year of planting and after that only one irrigation with good quality water in summer (June) month were given. No fertilizer was applied to the plants during the study period. The observations on survival percentage, plant height, diameter at stump height (DSH), diameter at breast height (DBH), crown diameter and number of stems and branches at 1/3rd plant height were taken and reported on average basis. Among the tree species planted *P. juliflora, T. arjuna, P. pinnata, A. nilotica, C. equisetifolia* and *P. dulce* found highly tolerant under sodic conditions and showed > 95 % survival after 10 years of planting. Maximum plant height was attained by *E. tereticornis* followed by *C. equisetifolia* and *P. juliflora.* Amongst all the species tried, the growth in case of *C. siamea* and *A. indica* was quite slow. Maximum crown diameter was recorded with *P. Juliflora* followed by *A. nilotica. However, m*aximum diameter at stump height (DSH) was in *E. tereticornis* (9.42 m) followed by *P. pinnata* (8.60 cm) and *P. juliflora* (8.43cm). Maximum tree areal biomass (140.0 kg/tree) was recorded with *P. juliflora* followed by *A. nilotica* (123.6 kg/tree) and *C. equisetifolia* (105.6 kg/tree) harvested at 10 years age (Table 3).

Table 3. Dry biomass production of 10 species and allocation into root and shoot components

Tree species	Biomass (kg/tree)			Shoot: root ratio	Lopped biomass (Mg/ha)	Biomassat harvest (Mg/ha)	Total biomass (Mg/ha)
	Shoot	Root	Total				
T. arjuna	83.25	24.48	90.75	3.40	11.33	41.62	52.95
A. indica	38.45	9.80	48.25	3.51	6.40	19.22	26.62
P. juliflora	113.00	27.03	140.03	4.18	13.77	56.50	70.27
P. pinnata	53.20	14.73	67.93	3.61	10.09	26.60	36.69
C. equisetifolia	84.20	20.28	104.48	4.15	11.01	42.10	53.11
P. alba	55.50	13.65	69.15	4.06	11.37	27.75	39.12
A. nilotica	101.50	22.14	123.64	4.50	12.34	50.75	63.09
E. tereticornis	63.54	16.60	80.14	3.82	7.90	31.77	39.67
P. dulce	64.50	19.65	84.15	3.28	8.20	32.25	40.45
C. siamea	43.30	14.80	58.10	2.92	2.71	21.65	24.36
LSD *(P=0.05)*	6.34	2.36	-	0.54	1.12	5.42	7.52

Source: Singh *et al.* (2011)

P. juliflora produced the highest energy harvest of 1267.75 GJ/ha followed by *A. nilotica* with 1206.0 GJ/ha and the lowest of *A. indica* (520.66 GJ/ha). Leaves had slightly higher heat of combustion (21.40-23.71 MJ/kg) whereas; it was lowest in stem (20.45–23.23 MJ/kg) (Table 4). The calorific values of stem and branches exhibited less variation, with *A. nilotica* having the highest heat combustion in both stem and branches (23.23 and 24.24 MJ/kg, respectively). The differences in total energy production and its allocation to different plant parts, led to variation between biomass yield and its allocation to stem, branch and leaves per hectare.

Table 4. Energy values of different plant components in ten tree species.

Species	Calorific values (MJ/kg)			Total energy (GJ/ha)
	Stem	Branch	Leaf	
Terminalia arjuna	22.57	21.60	23.24	933.53
Azadirechta indica	20.60	20.54	21.42	520.66
Prosopis juliflora	22.53	23.20	23.71	1267.75
Pongamia pinnata	21.60	21.60	22.34	576.85
Casuarina equisetifolia	22.20	22.14	22.21	934.11
Prosopis alba	21.46	22.20	23.21	607.13
A.nilotica	23.23	24.24	23.64	1206.32
Eucalyptus tereticornis	20.45	22.43	21.40	662.12
Pithecellobium dulce	21.50	21.60	22.64	696.26
Cassia siamea	21.40	21.68	22.58	466.89

Source: Singh *et al.* (2010)

Study indicated that tree plantations on sodic soils improved the soil health in terms of pH, EC, ESP and organic carbon. Results indicated that soil pH reduced from 10.50 (control) to 9.53, 9.70, 9.74, 9.80, 9.81, 9.84, 9.89, 9.95,

10.00 and 10.01 in the surface soil of *P. juliflora, A. nilotica, P. pinnata, E. tereticornis, A. indica, T. arjuna, P. alba, P. dulce, C. equisetifolia* and *C. siamea,* respectively after 10 years of plantations. Similarly, electrical conductivity (EC) declined significantly from 1.43 dS/m to 0.30, 0.33, 0.39, 0.61, 0.63, 0.69, 0.70, 0.77,0.86 and 1.26 dS/m in the surface layer under *P. juliflora, A. indica, T. arjuna, P. pinnata, C. siamea, P. alba, P. dulce, A. nilotica, E. tereticornis* and C. *equisetifolia,* respectively. There was a remarkable decrease in exchangeable sodium percentage at 0-15 cm soil depth after 10 years of plantation from its initial value of 85 to 34, 41, 43, 46, 48, 48, 52, 56 and 61 under *P. juliflora, A. nilotica, P. pinnata, E. tereticornis, A. indica, T. arjuna, P. alba, P. dulce,* C. *equisetifolia* and *C. siamea,* respectively. The increase in organic carbon content of the surface soil (0-15 cm) in a span of 10 years was about 4 fold under *P. juliflora* and *P. pinnata* and about 3 folds in other species. Available N content in the surface soil increased from 94.5 kg/ha to 138.5, 131.40, 126.30, 122.60 and 120.00 kg/ha under *P. juliflora, C. equisetifolia, A. nilotica, T. arjuna* and *P. pinnata,* respectively and smaller increase in other species. The available P and K in soil after 10 years of plantation was also higher in *P. juliflora* followed by *C. equisetifolia, A. nilotica, T. arjuna, P. pinnata* and *P. dulce.* However, in natural fallow where there was no plantation, the available N, P and K status was slightly improved (Fig. 4).

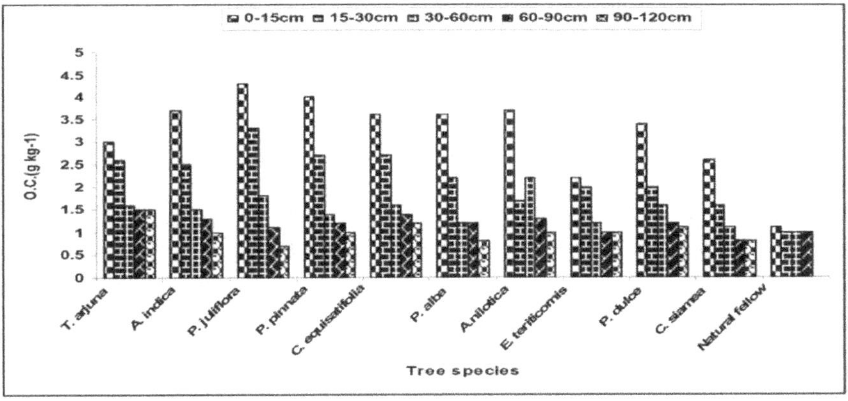

Fig. 4. Changes in organic carbon content after 10 years of silviculture system

Bulk density in 0-75 cm soil layer decreased, whereas, porosity and infiltration rate increased. Maximum reduction in bulk density was found under *C. equisetifolia* (1.21Mg/m^3) followed by *P. dulce* (1.25 Mg/m^3), *A. nilotica* (1.29 Mg/m^3), *P. juliflora* (1.32 Mg/m^3) and minimum under *A. indica* (1.48 Mg/m^3) over the initial value of 1.57 Mg/m^3. After 10 years of plantation cumulative infiltration rate increased to 26.30, 25.80, 24.30, 23.10, 21.90, 21.70 21.20, and 15.80 mm/day under *P. juliflora*, *C. equisetifolia*, *P. pinnata*, *P. dulce*, *A. nilotica*, *A. indica*, *T. arjuna*, *P. alba*, *E. tereticornis*, *and C. siamea*, respectively over the initial value of 2.10 mm/day whereas, in natural fallow the infiltration rate increased to the level of 11.80 mm/day (Table 5).

Table 5. Ameliorative effect of different tree species on physical properties of soil 10 years after plantation

Tree species	Bulk density (Mg/m^3)		Soil porosity (%)		Cumulative infiltration rate (mm/day)
	0-75 mm	75-150 mm	0-75 mm	75-150 mm	
T. arjuna	1.47	1.52	44.5	42.6	21.20
A. indica	1.48	1.56	44.1	41.1	21.70
P. juliflora	1.32	1.46	50.2	44.9	26.30
P. pinnata	1.36	1.57	48.6	40.7	24.30
C. equisetifolia	1.21	1.42	54.3	46.4	25.80
P. alba	1.37	1.61	48.3	39.2	20.00
A. nilotica	1.29	1.58	51.3	40.4	21.90
E. tereticornis	1.38	1.51	48.0	43.0	19.70
P. dulce	1.25	1.58	52.8	40.4	23.10
C. siamea	1.46	1.48	45.0	44.1	15.80
Natural fallow	1.50	1.57	43.4	40.7	11.80
Initial	1.57	1.64	40.7	39.6	2.10
LSD ($P=0.05$)	0.08	0.11	3.26	0.76	6.34

Source: Singh *et al.* (2011)

Silvi-pastoral systems: Silvipastoral model for rainwater conservation and production of fuel and forage from alkali lands has been developed by Grewal and Abrol (1986). With this model trees such as *A. nilotica*, *E. tereticornis* and *Parkinsonia aculeata* were planted on ridges and Karnal grass (*L. fusca)* was established in the trenches between ridges. The system conserves rain water during monsoon, which in turn increased the biomass of trees and inter crop of *L. fusca*. In addition to fire wood and forage production, this system was found useful in checking runoff and soil loss (Table 6). In addition, tree crops may be used in cattle production systems in order to provide live fences, windbreaks, and shade trees and for soil and water conservation purposes.

Table 6. Rainfall, run-off, soil loss and water balance in flat (FSS) and ridge-and-furrow planting systems

Year	Monsoon Rainfall	Run-off (mm)		Soil loss (Mg/ha) (mm)	
		FSS	RFS	FSS	RFS
1982	295.1	169.0	204.0	114.73	83.05
1983	584.6	401.3	469.6	23.86	9.39
1984	512.4	337.4	319.7	8.58	1.10

Component (mm)	1982		1983		1984	
	RFS	FSS	RFS	FSS	RFS	FSS
Rainfall	295	295	585	585	512	512
Run-off	0	169	0	401	0	337
Retention	295	126	585	184	512	175
Soil storage	88	58	132	79	216	95
Evaporation	207	69	453	105	269	80

Source: Grewal and Abrol (1986)

Planting of tree species with grasses and legumes in an integrated system and their utilization through cut and carry on forage in early years followed by in-situ grazing is known as silvipastoral system. Grasses, trees and trees + grasses mixtures were planted in different treatment combinations. The trees have been planted in auger holes of 45 cm diameter at the surface and 20 cm at the base and 120-140 cm deep. The pits were filled with a uniform mixture of original soil + 4 kg Gypsum + 2 kg UTK + 10 kg FYM + 20 kg silt before planting. The trees have been planted keeping a distance of 5 m between row to row and 4 m between plants. The grass species like Karnal grass (*L. fusca*), Gutton panic (*Panicum maximum*), Rhodes grass *(Chloris gayana)* and Berseem (*Trifolium alexandrinum*) were planted between interspacing of trees. From the study it is concluded that establishment of *P. juliflora* with planting of *L. fusca* for four years followed by *T. alexandrinum* for six years found more remunerative land use system than pure pastoral or silvicultural systems (Singh *et al*., 2014). This system improved the soil to such an extent that less tolerant but more palatable fodder species such as berseem (*T. alexandrinum*), shaftal (*T. resupinatum*), and senji (*Melilotus parviflora*) could be grown under *Prosopis* trees after 74 months (Table 7).

Table 7. Effect of *P. juliflora – L. fusca* silvi-pastoral system on soil properties

Soil properties	Depth (cm)	Initial	After 74 months
pH_2	0-15	10.3	8.9
	15-30	10.3	9.4
EC_2 (dS/m)	0-15	2.2	0.36
	15-30	1.5	0.60
Organic C (%)	0-15	0.18	0.58
	15-30	0.13	0.36
Available N (kg/ha)	0-15	79	165
	15-30	73	134
Available P (kg/ha)	0-15	35	30
	15-30	31	26
Available K (kg/ha)	0-15	543	486
	15-30	490	478

Source: Singh *et al.* (2014)

To find out a suitable tree and grass based agroforestry system for highly degraded sodic lands (pH_2>10.4, EC <2.0 dS/m and ESP >80.0), another field study was conducted at CSSRI, Regional Research Station, Research Farm, Shivri, Lucknow with three tree species namely *A. nilotica, C. equisetifolia* and *Eucalyptus tereticornis.* Three grass species *viz*. Rhodes grass (*Choloris gayana*), Gutton panic (*Panicum maximum*) and Napier grass (*P. purpureum*) were planted in between tree rows at a spacing of 50 × 50 cm, 50 × 50 cm and 75 × 30 cm respectively. Maximum survivable was recorded with *E. tereticornis* followed by *C. equisetifolia* and *A. nilotica, E .tereticornis* attains maximum plant height and diameter at stump height (DSH) over *A. nilotica* and *Casuarina equisetifolia.* Highest mortality was recorded with *P. purpureum* whereas minimum with *C. gayana.* From the data it is revealed that *C. gayana* performed well under both *C. equisetifolia* and *E. tereticornis* whereas *P. maximum* and *P. purpureum* shown good growth only under *A. nilotica.* Maximum annual fodder yield was recorded from *C. gayana* and minimum from *P. maximum, C. gayana* under *A. nilotica* yielded 68.71% and 46.94% higher yield over *C. equisetifolia* and *E. tereticornis* respectively. The yield of *C. gayana, P. maximum* and *P. purpureum* was very low where there were no trees because improvement in soil properties without trees was comparatively low. Maximum reduction in soil pH and improvement in organic carbon was recorded in *Chloris gayana* grown under *A. nilotica* followed by *Panicum maximum* and *Pennisetum purpureum.* (Table 8). The organic carbon content under *A.nilotica* and *C. equisetifolia* plantation with *Chloris gayana, Panicum maximum* and *Pennisetum purpureum* was at par and it was about 42% higher than the plantation of these grass species under *E. tereticornis.* Increment in organic carbon under *A. nilotica* and *C. equisetifolia* due to introduction of grass species was about 6% over the plantation of only trees.

Table 8. Interaction effect of tree and grass species on improvement in soil properties

Tree species	*Chloris gayana*			*Panicum maximum*			*Pennisetum purpureum*		
	pH_2	EC_2 (dS/m)	O.C. (%)	pH_2	EC_2 (dS/m)	O.C. (%)	pH_2	EC_2 (dS/m)	O.C. (%)
A. nilotica	9.32	0.67	0.37	9.47	0.47	0.36	9.52	0.97	0.37
C. equisetifolia	9.65	0.83	0.37	9.88	0.46	0.36	9.90	0.84	0.37
E. tereticornis	9.43	0.62	0.26	9.47	0.67	0.26	9.80	0.80	0.27

Source: Singh (2007)

Minimum bulk density was recorded under *C. gayana* with *C. equisetifolia* followed by the *Panicum maximum* with *C. equisetifolia* and *Pennisetum purpureum* with *A. nilotica*. The reduction in bulk density values also reflected the values of porosity under different systems. The maximum porosity (54.3%) was recorded under *Chloris gayana* with *C. equisetifolia* and minimum (46.79%) under *P. maximum* with *E. tereticornis.* The infiltration rate as indicated to know about the water transmission pattern within soil profile revealed that there were vide variation under different systems and varied from 20.70-23.60 mm/day. It was 14% faster under *C. gayana* with *A. nilotica* than *P. purpureum* with *E. tereticornis* (Table 9).

Table 9. Interaction effect of tree and grass species on physical properties of soil

Tree species	*Chloris gayana*			*Panicum maximum*			*Pennisetum purpureum*		
	Bulk density (Mg/m^3) (0-75mm)	Soil porosity (%) (0-75mm)	Cumulative infiltration rate (mm/day)	Bulk density (Mg/m^3) (0-75mm)	Soil porosity (%) (0-75mm)	Cumulative infiltration rate (mm/day)	Bulk density (Mg/m^3) (0-75mm)	Soil porosity (%) (0-75mm)	Cumulative infiltration rate (mm/day)
A. nilotica	1.27	52.07	23.60	1.31	50.56	22.10	1.28	51.69	22.00
C. equisetifolia	1.21	54.33	23.00	1.27	52.07	21.20	1.31	52.07	21.50
E. tereticornis	1.34	49.94	21.10	1.41	46.79	20.40	1.40	47.16	20.70

Source: Singh (2007)

Silvi-agriculture system: In this system the trees are grown for reasonable period of time, followed by growing agricultural crops. Prolonged occupation of sodic soils by trees results in their amelioration in terms of decreased pH and electrical conductivity and improved organic matter and fertility status. Singh *et al.* (1998) grew wheat and oat in pots filled with top (30 cm) soils collected from 24 year old plantations of *P. juliflora*, *A. nilotica*, *E. tereticornis*, *Albizzia lebbek* and *Terminalia arjuna* on a highly sodic soil and a reclaimed sodic soil. The organic carbon content and nutrient status of the soil under 24 years' old plantations was much higher than that of a farm soil reclaimed through gypsum. Soil amelioration was maximum under *Prosopis* and minimum under *Eucalyptus*. Grain and straw yield of both the crops were maximum under *Prosopis* and minimum under *Eucalyptus* (Table 10).

Table 10: Grain and straw yield of crops under different tree plantations

Species	Wheat		Oat	
	Grain (g/pot)	Straw (g/pot)	Grain (g/pot)	Straw (g/pot)
Eucalyptus tereticornis	32.2	25.3	42.7	58.5
A. nilotica	55.7	68.8	61.6	67.5
Albizzia lebbek	45.3	43.5	52.8	66.9
Terminalia arjuna	44.0	38.5	45.8	62.8
Prosopis Juliflora	61.7	87.5	87.9	111.1
Crop land	13.3	15.4	24.3	26.7
LSD ($P=0.05$)	2.8	2.0	7.0	9.4

Source: Singh *et al.* (1998)

Significant improvement in physico-chemical properties of the soil was observed due to combined effect of *Jatropha curcas* plantation and intercrops. Field study conducted at CSSRI, research farm, Shivri, Lucknow revealed that the bulk density of the control plot, where only Jatropha trees were planted and no intercrops were grown, was significantly higher over the plantation of *Jatropha curcas* and cultivation of crops in between the trees. However, soil porosity and infiltration rate increased significantly. Plant density had a significant effect on soil physical properties. Planting of *Jatropha curcas* at a spacing of 3 m × 2 m resulted in maximum reduction in soil bulk density and increase in soil porosity and cumulative infiltration rate because this treatment had the maximum litter fall and biomass yield (Singh *et al.*, 2015; Singh *et al.*, 2016). The winter months accounted for 80% of total litter fall that was composed of about 75–80% foliage, which in turn helped to increase organic carbon and reduce soil pH. Among the cropping systems evaluated, the sweet basil-matricaria (SB-M) cropping system had more positive changes in soil properties than sorghum-wheat (S-W), maize-linseed (M-L), and the control. After five years of plantation of *Jatropha curcas* and cultivation of crops in between *Jatropha curcas* rows (Fig. 5), the greatest improvement in terms of soil pH, electrical conductivity, and organic carbon, available N, P and K in the 0–15 cm soil layer was recorded under the plantation of *Jatropha curcas* at 3 m × 2 m spacing and growing of SB-M as the inter-crops (Table 11). The average reduction in soil pH under S-W, M-L and SB-M was 1.72%, 1.72% and 4.09% and the increments in organic carbon were 28.94%, 31.57% and 42.10%, respectively over the control. Baumert *et al.* (2015) also reported an 84% increase in soil organic carbon due to JCL plantation. The increase in organic carbon content of the surface soils in a span of five years was about three folds higher in the case of intercropping with JCL and about two-folds higher in JCL mono crops in sodic soils in comparison to barren sodic soils. The number of bacteria in the rhizosphere soils was about 40% higher than the non-rhizosphere soils. This is because of higher mineralization in the canopy area than outside the canopy due to an improved C : N ratio. Maximum microbial biomass carbon was recorded in soils under the SB-M cropping system followed by S-W and M-L and was lowest in the control where no intercropping was done. Increases in microbial biomass under S-W, M-L and SB-M were 17.81%, 8.75% and 24.68% over the control. This study shows that plantation of JCL in sodic soils stimulated the soil microbial population, and biological activities particularly in the rhizosphere zone of the soil (Singh *et al.*, 2016).

Fig. 5. Intercropping with Jatropha **(A)** maize in *kharif* and **(B)** linseed in *rabi*

Table 11. Ameliorative effect of Jatropha intercropping on physico-chemical properties of soil 5 years after plantation

Soil parameters	Initial (2006)	After 5 years (2012)								LSD (p=0.05)
		Control		S-W		M-L		SW-M		
		3×3m	3×2m	3×3m	3×2m	3×3m	3×2m	3×3m	3×2m	
Bulk density (g/cm^3) (0-75mm)	1.60±0.04	1.57±0.03	1.56±0.02	1.54±0.14	1.52±0.12	1.56±0.08	1.53±0.12	1.48±0.14	1.48±0.11	0.08
Soil porosity (%)	48.6±0.42	50.2±0.34	51.2±0.42	52.8±0.32	53.4±0.30	51.3±0.28	52.2±0.33	54.3±0.30	54.8±0.28	2.34
Cumulative infil-tration rate (mm/day)	8.62±0.12	14.2±0.14	15.4±0.21	18.75±0.16	21.2±0.18	16.62±0.20	18.4±0.16	21.2±0.13	22.1±0.14	4.23
Soil pH(1:2)	9.46±0.15	9.22±0.11	9.34±0.21	9.12±0.15	9.13±0.16	9.13±0.21	9.11±0.14	9.05±0.17	8.92±0.21	0.03
EC (µS/m)	177±8.62	210.2±9.11	450.1±10.13	670.3±10.25	210.3±9.65	130.5±8.66	620.2±9.20	450.3±10.32	300±9.68	0.08
ESP	40±2.23	36.2±1.15	32.4±1.68	32.4±1.85	35.4±2.12	32.5±2.21	35.2±2.10	30.2±2.13	35.3±2.43	3.2
O.C (g/kg)	1.0±0.02	1.8±0.03	2.0±0.12	2.4±0.20	2.5±0.14	2.5±0.13	2.5±0.16	2.7±0.20	2.7±0.08	0.002
Av. N (mg/kg)	41.9±1.12	103.6±1.15	110.1±1.16	95.2±1.23	107.8±1.42	95.9±0.86	114.8±1.12	103.6±1.21	111.9±1.32	34.2
Av. P (mg/kg)	11.1±0.26	9.6±0.23	10.1±0.26	12.3±0.31	13.5±0.22	10.9±0.25	12.1±0.26	13.6±0.30	14.5±0.26	1.12
Av. K (mg/kg)	173.5±22.50	137.6±17.62	146.9±16.50	166.9±18.10	171±18.21	148.9±16.42	199.2±17.52	171.8±16.16	189.7±18.21	13.6
MBC (mg/kg)	96.4±.3.23	96.4±2.56	100.1±2.48	112.2±3.12	119.3±3.22	102.5±2.86	111.2±3.10	120.5±2.80	124.5±3.10	2.32

Source: Singh *et al*. (2016)
LSD: Least significant differences, S-W: Sorghum-wheat, M-L: Maize-linseed, SW-M: Sweet basil-matricaria

Silvi-horti-pasture or horti-agricultural system: Horticultural species based agroforestry models for sodic soils have been developed by the Acharya Narendra Dev University of Agriculture & Technology, Ayodhya (Faizabad). In this model the growth rate of guava + Eucalyptus + Subabul was faster and production was higher in terms of fruit, fodder and fuel wood. Intercrops of bottle gourd, tomato, cabbage and spinach have been successfully grown in association with guava trees. The fruit species which can be cultivated successfully in alkali soils include: *Carissa carandus* (Karonda), *Zizyphus mauritiana* (Ber), *Emblica officinalis* (Aonla), *Psyzygium cumini* (Jamun) *and Psydium guajava* (Guava). Aromatic and medicinal crops such as Dill, Isabgol, Tulsi and matricaria can also be grown as intercrops between fruit trees in case of pH of soil is <9.5. The list of fruit crops suitable for developing horti-pasture system in sodic soils under different situations is given in table 12.

Table 12. Promising varieties of fruits for sodic soils

Plants	Promising varieties
Emblica officinalis (Aonla)	Chakaiya, NA-6, NA-7,NA-10
Zizyphus mauritiana (Ber)	BanarsiKarka, Gola
Psydium guajava (Guava)	Shweta, Allahabad Safeda
Punica granatum (Pomegranate)	Ganesh
Morus Alba (Mulberry)	K-2
Psyzygium cumini (Jamun)	

Source: Singh *et al.* (1996)

Planting techniques of trees in sodic soils

In addition to the sodium toxicity of soils and nutritional problems, the tree growth in sodic soils is impeded by limitations of roots proliferations through the hard kankar (calcite) pan existing usually at depths below 50-75 cm from the surface. Therefore, even the earlier afforestation attempts resorted to replacement of excavated alkali soil (50 cm deep pits) with normal soil to improve upon their drainage by digging holes (90-150 cm deep) and refilling the holes with a mixture of good soil, FYM and gypsum before planting tree saplings. The method was introduced in 1895 and named 'deep thala system' of plantation. Later, Yadav and Singh (1970), Yadav (1975) and Yadav (1980) concluded that addition of gypsum (50% GR) and FYM @ 25 kg per pit. (90 cm × 90 cm) was comparable to replacement of the original sodic soil (pH 10.0) with normal soil for the growth of saplings and their survival. The pit planting technique suffers from the disadvantage of higher requirements of amendments, laborious pit digging operation involving more earthwork and non-proliferation of roots through calcic horizon (hard pan). Keeping in these limitations in view, the planting technique has been improved through 'Auger-hole technique' at CSSRI, Karnal (Sandhu and Abrol, 1981). Here, 100-140 cm deep and 20-25 cm diameter auger holes are dug with a tractor-operated auger and saplings are planted after suitably amending the dugout soil. The performance of trees planted with this method has been highly satisfactory (Table 13). This method has picked up very well with the foresters because of reduced manual labour costs and speedy operations. In addition to piercing of hard Kankar layer, the advantage of this technique includes encouraging and training of deeper routing. Thus the trees are able to probe deeper soil layers for water and nutrients to sustain their growth.

Table 13. Comparative performance of trees planted with pit, trench and auger whole techniques in highly alkali soils

Dimension (cm)		*Eucalyptus tereticornis* (8 years)		*A. nilotica* (8 years)		*Prosopis juliflora* (6 years)	
Depth	Width	Height (m)	DSH (cm)	Height (m)	DSH (cm)	Height (m)	DSH (cm)
Pit							
90	90	9.2	17.3	7.5	19.6		
30	30					7.0	8.4
Trench							
30	30					6.9	7.9
Auger hole							
10	120	8.2	14.2	6.7	14.6		
10	120	8.9	16.2	7.7	16.2		
15	90					7.7	9.6
15	180	8.4	14.8	6.8	16.6		
15	180	9.1	16.6	7.4	17.6		

DSH: Diameter at stump height; *Source*: Gill and Abrol (1991); Singh *et al.* (1993)

Agroforestry for waterlogged sodic soils

Under saline soils the tree growth is adversely affected due to reduced water availability with excessive salts along with period water logging and poor aeration especially during the monsoon season. To improve upon aeration and reduce the water stagnation effect planting on high ridges are mounds showed good results. In water logged saline soils the salinity is usually higher in the surface layers and decreases with depth down to water table. Tomar *et al.* (1994) tried the subsurface planting of saplings (at a depth of 30 cm below surface) and compared it with ridge planting (40 cm high). Substantially, higher salts accumulated in the ridges that resulted in poor survival and sapling growth. The performance of trees was better when planted with subsurface method but need for spot irrigation was the main problem (Table 14). Minhas *et al.* (1996) observed the planting in the soul of furrow (60 cm wide and 20 cm wide) which was subsequently used for irrigating the tree saplings. Besides, uniform application of irrigation water and reduction of application cost, the subsurface planting and furrow irrigation method (SPFIM) helped increasing a low salinity zone below the sill of the furrows.

Table 14. Effect of planting methods on tree growth in a waterlogged saline soil

Tree species	Sub surface			Ridge trench		
	Height (m)	DSH (cm)	PS (%)	Height (m)	DSH (cm)	PS (%)
	After 9 years of planting					
A. nilotica	6.41	44.6	50			0
Acacia tortillas	5.31	34.3	56	3.11	10.8	25
Leucaena leucocephala	6.91	36.7	50			0
Prosopis juliflora	8.06	55.9	100	6.40	42.5	100
	After 27 months of planting					
	Surface			SPFIM		
Acacia auriculiformis	1.43		13	2.42		65
A. nilotica	3.21		69	2.89		95
Casuarina equisetifolia	2.13		46	3.00		95
Eucalyptus amanuensis	2.24		50	3.78		95
Terminalia arjuna	1.83		81	2.00		90

Source: Tomar (1994)

Another field study was conducted in waterlogged sodic soils of canal commands in Raebareli district of Uttar Pradesh, India. A vast area on either side of the canal is waterlogged coupled with sodicity. The area represents a semi-arid-sub-tropical climate, characterized by hot summers and a cool winter with mean annual rainfall of 984 mm, most of which occur during June to September. Average ten year maximum and minimum temperature varies between 21.36 to 38.53 ^{0}C and 7.72 to 26.81 ^{0}C, respectively. Geographical coordinates lies between 26^{0}30$'$18.90$''$ N latitudes and 81^0 6$'$40.18$''$ E longitudes at an elevation of 110 m above the mean sea level. The soil textural classes were observed as loam up to 30 cm, clay from 30 to 60 cm and sandy clay loam from 60 to 120 cm soil depth. Soil pH were observed to be 10.5, 10.3, 9.78, 9.43, 8.83 and 8.72; and EC were 2.60, 2.10, 1.02, 0.80, 0.54 and 0.55 dS/m for soil depths of 00 to 15, 15 to 30, 30 to 45, 45 to 60, 60 to 90 and 90 to 120, respectively. The soil pH and EC is high toward soil surfaces and decreases with increase in soil depths. Water table depth at site fluctuates from 0.00 m to 1.5 m below ground surface throughout the year. General field conditions of the area is shown in Fig. 6.

Fig. 6. Field conditions at experimental site

Plantation methods: One year old eucalyptus saplings were established using auger hole, pit method and raised bed cum auger techniques (Fig. 7 and 8) for studying the performance under waterlogged sodic soils. In case of raised bed method a platform of 600 mm × 600 mm × 300 mm was made by collecting soil from all around and an auger hole of 600 mm deep was made in middle for filling input mixture and sapling plantation. An input mixture of 5 kg gypsum, 5 kg farm yard manure and 10 kg canal sand was filled in holes. Three years field study reveled that the auger hole method performed best followed by raised bed and pit methods of plantation. This seems to be quite logical as auger hole was bored up to depth of 0.60 m and soil pH has decreasing trend with depth. Auger hole plantation was adopted for large scale plantation under waterlogged sodic conditions along Sharda Sahayak Canal.

Auger hole plantation technique was used for plantation of eucalyptus (*Eucalyptus camaldulensis*). Six month old eucalyptus sapling were planted and manually irrigated. The bio-drainage belt was established over a length of 400 m along with canal. The width of bio-drainage belt was 30 m. Spacing between row and plants were 1.5 m × 1.5 m. Total 267 row formed in bio-drainage belt. In each row, 20 eucalyptus plants were planted in bio-drainage belt. Four non-weighing type metallic lysimeters of one meter diameter and 2 m depth were installed inside the bio-drainage belt for measuring plant heights data at regular interval of time. Overall mortality level in entirely eucalyptus plantation was worked out 28.39%. The reason for high mortality is due to entry of salt laden surface runoff water to porous auger hole filled with input mixture during initial days of rainy period. Installation of lysimeters inside the bio-drainage belt was done with the sole objectives to avoid boundary effect and damage or uprooting by animals and passerby. Eucalyptus plant at an age of 18 months was uprooted and planted in the lysimeters for measuring evapotranspirative demand. Constant water table depths inside the lysimeters were maintained as it was observed outside the lysimeters by applying water every morning. The amount of water required to maintain the desired water level inside the lysimeters was considered as the total ET demands of the eucalyptus plant. Plant height of eucalyptus was measured on monthly basis.

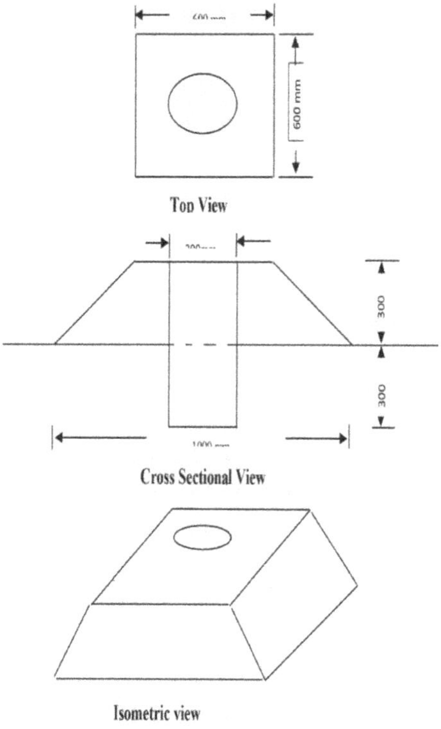

Fig. 7. Raised bed method of plantation.

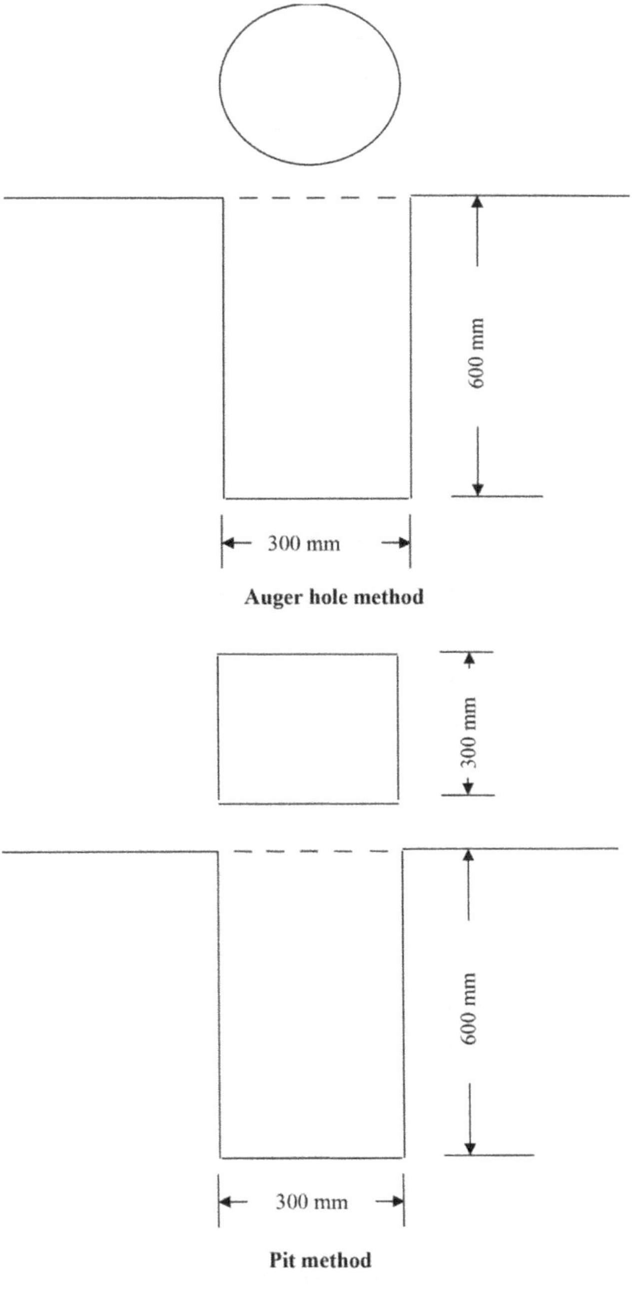

Fig. 8 Auger hole and pit methods of plantation.

On the basis of available information, a list of consistently better performing species recommended for water logged soils is given in table 15.

Table 15. Recommended tree species for waterlogged sodic soils

Soil EC (dS/m)	Tree species (Common name)
20-30	*Acacia farnesiana* (pissi babul), *Prosopis juliflora* (mesquite, pahari kikar), *Parkinsonia aculeate* (Jerusalem thorn, parkinsonia), *Tamarix aphylla* (faransh)
14-20	*A.nilotica*(desi kikar), *A. pennatula* (kikar), *A. tortilis* (Israeli Kikar), *Callistemon lanceolatus* (bottle brush), *Casuarina glauca* (casuarinas, saru*), C. obese, C. equisetifolia, Eucalyptus camaldulensis* (river-red gum, safeda), *Ferronia limonia* (kainth, kabit), *Leucaena leucoephala* (subabul), *Ziziphus jujube* (ber)
10-14	*Casuarina canninghamiana* (casuarinas, saru), *Eucalyptus tereticornis* (mysore sum, safeda), *Terminalia arjuna* (arjun)
5–10	*Albizzia caribaea, Darbergia sissoo* (shisham), *Gauzuma ulmifolia, Pongamia pinnata* (papri), *Samanea saman*
<5	*Acacia auriculiformis* (Australian kikar, akash mono), *A. deamii, A. catechu* (khair), *Syzygium cumini* (jamun), *Salix spp. (willow, salix), Tamarindus indica* (imli)

Source: Dagar and Singh (1994); Gupta *et al.* (1995)

Post planting management strategies

In sodic soils post planting irrigation during early establishment period is very essential. Marked response to irrigation in terms of survival and biomass yield of *Prosopis juliflora* was recorded in sodic soils (Singh *et al.,* 1990). In low rainfall areas (30-35 cm per annum), survival and biomass yield was significantly higher and plantation continued to respond to irrigation up to 4-5 years of planting. In experiments conducted in India from 1985 to 1988, Singh *et al.* (1989) found that, during the first two year, growth of *Prosopis* was far better when irrigated compared to the plantation which depend on rainfall alone. Within two year of planting, 36% of rainfed *Prosopis* died, while the mortality was only 9% among the irrigated plants. The water use efficiency also higher under irrigated conditions. Irrigation brought very little change in the chemical composition of different plant parts but significantly decreased root zone soil sodicity. Application of irrigation during first two years of planting is absolutely necessary. However, after two years, irrigation may be withdrawn as plant root can meet their requirement from ground water.

Conclusion

Agroforestry plays a vital role in biological amelioration of problem soils, wherein economically viable agriculture is possible in due course of time if standardized agroforestry systems are applied. While standardizing the suitable package of practices for degraded sodic lands major considerations are given on productivity, sustainability, economic utility and adaptability. We have developed sustainable agroforestry systems for salt affected soils involving multipurpose tree species (energy plantations and fruit trees), forage grass, and other species of economic importance. Many trees and grass species have been identified for salt affected soils, which provide not only food, fodder, fuel and other bio-energy resources but also several ecosystem services to maintain the quality environment around us including soil amelioration. Socio-economic development of the region is also carried out from the amelioration and judicious management of the degraded lands. Additionally this venture extends to continuous income generation, employment opportunity, food and nutritional security for small and marginal farmers.

References

Abrol, I.P., Yadav, J.S.P. and Massoud, F.I. 1988. Salt affected soils and their management. FAO Soils Bulletin 39. Food and Agriculture organization of the United Nations, Rome.

Ahmad, S., Ghafoor, A., Akhtar, M.E. and Khan, M.Z. 2016. Implication of gypsum rates to optimize hydraulic conductivity for variable-texture saline-sodic soils reclamation. *Land Degradation and Development*. Dev. doi:http://dx.doi.org/10.1002/ldr.2413.

Baumert, S., Khamzina, A. and Vlek, P.L. 2015. Soil organic carbon sequestration in *Jatropha curcas* systems in Burkina Faso. *Land Degradation and Development*. doi:http://dx.doi.org/10.1002/ldr.2310.

Bhargava, G.P. and Kumar, R. 2011. Characteristics, extent and genesis of sodic soils of the Indo-gangetic alluvial plain. In. Sharma, D.K., Rathore, R.S. Nayak, A.K. and Mishra, V.K. (Ed.) Sustainable Management of Sodic lands. Central Soil Salinity Research Institute, Regional Research Station, Lucknow. pp. 28-42.

Dagar, J.C. and Singh, N.T. 1994. Agroforestry options in reclamation of problem soils. In: Trees and Tree farming. P.K. Thampan (Ed.) Peekay Tree Crops Development Corporation, Cochin, India. pp. 65-103.

Drake, J.A., Cavagnaro, T.R., Cunningham, S.C., Jackson, W.R. and Patti, A.F. 2016. Does biochar improve establishment of tree seedlings in saline sodic soils? Land Degradation and Development. Dev. 27, 52–59. doi:http://dx.doi.org/10.1002/ldr.2374.

El-Mowellhey, N. 1998. Sustainable management of salt-affected soils in the arid ecosystem. Cairo, Egypt, 21-26 September 1997. pp 1-2.

Garg, V.K. 1998. Interaction of tree crops with a sodic soil environment: potential for rehabilitation of degraded environments. *Land Degradation and Development* 9: 81-93.

Gill, H.S. and Abrol, I.P. 1991. Salt affected soils, their afforestation and its ameliorating influence. *Int. Tree Crop J.*, **6**: 261-274.

Grewal, S.S., Abrol, I.P. and Singh, O.P. 1986. Agroforestry on alkali soils. Effect of some management practices on initial growth, biomass production and chemical composition of selected tree species. *Agrofor. Sys*. **4**: 221-232.

Gupta, R.K. and Abrol, I.P. 1990. Salt affected soils: their reclamation and management for crop production. *Adv. in Soil Sci*. **11**: 223-288.

Gupta, R.K., Tomar, O.S. and Minhas, P.S. 1995. Managing salt affected soils and waters for afforestation. P.23, Bull.7/95, CSSRI, Karnal, India.

Jaradat, A.A. 2003. Halophytes for sustainable biosaline farming systems in the Middle East. In: Alsharhan AS, Wood WW, Goudie WS, Fowler A, Abdellatif AM (eds) Desertification in the Third Millennium. Swets & Zeitlinger Publishers, The Netherlands, ISBN 90 5809 571 1, p. 187-204.

Kumar, A. and Abrol, I.P. 1986. Grasses in Alkali soils. Bull. No. 11, 95. Karnal. Central Soil Salinity Research Institute, ICAR.

Mandal, A.K. and Sharma, R.C. 2006. Computerized database of salt affected soils for agro-climatic regions in the Indo–Gangetic Plain of India using GIS. *Geocarto. Int.* **21**(2): 47-57. ·

Minhas, P.S., Singh, Y.P., Tomar, O.S., Gupta, R.K. and Gupta, R.K. 1996. Effect of saline irrigation and it schedules on growth, biomass production and water use by *Acacia* and *Dalbergia* on a highly calcareous soil. *J. of Arid Envt.* **36**: 181-92.

Mishra, A., Sharma, S.D. and Khan, G.H. 2002. Rehabilitation of degraded sodic lands during a decade of *Dalbergia sissoo* plantation in Sultanpur district of Uttar Pradesh, India. *Land Deg. and Devel.*, **13**: 375-386.

Mishra, A., Sharma, S.D., Pandey, R. and Mishra, L. 2004. Amelioration of a highly alkaline soil by trees in northern India. *Soil Use and Mangt.* **20**(3): 325–332.

NRSA and Associates. 1996. Mapping of salt affected soils of India, 1: 250,000 map sheets, Legend. NRSA Hyderabad, India.

Oo, A.N., Iwai, C.B. and Saenjan, P. 2015. Soil properties and maize growth in saline and nonsaline soils using cassava-industrial waste compost and vermicompost with or without earthworms. *Land Deg. and Devel.* **26:** 300–310. doi: http://dx.doi.org/ 10.1002/ldr.2208.

Pandey, V.C., Singh, K., Singh, B. and Singh, R.P. 2011. New approaches to enhance eco-restoration efficiency of degraded sodic lands; critical research needs and future prospects. *Ecological Restoration* **29**: 322–325.

Sandhu, S.S. and Abrol, I.P. 1981. Growth responses of *Eucalyptus tereticornis* and *A. nilotica* to selected cultural treatments in a highly sodic soil. *Indian J. of Agril. Sci.* **51**: 437-43.

Sharma, R.C., Rao, B.R.M. and Saxena, R.K. 2004. Salt affected soils in India–current assessment. In Advances in Sodic Land Reclamation, Proceedings; International Conference on Sustainable Management of Sodic Lands, Lucknow, India. 1-26.

Shukla, S.K., Singh, K., Singh, B. and Gautam, N.N. 2011. Biomass productivity and nutrient availability of *Cynodon dactylon* (L.) Pers. Growing on soils of different sodicity stress. *Biomass and Bioenergy*, **35**: 3440–3447.

Singh, B. 2006. Rehabilitation of alkaline wastelands on the gangetic alluvial plains of Uttar Pradesh, India through afforestation. *Land Deg. and Devel.* **1**(4): 305-310.

Singh, B. and Goel, L.V. 2012. Soil Amelioration through Afforestation. In: Restoration of Degraded Land to functioning forest ecosystem, Pp. 114-145.

Singh, G. 2009. Salinity-related desertification and management strategies: Indian Experience. *Land Deg. and Devel.* **20**: 367-385.

Singh, G. and Dagar, J.C. 2005. Greening sodic lands: Bichhian Model. Tech. Bull. No.2/2005, pp.51. Central Soil Salinity Research Institute: Karnal: India.

Singh, G.B., Abrol, I.P. and Cheema, S.S. 1990. Effect of irrigation on *Prosopis juliflora* and soil properties of an alkali soil. *Int. Tree Crops J.* **6**: 81-99.

Singh, G.B., Singh, N.T. and Tomar, O.S. 1993. Agroforestry in salt affected soils P.65, Res. Bull. 17, CSSRI, Karnal.

Singh, G.B. and Singh, N.T. 1993. Mesquite for the revegetation of salt lands. 24 pp. Bull. No.18, CSSRI, Karnal.

Singh, G.B., Abrol, I.P. and Cheema, S.S. 1989. Agroforestry system on an alkaline soil. Effects of spacing and lopping on mesquite Effect of spacing and lopping on mesquite (*Prosopis juliflora*) –Karnal grass (*L.fusca*). *Exp. Agric.* **25**: 401-408.

Singh, G.B., Dagar, J.C. and Singh, N.T. 1996. Growing fruit trees in highly alkaline soil: a case study. *Land Degradation and Rehabilitation* **8**:257-268.
Singh, G.B., Singh, H., Bhojvaid, P.P. 1998. Sodic soils amelioration by tree plantations for wheat and oat production. *Land Degradation and Rehabilitation* **9**: 453-462.
Singh, K. 2015. Microbial and enzyme activities of saline and sodic soils. *Land Deg. and Devel.* doi:http://dx.doi.org/10.1002/ldr.2385.
Singh, K., Pandey, V.C., Singh, B. and Singh, R.R. 2012a. Ecological restoration of degraded sodic lands through afforestation and cropping. *Ecol. Eng.* **43**: 70–80.
Singh, K., Singh, B. and Singh, R.R. 2012b. Changes in physico-chemical microbial and enzymatic activities during restoration of degraded sodic lands: ecological suitability of mixed forest over plantation. *Catena* **96**: 57–67.
Singh, K., Trivedi, P., Singh, G., Singh, B. and Patra, D.D. 2016b. Effect of different leaf litters on carbon, nitrogen and microbial activities of sodic soil. *Land Deg. and Devel.* **27**: 1215–1226. doi:http://dx.doi.org/10.1002/ldr.2313.
Singh, Y.P., Nayak, A.K., Sharma, D.K., Singh, G., Mishra, V.K. and Singh, D. 2015a. Evaluation of Jatropha Curcas genotypes for rehabilitation of degraded sodic lands. *Land Deg. and Devel.* **26**: 510–520. doi:http://dx.doi.org/10.1002/ldr.2398.
Singh, Y.P., Sharma, D.K., Singh, G., Nayak, A.K., Mishra, V.K., Singh, R. 2008. Alternate land use management for sodic soils. Tech. Bull. No. 2/2008. Central oil Salinity Research Institute, Regional Research Station, Lucknow, India, pp.16.
Singh, Y.P., Singh, G. and Sharma, D.K. 2014. Bio-amelioration of alkali soils through agroforestry systems in central Indo-Gangetic Plains of India. *J. For. Res.* **25**: 887–896.
Singh, Y.P., Singh, G. and Sharma, D.K. 2010. Biomass and bio-energy production of ten multipurpose tree species planted in sodic soils of Indo-gangetic plains. *J. of Forestry Res.*, **21**(1): 63-70.
Singh, Y.P., Singh, G. and Sharma, D.K. 2011. Ameliorative effect of multipurpose tree species grown on sodic soils of Indo-Gangetic alluvial plains of India. *Arid Land Restoration and Mangt.* **25**: 1–20.
Singh, Y.P., Singh, G. and Sharma, D.K. 2015b. Performance of pastoral: silvipastoral and silvicultural systems in alkali soils of Indo-Gangetic plains. *J. of Soil and Water Conservation.* **14**: 168–173.
Singh, Y.P., Mishra, V.K., Sharma, D.K., Singh, G., Arora, S., Dixit, H. and Cerda, A. 2016. Harnessing productivity potential and rehabilitation of degraded sodic lands through Jatropha based intercropping systems. *Agriculture, Ecosystems and Environment.* **233**: 121–129.
Suarez, D.L., Rhodes, J.D., Savado, R. and Grieve, C.M. 1984. Effect of pH on saturated hydraulic conductivity and soil dispersion. *Soil Sci. Soc. of America J* **48**: 50-55.
Sumner, M.E. 1993. Sodic soils: new perspectives. *Australian J. of Soil Res.* **31**: 683-750.
Szabolcs, I. 1979. Review of Research on salt-affected soils. *Natural Resources Res.,* **15**: 137 (UNESCO, Paris).
Tomar, O.S., Minhas, P.S. and Gupta, R.K. 1994. Potentialities of afforestation of waterlogged saline soils. (In) Agroforestry systems for degraded lands, Vol.1, pp.111-20. Singh, P., Pathak, P.S. and Roy, M.M. (Eds.) Oxford and IBH publishing Co.Pvt Ltd. New Delhi
Tripathi, K.P. and Singh, B. 2005. The role of revegetation for rehabilitation of sodic soils in semi arid subtropical forest. *Restoration Ecology* **13**(1): 29 -38.
Yadav, J.S.P. 1975. Improvement of saline alkali soils through biological methods. *Indian Forester* **101**(7): 385-395.
Yadav, J.S.P. 1980. Salt affected soils and their afforestation. *Indian Forester* **106**:259-272.
Yadav, J.S.P. and Singh, K. 1970. Tolerance of certain forest species to varying degree of salinity and alkalinity. *Indian Forester* **96**: 587-99.

12

Advances in Farm Mechanization in India

Sanjay K. Patel, B.K. Yaduvanshi and Prem K. Sundaram*

Introduction

Farm mechanization is application of machine power to work on land, usually performed by animate and mechanical power. It is also defined as an economic application of engineering technology to increase the labour efficiency and productivity. However, it not only includes production, distribution and utilization of a variety of tools, machinery and equipment but also planting, harvesting and primary process. Mechanization has a major impact on demand and supply of farm labour, agricultural profitability and a change in rural dynamics.

Aims of mechanization

- The main aim is to replace animal power on which agriculture has been based for very many centuries.
- It also aims at reducing the drudgery of certain operations which have to be performed either by human labour or by a combined effort of human beings and animals.
- Reducing cost of production.
- Enhancement of timeliness and profitability in farm operations.
- Reduction of labour requirements and enhancement of agricultural production through higher rates of work output.
- Improvement of work environment and enhancement of safety.
- Add value to primary products and so produce employment and income potential along the value chain.
- Increase cropping intensity.

**Corresponding author email: birjesh123@gmail.com*

Impact and status of farm mechanization

Crop yield increase is primarily a result of improved timeliness of operations. Lack of mechanization leads to inadequate and delayed seedbed preparation, high costs of land preparation and harvesting, increased harvest losses and loss of fodder all affect aggregate agricultural production output. Farm mechanization has positive effect on the livelihoods of farmers (Table 1). The status of farm mechanization for different operation is given in table 2.

Table 1. Examples of the positive impacts of mechanization

Factors affected farm mechanization	Improvements
Labour productivity	Farm family can cultivate land 1-2 ha by hand >2 ha by draft animal power or power tiller > 8 ha by tractor
Land productivity	Increased production through better placement of seed and fertilizer, better weed control through line-planting and improved timeliness
Timeliness of operations	Approximately 1% reduction of yield per day of delay in planting
Drudgery reduction	Reducing the need for women's muscle power, especially hand-hoeing and transport

Source: Sims and Kienzle (2016)

Table 2. Status of farm mechanization for different operations

Crops	% Operations mechanized			
	Seedbed preparation	Sowing/planting/ transplanting	Weed & pest control	Harvesting & threshing
Paddy	85-90	5-10	80-90	70-80
Wheat	90-95	80-90	70-80	80-90
Potato	90-95	80-90	80-90	70-80
Cotton	90-95	50-60	50-60	0
Maize	90-95	80-90	70-80	50-60
Chickpea	90-95	50-60	60-70	30-40
Sorghum	80-90	30-50	60-70	20-30
Millets	80-90	30-40	60-70	20-30
Oilseeds	80-90	30-40	60-80	20-30
Sunflower	80-90	40-50	80-90	60-70
Fodder crops	80-90	20-40	80-90	10-20
Vegetable crops	70-80	5-10	80-90	<1
Horticultural crops	60-70	30-40	40-50	<1

Women friendly equipment in agriculture

A lot of small tools and equipment have been designed by ICAR institutes, universities and other companies. Some of them which can be easily used by women are depicted below:

S. No.	Tools/ equipment	Remarks	Figures
1.	Dibbler	It is single row manually operated equipment for dibbling bold or medium seeds in row or gap filling into well prepared soil. It is suitable for drilling wheat, field pea, and maize in small plot.	
2.	Paddy drum seeder	The seeder consists of a seed drum, main shaft, ground wheel, floats and handle. Joining smaller ends of frustum of cones makes the seed drum. Nine metering holes of 10 mm diameter are provided along the circumference of the drum at both ends for a row-to-row spacing of 200 mm. Two floats are provided on either side to prevent the sinkage and facilitate easy pulling of seeder.	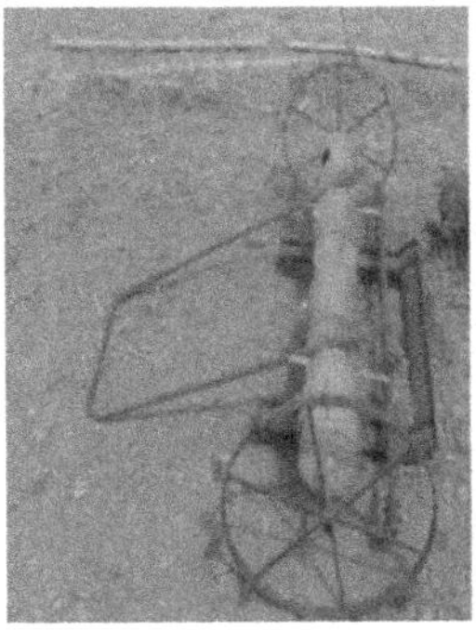
3.	V blade hand hoe	V blade hoe is used for weeding of the vegetable crop planted in the rows and earthing operation. It is a long handled weeding tool for operation in between crop rows. Consists of V blade, arms, ferrule and wooden handle. The pulling actions cause penetration of the blade into the soil and cut or uproot the weeds. Because of V shape , the blade creates small furrow between crop rows and also earthing of the plants	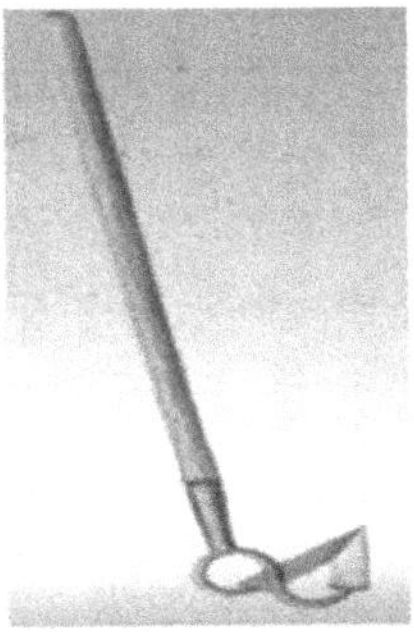
4.	Three tined hand hoe (grubber)	It is for weeding, inter-culture and breaking of the soil crust in vegetable gardens, flower crops and nurseries It is a simple and light weight, manually operated equipment for weeding and inter-culture in upland row crops in black soil. It consists of long handle, ferrule, three tynes and sweep type blades. The operator uses pull force to break the soil crust and uproot the weeds.	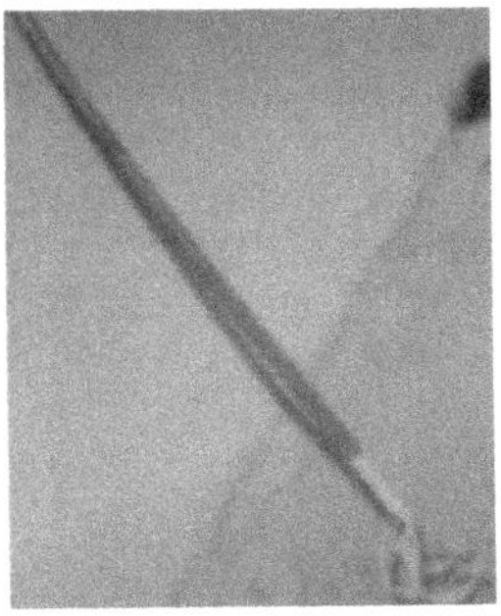

5.	Single wheel hoe	It is used for weeding and inter-culture of vegetables and other crops sown in rows. It is a widely accepted weeding tool for weeding and inter-culture in row crops. It is manually operated equipment for weeding and inter-culture in upland row crops spaced above 240 mm. It consists of wheel frame, V-blade with tyne and handle. Weeds cutting and uprooting are done through push and pull action of the unit.	
6.	Double wheel hoe	It is manually operated equipment for weeding and inter-culture in upland row crops in black soil region. It consists of twin wheels, frame, V-blade with tyne, U clamp, scrapper and handle. Weeds cutting and uprooting are done through push and pull action of the unit.	
7.	Cono-weeder	The cono-weeder is used to remove weeds between the rows of paddy crop efficiently. It is easy to operate, and does not sink in the puddle. The weeder consists of two rotors, float, frame and handle. The rotors are cone frustum in shape, smooth and serrated strips are welded on the surface along its length. The rotors are mounted in tandem in opposite orientation. The float controls working depth and doesn't allow rotor assembly to sink in the puddle. It is operated by pushing action.	

8.	Groundnut decorticator	It is a manually operated equipment to separate kernels from groundnut pods. The unit consists of frame, handle, oscillating arm sieve with oblong hole. The pods are feed in batches of 2 kg and crushed in between concave and oscillating arm having cast iron/ nylon shoe to achieve shelling.	
9.	Tubular maize sheller	It is a hand operated tool to shell maize from dehusked cobs. The unit consists of galvanized mild steel pipe with four tapered fins riveted to its inner periphery. The Sheller is held in left hand, a cob held in right hand is inserted into it with forward and backward twist, to achieve the shelling. Octagonal designs are also available.	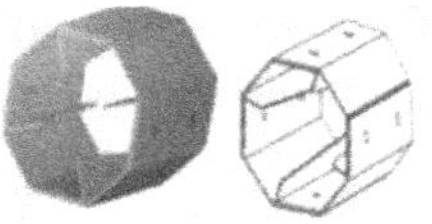 
10.	Fertilizer broadcaster for women	It is hand operated fertilizer broadcaster for women. It weighs only 3.5 kg. Its tank capacity is 7.5 kg and Swath width 5 m. an area of 1.1 ha can be broadcasted by it in one hour.	
11.	Hanging type grain cleaner	It is a simple hanging grain cleaner. Around 225 kg of grain can be cleaned per hour as against the conventional cleaning of 25 kg /hr.	

Land preparation, seeding/planting machinery

Manual operated

Mat type nursery sowing seeder: The mat type nursery sowing seeder is used for uniformly spreading of pre-germinated paddy seeds over the soil filled frames during sowing of mat types seedlings required for mechanical transplanting of paddy. Desired quantity of seed can be spread uniformly for sowing of seedlings. It saves about 80 per cent labour in comparison to manual spreading of seed.

Seed drill: Manually operated seed drill is used for seeding wheat and oilseed crops like rapeseed and mustard. The machine consists of seed hopper and fluted roller seed metering device. The power to metering mechanism is provided by chain and sprocket through a ground wheel. Shovel type furrow opener is used for opening the furrows. Machine is operated by one person. Machine is widely used for inter-row sowing of rape seed and mustard in wheat crop or moong in sugarcane crop. It can plant 0.3 to 0.4 ha/day.

Garlic planter/Multicrop planter: It is useful for sowing garlic and other bold seeds. In this machine, the planting mechanism has been mounted over the existing wheel hand hoe, which is used for inter-culture purposes and is already commercialized. The planting mechanism consists of a vertical plate with spoons. The capacity of the hopper is about 3.0 kg and two persons operate the machine. Plant spacing can be varied by varying the number of spoons on the periphery of vertical plate. Planting spoons are also available for sowing different crops like peas, sunflower, cotton, bhindi, maize, and soybean. It can plant 0.3 to 0.4 ha/day. Missing is only 4-5 per cent. Machine is getting popularity because of very simple design.

Tractor drawn seedbed preparation and sowing equipment

Tractor-drawn pulverizing roller attachment: Pulverizing roller is an attachment to the commercially available tine cultivators. It is suitable for puddling as well as dry seedbed preparation. The roller consists of 6 pulverizing members made of MS steel flats. These members pass thorough the slots in the star wheels, which are fixed on the central axle at a distance of 37 cm each. The quality of puddle as well as dry seedbed preparation is much better and less number of operations are required. The performance of the equipment at higher speed (4-5 km/hr) is better because of better churning action. The use of this machine can save 15-20 per cent of irrigation water because of better quality of puddle. The machine can cover 2.5-3.0 ha/day both in dry land as well as wet land conditions.

Multipurpose tool bar for mini tractor: The implement consisting of iron ploughs for tillage and clod crusher for breaking clods is useful for preparation of seed bed in a single pass with a saving of up to 50% in the cost of operation as compared to the cultivator (Fig. 1).

Fig. 1. Multipurpose tool bar
Source: (http://www.aau.in/sites/default/files/FMPE% 20Recommandantion.pdf)

Laser land leveller: Declining water table and degrading soil health are the major concerns for the present Punjab agriculture. The enhancement of water use efficiency and farm productivity at field level is one of the best options to redress the problem of water scarcity. Laser land levelling is one such technology which helped in using water efficiently, reduced irrigation time and enhanced productivity per unit input not only of water but also of other farm inputs. Laser leveller is trailed type equipment used for achieving precise levelling with desired grade. This two meter wide automatic levelling operation can be successfully carried with 50 hp or above tractor. It saves water to the tune of 25-30%, enhances efficacy of chemicals and fertilizers and improves productivity.

Post-hole digger: This equipment is used for digging pits of size ranging from 15 to 75 cm diameter and up to 90 cm depth. This machine is operated by the tractor PTO through gear box and is mounted on the 3-point linkage of the tractor. The field capacity of this machine depends on the type of soil and its moisture content. Under average conditions, it is capable of digging 60-70 hr of 90 cm depth.

No till seed-cum-fertidrill: The equipment is used for no-tillage system, requiring no previous seed-bed preparation after harvesting paddy and sowing of wheat crop effectively in one operation. This machine has inverted T-type furrow openers in place of shovel type furrow openers. The performance of the No-till drill was found to be most effective when operated in the fields where the loose straw after the combine harvesting of paddy has been dealt with. It can be operated by a 35 hp or above tractor. Its effective output is about 0.35 to 0.4 ha/hr. Its use saves 60-70 % diesel and time and cost of operation in comparison to the traditional method.

Strip-till drill: This machine is used for minimum tillage. It can sow wheat after paddy without any prior seedbed preparation. The machine consists of a standard seed drill with a rotary attachment mounted in the front of furrow openers. The rotary unit has C-type blades, which prepare a strip of 75 mm wide in the front of every furrow opener and hence only 40 per cent of area is tilled. Tilling and sowing is done simultaneously. Machine capacity is about 0.25-0.40 ha/h. Diesel saving with the use of this machine is 50-60 per cent as compared to conventional method whereas the time saving is 65-75 %.

Happy seeder: Happy seeder, combined the stubble mulching and seed drilling functions. The strip of stubble in front of the sowing tynes is cut, picked up and placed on the side of the drilled seed as mulch. The sowing tynes therefore engages bare soil. This PTO driven machine can be operated with 45 hp tractor and covers 0.2-0.3 ha/hr. Weed matter was nearly 50% lesser on happy seeder trial compared to conventionally sown plots. Happy seeder sows wheat directly in paddy residue in combine harvested field hence, prevents residue burning thus reduces air pollution. Mulched crops residue improved the soil hearth and added organic matter to the soil. This machine can also be used for sowing subsequent moong crop in wheat residue.

Inclined plate multi crop planter: It is suitable for sowing bold grains like maize, soybean, groundnut, cotton etc. In this machine, planting attachment has been added to commercially available seed-cum-fertilizer drills. It can plant 6-rows of groundnut at a spacing of 30 cm in addition to number of other crops like maize, cotton, soybean, sunflower etc. Seed metering mechanism in planting attachment is of inclined plate type with notched cells. Row to row spacing and plant to plant spacing is adjustable. It can cover 2.5-3.0 ha/day. It saves about 60% labour and time in comparison to manual planting.

Ridge planter for winter maize: It is suitable for planting maize on ridges. The machine consists of a 3-bottom ridger; a hopper with 2 units of inclined plate metering mechanism, driving wheel and furrow openers. The performance of the machine for sowing maize was found to be highly satisfactory. It can sow 2 rows and has a capacity of 0.2-0.35 ha/hr at a forward speed of 2.0-2.5 km/hr. Uniform plant to plant spacing can be achieved under good seedbed conditions. Machine saves about 80 per cent labour as compared to the traditional method.

Self- propelled single wheel riding type paddy transplanter: It is suitable for transplanting paddy seedlings in puddled soil. The machine consists of a 3.9 hp light weight diesel engine, power transmission system, main frame and mat type rice nursery tray, float and transplanting unit. It has a lugged wheel and the weight of the machine rests on the lugged wheel and float at the time of transplanting. Machine uses mat type seedlings and it can transplant 1.2-1.6 ha/day with the help of 4 persons. Hill population of 33 hills/m^2 can be achieved with 2-4 seedlings/hill. It saves about 80% labour as compared to manual transplanting.

Self-propelled walk behind type paddy transplanter: It is a 4 row walk behind type machine operated by a 4.3 hp petrol engine. It has only two lugged wheels and weight of the machine rests on the lugged wheel and float at the time of transplanting. The same lugged wheels were used for transportation. Row to row spacing is 30.0 cm and four settings are provided for plant top plant spacing i.e. 12, 14, 18 and 21 cm. There is a provision for adjusting the number of hills transplanted /m^2 and depth of transplanting. Machine uses mat type seedlings and it can transplant 0.8-1.2 ha/day with the help of three persons. It saves about 78% labour as compared to manual transplanting.

Self- propelled four wheel riding type paddy transplanter: Self propelled four wheel type paddy transplanter is a 6 row riding type machine operated by a 16.75 hp petrol engine. It has four lugged wheels and the weight of the machine rests on the lugged wheels. The same lugged wheels are used for transportation. The machine has five forward speeds. Row to row spacing is 30.0 cm and five settings are provided for plant to plant spacing i.e. 12, 14, 16, 18 and 21 cm. Five settings are provided for adjusting the number of hills transplanted per square meter. Transplanting depth is also adjustable. Machine uses mat type seedlings and it can transplant 4.0-4.8 ha/day with the help of five persons. It saves about 87% labour as compared to manual transplanting.

Sugarcane cutter planter: It is used for sowing of sugarcane. It cuts the seed sets of desired size in addition to doing other operations of opening the furrows placing the sets in the furrows, application of fertilizer, treatment of sets and covering of sets simultaneously. Machine is operated by a 35-hp tractor and has two rows. In this machine, two persons sitting on the machine feed the complete sugarcane one by one into the set cutting unit by picking from the seed hopper. The rotating blades cut the sets automatically before dropping the furrows. Machine can cover 1.0-1.2 ha/day with the help of 5 persons. Thus, machine saves about 75 per cent of labour in comparison to conventional method.

Trench digger for paired row sugarcane planting: A two row trench digger is designed and developed by the department for paired row sugarcane planting. The bottom width of trench was 30 cm and top width of bed was 90 cm. Depth of trench was 25 to 30 cm. The capacity of the machine is 0.3 to 0.4 ha/hr. About

8-10 per cent water saving was observed in trench planting of sugarcane crop. There is a possibility of inter-row cropping of wheat, chickpea, cauliflower/ cabbage, sarson, mentha on the beds.

Sugarcane bud planter: Sowing of sugarcane buds in field is cumbersome. Not maintaining uniform distances between the buds and varying the depth while sowing manually, may result in less productivity. As an attempt to address these issues, the sugarcane bud planter was born, which is a tractor operated bud planter. Using this planter the plantation cost is estimated to reduce to about ₹ 800/acre from about ₹ 6,000/acre using labour. The plant-to-plant sowing distance can also be adjusted according to the requirements. It can also be used for simultaneous application of fertilizer, pesticides or herbicides in the field. For this purpose, the machine is equipped with a sprayer pump. The machine can also be used for intercropping along with sowing of sugarcane buds. Pulses, wheat, and peanuts can be planted in between two rows of sugarcane. Apart from these, the machine can also be used for planting potatoes. The machine becomes helpful for those farmers, who face constant scarcity of farm workforce. The machine requires manpower only to fill the bud box whenever it gets empty. The innovator was awarded at the 8th National Award Function of the National Innovation Foundation – India (NIF) in 2015.

Semi automatic potato planters: These are two types i.e. belt & cup type and revolving magazine type potato planters. In the first type the seed tubers were placed in a hopper with its two sides slanting. The planter utilizes two endless canvas belts with cups riveted onto them. Each cup picks up a tuber. Two persons, one for each row, sit on the planter and observe the seed metering belts to ensure that each cup contains a tuber, if not then correction is made by removing/placing one of the tuber(s). Field capacity of the machine is 0.12-0.15 ha/hr. The second type of potato planter uses a revolving magazine having 10-12 compartments. Depending upon the number of rows (2/4) of the planter, one person is employed for each row to place (feed) the tubers into the seed compartments of each revolving magazine. A stationary plate is provided under the magazine with a slot directly over the delivery chute to drop the tuber. The output capacity of a two-row planter with revolving magazine varies from 0.12-0.15 ha/hr.

Revolving magazine potato planter:The second type of potato planter uses a revolving magazine having 10-12 compartments. Depending upon the number of rows (2/4) of the planter, one person is employed for each row to place (feed) the tubers into the seed compartments of each revolving magazine. A stationary plate is provided under the magazine with a slot directly over the delivery chute to drop the tuber. The output capacity of a two-row planter with revolving magazine varies from 0.12-0.15 ha/hr.

Automatic potato planter: Automatic potato planter is used for planting potatoes without the help of manual labour. It consists of a hopper, two picker wheels for picking tubers, seed tube, furrow openers, three bottom ridgers, a fertilizer metering mechanism and a frame. At the bottom of the hopper, there are agitators to improve the delivery of potato tubers to feeder. The metering mechanism is picker wheel type. The field capacity of the machine is 0.37 ha/h. Automatic potato planter saves about 70% of labour in comparison to semi-automatic potato planter. The performance of the machine is highly satisfactory when uniform size of the potato tubers is used.

Bed planter for vegetables: This machine makes beds and sow crop simultaneously on the beds. The machine is operated with a 45 hp tractor. This machine makes two beds each of 67.5 cm base width, 35 cm of top width and slant height of 20 cm. Machine also has a bed shaper after seeding to give proper shape to beds. The well-pulverised soil and proper moisture content are required for making of proper shape of the beds. The machine can cover 0.3-0.4 ha/hr and saves water to the tune of 15 – 20% for wheat crop. A single row one-meter bed width bed planter was also developed for planting onions, garlic and other vegetables.

Vegetable transplanter: A vegetable transplanter has been developed for transplanting seedlings of brinjal, cauliflower, cabbage, tomato and other crops like winter maize, African sarson etc. The machine consists of a frame, two lugged ground wheels, seedling tray, seat for the operator, furrow opener, compaction wheels, finger guide tunnel, picker wheel type metering mechanism, a water tank and a bed forming attachment. Picking forks has a spring mounted rubber flappers which opens before passing through the tunnel and close during its passage. Again, the flappers open at the bottom end of the tunnel to release the seedlings in a furrow. The wheel compacts the soil around the seedlings. The plant spacing in the machine is adjustable. Two persons one for each row sitting on the machine is required to places the seedlings in the flappers. The machine can transplant seedlings both on the beds and on flat surface. Machine can cover about 0.7-0.8 ha/day. Plant missing varies from 3-7 per cent. There is saving of labour up to 75-80 per cent in comparison to manual transplanting.

Inter-culture and spraying equipment

PAU wheel hand hoe: It is suitable for inter culture in row crops except paddy. For wider row crops, "*Trifali*" with 3 blades can also be used. It can cover 0.3-0.4 ha/day and one person can easily operate this equipment. The work rate of this equipment is about 4-5 times faster as compared to "*Khurpa*" without any additional load on the operator. Weed control with wheel hand hoe is only in the rows.

Rotary power weeder: It is a self-propelled engine operated power weeder for inter culture operation in horticulture and wider row crops. The depth of operation ranged from 4-7 cm. The machine can be operated at an average forward speed of 1.5 to 2.0 km/hr having average width of coverage 62.2 cm (in two pass). The weeding efficiency of the machine varies from 80-94% depending upon the type of crop. The field capacity ranged from 0.6 to 1.0 ha/day. The performance of the machine for weeding is found satisfactory on cotton, sugarcane, sunflower and chickpea.

Tractor operated rotary weeder: This machine is suitable for weeding in wider row crops like cotton, sugarcane etc. (Fig. 2). The machine consists of three rotary weeding blade assemblies. Power to these assemblies is provided from tractor PTO transmitted to main square shaft through gearbox mounted on main frame with the help of set of sprockets and chain. The machine has provision for adjustment of row to row spacing. Field capacity of the machine is 0.3-0.4 ha/hr. Weeding efficiency varies from 75-85%.

Fig. 2. Tractor operated rotary weeder
Source: (https://aicrp.icar.gov.in/fim/salient-achievements)

Fig. 3. Self-propelled power weeder
Source: (https://aicrp.icar.gov.in/fim/salient-achievements)

Self-propelled power weeder: The machine is useful in row crops, horticultural and vegetable crops for weeding and seedbed preparation. It consists of a 4.1 kW diesel engine mounted on the power tiller chassis, power transmission system, two MS wheels, a frame and a rotary tiller. The power from the engine is transmitted to the rotary with the help of belt and chain and through gear train to the ground wheels. The rotary tiller has been provided with 16 blades fitted on high-pressure pipe of 37.5 mm diameter with the help of nuts and bolts to the flanges. For depth adjustment, two skids made of flat are provided on both sides of power tiller. A power cut-off device is provided to engage or disengage the power to the rotary system. The wheels with lugs are provided for traction. The speed of power weeder ranges 2.3-2.5 km/hr with an effective working width of 550 mm giving field capacity of 0.10 to 0.13 ha/hr. The cost of self-propelled power weeder is about ₹ 40,000 and average cost of weeding is ₹ 1000/ha. The equipment saves 90% operating time and 30% in cost of weeding as compared to hand weeding by *khurpi*.

Fig. 4. Fertilizer band placement cum earthing up machine
Source: (https://aicrp.icar.gov.in/fim/salient-achievements)

Fertilizer band placement cum earthing up machine: The tractor operated (26 kW and above) fertilizer band placement cum earthing up machine has been designed and developed at GBPUAT, Pantnagar (Fig. 4). The machine is suitable for simultaneous placement of fertilizer, earthing up and cutting of weeds in crops such as maize, sugarcane, potato etc. having more than 0.50 m row to row spacing. The urea fertilizer application rate ranges from 60 to 250 kg/ha. It helps in top dressing of fertilizer at 50 to 100 mm from the plant. The field capacity of machine is 0.56 ha/hr with 82.4% field efficiency. The approximate cost of the machine is ₹ 50,000. There is considerable saving in fertilizer, time and labour over traditional method.

Tractor operated sprayer: It is suitable for spraying on cotton crop or any other wider row crops. It consists of a centrifugal pump, a tank, pressure regulator valve and a boom with nozzles and spray gun fitted on a frame. The sprayer is mounted on the 3-point linkage of the tractor and drive is given through from tractor PTO through a set of gears. Boom height can be adjusted from 10 to 225 cm from ground to suit different crop height. It can cover up to 1200 cm width and has a capacity of about 2.0 ha/hr at a field speed of 3.0 km/hr.

Self-propelled high clearance sprayer: It is most suitable for spraying on tall crop like cotton or wider row crops. The machine has a chassis with 120 cm ground clearance, four wheels, and 20 hp diesel engine, gearbox, water tank, seats for the operator, spray pump and boom with 18 nozzles. The boom height can be adjusted from 31.5 to 168.5 cm to suit different crops and can be folded during transport. The field speed is up to 5 km/hr and the road speed is up to 25 km/hr. The width of coverage is 1350 cm and it has a capacity of about 2.0 ha/hr at a field speed of 3.0-4.0 km/hr. Mechanical damage caused by the movement of high clearance sprayer in cotton crop is less in comparison to tractor operated sprayer.

Tractor operated fertilizer dibbler for ratoon sugarcane: The field after harvest of sugarcane is covered by a mat of trash up to a depth of 150 mm and punch application enables placement of fertilizer through crop residue (Fig. 5).

A tractor operated fertilizer dibbler for ratoon sugarcane has been designed and developed at TNAU, Coimbatore for placement of fertilizer without much soil disturbance and through crop residue. The principal components of the implement are revolving spade, fertilizer metering device, fertilizer placement funnel, soil covering and pressing device. The cost of the unit is ₹ 45,000/- and field capacity is 0.2 ha/hr. The cost of operation is ₹ 1550/ha and results in saving of 60% as compared to the conventional method.

GPS based variable rate granular fertilizer applicator: A GPS based variable rate granular fertilizers (NPK) applicator has been developed at IIT, Kharagpur and CIAE, Bhopal to ensure ideal application of fertilizers as basal dose (Fig. 6). It consists of a differential global positioning system (DGPS), micro-processor, micro-controller, DC motor actuator, power supply, threaded screw arrangement and fluted roller fitted metering mechanism. The fertilizer application rate is changed according to the prescribed application rate at the identified grid with coefficient of variation (CV) of 11.7-15.0%.

The RMSE and relative differences (RD) at different levels of application rates ranged from 1.3 to 4.6 and from 1.75 to 6.56, respectively. The fertilizer application accuracy ranges from 89.3% to 98.1% at various discharge rates. It was observed that the developed variable rate fertilizer applicator (VRFA) was effective and accurate to respond to the target application rates with small delay of time. It was concluded that the developed VRFA system closely met the target fertilizer application rate at the selected grid.

Harvesting and threshing equipment

Tractor front mounted vertical conveyer reaper windrower: It is suitable for harvesting and windrowing of wheat and paddy crops. The machine is mounted in front of the tractor and the power to the machine is provided through tractor PTO with the help of intermediate shaft running beneath the body of the tractor and a coupling shaft. Height of the machine above ground is controlled by tractor hydraulic with the help of pulleys and steel ropes. After the crop is cut by the cutter bar, it is held in a vertical position and delivered to one side of the machine by lugged belt conveyors and fall on the ground in the form of a windrow perpendicular to the direction of movement of machine. The machine is operated at a speed of 2.5-3.5 km/hr and has a field capacity of about 0.4 ha/hr. The use of this machine for harvesting can save about 60-70% labour and about 40-50% cost in comparison to manual harvesting.

Fig. 5. Fertilizer dibbler for ratoon sugarcane
Source: (https://aicrp.icar.gov.in/fim/salient-achievements)

Fig. 6. GPS based variable rate granular fertilizer applicator
Source: (https://aicrp.icar.gov.in/fim/salient-achievements)

Vegetable digger: This machine is used for digging various root crops like carrot, potato, garlic and onion. It consists of a digger blade made from carbon wear resistant steel. An elevator chain conveyor has been attached behind the blade. Two oval agitators are provided in the conveying system for separation of soil particles from the crop. The power to the elevator conveyor has been provided through a gear box. Two coulter discs are provided in front of the blade at the outer ends which help in easy slicing and lifting of soil by the blade. A roller behind the conveyor belt is provided for easy collection of dug material as it presses the soil before fall. The capacity of the machine varies from 0.21-0.28 ha/hr with harvesting efficiency ranging between 96-99%. Labour saving is about 60-70%.

Groundnut digger elevator: The machine is suitable for harvesting groundnut crop. It digs the groundnut vines below the pod zone and simultaneously elevates them by an elevator-picker reel (conveyor) for dropping on the ground. The soil attached to the vines is shaken off in the process and a window is formed with the help of deflector rods. The vines are dropped in such a manner that the pods get exposed to the sun for speedy drying. The field capacity of the machine varies from 0.3 to 0.4 ha/hr at tractor forward speed of about 2.5 km/hr with the help of three persons. Digging efficiency is around 97-98%. Per cent detachment of pods is about 3-4%. Machine saves up to 80-85% labour and 35-40% cost in comparison to tractor operated digger blade.

Fig. 7. Sugarcane chopper harvester

Sugarcane chopper harvester: The row spacing of header unit can be adjusted from 90 cm to 150 cm. This is a chopper type harvester (Fig. 7). Green top is decoupled and pushed aside, which is collected manually for feeding to the cattle. Cost of the machine is ₹ 1.5 crores.

Harambha (High capacity) thresher: Haramba thresher is suitable for threshing wheat crop and is highly popular. Threshing material pass through the concave. Light materials like chopped straw (Bhusa) is blown away with aspirator blower while the heavier materials like grains, nodes etc. fall on a set of reciprocating sieves. The sieves clean the grain and there is an optional attachment of an auger to elevate the grains and conveys directly on to a trolley. Feeding of crop is manual by standing on a platform provided with the thresher. Generally 3-4 persons are required for continuously feeding of the crop. Capacity of the thresher varies from 15-20 q/hr. The labour required for threshing with haramba thresher including the transportation of crop varies from 30 to 35 man-h/ha.

Axial flow paddy thresher: The paddy thresher had axial flow beater type threshing cylinder. The paddy thresher consists of feeding hopper, threshing cylinder, concave, cylinder casing, two sieves and screen for cleaning and blowers/aspirators. The crop is fed into the hopper and it is received by the threshing cylinder tangentially and then moves along the cylinder axially. For cleaning purpose, the thresher has two aspirators, one blower and one thrower. Threshing efficiency varies from 97-99%, cleaning efficiency varies from 90-97% and cylinder loss varies from 0.7 to 3.6%. It saves about 70% of labour as compared to conventional method of manual threshing by beating.

Maize dehusker-cum-sheller: This machine comprises of an axial threshing cylinder with a suitable concave and a thrower mechanism to eject empty stalk and husk. Grains fall on the cleaning sieves for cleaning. The machine can be operated by 25 to 35 hp tractor or 10 to 15 kW motor. The thresher can thresh the dehusked maize cobs having moisture content in the range of 12-24% successfully. The output capacity of the machine varies from 1200 -2400 kg/hr. The threshing and cleaning efficiency of the thresher is in the range of 96-98% and 94-98% respectively. Broken grains vary from 1-3%.

Fig. 8. Pedal operated maize sheller
Source: (http://www.aau.in/sites/default/files/FMPE%20Recommandantion.pdf)

Pedal operated maize sheller: The average capacity of machine to shell grains 61.50 kg/hr (6 times higher than hand operated maize sheller) and shelling efficiency 97.96%. The average cost of shelling for 100 kg grains is ₹ 44.50 and the saving in the cost of shelling is 71.08% as compared to hand operated maize sheller (Fig. 8).

Low cost power operated maize sheller: The machine is operated by 1hp electric motor. The highest through put rate of 1084 kg maize cobs/hr with highest shelling rate of 896 kg maize kernel and the lowest cost of shelling (₹ 5.0/q) as compared to hand (67 kg/hr, 55.71 kg/hr, and ₹ 160.25/q) and pedal (21.67 kg/hr, 18.97 kg/hr, and ₹ 68.17/q).

Groundnut thresher: The machine can detach the groundnut pods from the vines. The capacity of the thresher is about 200 kg/hr at moisture content of 35% (pods). The cleaning efficiency was found to be 94.2-97.0% whereas the threshing efficiency was of the order of 99%. Broken grains were observed to be 0.2-0.5%. There is a saving of about 31% in cost of operation and about 83% in labour requirement per hectare as compared to traditional methods of threshing.

Fig. 9. Low cost power operated maize sheller (*Source*: http://www.aau.in/sites/default/files/FMPE%20Recommandantion.pdf)

Sunflower thresher: Machine is suitable for threshing sunflower crop and has a capacity of 600-900 kg/hr of clean grain. Threshing efficiency of the machine is 100% where as cleaning efficiency is about 90%. Grain losses vary from 0.65 to 2.94 % depending upon crop moisture content. Machine requires 4-5 persons for crop feeding. It saves about 70% labour in comparison to traditional threshing.

Multi crop thresher: Equipment used for more than one crop with or without minor adjustment. This thresher has either spike tooth cylinder or rasp bar cylinder depending upon the manufacturer. It has cleaning and bagging attachments. Commercially available spike tooth type thresher has been used for threshing moong and mash after incorporating few modifications. The threshing cylinder has thirty six spikes placed six in each row. For threshing pulses, six spikes were retained on cylinder in six rows i.e. one in each row. The arrangement of spikes on cylinder periphery was axial. Threshing efficiency of the thresher is around 99% and cleaning efficiency is around 98-99%. Its capacity may be 300-2500 kg/hr. This thresher can be used for crops like paddy, wheat, sorghum, soybean, gram, millets, etc. It can be operated by 5-20 hp power depending upon the models.

Vegetable seed extraction machine: This machine is developed for extracting seeds from different vegetables and fruits like tomato, brinjal, chilli, watermelon, summer squash, cucumber, tinda etc. It consists of frame, feeding chute, a primary chopping chamber, crushing chamber, seed collecting chamber, rotor, concave screen, a seed out let, a waste (pulp) outlet, water pump, sprinkler and power transmission system. The fruits or vegetables are cut into small pieces in the primary chamber. Thereafter, these are further crushed by means of axially arranged blades attached to a rotor shaft. The shaft rotates at a speed of 250-300 rpm. Different concave screens for separating different seed sizes are used. Sufficient water should be available at the site of the machine. Three persons are required to operate the machine. The machine can extract seeds at the rate of 5.49, 3.78, 9.42, 4.68, 3.60, 6.60 and 1.42 kg/hr, respectively of brinjal, tomato, chilli, summer squash, watermelon, squash melon and cucumber.

Fodder harvesting machinery

Tractor-operated cutter bar type forage harvester and chaffer cum loader: The system consists of two separate units. First one is a mower and the second one is a chaff cutter cum loader. The mower has a working width of 196 cm and so designed that it can be folded easily during transportation. The chaffer cum loader has two powered feed roller and one compressing roller. It has two chaffing blades for chaffing of fodder and six thrower attachments are mounted on the periphery of the cutter. A reversing mechanism has also been provided for safety. The fodder is cut with mover and is collected in the field. The chaffer cum loader, chaff the fodder at site and the throwers mounted on the periphery of the cutter throws the fodder directly into the trailer. The field capacity of the side mover varies from 0.20-0.25 ha/hr and fuel consumption from 3.5-4.0 l/hr when operated at a speed of 1.5-2.0 km/hr. The throughput capacity of the chaffer cum loader varies from 12.0-18.0 tonnes/hr and fuel consumption from 4.0-4.5 l/hr, respectively. Three persons are required for feeding the crop to the chaffer. The chaffer cut the fodder pieces of uniform size of 18-20 mm. There is a saving of about 50 per cent in cost of operation and 65 per cent in labour cost.

Manually operated long handle scythe: It is suitable for harvesting barseem fodder. It has a long handle made of hollow pipe and a curved blade. One person in standing posture holds the handle from two points to operate the scythe. In one stroke a person cuts an area of about 1.2 m wide and 0.6 m long. In 2nd stroke, cut crop is swept with blade itself to make a sort of windrow. It has a higher work output in comparison to sickle as it has a better working posture. It saves about 50% labour.

Residue management machinery

Paddy straw chopper: This machine chops the paddy straw left in the field after combine harvesting into small pieces and spreads the chopped straw evenly in the field. This straw then can be incorporated into the soil and subsequent drilling of wheat can be done. This is an environment friendly technology as farmers do not need to burn the paddy straw left in the field and also improves the soil health. Paddy straw chopper consists of a rotary shaft mounted with blades named as flails for harvesting and chopping the paddy straw. Two counter rows having serrated blades are mounted on the concave of front portion of straw bruising which further assists in chopping the straw. The rotary shaft has two rows of flails. Each row consists of 20 number of flails. The shape of the flail is inverted gamma type. One counter row each having 21 serrated knives is fixed on the inside of the concave. Machine is operated by a 45-50 hp tractor. Field capacity of the machine is 0.3 ha/hr and approx. 70% straw is chopped in the size < 10 cm.

Straw baler: The straw baler is used for collecting and baling straw in the combine-harvested field. Before baling, first stubble shaver is operated to harvest the stubbles from base level. It can form bales of varying length from 40 to110 cm. The height and width of bales are generally fixed at 46 cm. The weight of bales varies from 15 to 45 kg depending on moisture content of straw and length of bales. The capacity of the baler varied from 0.34-0.38 ha/hr.

Straw combine: Straw combine is used to recover wheat straw after combine operation and is operated by a 35-40 hp tractor. Straw collected by straw combine is chopped into small size and collected in the trolley having a net to remove the dust. Also some grains are collected along with straw. The capacity of machine on an average is 0.5 ha/hr and straw recovery is about 55-60%. Machine is highly popular.

Fig. 10. Tractor operated banana stem shredder (*Source*: https://aicrp.icar.gov.in)

Tractor operated banana pseudo stem shredder: After the harvest of the banana bunch, the pseudo stem is manually cut and left in the rows (Fig. 10). After the harvest of the whole field, these are collected and left near the boundary for drying and subsequent burning. This process is tedious and time consuming.

The banana stem shredder helps in disposing of the stem immediately after harvest. Shredded material is suitable for mulching in the banana garden and also for vermi-compost. The banana stem shredder helps in disposing of the stem immediately. The banana stem consists of 95% of water and only 5% of fibre. The force required to cut the stem was calculated and accordingly the speed of the shredding drum (1060 rpm) were fixed. The shredding unit consisted of 4 blades placed perpendicular to each other at 2250 mm distance. Additionally, 12 nos. of spikes with flat cutting edge are fitted with a gap of 120 mm between the rows. The whole device is mounted on a frame made of MS angle. The blades are driven by the PTO of the tractor with a bevel gear box and the hopper is trapezoidal in shape with a height of 800 mm. During field trials it took 1.2 minutes to shred the stem having average height of 2400 mm. The stem was cut into small pieces and the water and fibre were separated.

Ergonomics and Safety

Slow moving vehicles emblem: Slow moving vehicle emblem (SMVE) was made sticker base rather than plastic base, which helps in simple fabrication and pasting on tractor trailer to avoid accidents during night.

Safety cage for tubewells: It is a ring shaped cage having a window type entry and is used for covering the wells and tubewells at farms and in villages to avoid accidents and saving human and animal from falling in tube well. The height of the ring is 75 cm and diameter is 210 cm.

Poisonous gas aspirator for tubewells: It is a simple aspirator and is operated by 35 hp or above tractor. Power to the aspirator is provided by tractor PTO and size of aspirator fan is 50 cm. The aspirator is connected to a flexible plastic pipe that can reach up to 30 ft deep well. The aspirator with the help of flexible plastic pipe sucks the poisonous gases produced in the tube wells and saves human lives especially during monsoon season. The above mentioned technologies developed by the department have resulted in impact on the economy of the state through reduction in the cost of cultivation, reduction in the losses, increase in the cropping intensity, decrease in labour requirement, increase in production and productivity because of timeliness of operations and reduction in drudgery of operator/worker.

Precision farming tools/machineries

Most farmers have practiced a form of variable-rate application (VRA) with a conventional sprayer for site-specific crop management (SSCM). A conventional sprayer applies a chemical that is tank-mixed with a carrier (usually water) using spray nozzles and a pressure-regulating valve to provide a desired volumetric application of spray mix at a certain vehicle speed.

Any change in the boom pressure or vehicle speed from that of the calibration results in an application rate different from the planned rate. Applicators have used this to their advantage at times. For example, when observing an area of heavy weed infestation, the applicator can manually increase the pressure or reduce the speed to apply a higher (but somewhat unknown) rate of herbicide.

Variable-rate application methods: One important technology-related question is: What methods of variable-rate application of fertilizer, lime, weed control, and seed are available? There are a variety of VRA technologies available that can be used with or without a GPS system. The two basic technologies for VRA are: map-based and sensor-based.

Map-based VRA adjusts the application rate based on an electronic map, also called a prescription map. Using the field position from a GPS receiver and a prescription map of desired rate, the concentration of input is changed as the applicator moves through the field.

Sensor-based VRA requires no map or positioning system. Sensors on the applicator measure soil properties or crop characteristics "on the go". Based on this continuous stream of information, a control system calculates the input needs of the soil or plants and transfers the information to a controller, which delivers the input to the location measured by the sensor. Because map-based and sensor-based VRA have unique benefits and limitations, some SSCM systems have been developed to take advantage of the benefits of both methods.

Map-based VRA: The map-based method uses maps of previously measured items and can be implemented using a number of different strategies. Crop producers and consultants have crafted strategies for varying inputs based on

(1) soil type (2) soil colour and texture, (3) topography (high ground, low ground), (4) crop yield, (5) field scouting data, (6) remotely sensed images, and (7) numerous other information sources that can be crop-and location-specific.

Some strategies are based on a single information source while others involve a combination of sources. Regardless of the actual strategy, the user is ultimately in control of the application rate. These systems must have the ability to determine machine location within the field and relate the position to a desired application rate by "reading" the prescription map.

Sensor-based VRA: The sensor-based method provides the capability to vary the application rate of inputs with no prior mapping or data collection involved. Real-time sensors measure the desired properties -usually soil properties or crop characteristics - while on the go. Measurements made by such a system are then processed and used immediately to control a variable-rate applicator.

The sensor method does n't necessarily require the use of a positioning system, nor does it require extensive data analysis prior to making variable-rate applications. However, if the sensor data are recorded and geo-referenced, the information can be used in future site-specific crop management exercises for creating a prescription map for other and future operations, as well as to provide an "as applied" application record for the grower.

Seeding VRA: Planters and drills can be made into VRA seeders by adjusting the speed of the seed-metering drive (Fig. 11). This will effectively change the plant population. VRA seeding is accomplished by separating or disconnecting the planter's seed-meter systems from the ground drive wheel. By attaching a motor or gear box (to change speed of the ground wheel input), the seeding rate can be varied on the go. Most of these devices will be matched with a prescription map and can have two or more rates. A two-rate scenario may be a system that reduces the seeding rates outside of the reach of a center pivot irrigation system, while multiple rates may be needed to adjust for soil types (water-holding capacity) and organic matter. it includes a hydraulic drive unit, processor, and ground speed sensor. A hydraulic motor (powered by tractor hydraulics) is attached to an electric stepper motor to control the speed delivered to the seed-meter shaft.

Fig. 11. Hydraulic motor to control seed meter (PAR-2 Variable Rate Drive; *www.trimble.com/agriculture/Variable-Rate-Application-Solution.aspx?dtID=overview*).

A controller receives a ground speed signal and coordinates the speed with planter width and seeding rate to send a signal to the hydraulic drive. On some planters/ drills, the seeding rates are matched with the application rates of fertilizer, herbicides, or insecticide units because they are driven by the same meter shaft (Fig. 12).

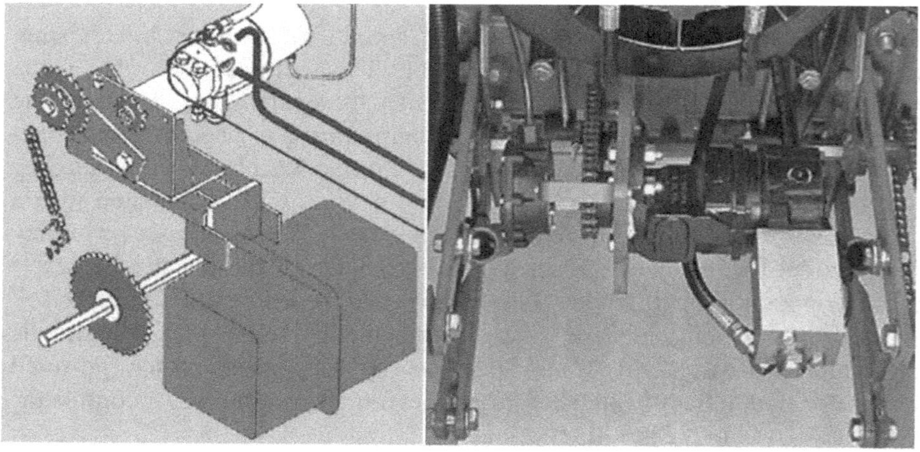

Fig. 12. Hydraulic motor attached to the seed-meter shaft (Trimble Variable Rate Controllers; *www.trimble.com/agriculture/Variable-RateApplication-Solution.aspx?dtID=overview*).

There is development of using on-the-go sensors to VRA seeding. There are soil organic matter (SOM) sensors that detect different levels of organic matter and adjust the plant population rate accordingly. Soil moisture meters that may be used for depth adjustment and for changing seeding rates are available.

Weed control VRA: For map-based weed control VRA systems, some form of "task computer" is required to provide a signal indicating the target rate for the current location (Fig. 13). Second, a system for physically changing the application rate to match the current prescribed rate is required. There are a number of different types of control systems on the market today that are adaptable to VRA. Three categories will be discussed:

1. Flow-based control of a tank mix.
2. Chemical-injection-based control, with the subset, chemical-injection control with carrier.
3. Modulated spraying-nozzle control system.

These systems eliminate much of the error in application that could occur if groundspeeds change from the calibrated setup. With the application rate managed by an electronic system, the ability to apply variable rates is a logical next step. This requires that the prescribed application rate, or "set point," be changeable according to the rate prescribed for that location.

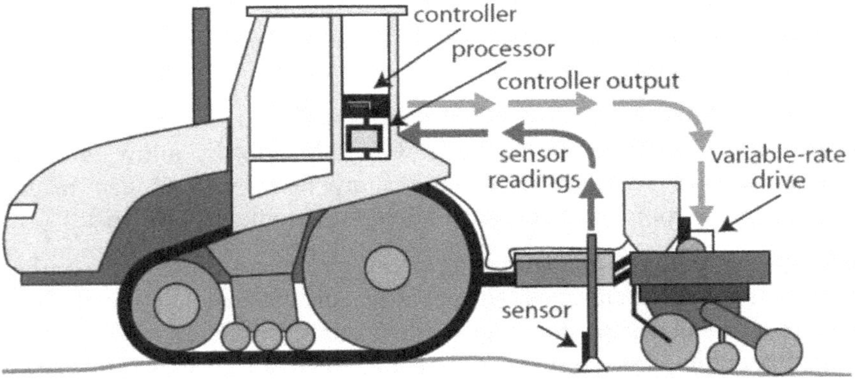

Fig. 13. "On-the-go" sensor (texture, electrical conductivity (EC), or soil organic matter (SOM)) measures soil characteristics before planting and adjusting the seeding rate (plant population) (*www.trimble.com/agriculture/Variable-Rate-Application-Solution.aspx?dtID=overview*).

Flow-based control systems: The flow-based control of a tank mix is the simplest of the three types discussed here (Fig. 15). These systems combine a flow meter, a groundspeed sensor, and a controllable valve (servo valve) with an electronic controller to apply the desired rate of the tank mix. A microprocessor in the console uses information regarding sprayer width and prescribed application rate to calculate the appropriate flow rate (gallons/min.) for the current groundspeed. The servo valve is then opened or closed until the flow-meter measurement matches the calculated flow rate. These systems have the advantage of being reasonably simple. They are also able to make rate changes across the boom as quickly as the control system can respond to a new rate command, which is generally quite fast (3 to 5 seconds).

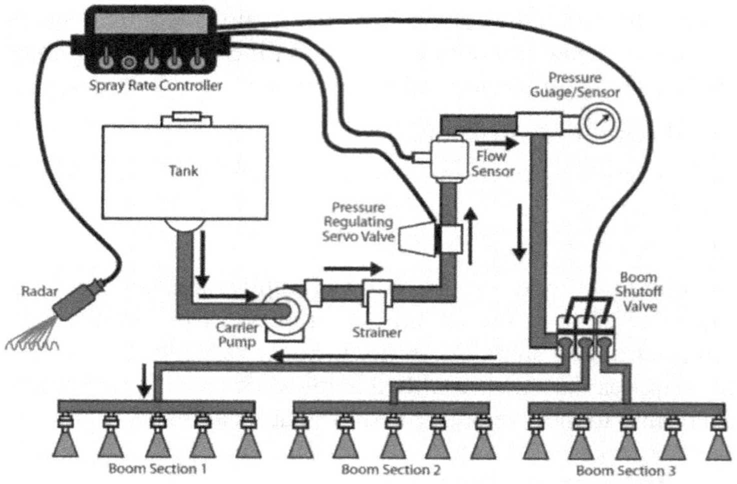

Fig. 15. VRA spraying system that is a flow-based control system of application rate (*Source*: Grisso et al., 2011)

As with any technology, flow-based controllers have limitations. The flow sensor and servo valve control the flow of tank mix by allowing variable pressure rates to be delivered to the spray nozzles. This can result in large changes in spray droplet size and potential problems with drift.

Some systems have warning provision for the pressure is out-side the optimum operating range for the nozzles. The operator can adjust vehicle speed to return the pressure to an acceptable range. This is the most widely used system. Its standard operating procedures specify that the operator must mix the chemical in the spray tank with the carrier and will generally have to deal with some leftover tank mix. However, this is a relatively simple system that should meet most needs while giving operators the capability of a single herbicide VRA.

Chemical direct-injection systems: An alternative approach to chemical application and control uses direct injection of the chemical into a stream of water (Fig. 16). These systems utilize the controller and a chemical pump to manage the rate of chemical injection rather than the flow rate of a tank mix. The flow rate of the carrier (water) is usually constant and the injection rate is varied to accommodate changes in groundspeed or changes in prescribed rate. Again, if the controller has been designed or modified to accept an external command (from a GPS signal and prescription map), the system can be used for VRA.

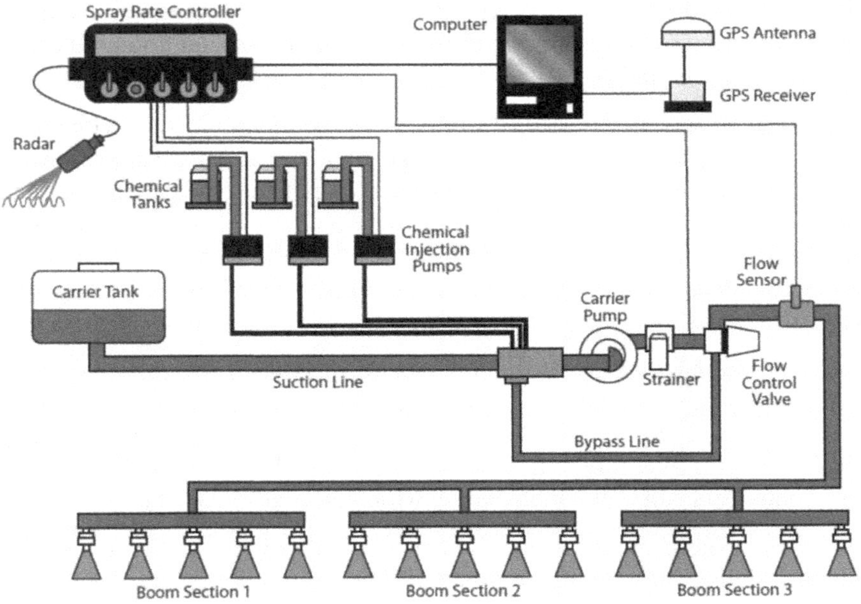

Fig. 16. VRA spraying system that incorporates chemical-injection technology. (*Source*: Grisso et al., 2011)

Chemical injection eliminates leftover tank mix and reduces chemical exposure during tank mixing. An additional advantage of this system is that the constant flow of carrier can be adjusted to operate the boom nozzles to provide the optimum desirable size and distribution of spray droplets. The principal disadvantage for variable-rate control is the long transport delay between the chemical-injection pump and the discharge nozzles at the ends of the boom. The volume within the spray plumbing (hoses and attachments) must be applied before the new rate reaches the nozzles. This can cause delays in the rate change and "Christmas tree" patterns of application as the new concentration of chemical works its way out through the boom.

For example, a simulation of a farmer owned broadcast sprayer (60-foot boom divided into five sections) indicated that nearly 100 feet of forward travel would occur before a newly prescribed rate would find its way to the end nozzles of that sprayer. However, a properly designed plumbing system and properly matched nozzles can shorten the reaction time. Some control systems will look forward (knowing location and speed) and make the required adjustments.

These limitations have led to systems that use both carrier and injection control. All manufacturers would recommend VRA be used in conjunction with carrier control as described below.

Direct chemical injection with carrier control: Chemical injection with carrier control requires the control system change both the chemical-injection rate and the water-carrier rate to respond to speed or application-rate changes. One control loop manages the injection pump while a second controller operates a servo valve to provide a matching flow of carrier. A perfect system of this type would deliver a mix of constant concentration as if it were coming from a pre-mixed tank.

The system can have many of the advantages of both of the earlier systems. Direct injection of chemicals means that there is no leftover mix to worry about, and the operator is not exposed to chemicals in the process of tank mixing. Change over from one rate to another occurs as quickly as both chemical and carrier controllers can make the change, which is usually very fast.

Disadvantages include a more complex system with higher initial cost and the problem of delivering varying amounts of liquid through the spray nozzles as rates change, with the resulting changes in droplet and spray characteristics. If a lot of spraying is done to avoid the hazards of tank mixing, these systems will give a great deal of control over the spraying operations and offer the capability of VRA of herbicides from a prescription map.

Modulated spraying-nozzle control systems: Modulated spraying-nozzle control (MSNC) systems permit VRA with spray drift control under a wide range of operating conditions (Fig. 17). MSNC controls the timing and duration of discharge from nozzles. High-speed valves are used to regulate the amount of time that spray is delivered from conventional nozzles. The systems offer the ability to change flow rate and droplet size distribution on the go. A brief description of the system follows.

MSNC-equipped sprayers use conventional sprayer nozzle assemblies that work in conjunction with direct-acting, in-line solenoid valves. Figure shows a schematic of a spraying system that incorporates modulated spraying-nozzle control. The system operates under the direction of a microprocessor and an application controller that responds to signals from flow and pressure sensors.

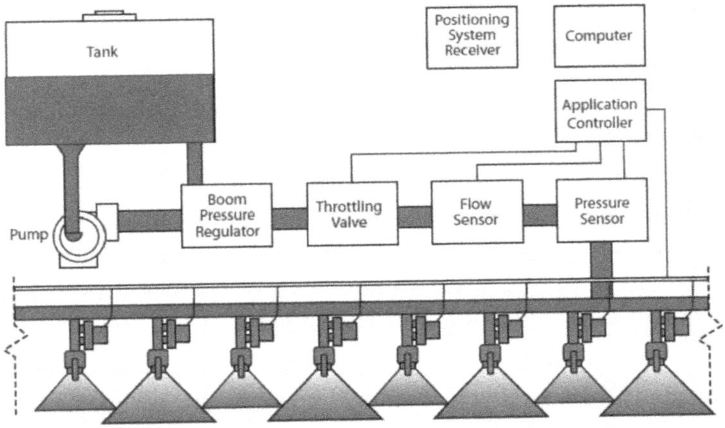

Fig. 17. VRA spraying system using modulated spraying-nozzle control (MSNC) technology. The controller can control individual nozzles or a single signal for the entire boom (*Source*: Grisso *et al.*, 2011)

The basic concept behind MSNC spraying is to operate each nozzle at full design pressure and flow during periods when a flow control valve is open. The key is to vary the amount of time that the valve stays open to produce variation in the flow rate (thus, application rate) without changing droplet size distribution or spray pattern. A fast-acting, electrical, solenoid-controlled nozzle assembly is mounted directly to a conventional nozzle assembly (Fig. 18).

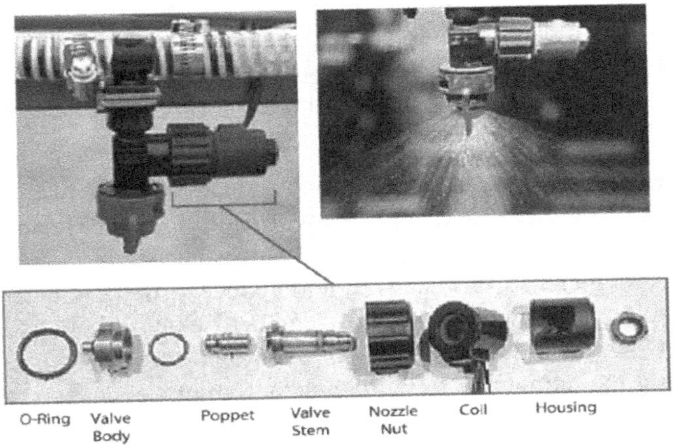

Fig. 18. Fasting acting, electrical, solenoid-controlled nozzle assembly (*Source:* Grisso *et al.*, 2011)

Systems are equipped with solenoids that operate at a frequency of 10 Hz. This means that solenoid position can be cycled between open and closed 10 times per second, as directed by a controller that responds to input from a computer and a set of sensors. A cycle of events (valve open/spray/valve close) takes place in one-tenth of a second.

In order for MSNC systems to operate most effectively, valve response must be quite rapid. An electrical signal to each valve is used to produce one of two flow conditions: full flow (completely open valve) or zero flow (completely closed valve). The solenoid-operated valves take only about 4 milliseconds (ms) or 0.004 second to respond to an electrical signal.

Changing valve position from open to closed and back (or vice versa) would take 8 ms during any 0.1-second cycle. In actual practice, this translates into a minimum duty cycle (amount of time the value is open for flow) of about 10 per cent and a maximum duty cycle of about 90 per cent if the control system is changing valve position during each 0.1-second time period. The MSNC system can also be operated at a full-open (100 per cent duty cycle) setting as well.

Because flow rate from each nozzle is governed by the amount of time (duty cycle) each flow-control valve stays open, the per centage of full, rated, nozzle flow would be equal to the duty cycle expressed as a per centage. This results in a range of flow rates from each nozzle of approximately 9-to-1, although the MSNC systems have been advertised with a more conservative rating of flow-control range at 8-to-1. For example, let's say that a standard nozzle has a rated capacity of 0.8 g pm at a pressure of 40 psi. The MSNC system is very effective at reducing nozzle flow rates while maintaining droplet size distribution and spray-pattern characteristics. Therefore, standard procedure/strategy is to install nozzles that will meet the maxi-mum flow demand in a particular spraying situation.

The MSNC system is then used to reduce rates as needed. A benefit of using larger nozzles is the reduced likelihood of plugging. In addition to controlling nozzle flow rates at a given system pressure, the MSNC system can be operated at reduced pressures to increase droplet size and reduce drift potential in locations and under atmospheric conditions in which drift would likely cause damage. Application rates could be maintained, even as system pressure is lowered, by increasing the amount of time the nozzle remains open during a minute.

Opening and closing nozzles as a sprayer travels through the field might appear to be a risky proposition. If a nozzle is held closed, even for an instant, no liquid will be discharged. Surely there will be areas of a field missed during normal operation of the sprayer. This is addressed by using a 1/20-second (1/2-cycle) "phase shift" of adjacent nozzles. When one nozzle is off, the nozzles adjacent to it are on. To increase spray-pattern overlap and minimize the effect of the "pulses and pauses" produced at the nozzles, these sprayers are equipped with wide spray-angle nozzles (110-degree angle versus the more-common 80-degree angle).

The potential benefits of using a chemical-application system that permits the tailoring of both application rate and droplet-size distribution throughout a field include the ability to:

- Produce a broader range in flow rates with much more consistent spray characteristics than conventional sprayers.
- Vary nozzle flow rates and/or travel speeds over a wide range without affecting spray pattern or drop-let-size distribution.
- Vary droplet-size distribution without changing application rate to minimize drift potential near sensitive areas or to increase spray coverage needed for some contact-type products.

MSNC technology can also be used to apply VRA nutrients. While drift control is not a major issue in fertilizer application, the MSNC provides yet another option for applicators wishing to take advantage of site-specific crop management methods.

Other useful devices

There are areas of the field that should not have chemicals applied. For example, a grass waterway is a best management practice (BMP) to reduce erosion impacts from a field, but if a non-selective herbicide is sprayed to the area while passing over the section, much damage to the BMP will result. By mapping these areas of the field, a boom control can automatically turn the boom (or sections of the boom) on and off to prevent application to selected areas. The controller can also automatically turn the boom section off if the boom section is in a previously applied area (Fig. 19) eliminating overlaps. It also eliminates skips by turning boom sections back on after leaving a previously applied area.

If the sprayer goes over an area that has already received an application, the controller detects the overlap and shuts off individual sections or nozzles of the implement to prevent the unnecessary usage of additional chemicals. When spraying oddshaped fields, grass waterways, or obstructions in a field, this boom control can have a tremendous benefit. Because an automated boom section control device requires a capital investment, applicators should weigh the cost of the machine against their potential savings on inputs before purchasing the equipment. However, one Virginia farmer using the technology indicated a 15 per cent savings in inputs (crop protection chemicals and liquid fertilizers) due to automatic boom control.

Sensor-based devices: Soil organic matter sensors (Fig. 20) can be used with VRA pre-plant herbicides because the amount of soil organic matter influences the effectiveness of some herbicides (often mentioned on the label). Such a sensor can be used to automatically adjust herbicide rates without prescription maps or other inputs. In this application, the sensor is pulled or pushed through the soil by the herbicide applicator.

Due to patchiness of weed infestations, uniformly treating entire fields can result in unsatisfactory weed control or unnecessary use of herbicides. Remote-sensing may be a technique that will improve weed scouting and result in better management decisions. Our eyes act as remote sensors. We can easily identify weed-free and weedy areas in a soybean field and distinguish between different weed species based on leaf shapes and sizes. When a remotesensing instrument collects reflectance at the field scale, reflectance values from individual features are averaged over the entire pixel area within the sensor. Using reflectance data of bare ground contrasted with green weeds growing between crop rows, some sprayers are equipped to switch the application device on and off. One example of a commercial unit is a Weed Seeker (Fig. 21), which has a reflectance sensor that identifies chlorophyll. The microprocessor interprets that data and when a threshold signal (when weeds are present) is crossed, a controller turns on the spray nozzle. The Weed Seeker system is built around closeproximity optical sensors using near-infrared (NIR), light-reflectance measurements to distinguish between green vegetation, bare soil and crop residue.

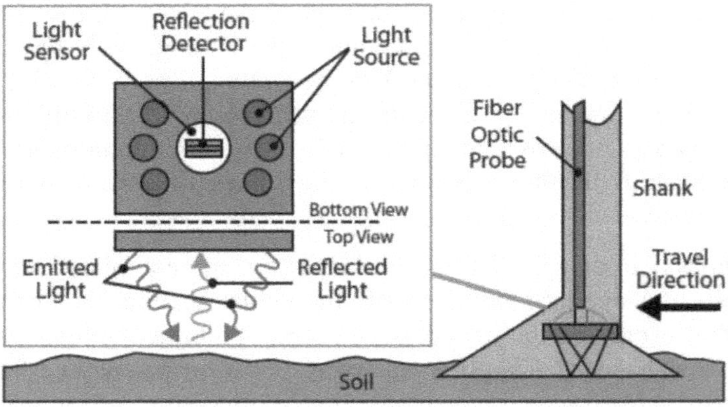

Fig. 20. Cross-section schematic of a subsurface, soil-reflectance optical sensor to measure soil organic matter (*Source*: Adamchuk and Jasa, 2002).

Each sensor unit consists of a light source and an optical sensor. The sensors are mounted on a bar or spray boom ahead of the spray nozzle and aimed at the ground. When a chlorophyll (green) reflectance signal exceeds a threshold (set during calibration by the operator) a signal is sent from a controller to a solenoid operated valve to release herbicide.

Fig. 21. Sensor-based Weed Seeker for herbicide control (*Source: www.ntechindustries.com/rowcrop.html.*)

The system is designed to turn on slightly before a weed is reached and stay on until slightly after a weed is passed. It can operate at travel speeds of 3 to10 mph. In areas where weed infestation levels are variable, the unit can significantly reduce chemical application amounts (compared to uniform, continuous applications). Because the Weed Seeker is not designed to distinguish between plant types (desirable crops versus unwanted weeds), its agricultural use is focused on between-the-row applications in standing crops or on-spot treatment of fallow ground.

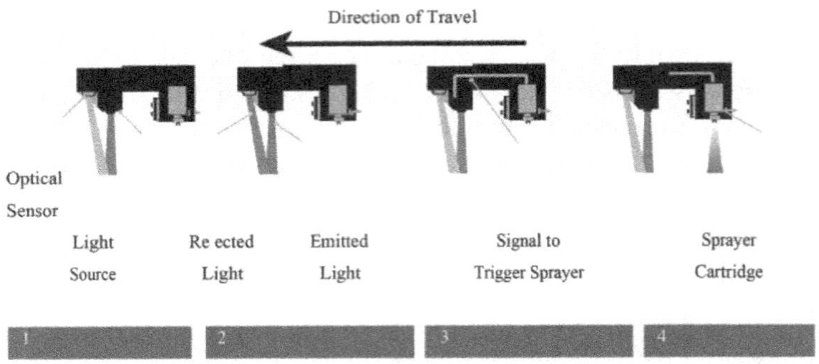

Fig. 21. The optical sensor control of the spray nozzle
(*Source*: Weed Seeker; *www.ntechindustries.com/rowcrop.html*)

Another device that is a sensor-based control is the boom-height control (Fig. 21). Even though this is not a VRA device, it does improve proper coverage from a spray boom, which will eliminate streaks and improper overlaps (Fig. 22). The ultrasonic sensors measure (40 times/second) the distance to the ground. This information allows the control system to make responsive height adjustments so that sprayer booms automatically follow the contours of the land. The system has shown reliable control with average speeds > 18 mph in all kinds of uneven terrain.

Fig. 22. Spray-boom control to eliminate streaks and improper overlaps
(*Source*: *www.norac.ca/products.php*)

Fertilizer VRA: Fertilizer applications can cover a wide area of application devices. Many of the VRA technologies for fertilizer applications are similar to weed control (liquid applications) and liming application (dry chemicals). Their effectiveness can be complicated based on weather impacts and the nutrient's availability and seasonal cycles. We will look at the major nutrients and why some are more likely to be applied with VRA.

Phosphorus VRA: VRA of phosphorus (P) is probably the second-most-profitable VRA activity. Soil phosphorus is not nearly as transient as soil nitrogen (N), meaning that grid soil tests can be used for a number of years. Also, there is evidence that long-term economic benefits might arise from building up soil-test phosphorus. This capital investment characteristic of soil-test phosphorus means that it is often profitable to uncover the intrinsic differences in soil-test phosphorus within a field at least at one point in time.

Nitrogen VRA: The adoption of VRA nitrogen (N) management by producers is low, despite the potential economic and environmental benefits of this practice. A major obstacle is the recommended nitrogen fertilizer rates based on yield goal are often poorly correlated with actual economically optimum nitrogen rates.

Nitrogen response patterns are often field and season specific and can vary widely within the same field, further complicating the development of prescription maps. Side-by-side comparisons of uniform and VRA-N management have revealed no consistent advantages for either strategy in yields achieved, profitability, whole field nitrogen usage, or nitrogen use efficiency by plants. In the future, a better understanding of temporal variation in nitrogen soil test levels, better crop simulation models, and improved nitrogen sensing and application equipment may assist producers in capturing the benefits of VRA-N management. Realtime sensors of crops offer the most potential for VRA-N, as these systems are designed to "sense" the nitrogen needs of the crop at the time of application. These systems require wellfertilized areas in the field to calibrate the sensor. Ongoing research will determine if these systems will be widely employed in the future.

Nitrogen application for grain crops: For nitrogen applications utilizing the Green Seeker, the concept is that the amount of fertilizer needed at a particular location within the field can be determined by implementing a nitrogen-rich strip at planting or shortly thereafter and comparing spatial variability of crop growth across the field to crop growth from the nitrogen strip. The nitrogen-rich strip provides an area in which nitrogen is not the yield limiting factor.

A nitrogen-rich strip is implemented by selecting one strip that transverses the field (typically one pass of the fertilizer application equipment) to receive a complete nitrogen application at planting. Then at side dress, NDVI readings are collected from the nitrogen-rich strip to calibrate the crop sensor system. Subsequently, as the fertilizer applicator covers the field, the sensors read NDVI values, compare them to the NDVI values from the nitrogen-rich strip, and apply an adjusted amount of nitrogen.

For example, if the NDVI value in the nitrogen-rich strip was 0.5 but was 0.6 at a particular location within the field, no nitrogen would be applied because the sensor determined that sufficient nitrogen is already available. Conversely, if the nitrogen-rich strip had an 0.5 NDVI reading but another location within the field had an 0.4 NDVI, then nitrogen would be applied in that area.

Recently, the use of a ramped calibration strip has been recommended. Instead of the nitrogen-rich strip consisting of one rate across the field, a range of nitrogen rates is applied across the field. This provides a benefit in that growers can see actual response to a range of nitrogen rates and when they are setting ranges for variable rate application, they have more information about how to appropriately establish the breaks for the assorted nitrogen rates.

Concept of custom hiring centres (CHCs) for farm equipment

Custom hiring of farm machinery (CHCs) was introduced in Indian agriculture in as early as 1912. Organised move to promote multi-farm use of agricultural machinery was made in mid1960 when Agro-Industries Corporations were established within the states.

CHCs are a unit comprising a set of farm machinery, implements and equipment for hiring by farmers. These centres give farm machinery on a rental basis to farmers who cannot afford to purchase high end agricultural machinery and equipment, apart from servicing old machinery. Farm machineries/equipment available at CHCs include tractor, rotavator, multi-crop thresher, MB plough, cultivator, leveller blade, blade harrow, seed cum fertilizer drill, knapsack sprayer, power weeder, winnowing fan, electronic balance, repairing tools.

These units are generally located in close proximities to large and small land holdings, which supply machinery and equipment to villages close to it reducing transport cost and transportation time. States such as Punjab, Haryana, Uttar Pradesh, Uttarakhand, Gujarat, Maharashtra, Karnataka and Tamil Nadu, which are highly mechanized, have maximum number of registered and unregistered CHCs catering to the machinery and equipment requirements of the farmers. Other states do have operational CHCs but farm mechanisation still remains a serious concern. Custom hiring centres, in order to work effectively and efficiently require a sound infrastructure setup for its operations, which include all weather roads, spatial distribution of machinery ownership and servicing network.

Conclusion

The farm mechanization is an important aspect of crop production system. Starting from the first operation of tillage to the post harvest stage. The machinery used in different farm operations tillage, seeding and planting, intercultural and weed management, harvesting and threshing, residue management, etc to improve the timelines and cost effectiveness. The equipments used through different power sources as well as equipment suitable for women are also developed to harness the spare women labour resources. Apart from these the recent developments in the mechanization sector i.e. equipment for precision farming systems are also being introduced in the farming system to reduce the wastage of important inputs like seed, fertilizer and water. The automatically guided sensor based instruments like laser leveller, image processing based precision sprayers, GPS based variable rate applicators for fertilizers and weedicides are also discussed here in details. These machines are costly and may be available on custom hiring bases to financially incapable small and marginal farmers.

References

A Saga of Progress: Compendium of 50 Years of Achievements, Punjab Agricultural University, Ludhiana, 2012, pp. 230.

Adamchuck, V.I. 2005. Characterizing soil variability using on-the-go sensing technology. Inter-national Plant Nutrition Institute, Site-Specific Management Guidelines, SSMG-44. Norcross, Ga. www.ipni.net/ppiweb/ppibase.nsf/b369c6dbe705d d13852568e 3000de93d/8bfeba411afe85e2852571 8000690a10/$FILE/SSMG-44.pdf

Adamchuck, V.I., and Jasa, P. 2002. On-the-go vehicle-based soil sensors. University of Nebraska- Lincoln Extension EC02-178. www.ianrpubs.unl.edu/epublic/live/ec178/build/ec178.pdf

Adamchuck, V.I., and Mulliken, J. 2005. Site-specific management of soil pH (FAQ). University of Nebraska-Lincoln Extension EC05-705. www.ian-rpubs.unl.edu/epublic/live/ec705/build/ec705.pdf

Anonymous, 2018. Recent technologies for production agriculture. ICAR-Central Institute of Agricultural Engineering, Bhopal.

Doerge, T.A. 2002. Variable-rate nitrogen management creates opportunities and challenges for corn producers. Crop Management doi:10.1094/CM-2002-0905-01-RS. www.plantmanagementnetwork.org/ pub/cm/review/variable-n/

Grisso, R.D., Alley, M.M., Thomason, W., Roberson, O.T. and Holshouser, D. 2011. Precision farming tools: Variable-rate application. https://www.researchgate. net/ publication /309121121

Humburg, D. 1990. Variable rate equipment — technology for weed control. International Plant Nutrition Institute, Site-Specific Management Guidelines, SSMG-7. Norcross, Ga. www.ppi-pic.org/ppiweb/ppibase.nsf/$webindex/article=A172CE4C852569610063 1668C0F666E3 or www.agri-culture.purdue.edu/ssmc/

Kapur, R., Chouhan, S., Gulati, S. and Saxena, V., 2015. Transforming agriculture through mechanization: A knowledge paper on Indian farm equipment sector. A report prepared by Grant Torton Pvt. Ltd and FICCI.

Lambert, D., and Lowenberg-DeBoer , J. 2000. Precision Agriculture Profitability Review. Site-SpecificManagement Center, School of Agriculture, Purdue University. www.agriculture.purdue.edu/ssmc/Frames/newsoilsX.pdf.

Sims, B. and Kienzle, J. 2016. Making mechanization accessible to smallholder farmers in Sub-Saharan Africa. Environments **3**(11):1-18. doi:10.3390/environments3020011.

13

Resource Conservation Techniques for Sustaining Crop Production in Rainfed Foothills Under Changing Climate

Sanjay Arora and Rajan Bhatt*

Introduction

The Indian Himalayan Region (IHR), with geographical coverage of over 5.3 lakh km^2, constitutes a large proportion of the hotspot and, therefore, contributes greatly to richness and representativeness of its biodiversity components at all levels. Out of this 5.3 lakh km^2, 33.13 million ha area is being constituted by the North-Western Himalayas. Further, most of the water used to grow maize crop in the sub-humid foothill region of northwest Himalayas, is derived from rainfall. Erratic rains, fragile ecosystems and traditional indigenous management practices are mainly responsible for the current situation (Kukal and Bhatt, 2010). IHR covers 11 states entirely (i.e. Jammu & Kashmir, Himachal Pradesh, Punjab, Uttarakhand, Sikkim, Arunachal Pradesh, Nagaland, Manipur, Mizoram, Tripura, Meghalaya), and two states partially (i.e. hilly districts of Assam and West Bengal). The region represents nearly 3.8% of total human population of the country and exhibits diversity of ethnic groups which inhabit remote terrains. Further it is reported that the Northwestern Himalayan region (NWHR) which spreads to an approximate area of 33.13 million ha, comprising of Jammu & Kashmir, Himachal Pradesh, Uttarakhand is 10.1% of country's total geographical area, supports 2.4% and 4% of human and cattle population of the country, respectively. This region has a diverse climate, topography, vegetation, ecology and land use pattern. The annual average rainfall varies from 80 mm in Ladakh to over 200 cm in some parts of Himachal Pradesh and Uttarakhand. The major natural resources are water, forests, floral, and faunal biodiversity. Forests constitute the major share in the land use of the region with only 15% of the net sown area and 162% cropping intensity (Bhatt, 2011). Due to hill and

**Corresponding author email: aroraicar@gmail.com*

mountainous topography, the region differs from plain in respect to weather and soil parameters, biodiversity, ethnic diversity, land use systems and socio-economic conditions. Growing concerns for deteriorating environment by stakeholders and others seem to have linkage with gigantic cause-and-effect arguments due to deforestation, landslides, large-scale downstream flooding, increasing poverty and the malnutrition. Recent estimates indicate that NW Indian Himalayas has considerable area under potential erosion rate which is really alarming (Bhatt *et al.*, 2013). In north-western Himalayas, on an average, 17% of the area falls in very severe category with erosion rates > 40 t/ha/yr, while about 25% area has erosion rate of > 10 t/ha/yr. The states of Uttarakhand in western Himalayas have maximum area (33%) under very severe category with erosion rates of > 40 t/ha/yr. It calls for serious efforts to employ appropriate conservation measures to check land degradation problems (ICAR & NASC, 2010). In Punjab around 10% area of the state is under sub-mountainous tract which is also severely affected by the problem of soil erosion in general and gully erosion in particular. Maize based cropping systems mainly dominated in the region even though water use efficiency of maize (a C_4 crop) is approximately double (Huang *et al.*, 2006). Therefore, moisture conservation techniques those are suitable for the area will serve the purpose and should be a paramount objective to attain sustainable maize yield in this region. Maize along with leguminous crops not only improves the soil fertility status but also improve the livelihood of the farmer of the region. Maize being widely spaced crop provides an opportunity to grow legumes without reducing its own yield. The main concept of this legume intercropping is to get increased land productivity and judicious utilization of land resources and farming inputs. Intercropping improved the availability of the resources such as water, light and nutrients. Besides, under rain fed conditions rain water use efficiency can be elevated in maize + black gram intercropping system (Solanki *et al.*, 2011). Mulching as it interfere the direct hitting of the solar radiations onto the bare soil surface, conserves the moisture, regulates the soil temperature and increased both the water as well as land productivity in the region (Bhatt and Khera, 2006; Bhatt *et al.*, 2006).

Maintenance of nutrient supply is, and can be one of the strategies to improve rain water and nutrient use efficiency. The most important management interaction in many drought stressed maize environments is between soil fertility management and water supply (Huang *et al.*, 2006). Reduced growth rate in nutrient deficit plants is generally associated with reduced water use efficiency (Bacon, 2004). Water use efficiency was increased by application of nitrogen (Ogola *et al.*, 2002). Adequate soil fertility management to remove nutrient constraints on crop production for every drop of water available through rainfall is also considered as one of the promising strategies for rainwater use efficiency. Efficient crop productivity under conditions of limited rainfall and erratic distribution is of great importance in rainfed agriculture (Rockstrom *et al.*, 2003). Hence, there is a need to identify and assess the different resource conservation technologies (RCTs) responsible for enhancing the water and nutrient use efficiency even in the era of water stressed conditions for attaining sustainable yield of crops in any soil and agro-climatic condition. Only then livelihood of the farmers of the region could be improved without adversely affecting the natural resources *viz.* soil and water (Maruthi Sankar, 1986; Vittal *et al.*, 2003).

The problems of small sized and fragmented holdings, difficult terrains, limited accessibility, and prolonged winters also proved dire necessity for following conservation approach for sustainable production of food grains. To meet the demand of increasing population and sustaining the resources for future, conservation approaches like storage of rainwater, conservation tillage, contour bunding, addition of organic matter in the soil, strip cropping, cover cropping, mulching, vegetative barriers, crop diversification, recycling of residues, sloping land management, adoption of improved varieties according to agro-climatic needs and construction and maintenance of water harvesting storage necessary to be adopted by every stake holder. In this chapter we are discussing the different resource conservation technologies which will mitigate the adverse effect of the global warming with a greater stress on the NW parts of the Himalayan region for the judicious use of the natural resources *viz.* soil and water along with improving the livelihood of the farmers of the region.

Extent of soil erosion in Himalayan region

The Himalayas due to steep slopes, fragile geology and intense high storms are intrinsically prone to soil erosion. The problem has been further aggravated by several developmental activities undertaken in the region like road construction, mining and hydro-power projects.

Recent estimates indicate that nearly 39% area of the Indian Himalayas has potential erosion rate of > 40 t/ha/yr, which is really alarming. In north-western Himalayas, on an average, 17% of the area falls in very severe category with erosion rates > 40 t/ha/yr, while about 25% area has erosion rate of > 10 t/ha/yr. The trend is similar in the north-eastern Himalayan states. The states of Uttarakhand in western Himalayas and Nagaland in (NEH) region have maximum area of 34% and 63.5%, respectively under very severe category with erosion rates of > 40 t/ha/yr. Overall, about 23% area in the Himalayan states has potential erosion rates > 40 t/ha/yr. It calls for serious efforts to employ appropriate conservation measures to check land degradation problems.

Soil loss tolerance limit (T) is defined as the threshold upper limit of soil erosion that can be allowed without degrading long term productivity of a particular soil. In India, a default soil loss tolerance limit (SLTL) of 4.5-11.2 t/ha/yr is adopted for planning soil conservation activities. If soil erosion rates are higher than the tolerance, they need to be reduced for maintain sustainable productivity. It is known that a given erosion rate is not equally serious on all soils. On shallow soils with a T-value of 5 t/ha, erosion at a rate of 12.5 t/ha could lead to relatively rapid loss of productivity. In contrast, on some deep soils with a T-value of 12.5 t/ha, erosion at the same rate would not be expected to reduce soil productivity to the same extent. The soil erosion rates in the NE region vary widely from < 5 t/ha to > 40 t/ha/yr. About 29.82% area in the region falls under the very severe erosion category, whereas very low, moderate and severe erosion classes are 4.47, 21.16, 16.79 and 12.9%, respectively. It was further revealed that due to strong resistance capacity, T-value were more or less consistent at around 7.5 and 12.5 t/ha/yr. As our main emphasis is on the climate change and its consequences on the agriculture, thus there is a need to have an idea regarding climate change.

Climate change and soil degradation

The potential changes in soil forming factors directly resulting from global climate change would be in organic matter supply, temperature regimes, hydrology and potential evapotranspiration. Soil organic matter declines in a warmer soil temperature regime. Drier soil conditions will suppress both root growth and decomposition of organic matter and will increase vulnerability to erosion. In areas where winter rainfall becomes heavier, some soils may become more susceptible to erosion. Not only does climate influence soil properties, it also regulates climate via the uptake and release of greenhouse gases such as carbon dioxide, methane and nitrous oxide. Soil can act as a source and sink for carbon, depending on land use and climatic conditions. Higher temperatures could increase the rate of microbial decomposition of organic matter adversely affecting soil fertility in the long run. Increased rainfall in regions that are already moist could lead to increased leaching of minerals, especially nitrates.

Large increases in fertilizer applications would be necessary to restore productivity levels. Decreases in rainfall, particularly during summer, could have a more dramatic effect, through the increased frequency of dry spells leading to increased proneness to wind erosion. Many soil related processes/properties like mineralization, immobilization, weathering of minerals, biological nitrogen fixation, root-microbe interaction etc., depend upon the prevailing weather/climatic parameters like temperature, carbon dioxide, humidity, rainfall, etc. In foothill region of NWHR, any minor change/

shift in the prevailing weather/climatic parameter affects the availability of essential nutrients to growing crops. For example, heavy precipitation results in excessive leaching and this in turn affects the availability of NO_3, SO_4, Ca and Mg. Similarly, high temperature coupled with drought alters biological nitrogen. It has been noted that in the foothills of Jammu in N-W region, the area under maize crop showed a decline in temperate as well as sub-tropical zone (Sharma and Arora, 2010; Fig. 1), which can be attributed to the comparatively low relative production efficiency index of these soils as a result of change in temperature and precipitation pattern.

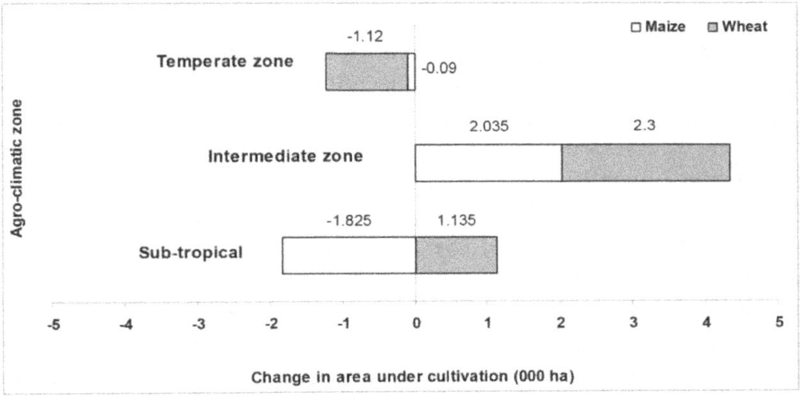

Fig. 1. Changes in area under cultivation of maize and wheat in the foothills of Jammu as impact of changing climate

Climate change and its consequences in Himalayan region

In Asia, more than a billion people could be affected by a decline in the availability of fresh water, particularly in large river basins, by 2050. Glacier melt in the Himalayas, which is projected to increase flooding and rock avalanches, will affect water resources in the next two to three decades. As glaciers recede, river flows will decrease. Climate change is expected to increase the frequency and intensity of current hazards and the probability of extreme events and new vulnerabilities with differential spatial and socio-economic impacts. This is expected to further degrade the resilience of poor, vulnerable communities. It is therefore important to understand a number of processes that are rapidly changing India's landscape, altering livelihood opportunities and wealth distribution, which in turn affect the vulnerability of many communities and stakeholders and their capacity to adapt to long-term risks. India faces a turbulent water future. The country has a highly seasonal pattern of rainfall, with 50% of precipitation falling in just 15 days and over 90% of river flows occurring in just four months. The Indian mainland is drained by 15 major (drainage basin area >20,000 km^2), 45 medium (2,000 to 20,000 km^2) and over 120 minor (<2,000 km^2) rivers, besides numerous ephemeral streams in the western arid region. The Himalayan glaciers feed India's most important rivers. But rising temperatures means that many of the Himalayan glaciers are melting fast, and could diminish significantly over the coming decades with catastrophic results. In the long run, the water flow in the Ganges could drop by two-thirds, affecting > 400 million people who depend on it for drinking water. In the short term, the rapid melting of ice high up in the Himalayas might cause river swelling and floods. The formation of glacial lakes of melt-water creates the threat of outburst floods leading to devastation in lowland valleys. According to a report by the Intergovernmental Panel on Climate Change (IPCC) looking at the threat from climate change to human development and the environment, "only the polar ice caps hold more fresh water than the Himalayan glaciers": "If the current trends of climate change continue, by 2030 the size of the glaciers could be reduced by as much as 80%". The adverse impact of climate changes includes water crisis and an increased risk of extinction for an estimated 20 to 30 per cent of plant and animal species in India if the global average temperature exceeds 1.5 to 2.5 ^{0}C. Climate change will also significantly impact health in India. The most vulnerable will be the poor, the disabled, the youngest and oldest members of the population as they already face limited access to health facilities and have limited disposable income to cover additional medical costs. The Ganges river basin runs from the central Himalayas to the Bay of Bengal, and covers parts of Nepal, India, China and Bangladesh. The basin occupies 30% of the land area of India and is heavily populated, increasing in population density downstream to Bangladesh, the most densely populated country in the world (Rashid and Kabir, 1998). Water withdrawal poses a serious threat to the Ganges. In India, barrages control all of the tributaries to the Ganges and divert roughly 60% of river flow to large scale irrigation (Adel, 2001). India controls the flow of the Ganges into Bangladesh with over 30 upstream water diversions. The largest, the Farraka Barrage, 18 km from the border of Bangladesh, reduced the average monthly discharge of the Ganges from 2,213 m^3/s to a low of 316 m^3/s [14%] (Goree, 2004). Climate change will exacerbate the problems caused by water extraction. The Himalayan glaciers are estimated to supply 30-40% of the water in the Ganges, which is particularly critical in the dry season prior to the monsoon rains. The projected annual renewable water supply for 2025 indicates water scarcity (WWF, 2007). Although the Ganges catchment drains virtually all of the Nepal Himalayas and water supply per person in the basin ranges from adequate to ample, its dry season outflow (from December to February) to the sea is non-existent. Overall, excessive water diversions threaten to eliminate natural flows and severely damage people's livelihoods in the Ganges. The Indus river basins pans parts of four countries (Afghanistan, Pakistan, India and China) in an area that is > 30% arid, and much drier than the nearby Ganges river basin (WRI, 2003). The Indus river is critical for Pakistan's 160 million people and irrigates 80% of its 21.5 million ha of agricultural land (Rizvi, 2001). The watershed is also an area of rich biodiversity, particularly where it opens to the Arabian sea. The Indus river delta is a highly productive area for freshwater fauna and an important region for water birds (Ramsar Convention on Wetlands, 2003). The Indus river is extremely sensitive to climate change due to the high portion of its flow derived from glaciers. Temperature controls the rate of glacier melt, which in turn, provides more water in dry, warm years and less water in cool years. River catchments with a large portion of glacial melt water experience less variability in water flows. With climate warming, many glaciers will no longer exist to moderate the flow of these rivers. Thus communities which depend on glacier water will face more severe water shortages, variability and potentially greater flooding too (Rizvi, 2001). The Himalayan glaciers provide the Indus with 70-80% of its water (Kiani, 2005), the highest proportion of any river in Asia. This is double the proportion of water that they provide the Ganges (30-40%). Himalayan glaciers provide 44.8% of the water in the Upper Indus in China alone (Yang, 1991).The Indus basin is already suffering from severe water scarcity due to over extraction for agriculture, causing salt-water intrusion in the delta (WRI, 2003). In 1995, the Indus river already supplied much less water per person than the minimum recommended by the United Nations (UN) and by 2025 is predicted to suffer even more severe waters scarcity (Revenga *et al.*, 2000). Well-managed riparian forests are especially important in minimizing the impacts of climate change on river biota. They provide shade and temperature regulation, can moderate the effect of frequent, short duration storm event sand can support natural water flow regimes. Climate change will exacerbate the impact of deforestation on water regulation. Although the Indus system is currently robust enough to cope with shortages of 10-13% in river flows, when the rivers flow drops to15-20% below the average, irrigation shortages occur. Climate change will surely exacerbate the problems of irregular and low flow.

Effect of tillage and mulching

In-situ soil moisture conservation: Mulch material was found to be quite effective in conserving soil moisture. However, it's role differed to a significant extent under different tillage treatments and with different mode of mulch application. Observations of periodic soil moisture content of surface and sub surface soil reveal that the minimum tillage (T_m) was more effective in conserving moisture than the conventional tillage (T_c) under all modes of mulch application. As compared to the control the application of straw mulch @ 6 t/ha in whole plot had 4.0 to 5.1% higher soil moisture content in the 0-15 cm soil depth and 3.5 to 7.1% high soil moisture content in 15-30 cm soil depth under minimum tillage treatment. However, under the conventional tillage treatment the effect of mulch applied on soil for conserving the moisture was relatively less. The other three modes of mulch application helped in conserving the soil moisture but the amount of water conserved was relatively small as compared to that under fully covered treatment. The minimum tillage (T_m) was more effective in conserving moisture than conventional tillage (T_c). Greater the surface area covered by that of the mulch material in a particular mode, greater is it's effectiveness in conserving the soil moisture. Thus among different mode of mulch application the trend varied as $M_w > M_{1/3} > M_s > M_v > M_o$ as area covered varied as 98% > 33% > 18% > 1.8% > 0%.

It has been reported earlier (Hillel, 1980) that mulches are most effective in conserving moisture in the wet range when the evaporation is in the constant rate range and controlled by the atmospheric evaporativity. Use of crop residue as surface mulch help in reducing the soil surface sealing and maintaining high infiltration rate (Wagger and Denton, 1992) and reducing vapour losses. Vegetation cover is thus a very important factor, as it absorbs the impact of raindrops and protects the aggregate breakdown which further promotes the greater soil moisture content (Reddy and Reddi, 1995).

Soil temperature: Straw mulch provides surface cover on the bare soil and thus acts as a barrier for the sun rays to reach the soil surface. Mulch also reduces the soil temperature by reflecting the sun rays. As compared to control, minimum soil temperature in fully covered plots was 1.4 to 2.0°C higher in T_m and 1.4 to 2.4°C higher in T_c plots. Thus tillage did not significantly affect the minimum soil temperature. However, as compared to control, the maximum soil temperature in the fully covered-mulched plots was 1.9 to 2.8 °C and 2.5 to 4.0 °C lower under minimum and conventionally tilled plots, respectively. The straw mulched plots showed lower temperature than the bare plots and the trend observed was $M_o > M_{1/3} > M_w$. However, there was no significant effect of the tillage treatment on maximum soil temperature.

Runoff: During the study period, 31 rainstorms were received. However, only 12 out of them produced significant runoff and soil loss. On perusal of data in Table 1, runoff in conventionally tilled plots (T_c) under all the modes of mulch applications was 5.1% higher than under minimum tilled plots (T_m). This may be observed to the fact that minimum tillage promotes aggregation, maintains the soil structure and thus increases the infiltration (Mcgregor *et al.*, 1998) and helps in reducing runoff. As compared to the control, straw mulch was quite effective in decreasing the runoff. The mean runoff decreased from 50.2% under controlled plots (M_0) to 16.8% under fully covered plots (M_w). The per cent runoff was observed to be 44.7%, 30.0% and 24.3% in vertical mulching (M_v), strip application (M_s) and $M_{1/3rd}$ plots where the lower 1/3rd of plots was applied with mulch. Differentiating behavior of different modes of mulch application may be due to the different amount of surface cover provided by these treatments. It was observed that M_w, $M_{1/3rd}$, M_s, M_v and M_o had 98%, 33%, 18%, 1.8% and 0% surface covered. The role of surface cover in decreasing the runoff had been reported earlier by several scientists (Mannering and Meyer, 1963; Khera and Singh 1998).

Mulch also provided hydraulic resistance to the running water. Greater the surface covered, greater would be the resistance offered by the mulch material to the runoff. However as far as interactive effect of tillage and mulch was concerned it was observed that M_w, $M_{1/3}$, M_s, M_v and M_o had 16.9%, 25.9%, 31.2%, 44.1% and 47.0 % in T_m plots while these values were increased to 21.0 %, 29.1%, 34.0%, 51.2% and 54.9% in T_c plots. It meant that $T_m M_w$ proved to be best while $T_c M_o$ was the worst treatment in checking the runoff. Percent runoff under all the modes of mulch application with T_m was significantly lower than T_c treatment. The application of straw mulch avoided the direct raindrop impact, decreased surface sealing and entrapped the soil particles and enhanced residence time for water to infiltrate into soil (Khera and Singh, 1998).

Soil loss: The soil loss was differed significantly affected by both the tillage treatments as well as by the different modes of straw mulch application. As far as the tillage is concerned, it was observed that on an average T_c (conventionally tilled) plots had 40.1% higher soil loss than the T_m (minimum tilled) plots. The conventional tillage resulted in the breakdown of larger aggregates into the smaller ones which were found to be more erodible (Bhatt, 2015). Mulch was quite effective in decreasing the soil loss by providing proper surface cover to the bare soil. The mean soil loss (averaged over all levels of tillage) in M_o, M_v, M_s, $M_{1/3}$ and M_w was observed as 5554, 4691, 2073, 1425 and 548 kg/ha, respectively. It was due to the fact that different modes of straw mulch application provided variable ground cover. The efficiency of this ground cover in decreasing the soil loss was also reported earlier by Khera and Singh (1995). Among different modes of mulching M_w had highest ground covered area followed by $M_{1/3}$, M_s, M_v and M_o.

However, as far as interactive effect of tillage and different mode of mulch application was concerned, it was observed that the soil loss was 2063.7, 6258.4, 8058.1, 15498.9 and 17152.7 kg/ha in M_w, $M_{1/3}$, M_s , M_v and M_o under minimum tilled plots (T_m) as compared to the 3118.5, 6769.6, 9869.5, 21356.4 and 24947.2 kg/ha under the T_c (conventional tilled) plots, respectively.

Table 1. Effect of tillage and mulching on the runoff (%)

Treatments	Dates												
	2/8	3/8	4/8	6/8	10/8	12/8	13/8	2/9	3/9	7/9	8/9	12/9	13/9	Mean
T_m	30.1	31.0	33.0	30.02	28.3	30.0	30.3	28.4	33.1	31.3	31.0	33.7	30.8	33.4
T_c	33.9	35.6	38.06	34.63	35.6	34.3	34.8	32.7	37.7	35.9	35.3	38.7	35.7	38.5
CD (5%)	1.7	2.3	2.02	2.4	NS	2.1	1.78	3.49	2.38	1.63	2.48	2.0	2.7	-
M_w	16.2	16.4	18.9	15.4	16.6	15.3	15.8	16.6	18.2	16.9	16.4	18.7	17.0	16.8
$M_{1/3}$	23.4	23.6	27.5	22.8	23.7	22.6	22.9	24.2	26.2	24.4	23.7	26.9	24.0	24.3
M_s	29.2	29.3	32.6	28.6	29.4	29.5	28.8	29.5	31.6	29.8	29.5	32.1	29.6	30.0
M_v	41.8	47.1	47.6	45.8	39.9	45.4	46.1	46.5	48.7	46.6	46.4	49.9	45.6	45.9
M_o	49.5	49.9	51.0	49.1	50.0	48.8	49.3	50.2	52.4	50.3	49.6	53.1	50.3	50.2
CD (5%)	1.2	1.24	1.73	1.3	8.6	1.15	1.33	13.3	1.33	1.1	1.26	1.52	1.24	-

(*Tm is minimum tillage, Tc is conventional tillage, Mw is mulch applied on the whole plots, M1/3rd is mulch applied on lower 1/3rd of the plots, Ms is mulch applied in the strips, Mv is the vertical mulching and Mo is the control plots*). *Source*: Bhatt and Khera (2006)

The interaction between modes of mulch applications and tillage was observed to be significant. Except under M_w and $M_{1/3rd}$ treatments, soil loss under the other three modes of mulch application significantly higher under conventional tillage as compared to minimum tillage. Application of mulch reduces the soil loss as it decreases the erosivity of rain by intercepting the rain drops, increases surface detention and residence time for water to infiltrate and helps in entrapment of soil particles (Mannering and Meyer, 1963).

Gully erosion and its dynamics in sub-mountainous tracts

Foothills of lower Shiwaliks occupies an area of 2.14 m ha and represents the most fragile eco-system of the Himalayan mountain range because of its peculiar geological formations. It lies mainly in the states like Jammu and Kashmir (Jammu, Udhampur and Kathua) (0.80 m ha), Punjab (Hoshiarpur, Ropar, Nawanshahr and Gurdaspur) (0.14 m ha), Haryana (Ambala and Yamuna Nagar) (0.06 m ha) and Himachal Pradesh (Kangra, Una, Bilaspur, Hamirpur, Chamba, Solan and Southern parts of Sirmaur) (1.14 m ha) (Kukal *et al.*, 2006). Prior to the middle of 18th century, the Shiwalik hills were strictly preserved for hunting and no cultivation, grazing or exploitation of timber was permitted. The increased population of mankind along with the livestock density far exceeding the current carrying capacity of the land, frequent forest fires and mismanagement of land resources, resulted in the steady but obvious natural resource degradation and presently the garden of Punjab has changed to more or less a desert, which is still highly exploited by various anthropogenic activities (Kukal and Bhatt, 2006).

Ecological degradation in Shiwalik hills is the outcome of the overexploitation and mismanagement of soil resources through deforestation, overgrazing and clearance of the vegetation for the agricultural purposes disregard to their slope and topography (Bhatt *et al.*, 2004; Arora and Saygin, 2011). The whole belt experiences severe problem of soil erosion because of undulating slopes and highly erodible soils coupled with highly intensive rain storms and represents the most fragile ecosystem of Himalayan mountain range because of its peculiar geological formations, and highly erodible soils (Thakur *et al.*, 2013). It is reported that runoff and soil loss in the region varies from 35-45% and 25-225 t/ha/yr (Sur and Ghuman, 1994). Minimum tillage coupled with rice straw mulching @ 6 t/ha reported to mitigate the losses of soil erosion to a significant extent (Bhatt and Khera, 2006; Arora and Bhatt, 2006). Assessing the impacts of climatic and land use changes on rates of soil erosion by water is the objective of many projects (Nearing, 2001). Among different types of soil erosion, gully erosion is the most serious one in the region as around 20% of the area is already under gullies (Kukal and Sur, 1992) and spatially advances at an increased rate on upslope the catchment as organic carbon and clay content decreased up the slope (Kukal *et al.*, 2005). Thus, the problem of soil erosion is quite serious in the region and farmers mostly depend on their indigenous technical knowledge (ITK) to grow their crops sustainably in the region and many of these techniques are scientifically sound also (Kukal and Bhatt, 2006; Bhatt, 2013).

Ephemeral gully erosion has been reported to account for 48.5 to 72.8% of the total soil loss (Zheng *et al.*, 2009). About 70-80% of the gully erosion control structures have failed in the region (Kukal *et al.*, 2002). The reasons attributed for the failure of gully control structures in the region include lack of information on gully network including distribution and extent of different-ordered gullies, gully density, gully texture, behaviour and development of gullies in the region (Kukal *et al.*, 2006). Secondly, the installation of gully control structures is generally done in the highest-ordered gully. After some time, these structures gets silted up along the upstream side, after which the runoff water starts falling down from the crest height of the structure and causes higher erosion losses. The lower ordered gullies are seldom tackled in the region while controlling the runoff and soil loss and are generally ignored in all the soil conservation programmes (Kukal *et al.*, 2006). To evaluate the temporal advancement of gullies both in terms of their number and length, two surveys were being conducted in the four catchments of Hoshiarpur district *viz*. firstly in 2003 and secondly five years thereafter in 2008 this detail survey was conducted in 2003 and then on 2008 in the four catchments of Hoshiarpur district of Punjab with the objectives:

- To evaluate the spatial variation of vegetation all along the slopes.
- To study distribution of gullies on either side of main gully.
- To study temporal advancement of gullies from 2003 to 2008 both in terms of gully density and texture.
- To formulate strategies for the assessment of gullies and brought out effective techniques for mitigating the adverse effects of gully erosion in particular and soil erosion in general.

Study area and its characteristics

The survey was carried out in four catchments of Hoshiarpur district of Punjab in the Shiwaliks region of lower Himalayas in North India. The region lies between 30^0 10' to 33^0 37' N latitude and 73^0 37' to 77^0 39' E longitude and stretches to about 530 km length wise and 25-95 km width wise. The climate of the region varies between semi-arid to sub-humid (Fig. 2). The maximum temperature (41-42 ^{0}C) is recorded in the first fortnight of June, whereas minimum temperature (5-6 ^{0}C) is recorded in the month of January. Majority of the soils range from loamy-sand to sandy loam in texture and have low to medium moisture retention capacity and are highly erodible. It is reported that majority of soils (67%) are loam followed by loamy skeleton (28%), sandy skeletal (1.0%) and sandy (0.7%) (Sidhu, 2002). Soils vary widely in their (Shallow to very deep), texture (Sand to clay loam), organic carbon (0.1 to 1.1%) and pH (5.3 to 8.4) depending upon the physiography, parent material, vegetation cover and climatic conditions. The area receives an average annual rainfall of about 800-1,400 mm with a high coefficient of variation. About 80% of this rain occurs in the three months of monsoon season (July to September) with highly intensive rainstorms.

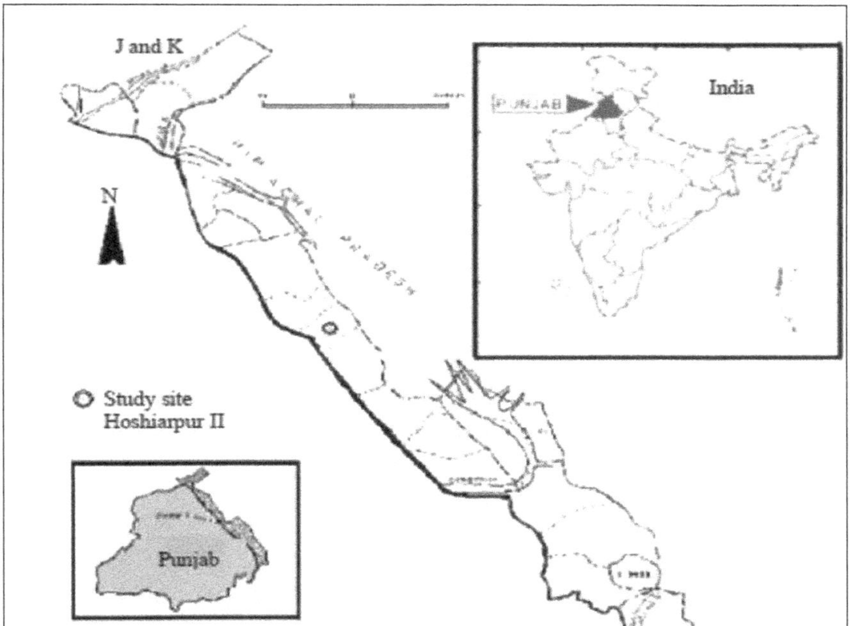

Fig. 2. Location map of the study map

In Shiwaliks of Punjab, rainfall aggressiveness (ratio of highest monthly rainfall to total annual rain) an index of rainfall concentration in a period, varies from 55.9 to 502.4 with an average value of 218±121.7 (Singh, 2000). Rainfall constitutes the major water resource and is sufficient to take two crops annually, but because of its ill distribution in time and space, the farmers are unable to utilize it properly for agricultural purposes. Sidhu *et al.* (2000) reported that in the region 40% land has steep slopes (>15%), 10% moderate (8-15%), 17% gentle (3-8%) and 25% very gentle slopes (10-13%). Further, Kukal *et al.* (2006) reported that convexo-concave (NE side) and concave slopes (SW sides) dominate in the region with a conclusion that SW sides of a catchment exposed to greater sunshine hours which resulted in greater evaporation, lesser moisture, lesser vegetation, lesser organic matter and greater erosion intensity. In the present study, we surveyed four catchments of Hoshiarpur district of Punjab to study the temporal advancement of gully texture and density from 2003 to 2008 (Fig. 3).

Various factors controlling gully growth are catchment characteristics *viz.* area (Burkard and Kostaschuk, 1997), slope shape (Meyer and Martinez-Casasnovas, 1999), gully development parameters, slope steepness (Kukal *et al.*, 1991), surface runoff, precipitation, soil moisture and piping (Stocking, 1980). In the present study, detailed field survey for gully erosion was carried out by dividing catchments into grids of 50 × 50 m^2 each and then each gully line was sketched on the contour maps (at a scale of 1: 1000) manually after measuring the distance between wooden pegs laid out in the grids.

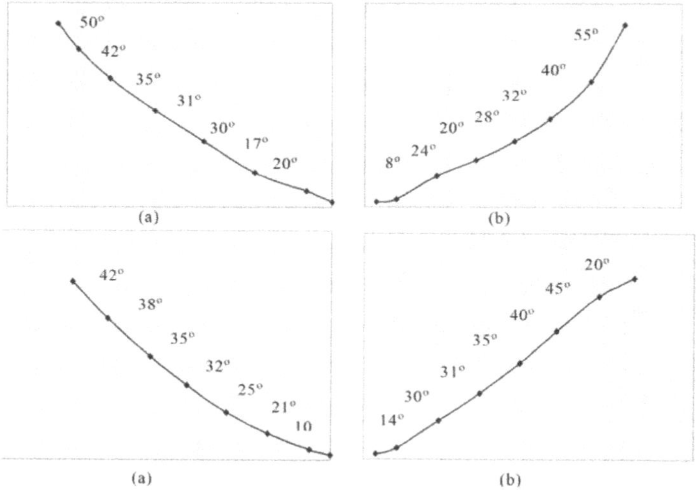

Fig. 3. Typical slope profiles in catchments **(a)** facing NE and **(b)** facing SW in foothill region of Punjab

The gullies up to the first order were marked on the maps. Slope angles across the slope transacts were measured at certain intervals in the selected catchments with an Abney level and measuring tape to determine the slope shape profiles. Gully heads were also marked in each catchment on the base maps and their status of activeness found out. Gullies were classified as 1st, 2nd, 3rd, 4th and 5th order gullies, depending upon extent of their bifurcation. The length of different ordered gullies was measured in each catchment from the gully erosion map. The total length of all the gullies in the catchment were expressed as "gully density" (km/km^2). The number of first-order gullies per unit area was expressed as "gully texture" (number/km^2).

Temporal variation in gully erosion

A detailed survey was firstly made in the year, 2003 in the Saleran catchments of Hoshiarpur district and then after 5 years, similar catchments were surveyed again to have an idea regarding the advancement in gullies both from their length as well as their number. A comparison of "gully density" (km/km^2) and "gully texture" (number/km^2) surveyed in 2003 and 2008 shows that both gully texture and gully density has increased over the years (Table 2).

Table 2. Temporal variation of gullies in the study catchments (2003-08)

Catchment	Gully texture			Gully density		
	2003	2008	% increase	2003	2008	% increase
I	758.0	1453.8	91.8	31.7	35.8	12.9
II	439.9	1174.8	167.0	15.5	19.4	25.2
III	722.0	1151.2	59.5	15.8	17.0	7.6
IV	251.2	464.8	85.0	8.6	9.4	9.3

The catchment II experienced highest increase in gully texture (167%) and gully density (25.2%) followed by catchment I (91.8% and 12.9%), catchment IV (85% and 9.3%) and catchment III (59.5% and 7.6%) (Table 2). Interestingly, the per cent increase in gully texture was significantly higher than gully density in all the catchments. This indicates that more number of first-order gullies was added every year than the addition in the gully length. In fact the rainfall aggressiveness (ratio of highest monthly rainfall to the average annual rainfall) recorded to be higher (Matharu *et al*., 2003) in the region leads to the creation of new gullies (Morgan, 2005). It is thus clear that gully networks have been expanding with time in the region. To check this expansion of gully network, gully control strategy as discussed previously needs to be adopted at a wider scale in the region.

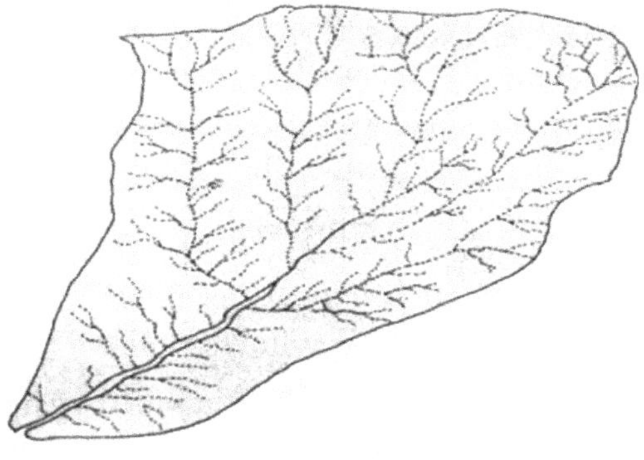

(a) Saleran, Hoshiarpur, IVth catchment (b) Saleran, Hoshiarpur, Ist catchment

Fig. 4. Gully erosion network in the 1st and 4th catchment of Saleran watershed, Hoshiarpur, Punjab

First ordered gullies were significantly higher both in number and length (Fig. 4), and these are the main feeding braches to the higher ordered gullies. Thus, there is an urgent need to control these gullies rather than to construct the check dams in the highest/main gully which also got silted up after some time.

Vegetation density

The average tree density in different catchments decreased significantly by 76% from top-slope to the toe-slope segment (Table 3).

Table 3. Vegetation density on different slope segments in the study catchments

Slope segment	Catch I	Catch II	Catch III	Catch IV	Mean
Tree density (%)					
Top slope	25.2	31.1	41.1	25.4	30.7
Mid slope	18.4	17.3	18.1	16.0	17.5
Toe-slope	7.7	6.8	6.8	7.9	7.3
Mean	17.1	18.4	21.8	16.4	
LSD (***P***=*0.05*)	Catch = 0.67	Slope segment (SS) = 0.58		Catch × SS = 1.16	
Bush density (%)					
Top slope	28.4	19.1	27.4	16.8	22.9
Mid slope	27.8	27.6	27.4	20.3	25.8
Toe-slope	22.4	26.1	31.7	29.6	27.5
Mean	26.2	24.3	28.9	22.2	
LSD (***P***=*0.05*)	Catch = 0.83	Slope segment (SS) = 0.72		Catch × SS = 1.45	
Grass density (%)					
Top slope	13.5	14.6	9.2	11.5	12.2
Mid slope	9.1	13.2	16.1	8.7	11.8
Toe-slope	5.5	14.7	2.7	5.4	7.1
Mean	9.4	14.2	9.4	8.6	
LSD (***P***=*0.05*)	Catch = 0.53	Slope segment (SS) = 0.46		Catch × SS = 0.92	

The decrease was significantly higher in catchment III (21.8%) compared to catchment I (17.1%), II (18.4%) and IV (16.4%). The decrease in tree density down the slope was also highest in catchment III (83%) compared to 69-78% in other catchments. This decrease in tree density down the slope could be due to the reason that the trees growing on or near the toe-slope segment were more prone to be cut by the local population for timber or fuel purposes. The catchment III being smaller in size and more easily accessible to the human beings and animals could have faced more destruction of vegetation.

The average density of bushes increased down the slope by 20%. However, the increase was 76% in catchment IV compared to 14.6% in catchment III and 36.6% in catchment II. In catchment I, the density of bushes rather deceased down the slope. As in case of tree density, the density of bushes was highest (28.9%) in catchment III, followed by catchment I (26.2%), II (24.3%) and

IV (22.2%). The overall increase in bush density down the slope could be due to the dominance of *Lantana* spp. which is not liked by animals as fodder. Also this, being thorny and of bad odour is not preferred by the local people for any purpose. Moreover, the higher density of these bushes near the toe-slope segment could be due to higher soil moisture content at the lower slope segments.

Unlike trees and bushes, the density of grasses was higher in catchment II (14.2%) compared to about 8.6-9.4% in other catchments. The average grass density also decreased down the slope by about 42 per cent. However, the extent of decreased varied in different catchments with 70.6% in catchment III, followed by 59% in catchment I and 53% in catchment IV. However, in catchment II, it did not vary down the slope.

The density of vegetation was affected by the slope aspect. In general the slopes facing north-east direction had significantly higher tree density than the slopes facing south-west direction at top and toe slope. This is due to the fact that slopes, facing south-west direction experience longer sunshine hours and hence more evaporation losses and greater aridity resulting in lower tree density. However, the average bush and grass density was similar on the two slope aspects, in contrast to the previous studies carried out in the region (Kukal *et al.*, 1999).

The density of trees decreased from top-slope to toe-slope segment on both the slope aspects. However, the decrease was more on slopes facing south-west direction (78.8%) than on slopes facing north-east direction (71.8%). The density of bushes unlike trees increased significantly down the slope. The grass density deceased from top-slope to toe-slope segment by 18.3 per cent on slopes facing north-east in comparison to 60 per cent on slopes facing south-west direction. The highest decrease (95%) in tree density from top-slope to toe-slope segment was observed on south-west facing slopes in catchment III, followed by catchment IV (82%), catchment II (78%) and lowest in catchment I (72%). The trend of variation in bush density down the slope was opposite to that of tree density. The bushes increased down the slope in almost all the catchments except the north-east facing slopes in catchment I, whereas tree density decreased significantly down the slope. The increase in bush density down the slope could be due to the higher prevalence of *Lantana* spp. as discussed earlier.

The grass density as in case of trees, decreased down the slope in all the catchments. As in tree density, the decrease in grass density was more on the slopes facing south-west than those facing north-east. The highest decrease (89.3%) was observed in south facing slopes of catchment III followed by catchment II (76%), catchment IV (70%) and lowest in catchment I (14%).

The temporal surveys conducted in 2003 and 2008 in the four catchments of the Hoshiarpur district showed that (1) First ordered gullies are the main culprit which collect the rainwater from each nook and corner of the catchment and supplied it to the higher ordered gullies. Governments spend huge amounts to install check dams in the higher ordered gullies which prove to be failure to control the soil erosion at a long time scale. Thus, if we control the first ordered gullies in the catchment, then no water supplied in the higher ordered gullies and thus extent of soil erosion could be significantly reduced. (2) Catchment sides facing SW side reported to have intense gully network as compared to NE because of greater sunshine hours, more evaporation, lesser vegetation, lesser organic matter, poor aggregation and finally, higher erodibility. (3) The severity of gully erosion was observed to be a function of average relief and lamniscate ratio of the catchments as increased steepness results in increased runoff speed which aids deepening and widening of gullies. Lower value of lamniscate ratio indicates more erosion in the catchment due to its more compact shape with reduced time of concentration. (4) Concave slopes in a particular catchment have been shown to be more prone at the lower slope segments for initiation of gullies as the runoff water concentrates at the lower slope segment where the slope steepness decreases and runoff water starts concentrating. The slope angles were steeper on the upper slope segments and it decreased in the down slope direction.

Erosion

Flood damages are generally viewed to occur in plain areas. However, rivers and torrents (seasonal streams with flash flow during monsoon) in the Himalayan region also cause heavy damage to life and property. A recent example is that of flood induced deluge occurring in Kedranath valley of Uttarakhand state during mid June, 2013 which took a toll of thousands of lives and destroyed property including agricultural lands which are not retrievable. This has necessitated us for taking adequate flood control measures in the Himalayan rivers/streams.

Bio-engineering technology has been evolved at CSWCRTI, Dehradun for treatment of torrents in Shiwaliks where mechanical measures like spurs, retaining walls and earthen embankments have been used in conjunction with suitable vegetable species for torrent training. Species like *Arundo donax* (Narkul or Nada), *Vitex negundo* (Shimalu), *Ipomoea* (Besharam), Bamboo, *Napie*r (Hathi ghas), *Saccharum munja* (Munj ghas) have been found suitable for bank protection and vegetative reinforcement of structures. The benefit cost ratio of river training works has been found over 2.65:1.

Extreme rainfall induced disaster in Uttarakhand

A disaster had occurred in Uttarakhand as a result of extreme rainfall occurring during mid June, 2013 resulting in a huge loss to life and property. Under a joint initiative of NARS and Uttarakhand state government, a survey conducted to observe the damage to natural resources *inter alia* revealed the following:

- The agricultural fields/habitations situated within the high flood level of rivers/streams were washed away and damage was noticed to adjoining lands also wherever flood water entered.
- The intensity of damage was more in untreated watersheds compared to the treated ones.
- Maximum mass erosion problem observed was that due landslides/slips, especially along roads. Landslides/slips were more at places where no retaining walls, toe drains were provided and slopes were without vegetation.
- The drainage lines (Nalas/gullies) treated with proper bio-engineering measures (gabion check dams etc.) even before 20-30 years back, were not much affected by extreme rainfall events (14 – 17 June, 2013).
- The diversion drains constructed by some farmers (at their own initiative) for safe disposal of runoff water saved valuable agricultural land and crops.
- The degraded hill slopes and landslides/slips treated earlier (12 years before) with Geo jute technology were found stable and lush green with vegetation.
- Erosion problem was less or absent and damage were minimised with good agroforestry practices.

Thus, there is a need to control this menace of soil erosion by managing the watershed approach which on integrated basis will serve the purpose. Following are some of the case studies of the Shiwaliks.

Watershed management programme

The increasing anthropogenic pressure on the Himalayan resources to meet the ever-increasing demands for material supplies is leading to their widespread degradation. It was established that ecologically relevant destruction took place in rainfed old croplands within mid-slope and high landscape positions. The continued degradation of the fragile Himalayan region would affect adversely the socio-economic and environmental stability of the region. Major part of the land area of north-western hill region is hilly terrain and considerable part of this is under forest cover. Thus, very small area is available for cultivation and considerable part of it is under rainfed having low productivity. Undulating topography, varied climate, scanty cultivated land, overwhelming percentage of small and marginal holding, difficult conditions, high cost and low returns on food grain crops, poor economic condition of the farmers etc. are main causes responsible for this situation. Majority of the hill population resides in the ranges of the middle Himalayas having elevation between 600 to 2000 m and is mostly dependent upon rainfed agriculture. Farmers are in a habit of using traditional agricultural techniques and methods for crop production and are getting low productivity almost from all the crops being grown in the region. Therefore, more holistic approach to land use and management is needed to cope with increased pressure on soil resources for sustainable food and fiber production while reducing the adverse off-site environmental impacts of agricultural practices. Besides regular watershed programmes launched by the government in the country such as NWDPRA, IWDP some programmes specially tailored for the hilly regions were implemented as given below:

Integrated watershed development project **(IWDP)**: The World Bank assisted Integrated Watershed Development Project (IWDP) was implemented in the Shiwalik belt in two phases since 1980 to 2005 covering the Shiwalik region in States of Himachal Pradesh, J&K, Uttarakhand, Punjab and Haryana. The project under IWDP hills I & II covered a total area of 5,84,564 ha at an investment of ₹108347.1 lakhs.

Uttaranchal decentralised watershed development project **(UDWDP)**: The World Bank aided Uttaranchal Decentralised Watershed Project (UDWDP) also known as *"Gamya"* was operationalised in Uttarakhand state since 2004 for a period of 7 years (2004-2012) covering a total of 461 Gram Panchayats (GPs) spread over 77 micro-watersheds and covering a total area of 2348 km^2 (part of 11 districts of Uttarakhand state). The project aimed at improving the productive potential of natural resources and increase income of rural inhabitants in selected watersheds following socially inclusive and sustainable approaches.

Watershed development project in shifting cultivation areas **(WDPSCAP)**: Shifting or *Jhum* cultivation is a serious problem in the NE hills where an area of about 43.57 lakh ha is affected by this problem which has been further aggravated due to reduction of earlier *Jhum* cycle of 20-30 years to 3-6 years now. The watershed development project in *Shifting* cultivation area (WDPSCAP) with 100% financial assistance was implemented in all NE states with the objectives of protecting the hill slopes of *Jhum* areas from excessive soil erosion and to create livelihood support activities and introduce appropriate land uses.

Integrated watershed management programme **(IWMP)**: The Department of Land Resources (DoLR), Ministry of Agriculture, Govt. of India was implementing Drought Prone Areas Programme (DPAP), Desert Development Programme (DDP) and Integrated Wastelands Development Programme

(IWDP) till 01.04.2008. All these programmes were constructed into a single modified programme called Integrated Watershed Management Programme (IWMP) *w.e.f.* 26.02.2009. The provision of treatment of micro-watersheds on cluster basis, reduced number of instalments, delegation of power of sanction of projects to the states, dedicated institutions, capacity building of stakeholders, monitoring and evaluation, specific budget provision for detailed project report (DPR) preparation, livelihood for asset less people and productivity enhancement are major features of IWMP.

Physical and financial achievements

The area treated/reclaimed in different Himalayans states under various watershed development programmes upto Xth Plan is presented in table 2 (Sharda *et al.*, 2012). It is evident from the table 4 that only 13.46% of TGA of north-west states and 15.5% of TGA north-east Himalayan states has been covered. In comparison to overall situation at national level, the treated area in Himalayan states is only 14.36% against 17.20% of TGA at country level. As the hilly regions are more fragile and ecologically sensitive, they need priority in treatment.

The soil conservation measures taken up in watershed development include mechanical measures like bunding, terracing, check dams, agronomic measures, vegetative barriers, alternate land use systems, runoff harvesting and recycling systems etc. These measures were under taken on agricultural lands (arable/non-arable) for enhancement of productivity and profitability.

Table 4. Physical and financial status of treated area in different Himalayan states of India under various watershed programmes (up to Xth plan)

States	Total geographical area (TGA)	Area treated (M ha)	Expenditure (₹ in lakh)	% of TGA treated
Jammu & Kashmir	22.22	1.19	57,609.29	5.34
Himachal Pradesh	5.57	1.45	67,688.94	25.94
Uttarakhand	5.35	1.82	76,316.63	33.93
NW Himalayas	33.14	4.46	2,01,614.86	13.46
Arunachal Pradesh	8.37	0.5	11,639.43	5.97
Assam	7.84	1.09	27,321.18	13.97
Manipur	2.23	0.44	16,227.47	19.69
Meghalaya	2.24	0.33	11,534.61	14.76
Mizoram	2.11	0.66	27,334.00	31.37
Nagaland	1.66	0.58	31,677.20	35.19
Sikkim	0.71	0.23	10,167.69	32.44
Tripura	1.05	0.23	11,109.52	21.57
NE Himalyas	26.21	4.06	1,47,011.10	15.49
Himalyas hill total	59.35	8.52	3,48,625.96	14.36
All India total	328.66	56.54	19,47,057.24	17.20

Under IWMP (2009-10 to 2012-13), a total number of 862 watershed projects were undertaken covering an area of 3.633 M ha with funding of ₹ 680.56 crores. Under Integrated Wastelands Development Programme (IWDP) a total of funds released amounted to ₹ 1514.81 crores (2007-08 to 2012-13).

Impact of watershed development programme

There is ample evidence of positive impacts of watershed programmes in terms of reduction in soil and water losses and improved agricultural productivity in normal rainfall years in regions that were bypassed in the conventional green-revolution era. Several reviews on the performance of watershed development projects (Palanisami *et al.*, 2002; Joy *et al.*, 2005) in India have diagnosed various limitations of watershed programmes. Participatory integrated watershed management (IWSM) approach being adopted in the recent past has shown encouraging results over the previously adopted commodity based or sectoral approaches. Operational research project on watershed management at Fakot in outer Himalayas which was implemented by ICAR-Central Soil & Water Conservation Research and Training Institute (CSWCRTI), Dehradun, Uttarakhand during 1975-86 is a successful example of this participatory approach. Similar trends in production increase and environmental benefits have been observed in other watersheds developed under NWDPRA, DPAP, RVP and other bilateral projects implemented during 1990s.

Landslide and mine spoil rehabilitation

It was observed that as a result of the bioengineering treatment of landslide affected area (Nalotanala watershed – 60 ha) and mine spoil affected area (Sahastradhara watershed – 64 ha), heavy soil erosion was checked and brought within permissible limit, surface runoff was drastically reduced with attenuation of flood peaks, the lean period flow increased due to groundwater recharge and biodiversity improved (Table 5). Restoration of lime stone mine spoil area which had excess of calcium resulted in improving the water quality through reduction in calcium content.

Table 5. Effect of bioengineering measures on landslide (1964-1994) and mine spoil rehabilitation (1984-1996) project

Particulars	Landslide project		Mine spoil project	
	Before treatment	After treatment	Before treatment	After treatment
Runoff (mm)	55	38	57	37
Sediment load (t/ha/yr)	320	5.5	550	08
Dry weather flow (days)	100	250	60	240
Vegetative cover (%)	<5	>95	10	80

Recommendations

From the foregoing discussion it can be inferred that treatment of mass erosion affected areas need to be brought within the purview of watershed development programmes. Some of the priority areas for treatment are given under.

- *Torrent/river training:* Rivers/torrents/streams which cause flood damages in watersheds should be treated through cost effective bioengineering technology using mechanical measures (spurs etc.) and locally adaptable suitable vegetative species which possess requisite hydraulic properties to withstand and dissipate the impact force of flood waters. Technique for *Katta crate* structures using locally available river bed material has been evolved by CSWCRTI, Dehradun. The adjoining lands should be put under permanent tree cover. The area up to 300 m from the stream banks should be banned for habitation and constructional activities.
- *Mine spoil rehabilitation:* Areas wherever mining/quarrying has been done must be put back under green cover after mining is over. Use of natural geotextiles (jute/coir nets) has been found effective for quick revegetation.
- *Landslides/slips control:* Landslides/slips particularly occurring along roads, agricultural lands, habitations etc. should be treated through bio-engineering technologies on watershed basis already available. Green technology for road construction should be followed. Treatment of the eroded/degraded/cut slopes along the roads should be mainstreamed into the road construction work. A separate soil conservation wing with the road construction agencies should be created for the purpose. Training/awareness programmes on bio-engineering aspects for the highway engineers need to be conducted. However, landslide/slip treatment is possible in areas which are not geologically unstable.
- *Cost estimates:* As mass erosion control requires more intensive measures particularly engineering ones, the cost of treatment per hectare naturally increases. From the past experiences, on an average, the landslide/slip affected watersheds may need an investment of ₹1-1.5 lakh/ha. For treatment of torrents an amount of ₹15-20 lakh/km of torrent length may be required.

In the watershed programmes so far the focus has been on resource conservation and productivity enhancement on agricultural lands. But in the Himalayan region due to their geological, topographical and climatic setting, the problem of mass erosion such as landslides/slips, river bank erosion and high sedimentation are common. In the changing climate scenario, such problems are expected to increase due to forecasting of high intensity storms and denudation of forest cover. Hence, there is a need to mainstream treatment of such problems areas into the watershed programmes.

Integrated approach to mitigate the adverse effect of the climate change

Delineation of the process or evidence showing control will acts as a booster for the farmer and encourage them to come out of this critical condition of poor livelihood. Here we are going to discuss some of the pioneer techniques which proved to be effective in controlling the problem of the soil erosion. North-western Himalaya region is severely suffering from the problem of the soil erosion. Hence, soil conservation measures suitable to an area must be adopted to improve the livelihood of the farmers. Soil and water conservation includes conservation farming, storage of rain water in soil profile, modification of soil profile and addition of organic matter etc. The various soil and water conservation technologies are land levelling, contour bunding, trenching, contour cultivation, strip cropping, cover crops, mixed cropping, conservation tillage, cultivation of fast growing and early maturing crops, recycling of crop residues, use of mulches, vegetative grass barriers, agro-forestry systems, sloping land management etc. (Sharma and Singh, 2013). Among these mulching found to be quite effective but its rate and mode are equally effective (Bhatt and Khera, 2006). Recognizing the importance of these techniques we are going to discuss them each one by one for delineating their role in conserving natural resources in hill agriculture.

Contour farming: In regions of low rainfall, the primary objective of contour farming is to provide the greatest possible conservation of rain water, in humid regions on the other hand, the primary purpose is to reduce soil loss by water erosion, although water conservation is also highly important. Purpose of contour farming is to place rows and tillage lines at right angles to the normal flow of surface water. The resistance developed by the crop rows and by furrows, between the ridges, to the water flow thus, reduces runoff velocity and gives more time to water to infiltrate in to the soil. Contour tillage operations, thus, hold part of the rain water and store it *in-situ*, thus reducing runoff and soil erosion and bring about a more uniform distribution of water received through rainfall. Contour farming, in addition to reducing soil loss and increasing yields. The practice of up and down method of cultivation in many parts of the country is one of the causes of encouraging man made erosion. On steep slopes, this practice enables rain water to gain velocity, facilitating runoff water to erode soil as time of concentration decreased to manifolds. Thus, the upper layer of the fertile soil is washed away and silted up in the dams and bunds installed in the highest ordered gully and decreased its lifespan. Contour farming is a system of crop husbandry in which all cultivation operations are done on contour e.g. preparatory tillage, sowing, inter-culture etc. water loss and soil erosion are reduced by contour farming as erosivity of the rainwater decreased to marginal levels.

Soil loss decreased to 32.2% when the site was brought to contour farming as compared to the conventional methods in Dehradun (Anonymous, 1972). Grass gave the least runoff and soil loss and cultivated fallow gave the maximum soil loss (Table 6).

Table 6. Soil and water loss on 8 per cent slope under different cropping systems at Dehradun

Treatment	Rainfall (mm) (June-October)	Water loss (mm)	Water loss as % of rainfall	Soil loss (t/ha)
	(Average of 4 years) (8 per cent slope - Dhulkot silty clay loam)			
Up and down cultivation maize – wheat	1239	670	54.1	28.5
Contour cultivation maize + cowpea – wheat	1239	511	41.2	19.3
Cowpea – wheat	1239	405	32.7	28.3
Giant star grass	1239	31	2.5	1.3
Cultivated fallow	1239	445	35.9	44.0

Use of cover crops or mulching: The effect of mulching in improvement of water productivity has been ascribed to restricted evaporative losses because of decrease in radiant energy reaching to soil to cause phase change from liquid water to gaseous phase, decrease of vapour pressure difference within soil and ambient air and finally because of decrease in vapour lifting capacity of the air (Fig. 5) and thereby causing yield augmentation (Bhatt and Khera 2006; Bhatt *et al*., 2004).

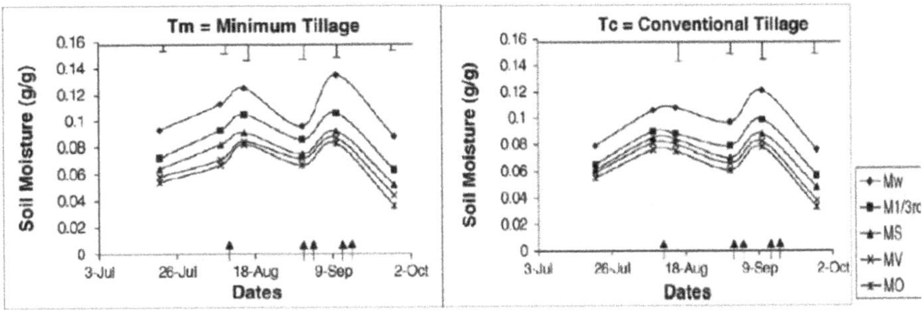

Fig. 5. Soil moisture content of surface soil as affected by tillage and different modes of mulch application (Bhatt and Khera, 2006)

Cover crop is a close growing crop raised mainly for protection and maintenance of soil. Effectiveness of the cover crop depends on close spacing and development of good canopy for interception of rain drops so as to expose minimum soil surface for erosion. Good ground cover canopy gives protection to the land like an umbrella. Erosion from cultivated fields can be reduced if the land has enough crop canopy during the peak season (Sharma *et al*., 2012). Study at Dehradun show the canopy development of various legumes crops during rainy season (Fig. 6).

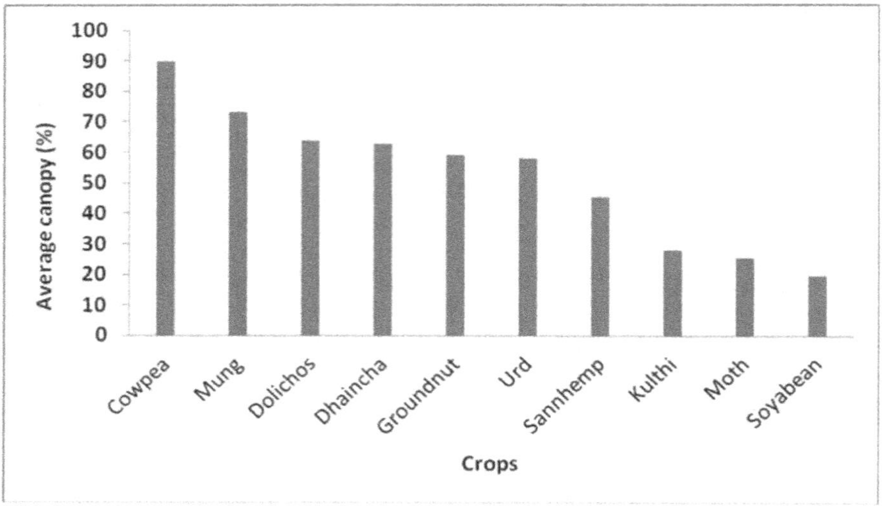

Fig. 6. Average crop canopy of different legumes

Strip cropping and hedgerow cultivation: Strip cropping is a system under which ordinary farm crops are planted in relatively narrow strips, across the slope of the land and are arranged in such a manner that the strips of erosion permitting crops are always separated by those of close growing or erosion resisting crops. Seven strip mixture treatments of maize and soybean were tested by Singh and Bhushan (1976) in high rainfall area at Dehradun. Crops behave differently in their capacities to produce vegetative cover and root development and consequently affect splash erosion, runoff and soil loss. Legumes, in general, as they produce good cover, are erosion resisting whereas open tilled crops like maize, cotton are erosion permitting. Farmers cultivate the lands to satisfy his food requirement and at the same time produce minimum erosion. Strip cropping is the system, which meets such requirement. The analysis indicated that with 2 rows of maize (60 cm apart) and 8 rows of soybean (30 cm apart), maximum yield was obtained. Since, there was a major contribution of soybean in this production system (3708 kg/ha), maximum economic returns were also obtained under this combination (Table 7). Mittal *et al.* (1998) found that maize and cluster been in alternate strips of 9 m each at 1.5% slope in Shiwalik region reduced the runoff and soil loss by 24 and 43%, respectively as compared to maize cultivation along the slope.

Table 7. Yield (kg/ha) of maize and soybean under different crop mixtures (4 years average)

Crop	No. of maize rows				No. of soybean rows		
	10:0	8:2	6:4	5:5	4:6	2:8	0:10
Maize	3690	3254	2722	2502	2557	2054	-
Soybean	-	285	358	632	900	1654	2071
Total grain production	3690	3539	3079	3215	3457	3708	2071

Mixed/inter cropping: Mixed/Inter cropping is a practice of growing more than one crop in the same field simultaneously. This system of cropping is very extensively adopted by the farmers in India. The most important mixture consists of legume and non-legume crops. System gives better cover on the land surface, good protection to soil from beating action of rain drops. Study conducted at different places in the country has shown that mixed/inter-cropping gives less runoff and soil loss. Studies in maize + cowpea intercropping show that splash erosion reduced from 64 g/57.6 m^2/77 days under pure crop of maize (with 68% canopy) to 22 g/57.6 m^2/77 days under intercrop (with 94% canopy). The results conclusively suggest that intercropping of cowpea (double row) for vegetable purposes is more advantageous both in terms of monetary benefits and soil conservation (Bonde and Mohan, 1984). Intercropping of legumes in maize not only advantageous in reducing runoff and soil loss by 15 and 27 per cent but also resulted in higher yield of succeeding wheat crop by 61 per cent (Tomar, 1992; Ghosh *et al.*, 2006).

Tillage and crop residue management: Tillage is well-known soil and water conservation technology. It results in improved soil organic matter (SOM), which positively influences physical, chemical and biological properties of soils. Conservation tillage with the maintenance of crop residues cover on 30% of the soil surface is soundly based within the frame work of conservation of natural resources *viz.* soil, water, nutrients and SOM and sustained production. The impact of a zero tillage (ZT) and crop residues management associated with soil health and quality by improving soil properties, minimizing soil erosion, soil water evaporation and conserving soil moisture is well documented. Hence, reduced tillage practices have been widely used in the last decades as an attractive alternative over conventional tillage practices because of their potential to reduce production/ operating costs and benefit for the environment and can save considerable time with seedbed preparation compared with conventional tillage practices (Al-Kaisi and Yin, 2004). Bhardwaj and Sindhwal (1998) found that ZT with plot weeded in 30cm strips along the maize rows leaving 60 cm unweeded used as mulch 4.5 to 5 t/ha dry weeds cut and spread over inter-row spaces as live mulch prior to seeding reduced runoff from 42 to 22 % and the soil loss from 12 to 3 t/ha at 4% slope. Reduction in runoff and soil loss was also reported by Ghosh *et al.* (2013) by using weed as mulch along with a practice of minimum tillage at 2% sloppy lands. Covering the land surface through crop residues, organics manures, leaf litter and other vegetative materials and its proper management is the most effective measure to reduce evaporation, keep down weeds, improve soil structure, infiltration rate and ultimately increase crop yields. In general, mulches used in *rabi* season crops have always conserved soil moisture and increased considerable yield in many cases. In hilly region, farmers use heavy mulching through organics and FYM for moisture conservation. The results of a study conducted for three years to assess the effect of four mulching treatments on yield of wheat crop reported by Tejwani *et al.* (1975) revealed that maize residues ploughed under and seed bed prepared by sub-surface tillage was beneficial as it gave higher grain yield of wheat as compared to other treatments. In Doon valley after the harvest of *kharif* crops, *rabi* crops suffer from moisture deficit at sowing and also during crop growth period. Therefore, Sharma *et al.* (2013) suggested that in maize-wheat cropping system, maize field should be ploughed immediately after maize harvest followed by planking and covering soil surface by maize stover mulch @ 5 t/ha up to wheat sowing to reduce soil moisture losses (up to 33% at surface layer) and increased germination (43%), water use efficiency (63%) and yield of wheat crop as compared to control. Advantage of mulch along with conservation tillage has also been described by Sharma *et al.* (2000) for wheat sowing in maize based cropping system. In a study in middle Himalayan region, Bhattarcharyya *et al.* (2009) reported that management strategies, such as reduced tillage (ZT and MT) and continuous leguminous cropping sequences (soybean-wheat/lentil/garden pea) played significant roles in total SOC sequestration in soil in the rainfed hilly agro-ecosystem. On an equivalent initial soil mass basis, SOC storage to about 15 cm soil depth after four years was 26.0 Mg/ha in continuous NT plots, but 23.9 Mg/ha in continuous CT (Bhattarcharyya *et al*., 2009).

Crop rotations: One of the basic components of the conservation agriculture is the rotation of the crops as it changed the rhizosphere depths. Rotation of crops is an order in which the chosen cultivated crops follow one another in a set cycle on the same field over a definite period for their growth and maturity with an objective to get maximum profit from least investment without impairing soil fertility. Crop rotations can be an important soil and water conservation practice. In a good crop rotation, the effects of soil improving crops offset the effects of soil depleting crops. With the increased availability and use of fertilizers, it is now possible to take more effective care of the fertility aspects of the soil under different rotations. Inclusion of thick growing crops in the rotation protects soil from the impacts of rain drops, besides the interception of runoff.

Plant population: Greater the plant population greater is the hindrance in running water even at the steeper slopes which finally decreases its erosive potent to cause the soil erosion. Further, it is reported that increasing plant population to 1.5 lakh/ha and using 50% plants as mulch after one month of sowing considerably controlled soil and water losses in maize. But considering increase in yield and reduction in runoff and soil, 1.0 lakh population should be preferred over 55,000 plants/ha (Narain and Singh, 2000). Comparing maize plant populations of 40,000; 55,000; 70,000 and 85,000 plants per ha as a sole crops and in combination with cowpea and blackgram as intercrops concluded that maize (70,000 plants) + cowpea gave highest maize equivalent total production over six years of 4,270 kg/ha (Tyagi *et al*., 1998). Runoff and soil and nutrient loss followed the trend of canopy cover and plant population i.e. decreased with increase in plant population of maize or intercropping with legumes. However, Khola *et al*. (1999) concluded that maize grain yield was higher by 144 kg/ha under 60 × 20 cm (83,000 plants/ha) spacing than 60 × 30 cm (55,000 plants/ha) but succeeding wheat grain yield was slightly lower by 58 kg/ha under the farmer. Thus, having higher plants will certainly improves both the land and water productivity as on one side plants increased the organic carbon status which further improved the aggregation strength with which primary particles are bound and thus finally decreased the erodibility of the soil even at steeper slopes.

Green manuring: Green manuring can be defined as a practice of ploughing or turning into the soil undecomposed green plant tissues for the purpose of improving physical structure as well as fertility of the soil (Singh *et al*., 2013). In *in-situ* green manuring, crops are grown and buried in the same field which is to be green manured. The most common green manure crops grown under this system are sunhemp (*Crotalaria juncea*), *dhaincha* (*Sesbania aculeata*) etc. Studies conducted at Dehradun indicated that sunnhemp is an excellent green manure crop, which has the inherent ability for improving the fertility status of the soil besides reduction in runoff and soil loss (Sharda *et al*., 1999). Considerable reduction in runoff amounting to 18.6 per cent in green manuring (GM)-wheat system has been recorded as compared to the fallow-wheat system. Reduction in soil loss was highly significant (33.1%), mainly due to fast and vigorous growth of sunhemp after the onset of monsoon which provides dense cover to the soil. Study at Dehradun indicated that growing of sunhemp for green manuring in 60 cm wide contour strips between maize rows and recycling after one month as surface mulch reduces erosion and controls weed growth (Anonymous, 2006). Incorporation of mulch after maize also improved the wheat yield due to moisture conservation and nutritional built up (Table 8).

Table 8. Effect of green mulching with green manuring and land slope on runoff (%) and soil loss (t/ha)

Land slope (%)	With green manuring		Without green manuring	
	Runoff	Soil loss	Runoff	Soil loss
0.5	6.0	3.07	11.2	4.1
2.5	12.2	6.83	15.7	8.6
4.5	18.3	12.80	21.2	17.7
9.5	24.9	23.19	28.2	28.2

Soil fertility approach: Sustenance of soil fertility is important through balanced and integrated plant nutrient management to enhance the productivity and sustainable agriculture in the region. Preference should be given to locally available organic nutrient sources like FYM, green manuring etc. Four technologies namely, (i) maize + cowpea (1:2)-wheat, (ii) maize–wheat + mustard (9:1), (iii) paddy-wheat, paddy (SRI) - wheat and (iv) maize-potato-onion with INM to each crop rotations were demonstrated among 19 farmer's fields under Farmers Participatory Action Research Programme (FPARP) on more crop per drop of water in the agro-climatic zone of valley land of NW Himalayas (Agro-ecological region 14) in the districts of Dehradun (Uttarakhand) and Sirmour (Himachal Pradesh) during 2007-09. Results revealed that irrespective of technology, water use efficiency (WUE) increases to the tune of 44.7% with yield increase of 40.7% and the corresponding net return increases to 90.1%. It is inferred from the study that of the crops rotations practiced in this region, maize-potato-onion with INM is an important option for best management of land, water resource; net return and improving the soil quality under limited irrigation condition whereas maize-wheat + mustard under rainfed conditions (Ghosh *et al*., 2011).

Mulch farming: Mulch farming is a system in which organic residues or other materials are neither ploughed into the soil nor mixed with it, but are left on the surface to serve as mulch. Mulch farming is not only useful for reducing soil and water losses but is also useful for maintaining high soil moisture in the field. Thus, mulching can be used in higher rainfall area for decreasing soil and water losses and in low rainfall region for improving soil moisture. Studies conducted at various places have amply demonstrated that mulching increases both soil moisture and yield of crops. Khola *et al.* (1999) in their study observed that intercropping of cowpea or soybean with maize either for fodder or *in-situ* mulching reduced maize grain yield by 101-451 kg/ha compared to sole counterpart. However, succeeding wheat grain yield was higher under mulching of cowpea (2744 kg/ha) or sorghum (2683 kg/ha) in maize and lower under sole stands of maize (2291-2349 kg/ha). Tyagi *et al.* (1998) observed that planting of perennial pigeon pea as a vegetative barrier and incorporation of its non-woody material, cut at 50 cm height in the soil registered minimum runoff (24%) and soil loss (5.3 t/ha). Sharma *et al.* (2010)

concluded that legume mulching is a highly beneficial practice for enhanced moisture and nutrient conservation, leading to increased productivity and soil health of maize-wheat cropping system under Doon valley conditions of north-western India. Beneficial effects of mulching of intercropped legumes with maize for mulching/seed production/green pods on canopy cover, soil and water losses and productivity have also been reported by Mohan (1992), Narain and Singh (2000) and Singh *et al.* (2003).

As mulching can be used in higher rainfall period/region for decreasing soil and water loss similarly is can be used in low rainfall period/region for increasing soil moisture. Studies conducted at various places have demonstrated that mulching increases soil moisture and yield of crops. Wheat crop sown in *rabi* season at Dehradun, gave significantly higher yield with surface mulch. Study conducted at Dehradun show that mulches applied immediately after the harvest of maize increased the grain yield of wheat significantly. Straw mulching before sowing was superior giving 36% higher yield than control followed by grass mulch (Singh and Bhushan, 1978). However, dust mulch was found as most practical way of moisture conservation, which produced 22% higher yield than control. Sharma *et al.* (1997) conducted a study to find out the optimum quantity of air dried *Leucaena* leaves mulch and its time of incorporation preceding to rainfed wheat sowing. The equation $Y= a+bM+cT+dM^2 +eT^2+ FMT$ was fitted between wheat grain yield (Y) and combination of doses of mulch (M) and timings of its incorporation (T). The results revealed that in addition to recommended dose of chemical fertilizers, air dried *Leucaena* leaves mulch @ 2.25 t/ha may be incorporated into the soil within a period of 30-40 days of maize harvesting. The wheat yield can be obtained about 32q / ha, which is about 25% higher than the control (Table 9).

Table 9. Effect of different levels and time of incorporation of *leucaena* mulch on grain and straw yield of wheat

Treatment	Grain yield (kg/ha)	Straw yield (kg/ha)
Leucaena mulch level (t/ha)		
0	2607	4548
2	3207	5287
4	3435	5380
6	3623	5560
LSD (*P*=0.05)	236	333
Days of incorporation after spread of mulch		
0*	3085	5141
15	3292	5176
30	3231	5199
45	3264	5259
LSD (*P*=0.05)	NS	NS

NS = Non-significant; * Incorporation immediately after maize harvest

Water harvesting and recycling: In region, particularly in high rainfall areas, the water harvesting structure can be a source of supplemental irrigation during the time scarcity for cash crops for high productivity and profitability (Sharda *et al.*, 1986). The harvested water in dugout or embankment type ponds can be efficiently utilized for supplemental irrigation during lean period to boost crop production. In the middle of Himalayas, where large water harvesting structures are not feasible owing to topographical limitations, small tanks locally called "*tankas*" of capacity ranging from 10 cum to 20 cum have been successfully constructed at farmers' fields. The top length varying from 5 to 6.5 m and width 3.5 to 5 m have been found suitable with side slopes of 1:1 and depth of 1.6 m (Juyal and Gupta, 1985). They suggested that "*tankas*" can be suitably lined with 1,000 gauge (250 micron) LDPE sheet. Before laying the sheet in excavated tank, protruding stone pieces and roots should be removed. For Kumaon region two water harvesting systems have been found feasible, namely (i) intra-terrace and (ii) micro-watershed. In the former, runoff is collected from a single terrace of about 0.025 ha and recycled back to the same terrace, whereas in the later, water is collected from a micro-watershed of 0.08 ha and recycled to a command area of about 0.06 ha in the downstream. Both the systems are found to reduce the peak flows (Srivastava, 1983).

The harvested water in dugout or embankment type ponds can be efficiently utilized for supplemental irrigation during lean period to boost crop production. In Doon valley, there are three options available for utilizing harvested water *viz;* (i) pre-sowing irrigation (5 cm) (ii) application at crown root initiation (CRI) stage (5 cm), and (iii) pre-sowing + CRI (total 10 cm). It has been found that between the first two options, increase in crop yield was almost same. However, owing to excessive storage losses, delayed application in the second case results in reduced command area. In case of the third option, though two irrigations boost the crop yield compared to first two methods (Table 10), but, corresponding reduction in command area is more (Singh *et al.*, 1981; Singh and Bhushan, 1980).

Table 10. Effect of supplemental irrigation on wheat yields

Particulars	Pre-sowing (5 cm)	CRI (5 cm)	Pre-sowing + CRI (10 cm)	Control
Yield (kg/ha)	3203	3236	3898	2045
% increase due to supplemental irrigation	80	50	25	-

In a study at Dehradun on low cost water harvesting and micro-irrigation, this year 3 silpaulin lined (200 GSM) water harvesting tanks of 7-8 cum capacity were constructed at Ashti integrated with low cost drip irrigation system operating on low head (3-5 m). Integrated system is more convenient in irrigating the crops for higher productivity and saving in labour as compared to hand watering normally practiced in hilly region (Anonymous, 2012).

Sub-surface water harvesting: In Himalayan foothills, seepage or oozing/trickling of water from the hill slopes as a surface/sub-surface flow is a common phenomena. Duration and discharge of flow depends on the local geological set up which forms perched type of aquifers/cracks allowing short duration or sustained release of water from the permeable faces of the hills. Such type of flow can be observed in most of the parts in the Himalayan region. The water can be stored in the harvesting structures which are specially designed so that it allow the flow into the tank from upper walls (pervious wall) and stored against the lower wall (impervious). Such structure (21 × 8 × 2 m) was constructed in Kalimati village in order to harvest the sub-surface flow. The upper long wall of the tank was constructed with pervious gabion work and, other walls were constructed with masonry work, i.e. impervious so that the incoming water from the upper pervious wall could retain in the tank.

The idea was that pore space available in gabion walls between boulders would act as inlet and allow sub-surface water flow to be collected in the tank. In all, 33 families from Kalimati village and 17 from Badasi village participated in developing the water resource for irrigation to 16 ha of land. Stakeholders of Kalimati and Badasi villages (50) had the right to share the resource for crop production. The yield of wheat in limited irrigation ranges from 2,800 to 3,600 kg/ha with an average 3,240 kg/ha over the farmers practice as control. The increase in yield was recorded 44% which had been achieved because of irrigation and generation of water resource in the village (Samra *et al*., 2000).

***Conservation bench terrace* (CBT)**: CBT system comprises 3:1 ratio of contributing and receiving area and 20 cm depth of impoundment at the end. A study was initiated with a view to study the efficacy of conservation bench terraces system, under recommended cropping sequence of CSWCRTI, Dehradun. To store the excess runoff from these treatments, a brick masonry tank of size 30 × 15 × 2.5 m was constructed at the experimental site at S.C. Farm, Selakui of CSWCRTI. From the analysis, it has been concluded that the CBT system is most effective in reducing runoff by over 80% and soil loss by more than 90% compared to conventional system i.e. sloping boarder with zero depth of impoundment; crop rotation: maize + cowpea-wheat + mustard. In the levelled bench of CBT system, paddy was taken during *kharif* season and intercrop of mustard with wheat was taken during *rabi* season. The paddy yield under the CBT system makes the system more remunerative compared to the traditional system of maize-wheat rotation. The CBT system also recorded highest yield of wheat and mustard crops on an average. The nutrient loss from CBT system was observed to be minimum. The system has a great scope for harvesting the inevitable runoff particularly in Doon valley and its subsequent utilization during intervening droughts for sustaining the crop yields (Sharda *et al.*, 1999).

Gravity fed participatory water resource development and management: Low productivity and cropping intensity, substance level of farming is attributed to poor water resource development in the hilly areas. A major part of rainfall goes unutilised as surface runoff which is not being used for productive purposes. Community is unable to invest huge capital on water resources development due to their poor resource conditions particularly in hilly area. ICAR-Central Soil and water Conservation Research and Training Institute (CSWCRTI), Dehradun initiated water resource development and its management for enhancing livelihood security of hilly farmers in villages *viz.,* Pasauli, Devthala in Vikasnagar block of district Dehradun, Uttarakhand situated in N-W Himalaya.

Water is being tapped from a perennial spring located at 771 m msl in outer Himalaya. An intake well (IW) was constructed at the source to convey water from source to distribution tank (DT) through galvanized iron (GI) pipe (dia: 110 mm) of length 2080 m, by gravity flow with hydraulic gradient 1.94 %. Distribution tank having capacity of 50,000 litres and inflow at DT is 10 LPS. The tank is located at the head of the command area (25 ha) in order to convey the water through underground PVC pipelines to the farmers' fields, by gravity flow with maximum 26.5 m gravity head. PVC pipeline (1360 m) was laid across the command area in such a way that it can supply the water to the farmers' fields spread in the command area with 10-12 LPS discharge. The command area is divided into 15 unit command areas (UCA) and each UCA covers 1.5 ha. At the head of each UCA, a riser is fitted with the PVC pipeline to uniformly deliver the water. In order to enhance the field application efficiency, collapsible LLDPE pipes are being used by the farmers. Distribution of water for the irrigation to each beneficiary farmer is regulated by the committee constituted by the water users association (WUA). Water being shared among the farmers' fields on the rotation basis as decided by the WUA and farmers get his turn in 15 days (irrigation period). In one season, delivery of water follow the sequence from head to the tail riser and in next season delivery sequence gets reversed order to negate the clash among the farmers. Productivity of major crops *viz.,* maize, paddy, wheat and *toria* were increased by 48%. Enhanced cropping intensity by 29% due to intervening crop of *toria* in between maize-wheat sequence and cultivation of *rabi* wheat in fallow land (Dhyani *et al.*, 2011).

Conclusions

Finally, it could be concluded that NW Himalayan region of the India form the suffering severely from the problem of soil erosion and the farmers of the region bound to live with a poor living status because of lower water and land productivity. However, this problem could be tackled in a smart way by adopting certain integrated approach as a single approach might not be so effective. For controlling gully erosion, tackling 1st ordered gullies will serve the purpose. With integrated approach, soil health and livelihood of the farmers certainly improved by mitigating adverse effect of the global warming and the soil erosion. Following are some also important conclusions:

- First ordered gullies are the main culprit which collect the rainwater from each nook and corner of the catchment and supplied it to the higher ordered gullies. Governments spend huge amounts to install check dams in the higher ordered gullies which prove to be failure to control the soil erosion at a long time scale. Thus, if we control the first ordered gullies in the catchment, then no water supplied in the higher ordered gullies and thus extent of soil erosion could be significantly reduced
- Catchment sides facing SW side reported to have intense gully network as compared to NE because of greater sunshine hours, more evaporation, lesser vegetation, lesser organic matter, poor aggregation and finally, higher erodibility.
- The severity of gully erosion was observed to be a function of average relief and lamniscate ratio of the catchments as increased steepness results in increased runoff speed which aids deepening and widening of gullies. Lower value of lamniscate ratio indicates more erosion in the catchment due to its more compact shape with reduced time of concentration.
- Concave slopes in a particular catchment have been shown to be more prone at the lower slope segments for initiation of gullies as the runoff water concentrates at the lower slope segment where the slope steepness decreases and runoff water starts concentrating. The slope angles were steeper on the upper slope segments and it decreased in the down slope direction.

References

Al-Kaisi, M.M. and Yin, X. 2004. Stepwise time response of corn yield and economic return to no tillage. *Soil and Till. Res.* **78**: 91–101.

Arora, S. and Saygın, Selen D. 2011. Soil and water conservation in hilly terrain in north India and central turkey: issues and challenges. *J. of the Soil and Water Cons.* **10**(2): 87-93.

Bhardwaj, S.P. and Sindwal, N.S. 1998. Zero tillage and weed mulch for erosion control on sloping farm land in doon valley. *Indian J. Soil Cons.* **26** (2): 81-85.

Bhatt, R. 2015. Soil water dynamics and water productivity of rice-wheat system under different establishment methods. Dissertation submitted to the Punjab Agricultural University, Ludhiana.

Bhatt, R., Khera K.L. and Arora, S. 2004. Effect of tillage and mulching on yield of corn in the submontaneous rainfed region of Punjab, India. *Int. J. of Agric. and Biol.* **6**:1126-28.

Bhatt, R., Kukal, S.S. and Arora, S. 2013. Resource conservation technologies for improving water productivity. *J. of Soil and Water Cons.* **12**(4): 313-320.

Bhatt, J.C. 2011. Crop production system in hills: status, trend analysis, constraints, potential and strategies. Proceedings of workshop 'Mountain Agriculture in Himalayan region: status, constraints and potentials". April 2-3, 2011. Pp 41-56.

Bhatt, R. and Khera, K.L. 2006. Effect of tillage and mode of straw mulch application on soil erosion losses in the submontaneous tract of Punjab, India. *Soil Till. Res.* **88**: 107-115.

Bhattacharyya, R., Ved-Prakash, Kundu, S., Srivastva, A.K., and Gupta, H.S. 2009. Soil aggregation and organic matter in a sandy clay loam soil of the Indian Himalayas under different tillage and crop regimes. *Agric., Ecosys. & Env.* **132**: 126-134.

Bicon, M.A. 2004. Water use efficiency in plants biology. CRP Press.

Bonde, W.C. and Mohan, S.C. 1984. Intercropping in maize for effective soil conservation, moisture utilization and maximizing production in Doon valley. Annual report, CSWCRTI, Dehradun, pp. 72-73.

Bouyoucos, G.J. 1935. The clay ratio as a criterion of susceptibility of soils to erosion. *J. Am. Soc. Agron.* **27**: 738-41.

Burkard, M.B. and Kostachuk, R.A. 1997. Patterns and control of gully growth along the shoreline of lake Huron. *Earth Surf Process Landforms* **22**: 901-911.

Dhyani, B.L., Madhu, M., Singh, C., Kumar, A., Bankey, B., Mandal, D. and Muruganandam. 2011. Participatory dissemination and assessment of land and water management technologies for livelihood security in rainfed areas of north western Himalayas. Annual Report, 2011-12, CSWCRTI, Pp.99-100.

Ghosh, B.N., Khola, O.P.S. and Dadhwal, K.S. 2006. Impact of organic matter build up on resource conservation under different crops on 2% slope. Annual Report, 2005-06, CSWCRTI, Dehradun, pp.29-30.

Ghosh, B.N., Sharma, N.K. and Dadhwal, K.S. 2011. Integrated nutrient management and cropping systems impact on yield, water productivity and net return in valley soils of north-west Himalayas. *Indian J. of Soil Cons.* **39**(3): 236-242.

Ghosh, B.N., Sharma, N.K. and Dogra, P. 2013.Yield maximization and resource conservation through organic input management Annual Report, 2013, CSWCRTI.

Goree, K. ed. 2004.Water-L News. International Institute for Sustainable Development in Collaboration with the Third World Water Forum Secretariat and the World Water Council. Issue 9 (February 24-March 16) 2004. Bangladesh.

Hadda, M.S. and Sur, H.S. 1987. Effect of land modifying measures on erosion, nutrient, water storage and yield of pearl millet fodder. *J. Indian Soc. Soil Sci.* **35**: 480-486.

Hillel, D. 1980. Applications of soil physics. Academic Press, New York, 385pp.

Hudson, N. 1984. The erosivity of rainfall. *Soil Cons.* 62-80 pp.

ICAR & NASC. 2010. Degraded and wastelands of India -status and spatial distribution, Directorate of Information and Publications of Agriculture Indian Council of Agricultural Research New Delhi, pp158.

Joy, K.J., Parnjpe, Suhas, Shah, Amita, Badigar, Shrinivas and Lele, Sharachchandra. 2005. Scaling up of watershed development projects in India: learning from the first generation projects. Fourth IWMI-Tata Annual Partners Meet, International Water Management Institute, Anand, India.pp.133-134.

Juyal, G.P. and Gupta, R.K. 1985. Construction of LDPE lined 'TANKAS' in hills – A case study. *Indian J. Soil Cons.* **13**(1): 10-13.

Khan, A.R. 1999. An Analysis of the Surface Water Resources and Water Delivery Systems in the Indus Basin. International Water Management Institute, Consultative Group on International Agricultural Research. Report 54.Battaramulla, Sri Lanka.

Khera, K.L. and Singh, G. 1998. Effect of crop cover and field slope on soil erosion in northern plain hot sub humid Punjab. *Indian J. Soil Cons.* **26** (1): 19-21.

Khera, K.L. and Singh, G. 1995. Effect of paddy straw mulch and rainfall intensity on runoff and soil loss under stimulated rainfall. *Indian J. Soil Cons.* **23** (1): 20-23.

Khola, O.P.S., Sharma, N.K. and Khullar, A.K. 1999. Evaluation of different maize growing systems for resource conservation and biomass production under varying fertility conditions. Annual Report, CSWCRTI, 1999-2000, Dehradun, p.32-33.

Kiani, K. 2005. Water-related crisis feared in 20 years. Dawn: the Internet Edition. 04 January 2005. The Dawn Group of Newspapers: Islamabad, Pakistan.

Kukal, S.S. 1987. Soil erosion risk assessment survey of a representative area in submontaneous tract of Punjab. M.Sc. thesis, Punjab Agricultural University, Ludhiana, India.

Kukal, S.S., Bawa, S.S. and Khera, K.L. 2002. Evaluation of gully control measures in foothills of lower Shiwaliks-A Compendium, Department of Soils, Punjab Agricultural University, Ludhiana, Punjab, India.

Kukal, S.S., Bawa, S.S., Bhatt, R. and Kamboj, A. 2006. Behaviour and patterns of gully erosion in foothills of lower Shiwaliks. Final report submitted to Department of science and Technology (DST), New Delhi, India.

Mannering, J.V. and Meyer, L.D. 1963. The effect of various rates of surface mulch on infiltration and erosion. *Soil Sci Soc. Am. Proc.* **27**: 84-86.

Maruthi Sankar, G.R. 1986. On screening of regression models for selection of optimal variable subsets. *J. of Indian Soc. of Agril. Stat.* **38**: 161-168.

Matharu, G.S., S.S. Kukal and S.S. Bawa, 2003. Rain characteristics in relation to runoff in submontane Punjab. *J. Indian Soc. Soil. Sci.* **51**: 288-290.

McGregor, K.C., Cullum, R.F. and Mutchler, C.K. 1999. Long term management effects on runoff, erosion and crop production. *Trans ASAE* **42**(1): 99-105.

Meyer, A. and Martinez-Casasnovas, J. A. 1999. Prediction of existing gully erosion in Vineyard parcels of the NE Spain: a logistic modelling approach. *Soil Till. Res.* **50**: 319-331.

Mohan, S.C. 1992. Effect of green manuring with intercropping on soil fertility and moisture conservation under rainfed conditions, Annual Report, 1992-93, CSWCRTI, Dehradun, p.44.

Morgan, R.P.C. 2005. Soil Erosion and Conservation. 3rd Edn., Blackwell Publishing ISBN: 1-4051-1781-8 pp: 324.

Narain, P. and Singh, R.K. 2000.Quantification of ground cover effect on runoff and soil loss through artificial and natural rainfall. Annual Report, CSWCRTI, Dehradun, p.23-24.

National Academy of Agricultural Science (NAAS). 2010. Degraded and wastelands of India: Status and spatial distribution, ICAR, New Delhi.

Nearing, M.A. 2001. Potential changes in rainfall erosivity in the US with climate change during the 21(st) century. *J. of Soil Water Cons.* **56**(3): 220–232.

Ogola, J.S., Mutuilah, W.V. and Omulo, M.A. 2002. Impacts of gold mining on the environment and human health: A case study in the Migori Gold Belt, Kenya, *Environmental Geochemistry and Health* **24**: 141 -158.

Palanisami, K., Suresh K., D. and Chandrasekharan, B. (Eds.). 2002. Watershed Management: Issues and Policies for 21st Century. Associated Publishing Company, P.O. Box No. 2679, New Delhi, India.341 pp.

Reddy, T.Y. and Reddi, G.H.S. 1995. Soil Conservation. *Principles of Agronomy*: 335-67.

Rockstrom J., Barron, J. and Fox, P. 2003. Water Productivity in Rain-fed Agriculture: Challenges and Opportunities for Smallholder Farmers in Drought-prone Tropical Agro ecosystems; In Braker R., Molden D. (eds.), Comprehensive Assessment of Water Management In

Agriculture Series; Water productivity in Agriculture: Limits and Opportunities for Improvement; CAB International.

Samra, J.S., Mishra, A.S., Mohan, S.C., Dhyani, B.L. and Tomar, D.S. 2000. Participatory generation, assessment refinement and dissemination of rural technology (1996 to 1999). CSWCRTI, Dehradun, Pp. 206.

Sehgal, J.L. and Sys, C. 1970. The soils of Punjab–India. III Classification and Geographic Distribution. *Pedologic* 20: 357-80.

Sharda, V.N. 2011. Land degradation and watershed management related issues in the Himalayan Region: Status and strategies. Proceedings of workshop "Mountain Agriculture in Himalayan region: status, constraints and potentials. April 2-3, 2011. Pp 1-22.

Sharda, V.N., Khybri, M.L., Sharma, N.K., Mohan, S.C. and Juyal, G.P. (1999).Green manuring for conservation and production in Western Himalayan Region : (I) effect of runoff, soil loss and fertility improvement. *Indian J. of Soil Cons.* **27**(1): 26-30.

Sharda, V.N., Sastry, G. and Joshi, D.P. 1986. Studies on control of seepage losses. Annual Report, pp. 34-36, CSWCRTI, Derhadun.

Sharda, V.N., Shrimali, S.S. and Khola O.P.S. 1999. Hydrological evaluation of recommended conservation measures on mildly sloping lands. Annual report, 1999-2000, CSWCRTI, Dehradun.

Sharda, V.N., Sikka, A.K. and Juyal, G.P. 2012. Participatory integrated watershed management–A field manual (Second Edition), CSWCR&TI. Dehradun.

Sharma, A. and Arora, S. 2010. Soil quality indices and relative production efficiency for maize and wheat crops in different agro-climates of North-West India. *Soil Sci.* **175** (1): 44-49.

Sharma, B.R., Rao, K.V., Vittal, K.P.R., Ramakrishna,Y.S. and Amarasinghe, U. 2010. Estimating the potential of rainfed agriculture in India: Prospects for water productivity improvements. *Agril. Water Mangt.* **97**: 23-30.

Sharma, N.K. and Singh, R.J. 2013. Agronomic Practices for Erosion Control. *Popular Kheti.* **1**(3): 57-60. (Available online at www.popularkheti.info)

Sharma, N.K., Ghosh, B.N., Khola, O.P.S. and Dubey, R.K. 2013. Residue and tillage management for soil moisture conservation in post maize harvesting period under rainfed conditions of north-west Himalayas. *Indian J. of Soil Cons.* **41**(3): 287-292.

Sidhu, P.S., Hall, G.F. and Sehgal, J.L. 1976. Studies on some soils at varying stages of pedogenic development in the Central Punjab-1.Morphology and Physiochemical characterization. *J. Res PAU* **13** : 23-24.

Sidhu, H.S. 2002. Crisis in agrarian economy in Punjab: some urgent steps. *Economic and Political Weekly*. **37**(30): 3132-3138

Singh S., Singh, R.J., Kumar K., Singh B., Shukla L. 2013. Biofertilizers and green manuring for sustainable agriculture. In book: Modern Technologies for Sustainable Agriculture, Edition: 1, Publisher: New India Publishing Agency, New Delhi, Editors: Sunil Kumar, Birendra Prasad, pp.129-150. ISBN: 978-93-81450-61-1.

Singh, G. and Bhushan, L.S. 1980. Supplemental irrigation to wheat crop from runoff harvested water under dry farming conditions. *Annl. of Arid Zone* **19** (3): 215-220.

Singh, G., Ram Babu, Narain, P., Bhushan, L.S. and Abrol, I.P. 1992. Soil erosion rates in India. *J. of Soil and Water Cons.* **47**: 97-99.

Singh, R., Dhyani, S.K. and Dubey, R.K. 2003. Conservation tillage and green manure mulching for optimizing productivity in maize-wheat cropping system in the sub-mountainous Himalayan region. Annual Report, CSWCRTI, Dehradun, p.21-22.

Solanki, R., Anju, Poonum and Dhankar, R. 2011. Zinc and copper induced changes in physiological characteristics of *Vigna mungo* (L). *J. Environ Biol.* **32**: 747–751.

Srivastava, R.C. 1983. Film lined tanks for hilly regions. *Indian Farming* XXXIII (II): 27-31.

Stocking, M.A. 1980. Examination of the factors controlling gully growth. In: De Boodt, M., Gabriels, D. (Eds.), Assessment of Erosion. Wiley, Chichester, UK, pp. 505-520.

Sur, H.S. and Ghuman, B.S. 1994. Soil management and rain water conservation and use-IV. Alluvial soils under medium rainfall. *Indian Soc. Soil Sci. Bull.* **16** : 56-65.

Sur, H.S. and Ghuman, B.S. 1992. Soil and rainfall water management in medium rainfall alluvial soils. Paper presented at the 57th convention of Indian Soc Soil Sci. held at Hyderabad from 26-29 Nov.

Tejwani, K.G, Bhaskaran, A.R. Thomus, P.K. Verma, Bdyal, R., Khybri, M.L. and Bains, K.S. 1975. Research in agronomy for soil conservation. In :Tejwani, K.G., Gupta, S.K. and Mathur, H.N. (Eds). Soil and water conservation research, ICAR, New Delhi.Pp106-145.

Thakur, K.K., Arora, S., Pandita, S.K. and Goyal, V.C. 2013. Land use based characterization of soils in relation to geology and slope in micro-watershed of Shiwalik foothills. *J. of Soil and Water Cons.* **12**(4): 348-354.

Tomar, D.S. 1992. Effect of intercropping of maize on soil erosion and yield. Annual Report 1992-93, CSWCRTI, Dehradun, p.38.

Tyagi, P.C., Mohan, S.C. and Joshi, B.P. 1998. Crop cover management in maize legume cropping system for sustainable production. Annual Report, 1998-99 CSWCRTI, Dehradun, p.26-27.

Vittal, K.P.R., Singh, H.P., Rao, K.V., Sharma, K.L., Victor, U.S., Ravindra Chary, G., Sankar, G.R.M., Samra, J.S. and Singh, G. 2003. Guidelines on drought coping plans for rainfed production systems. All India Coordinated Research Project for Dryland Agriculture, Central Research Institute for Dryland Agriculture (CRIDA), Indian Council of Agricultural Research, Hyderabad, A.P., India, p.39.

Wagger, M.G. and Denton, H.P. 1992. Crop and tillage rotation. Grain yield, residue cover and soil water. *Soil Sci. Soc. Am. J.* **56**: 1233-1237.

World Resource Institute (WRI). 2003. Watersheds of the World_CD. The World Conservation Union (IUCN), the International Water Management Institute (IWMI), the Ramsar Convention Bureau, and the World Resources Institute (WRI): Washington, DC.

WWF. 2007. World's top 10 rivers at risk. WWF International. Gland, Switzerland.

Zheng, F., Huang, C. and Romkens, M. 2009. Ephemeral gully erosion research at the Loess Plateau of China. Paper presented at the annual meeting of the soil and water conservation society, Saddlebrook Resort, Tampa, Florida, USA http://www.allacademic.com/meta/p202367_index.html.

14

Advances in Reclamation and Management of Salt Affected Soils for Sustainable Crop Production

Sanjay Arora, Y.P. Singh and Atul K. Singh*

Introduction

Soils are formed by weathering of rocks and minerals and all soils contain some amount of soluble salts. Many of these act as a source of essential nutrients for the healthy growth of plants. However, when quantity and quality of salts in the soil near rhizosphere exceeds a particular value, growth, yield and/or quality of most crops is adversely affected. Such a soil is called salt-affected. The degree of adverse effects depends upon the type and quantity of salts, crop and its variety, stage of growth, cultural practices and environmental factors *viz.* temperature, relative humidity, and rainfall etc. Development of salinity and waterlogging is a serious problem in arid and semi-arid regions of the world and threatening the sustainability of irrigated agriculture.

Salt-affected soils occupy an estimated 952.2 million ha of land in the world that constitutes nearly 7% of the total land area and nearly 33% of the potential arable land (Dudal and Purnell, 1986). In India, the salt affected soils account for 6.727 million ha i.e. 2.1% of geographical area of the country. These soils are mostly found in the states of Uttar Pradesh, Haryana, Punjab, Madhya Pradesh, Bihar and Andhra Pradesh. In the last 25 years 1.1 million ha of alkali soils have been reclaimed in the states of Haryana, Punjab and Uttar Pradesh. These have contributed to the additional food grain production of 10 million tonnes annually. Reclamation of alkali soils have also decreased the incidents of floods and malaria and increased the groundwater recharge.

Salt affected soils differ from arable soils with respect to two important properties, namely, the soluble salts and the soil reaction. A build-up of soluble salts in the soil may influence its behaviour for crop production through changes in the proportions of exchangeable cations, soil reaction, physical properties

**Corresponding author email: aroraicar@gmail.com*

and the effects of osmotic and specific ion toxicity. Salt related properties of soils are subject to rapid change. Therefore, to facilitate discussion on soil management and the influence of the two common kinds of salts (neutrals and alkali salts) on soil properties and plant growth, salt affected soils are broadly grouped as either saline or alkali soils (Szabolcs, 1974; Abrol and Bhumbla, 1979).

Extent of salt affected soils in India

Remote sensing techniques and GIS tools have proven to provide useful information on monitoring, delineating and mapping salt affected soils (Arora, 2013). Remote sensing based identification of salt-affected soils based on differences in spectral signatures using multispectral and hyper spectral data is the most common approach (Sharma and Mondal, 2006; Singh *et al.,* 2017). National Remote Sensing Agency (NRSA), Hyderabad in association with other national and state level organizations like ICAR-Central Soil Salinity Research Institute (CSSRI), Karnal; National Bureau of Soil Survey & Land Use Planning (NBLSS&LUP), Nagpur; All India Soil Survey & Land Use (AISS&LU), Delhi; and state government agencies conducted survey and used remote sensing data to prepare the maps of salt affected soils of India in 1996. The Landsat satellite images were used in mapping salt affected soils at 1:250,000 scale. Satellite images were interpreted for broad categorization of different types of salt-affected soils, sample areas for field verification (ground truthing) were identified and surveyed for soil sampling and characterization. The salt affected soils were classified according to norms for pH, electrical conductivity (EC) and exchangeable sodium percentage (ESP). The state wise extent of salt affected soils in India is given in table 1. It shows that maximum area of salt affected soils occur in Gujarat followed by Uttar Pradesh and Maharashtra which account for about 62.4 per cent of the total. Due to the limitation of small scale some very small and isolated patches of salt affected soils occurring in the states of Delhi and Himachal Pradesh could not be detected.

Out of the total 6.727 million ha of salt affected soils, 2.956 million ha are saline and the rest 3.771 million ha are sodic. Out of the total 2.347 million ha salt affected soils in the Indo-Gangetic Plains, (IGPs) 0.56 million ha are saline and 1.787 million ha are sodic.

However, it has been noticed that large tract of the salt affected soils have been reclaimed through interventions of various techniques and also vast areas have been reported to have develop soil salinity or sodicity through anthropogenic activities and improper management. It has been observed that $>$ 7,500 ha of salt affected soils have been developed in Ravi-Tawi canal command area of

Jammu and its improper management has lead to more severity in recent years (Jalali *et al.*, 2005, 2006; Sharma *et al.*, 2012).

Table 1. Extent of salt-affected soils India ('000 ha)

State	Saline	Sodic	Total
Andhra Pradesh	77.598	196.609	274.207
Andaman & Nicobar Island	77.000	0	77.000
Bihar	47.301	105.852	153.153
Gujarat	1680.570	541.430	2222.000
Haryana	49.157	183.399	232.556
Karnataka	1.893	148.136	150.029
Kerala	20.000	0	20.000
Madhya Pradesh	0	139.720	139.720
Maharashtra	184.089	422.670	606.759
Orissa	147.138	0	147.138
Punjab	0	151.717	151.717
Rajasthan	195.571	179.371	374.942
Tamil Nadu	13.231	354.784	368.015
Uttar Pradesh	21.989	1346.971	1368.960
West Bengal	441.272	0	441.272
Total	2956.809	3770.659	6727.468

Source: NRSA & Associates (1996)

Characterization of salt affected soils

Salt-affected soils in India are broadly placed into two broad groups; sodic (alkali) soils and saline soils. There are certain specific situations where saline-sodic soils also do exist. Since the management of saline sodic soils will be more similar to that of the sodic soils, they are generally grouped with the sodic soil category. The only management difference for their reclamation is that such soil needs ponding of water for leaching of soluble salts before amendment application. The sodic soils have higher proportion of sodium in relation to other cations in soil solution and on the exchange complex. Growth of most crop plants on sodic soils is adversely affected because of impairment of physical conditions, disorder in nutrient availability and suppression of biological activity due to high pH, exceeding even 10 in severe cases, and exchangeable sodium percentage (ESP) of up to 90% or so (Kanwar and Bhumbla, 1969). Salt solutions contain preponderance of sodium carbonates and bicarbonates capable of alkaline hydrolysis, thereby saturating the absorbing complex with sodium. The sodic soils of the IGP are generally gypsum ($CaSO_4.2H_2O$) free but are calcareous, with $CaCO_3$ increasing with depth, which is present in amorphous form, in concretionary form, or even as an indurate bed at about 1 m depth. The accumulation of $CaCO_3$ generally occurs within the zone of fluctuating water table. The dominant clay mineral is illite. The processes which target the dissolution of $CaCO_3$ have significant role in reclamation of alkali or sodic soils. Crops like rice helps in reclaiming sodic soils (Chhabra and Abrol, 1977). These soils are deficient in organic matter, available N, Ca, and Zn. Certain micro-nutrients present problems of either deficiency or toxicity. Toxicities of Al, Mn, and Fe sometimes pose problems for wheat when over- irrigated and results in yellowing of the crop.

The inland saline lands are widespread in the canal irrigated, arid and semi-arid regions. These soils are characterized by the presence of excess neutral soluble salts like chlorides and sulphates of sodium, calcium and magnesium. Sodium chloride is the dominant salt. High soil salinity is often accompanied by high water table, often within 2 m of soil surface. Sub-soil waters are generally salty and, therefore, their use for irrigation presents major constraints to crop production. In general, these soils have good physical properties but poor natural drainage. The formation of saline soils is generally associated with the rise in water table due to introduction of irrigation and inadequate drainage.

Saline soils: Saline soils, also known as solonchaks, are those which contain appreciable amounts of soluble salts, so as to interfere with plant growth. Saline soils are often recognized by the presence of white salt encrustation on the surface and have predominance of chloride and sulphate of Na, Ca and Mg in quantities sufficient to interfere with growth of most crops. Soil with neutral soluble salts has saturation paste pH of < 8.5. The electrical conductance of saturation extract of saline soils is > 4 dS/m at 25 °C and ESP < 15. The sodium adsorption ratio (SAR) of the soil solution is generally < 15. However, soil salinisation with neutral soluble salts of Na invariably results in soil solution with SAR > 15. Such soils are termed saline-sodic. Chemical characteristics of some of the saline soils are given in table 2.

Table 2. Chemical properties of some saline soils

pHs	ECe (dS/m)	SAR	Concentration of ions in the saturation extract (meq/l)							
			Na^+	K^+	Ca^{2+}	Mg^{2+}	HCO_3^-	CO_3^{-2}	Cl^-	SO_4^{-2}
7.30	2.2	4.14	11	2.2	6.9	7.2	2.9	Nil	12	13
7.60	26.6	54.42	260	5.5	16.1	24.0	4.2	Nil	110	249
7.20	32.3	37.75	252	6.1	36.6	62.8	3.8	Nil	285	114
7.45	27.9	22.76	174	17.3	44.9	72.0	4.8	Nil	224	104

Source: Chhabra (1996)

Due to the presence of excess salts saline soils remain flocculated and their hydraulic conductivity is equal to or slightly more than that of similar non-saline soils. Poor plant growth in saline soils results from high osmotic pressure of soil solution causing low physiological availability of water to the plant. Direct toxic effects of individual ions and complex interaction between sodium, calcium and magnesium lead to disturbed equilibrium of these ions and hampers plant's ability to absorb water and nutrients in required amounts.

Alkali soils: Alkali soils also known as sodic or solonetz are those which have high pH, high ESP and contain low concentration of soluble salts. Soils containing excessive salts of sodium carbonate and sodium bicarbonate and having sufficient exchangeable sodium to interfere with growth of most crops plants are called alkali. These soils have pHs > 8.5, ESP >15 and ECe < 4 dS/m. Chemical characteristics of a typical alkali soil of the IGPs of India are given in table 3 (Chhabra, 2008). Due to these characteristics, alkali soils are highly dispersed and have poor physical properties. Some times distinction is made between alkali and sodic soils especially for the Vertisols where the term sodic is preferred because in those soils in contrast to the alfisols, pH increases slowly with increase in ESP.

Table 3. Chemical characteristics of a typical alkali soil

pHs	ECe dS/m	SAR	Concentration of ions in the saturation extract meq/l							
			Na^+	K^+	Ca^{2+}	Mg^{2+}	HCO_3^-	CO_3^{-2}	Cl^-	SO_4^{-2}
10.6	22.34	96.0	248.3	0.70	0.20	0.40	141.6	136.2	6.60	3.90

Source: Chhabra (2008)

Management of alkali soils present difficulties due to their physical, chemical and hydrological properties, which affect the field preparation, irrigation practices, drainage and choice of crops. Most alkali soils exhibit impervious characteristics with nil to slow infiltration rates. The surface horizon with the highest pH, ESP, high content of sodium carbonate and platy structure is the limiting horizon. At about1m depth 30-60 per cent calcium carbonate concretion 30 to 70 cm thick layer is invariably found. Roots have to find their way through the nodules and, therefore, find less space for growth. Total nutrient reserve in this horizon is accordingly low. The underground water in most of the alkali soils is non-saline and non-sodic.

In alkali soils chemical environment is unfavorable for plant growth. Crop failures on alkali soil results largely from toxicity of sodium carbonate and bicarbonate and osmotic effects of other salts present. Soluble and exchangeable calcium and magnesium precipitate as calcium carbonate rendering soil deficient in these elements. Availability of trace elements except molybdenum and boron decreases due to reduced solubility. Weathering under alkaline conditions results in the release of high quantities of potassium, silica and iron. Alkali soils have flat, very gently sloping surface and negligible infiltration rates. Soil reclamation through application of amendments improves infiltration rates leading to reduced surface run-off losses and increased ground water recharge. Drainage of excess water accumulating on surface is accomplished through shallow ditches dug along the natural slope gradients. In those areas where water table prevails within 1 to 2 m depths and is not utilized noticeably for irrigation, vertical drainage through a net work of deep tube-wells is necessary.

Under field conditions, plant growth is adversely affected due to combination of many factors. The extent depends upon the amount of exchangeable sodium, pH, nature and stage of crop growth, environmental conditions and the overall management levels (Table 4).

Table 4. Exchangeable sodium percentage of soil and sodicity hazards to the plants

Approximate ESP	pHs	Sodicity hazards
< 15	8.0-8.2	Non-slight
15-35	8.2-8.4	Slight to moderate
34-50	8.4-8.6	Moderate to high
50-65	8.6-8.8	High to very high
> 65	>8.8	Extremely high

Source: Chhabra (1996)

Saline-alkali soils: Theoretically, soils having pH > 8.2, ESP > 15 and ECe > 4 dS/m are referred to as saline-alkali soils. Many a times saline-sodic soils are grouped with sodic soils because of several common properties and management approaches (Qadir *et al.,* 2007). However, they suffer from not only high sodium content but also high levels of neutral salts. When cropped directly i.e. without adopting any soil amelioration measures, plants in these soils sufferdue to adverse effects of high pH, high ESP/SAR and high concentration of soluble salts. Such soils are formed under a situation when irrigation is given with high RSC (Residual sodium carbonate) water leading to precipitation of soluble and exchangeable-Ca^{2+} of the soil as $CaCO_3$. This enables Na^+ to enter exchange complex leading to increase in ESP resulting in rise in pH, especially of the surface layers. Mostly such soils occur in the geographically transitional zone of alkali and saline soils. Further, soils having high pH and high ECe as in alkali soils formed *in-situ*; soils having high pH and high ECe formed due to use of high RSC waters and soils with moderate pH but high SAR and high EC formed due to shallow saline water table high in SAR are all classified as saline-alkali soils. All the three categories need different reclamation measures which create a lot of confusion in the mind of soil survey officials, planners and developing authorities.

These soils are mostly confined to regions with around 550 mm mean annual rainfall in the form of a narrow band separating the alkali and saline soils. These have a preponderance of neutral salts but contain sizeable quantities of sodium carbonates and bicarbonates. These generally have sandy to loamy textural gradation and may have a calcic or a petrocalcic horizon in the substratum thereby resulting in reduced water intake.

Results of several field experiments in India and Iraq on sandy loam soils suggest a limited value of amendments in the reclamation of saline-sodic soils. However, significant responses to application of amendments can be expected in soil with inherent low permeability. When salinity and sodicity occur together, limited evidence suggests that the effect of the two factors on plant growth is non-additive and non-interactive, and primarily salinity effects limit the growth. Many saline-sodic soils contain soluble carbonates besides the excess of neutral salts. Such soils manifest alkali soil properties. It is, therefore prudent that the saline-sodic soils that do not contain soluble carbonates be grouped and managed as saline soils and the rest of them grouped and managed like alkali soils.

***Saline vertisols*:** Vertisols and associated soils cover nearly 257 million ha of the earth's surface (Dudal and Bramao, 1965) of which about 72 million ha occur in India. This shows that nearly 22% of total geographical area of the country is occupied by Vertisols. In the central part of India known as the Deccan Plateau, the soils are derived from weathered basalts mixed to some extent with detritus from other rocks. In other areas, particularly in the south, the soils are also derived from basic metamorphic rocks and calcareous clays. Similarly, in the western region, these are derived from marine alluvium that account for nearly 19.6 million ha. These soils are generally deep to very deep and heavy textured with clay content varying from 40-70%. Further, these are also low in organic carbon content, high in cation exchange capacity, slight to moderate in soil reaction and are generally calcareous in nature. Vertisols, when kept fallow during *kharif* season are exposed to soil erosion hazards. Their inherent physicochemical characteristics such as poor hydraulic conductivity, low infiltration rates, narrow workable moisture range, deep and wide cracks pose serious problems even at low salinity level. However, the Vertisols of Bara tract in Gujarat are generally very deep (150 to 200 cm), fine textured with clay content ranging from 45 to 68% with montmorillonite as dominant clay minerals. The soils exhibit high shrink and swell potential and develop wide cracks of 4-6 cm extending up to 100 cm depth. The soils are calcareous in nature having calcium carbonate ranging from 2 to 12% in the form of nodules, kankar and powdery form.

Reclamation and management of saline soils

Establishing a good crop stand in saline soils or with saline water irrigation is a challenging task. Unlike normal soils, the agronomic practices for crop production in saline water irrigated soils are different. An ideal package of cultural practices besides soil fertility and irrigation water management can ensure a good crop stand vis-à-vis good yield. No single management option in isolation can be fully effective in controlling the salinity of irrigated soils, but several practices interact with each other in an integrated manner but for better understanding each option should be understood separately.

Salt leaching: Reduction in crop yield depends on factors such as crop growth, the salt content of the soil, climatic conditions, etc. In extreme cases where the concentration of salts in the root zone is very high, crop growth may be entirely affected. To improve crop growth in such soils the excess salts must be removed from the root zone. Leaching is the most effective procedure for removing salts from the root zone. Leaching is accomplished by ponding fresh water on the soil surface and allowing it to infiltrate and it is effective only when the salty drainage water is discharged through drains out of the area under reclamation. Process may reduce salinity levels in the absence of artificial drains when there is sufficient natural drainage, i.e. the ponded water drains without raising the water table. Leaching should preferably be done when the soil moisture content is low and the groundwater table is deep. Leaching during the summer months is, as a rule, less effective because large quantities of water are lost by evaporation. The actual choice will however, depend on the availability of water and other considerations. Leaching requirement of a saline soil is usually dependent on depth of soil to be reclaimed, the initial salinity, type of salts and soils inherent characteristics such as texture, structure, permeability and infiltration (Rhoades, 1982). In some parts of India for example, leaching is best accomplished during the summer months because this is the only time when water table is deepest, soil is dry and large quantities of fresh water can be diverted for reclamation purposes. The term reclamation of saline soils refers to the methods used to remove soluble salts from the root zone. Methods commonly adopted or proposed to accomplish this include the following:

Drainage: Irrigation is the most effective means of stabilizing agricultural production in areas where the rainfall is either inadequate for meeting the crop requirements or the distribution is erratic. Smedema and Ochs (1998) highlighted the importance of drainage in arresting land degradation due to waterlogging and salinity in irrigated areas of arid zone. Before the introduction of large quantities of water through irrigation to an area, water balance between rainfall on the one hand and stream flow, groundwater table, evaporation and transpiration on the other exists. This balance is seriously disturbed when additional quantities of water are artificially supplied to grow agricultural crops, introducing additional factors of groundwater recharge by seepage from unlined canals, distributors and field channels, and from the irrigation water let on the fields more than the quantities actually utilized by the crops, causing rise in groundwater table. Once the groundwater table is close to the soil surface, its evaporation from the surface cause appreciable movement of groundwater and salts in the root zone.

Irrigation management: Irrigation should meet both water requirements of crops and the leaching to maintain a favourable salt balance in the root zone for optimum yield. Multi-quality irrigation water, methods of irrigation, frequency and quantity of water applied could be possible interventions. Successful operation of field application must supply water to the plants at the right time in appropriate quantity and quality with minimum waste and at the place where the plants need it for optimum growth. On-farm water management technology includes proper land leveling and shaping, efficient design and layout of irrigation methods, scientific scheduling of irrigation under both adequate and deficient water supply conditions, irrigation management under high water table depths and crop planning for optimum water use.

Pre-sowing irrigation: In saline soils or saline water irrigation, salt often accumulate in the top few centimeters of the soil during non-crop periods. Where high water table exists, fallowing may result in excessive salt accumulation in the root zone particularly in arid and semi-arid regions. Under these conditions both germination and yields are adversely affected. A heavy pre-sowing irrigation to leach the accumulated salts from the root zone is very useful and essential to improve germination and early growth. Wherever available, pre-sowing irrigation should be given with good quality canal water.

Irrigation intervals: During intervening periods between two irrigation cycles, crop evapo-transpiration reduces soil water, matric potential and solute potential at a rate which is governed by rate of ET and soil moisture characteristics causing variable effects on crop yields. These effects are more pronounced and set very aggressively in saline environments, so, irrigation in saline soils should be more frequent to reduce the cumulative water deficits. As the soil progressively dries out due to evapo-transpirational losses, the concentration of salts in the soil solution and, therefore, its osmotic pressure increases making the soil water increasingly difficult to be absorbed by the plants. Thus infrequent irrigation aggravates salinity effects on growth. More frequent irrigations, by keeping the soil at higher soil moisture content prevent the concentration of salts in the soil solution and tend to minimize the adverse effects of salts in the soil. For these reasons crops grown in saline soils must be irrigated more frequently compared to crops grown under non-saline conditions so that the plants are not subjected to excessively high soil moisture stresses due to combined influence of excess salts and low soil water contents.

Irrigation method: Sprinkler irrigation is an ideal method for irrigating frequently and with small quantities of water at a time which results in leaching of soluble salts more efficiently. The trickle or drip irrigation method is suitable for perennial or seasonal row crops; it has been found particularly useful when irrigating with water of high salinity. The method has the advantage that it keeps the soil moisture continuously high in the root zone, therefore maintaining a low salt level. Although sprinkler and trickle irrigation methods are highly efficient, both from the view of water use and salinity control, their high initial costs often preclude their use.

***Mulching*:** During periods of high evapo-transpiration between the two irrigations and during fallow periods there is a tendency for the leached salts to return to the soil surface. Soil salinization is particularly high when the water table is shallow and the salinity of groundwater is high. Practices that reduce evaporation from the soil surface and/or encourage downward flux of soil water can help to control root zone salinity. Studies on soil salinity changes under bare fallow and straw mulch on fallow have shown that mulch reduced salinity on bare fallow soil. Under straw mulch a significant reduction in soil salinity results in favourable edaphic environments and thus increased crop yields. Periodic sprinkling of mulched soils result in greater salt removal and higher leaching efficiency than flooding or sprinkling of bare soil.

Crops and cropping sequences: Since crops vary in salinity tolerance from very sensitive or semi-tolerant to tolerant (Table 5), so the selection of crops and cropping sequences for saline water irrigation and saline soils is of paramount importance. These inter and intra-generic variations in salt tolerance of plants can be exploited for selecting crops or varieties that can produce satisfactory yields under given salt stress.

Table 5. Soil salinity classes and effects on crop plants

Class	ECe (dS/m)	Effect on Crop Plants
Non saline	0 - 2	Salinity effects negligible
Slightly saline	2 - 4	Yields of sensitive crops may be restricted
Moderately saline	4 - 8	Yields of many crops are restricted
Strongly saline	8 - 16	Only tolerant crops yield satisfactorily
Very strongly saline	> 16	Only few very tolerant crops yield satisfactorily

Reclamation and management of alkali (sodic) soils

The chemical amendment based technology has been developed to reclaim the alkali/sodic soils (Singh *et al.,* 2003). Various components of this technology includes; field leveling, bunding, soil sampling to know the sodicity status for working out amendment dose, application of gypsum/pyrite as per requirement of the soil and its mixing in upper 10 cm of soil, keeping water ponding for 5-7 days, following rice-wheat rotation for the first 3 to 4 years and growing sesbania during summer as green manure crop after wheat harvest in April. By adopting this technology about 1.3 m ha area has been reclaimed in the states of Punjab, Haryana and western U.P. (Singh *et al.,* 2007). The reclaimed area is contributing 8-10 m tonnes additional food grains to the national food grains pool. In addition to food production and employment generation, the reclamation programmes have helped in minimizing flood hazards, increasing ground water recharge, reducing incidence of malaria and water borne diseases, growth in agro-based and auxiliary industries and increasing forest cover. Some of the constraints being experienced in further adoption of this technology includes; increased cost of amendments and withdrawal of subsidy, requirement of repeat application of gypsum in areas with high residual sodium carbonate waters or with shallow brackish water.

Crops and cropping pattern: Crops differ in their tolerance to soil sodicity (Abrol and Bhumbla, 1979). The relative tolerance of crops and grasses to soil ESP is given in table 6. In general, cereal like rice is more tolerant than legumes as they require less Ca, availability of which is a limiting factor in alkali soils. Crops which can withstand excess moisture conditions are generally more tolerant to alkali conditions. Among the cultivated crops, rice is most tolerant to soil sodicity. It can withstand an ESP of 50 without any significant reduction in yield. It is followed by sugar beet and teosinte. Crops like wheat, barley and oats etc. are moderately tolerant. Legumes like mash, lentil, chickpea and pea etc. are very sensitive and their yield decreases significantly even when the soil ESP is <15. Sesbania is an exception among the leguminous crops as it can grow at ESP up to 50 without any reduction in yield. Due to this it is an excellent crop for green manuring in alkali soils. Some of the natural grasses like Karnal grass and Rhodes grass are very tolerant to soil sodicity, and in fact grow normally under high alkali conditions. Karnal grass likes both waterlogging and high ESP in its growing environment.

Salt tolerant varieties: A sizable part of the salt-affected area in India is in possession of small and marginal farmers who are themselves poor. Under such situations, chemical amendments based reclamation technology without government subsidy is not feasible. Development of salt tolerant varieties of important field crops is an option of great promise for utilization of such areas. As most of these varieties give significant yields with or without little application of chemical amendments. Several varieties of field crops like rice, wheat and mustard have been developed which have potential to yield reasonable economic return both in high pH alkali soils and also in saline soils (Singh and Sharma, 2006). In case of rice, the most promising varieties include CSR10, CSR13, CSR19, CSR23, CSR27, CSR30, CSR36 and CSR43. These varieties can be cultivated in soils with pH and EC ranging from 9.4 to 9.8 and 6-11 dS/m, respectively. For wheat, KRL 210, KRL19, KRL1-4, Raj 3077 and WH157 are suitable for soils with pH and EC range from 8.8 to 9.3 and 6-10 dS/m. Pusa Bold, Varuna, CS52 and CS54 are salt tolerant mustard varieties.

Table 6. Relative tolerance of crops and grasses to soil ESP

Tolerant (ESP: 35-50)	Moderately tolerant (ESP: 15-35)	Sensitive (ESP: < 15)
Karnal grass (*Leptochloa-fusca*)	Wheat (*Triticum aestivum*)	Mash (*Phaseolus mungo*)
Rhodes grass (*Chloris gayana*)	Barley (*Hordeum vulgare*)	Chickpea (*Cicer arietinum*)
Para grass (*Brachiaria mutica*)	Oat (*Avena sativa*)	Lentil (*Lens esculenta*)
Bermuda grass (*Cynodon dactylon*)	Shaftal (*Trifolium resupinatum*)	Soybean (*Glycine max*)
Rice (*Oryza sativa*)	Lucerne (*Medicago sativa*)	Groundnut (*Arachis hypogea*)
Dhaincha (*Sesbania aculeata*)	Turnip (*Brassica rapa*)	Sesamum (*Sesamum* sp.)
Sugarbeet (*Beta vulgaris*)	Sunflower (*Helianthus annus*)	Mung (*Phaseolus aureus*)
Teosinte (*Euchlaena maxicana*)	Safflower (*Carthamus tinctorius*)	Pea (*Pisum sativum*)
	Berseen (*Trifolium alexandrinum*)	Cowpea (*Vigna unguiculata*)
	Linseed (*Linum usitatissimum*)	Maize (*Zea mays*)
	Onion (*Allium cepa*)	Cotton (*Gossypium hirsutum*)
	Gralic (*Allium sativum*)	
	Pearl millet (*Pennisetum typhoides*)	

Source: Abrol and Bhumbla (1979)

Alternate land use systems: Reclamation of salt affected soils in village community lands, lands along the roads, railway tracks and other government lands, offer ample opportunities to raise salt tolerant trees, bushes and grasses to produce fuel wood, fodder and energy. An alternate technology of raising multipurpose forest tree plantation, fruit trees, agro-forestry systems and other high value medicinal and aromatic crops seems quite feasible. Several salt tolerant forest and fruit species have been identified which can be grown in highly sodic and saline soils (Singh *et al.,* 1994; Minhas *et al.,* 1997; Tomar *et al.,* 2003). The promising forest species include *Prosopis juliflora, Acacia nilotica, Tamarix articulata* and *Casuarina equisetifolia*. Long term field trials in a highly sodic soil revealed that *Emblica officinalis, Psydium guajava* and *Carissa carandus* proved highly promising in terms of growth performance and fruit production (Singh *et al.,* 1997).

Agroforestry and grasses: Several grass species have mechanism to tolerate high salt concentration in the root zone soil. Some of these highly tolerant grasses either exclude the absorption of salts from the soil and/or deposit the absorbed/translocated salts at points within the plant system which do not allow them to interfere in metabolic processes. Grasses like *Leptochloa fusca* has the potential to yield high biomass even at pH level of 10.4 and more. Similarly, *Bricharia mutica* is another salt tolerant grass, which can be grown even under prolonged waterlogged and salt situation. Several experiments have been conducted at CSSRI, Karnal and elsewhere to study the performance of these grasses in association with salt tolerant trees like *Prosopis juliflora* and *Acacia nilotica* in a unified agro-forestry system (Singh, 1995; Singh *et al.*, 2017). The results of the experiments indicated that sodic soils can be reclaimed by growing *Prosopis juliflora* and *Leptochloa fusca* for five years. Also *Jatropha* based intercropping helps in reducing soil pH of sodic soil in IGPs thereby rehabilitating these degraded lands (Singh *et al.*, 2016a).

Medicinal and aromatic crops: A number of medicinal and aromatic crops have been screened for salinity and sodicity tolerance in India. Crops like Isabgol (*Plantago ovata*) and Matricaria can be successfully cultivated in soils having pH of 9.5 and EC between 8-10 dS/m (Dagar *et al.*, 2004; Dagar *et al.*, 2006). Similarly, dill (*Anethum graveolens*), a spice crop and Salvadora, a non-edible oil tree can be grown in salt affected Vertisols very successfully (Rao *et al.*, 2000; Rao *et al.*, 2003). Industrial species like *Euphorbia* and mulethi (*Glycyrrhiza glabra*) also have good scope for cultivation in salty environments. Halophyte plant species are very useful for bio-saline agro-forestry and phyto-remediation of coastal saline lands (Arora *et al.*, 2013).

On-farm development: For successful adoption of reclamation technology the farm should be properly developed. The on–farm development activities include land shaping (leveling), bunding and making provisions for irrigation and drainage. To ensure proper water management and uniform leaching of salts, the field should be leveled properly. To avoid major earthwork, the big fields should be divided into small parcels and leveled. Drastic removal of the surface soil will expose the subsoil containing $CaCO_3$, which can pose difficulties in reclamation, and cropping of the area. Best results are achieved when the moisture level in the field is 12 to16% on dry weight basis. Strong bunds are essential to store rainwater for leaching and irrigation in the field and to prevent its loss through surface runoff. These also prevent the entry of salt rich water from the adjoining areas to avoid resodication. Drainage of excess water may be needed under high rainfall conditions to prevent damage to young rice seedlings and to arable crops like wheat. Provision of good quality irrigation water is a must for successful reclamation of alkali soils. For this a shallow cavity tube well should be installed in the area or supply of canal water may be ensured.

***Application of amendments*:** To have successful crops, ESP of the soil must be lowered, which can be achieved by application of amendments. Amendments are materials which i) directly supply Ca for the replacement of exchangeable Na or ii) furnish Ca indirectly by dissolving calcite, natively found in the alkali soils due to their acidulating effect and or iii) dissolve native $CaCO_3$ by increasing the partial pressure of CO_2. The Ca so mobilized is used to replace Na from the exchange complex and this reclaims the alkali soil. Chemical amendments like gypsum ($CaSO_4.2H_2O$) and calcium chloride ($CaCl_2.2H_2O$) directly supply Ca to replace exchangeable Na there by lowering ESP and pH. Mine gypsum is the most commonly used amendment for alkali soil reclamation. The agricultural grade gypsum is only 75 to 80 % pure. The amount of amendment needed for reclamation depends upon the amount of exchangeable sodium to be replaced, which in turn is governed by the amount of sodium adsorbed on the soil (ESP & CEC), sodicity tolerance and the rooting depth of the crop to be raised. Optimizing gypsum requirement and following total protocol for reclamation with surely rehabilitate the degraded soils (Singh *et al.,* 2016b).

Alkali soils once reclaimed do not need repeated application of amendments provided these are continuously cropped. This is due to the fact that these soils contain high amounts of $CaCO_3$, which together with Ca added through fertilizers like CAN (calcium ammonium nitrate) and SSP (single super phosphate), and irrigation water is sufficient to meet the Ca needs of the plants and to keep Na out of the exchange complex.

Addition of crop residues and other organic materials in the soil is beneficial as these help to improve and maintain soil structure, supply needed plant nutrients, prevent soil erosion and hastens reclamation of alkali soils. Most commonly used organic amendments are materials like straw, rice husk, poultry droppings, groundnut and sunflower hulls, farm yard manure (FYM), compost, green manure, tree leaves, saw dust etc. Various organic amendments have been found to ameliorate the saline and sodic soils and enhance crop productivity of these degraded lands apart from improving the physico-chemical and microbial properties of the soil (Gupta and Arora, 2016; Arora *et al.,* 2016).

Industrial byproducts like phospho-gypsum, pressmud, molasses, acid wash and effluents from milk plants may be used to provide soluble Ca directly or indirectly by dissolving soil lime, for reclamation purposes. As these materials can be cheap and locally available, their use should be encouraged. However, care should be taken not to introduce toxic elements like F, which may be present in large amounts in products like phospho-gypsum (Chhabra *et al.,* 1980).

Microbial approach for bio-remediation

Both physical and chemical methods for saline/sodic soil reclamation are not cost- effective. The biotic approach 'plant-microbe interaction' to overcome salt stress has recently received a considerable attention from many workers throughout the world. Plant-microbe interaction is beneficial association between plants and microorganisms and also a more efficient method used for the reclamation of salt affected soils (Arora *et al.,* 2014b). Bacteria are the most commonly used microbes in this technique. Rhizosphere bacteria improve the uptake of nutrients by plants and /or produce plant growth promoting compounds and regenerate the quality of soil (Arora and Vanza, 2017). These plant growth promoting bacteria can directly or indirectly affect plant growth. Indirect plant growth promotion includes the prevention of the deleterious effects of phytopathogenic organisms by inducing cell wall structural modifications, biochemical and physiological changes leading to the synthesis of proteins and chemicals involved in plant defense mechanisms.

The biotic approach 'plant-microbe interaction' to overcome salinity problems has recently received a considerable attention throughout the world. The halophilic microbes have potential for bio-remediation of salt dominant soils. Native halotolerant bacteria were isolated from the rhizosphere of halophyte plants. The plant growth promoting halophilic bacteria having salt tolerance upto 15% NaCl concentrations were selected for bio-remediation. Halophilic bacteria *Planococcus maritimus* (CSSRO2) was found to be more efficient in reducing sodium concentration from 1,12,230 ppm to 1,00,190 ppm at 24 hour while bacterial strain *Nesterenkonia alba* (CSSRY2) reduced 92,730 ppm of sodium at 24 hrs (Fig. 1). The halophilic bacteria CSSRY1 and CSSRO2 have high potential to remove sodium ions from soil and increase metabolic and enzymatic activities thereby reducing salt stress and thereby enhancing flow of solute to enhance plant growth.

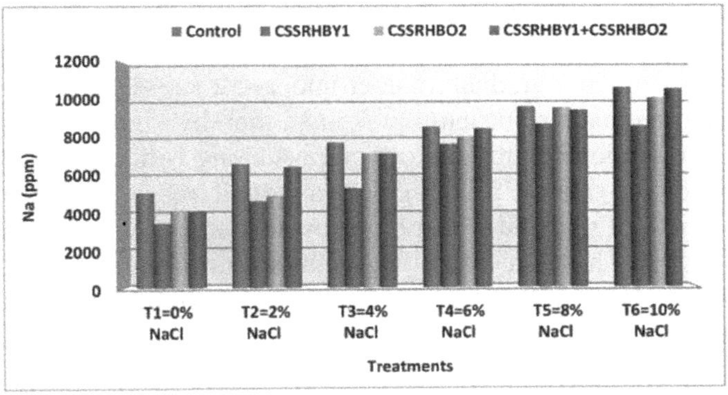

Fig. 1. Sodium removal efficacy of halophilic soil bacteria from saline soil

The phosphate solubilising bacteria and *Rhizobium* strains have been isolated and found to grow in soils with 10% salt (EC, 156 dS/m). The pure culture of the strains were propagated in presence of biological amendments to 10^8 colony forming units and lower down the Na content of soil. It was observed that there was 10-12 per cent increase in yield attributes and yield of wheat at 6% NaCl as compared to 2% NaCl (Arora *et al.,* 2014a). Introduction of halophilic plant growth promoting bacteria not only bio-ameliorates the saline and sodic soils but also enhances bio-chemical activity which enhances crop productivity under salt stress conditions (Arora *et al.,* 2016a,b). The applications of halophilic bacteria include recovery of saline soil by directly supporting the growth of vegetation thus indirectly increasing crop yields in saline soil.

Further, for bioremediation of sodic soil of IGPs, halophilic plant growth promoting bacteria were isolated from native sodic soil and screened for salt tolerance and plant growth promoting traits. The efficient halophiles among N-fixers and P-solubilizers were tested for crop production at different soil pH and were prepared as liquid bio-formulations 'Halo-Azo' and 'Halo-PSB' (Arora *et al*., 2016a,b). These bio-formulations can be used either used for seed/seedling treatment or soil application.

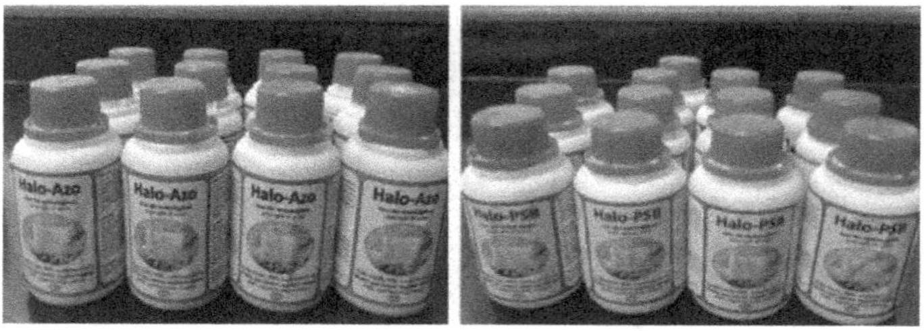

Fig. 2. Liquid Bio-formulations: *Halo-Azo* **and** *Halo-PSB*

Application of these bio-formulations helps to supplement plant nutrients like nitrogen and phosphorous through their activities in the rhizosphere and make available to plants in a gradual manner under salt stress. It ensures better root development and better nutrient uptake and thereby vigorous crop growth. Application of organic manures and/or compost along with liquid bio-formulations 'Halo-Azo' and 'Halo-PSB' (Fig. 2) was found to be effective to significantly enhance yield of rice and wheat crops by an average of 11.5% on sodic soils upto soil pH 9.8. These are also effective in mustard, vegetable and fodder crops under sodic stress. These formulations also help in maintenance of soil health, minimize environmental pollution and cut down on the use of chemicals to some extent. They are affordable for most of the farmers who are small and marginal. Bio-formulations are also ideal input for reducing the cost of cultivation and for promoting organic farming in salt affected soils.

Conclusion

The problematic soils need scientific management for optimizing crop yields. Before any reclamation program can be initiated, it is imperative to assess the kind and degree of the problem. There is need for visual identification along with sampling of soil/water for proper diagnosis and suggesting effective remedial measure. Various options suggested for crop production in salt stress environment aim at preventing the salt build up to the levels that limit productivity of soils and control salt balances in the soil-water system and minimizing the damaging effects of salts on crops at plant level or field scale. Soil, crop, climate and social factors are the deciding criteria for adoption of specific agronomic practices. Bio-remediation through halophilic plant growth promoting microbes has shown positive effect in management of salt affected soils and useful for successful application in agriculture.

References

Abrol, I.P. and Bhumbla, D.R. 1979. Crop responses to differential gypsum applications in a highly sodic soil and the tolerance of several crops to exchangeable sodium under field condition. *Soil Sci.* **127**:79-85.

Arora, S. 2013. Remote sensing and geographic information system approach for mapping and reclamation of salt-affected soils. In: Singh *et al.* (eds) *Participatory approaches to enhance farm productivity of salt affected soils*. IRRI-CSSRI, RRS, Lucknow, pp. 9-32.

Arora, S., Bhuva, C., Solanki, R.B. and Rao, G.G. 2013. Halophytes for bio-saline agro-forestry and phyto-remediation of coastal saline lands. *J. of Soil and Water Cons. India.* **12**(3): 252-259.

Arora, S., Patel, P., Vanza, M. and Rao, G.G. 2014. Isolation and characterization of endophytic bacteria colonizing halophyte and other salt tolerant plant species from coastal Gujarat. *African J. of Microbiol. Res.* **8**(17): 1779-1788.

Arora, S., Singh Y.P., Singh A.K., Mishra V.K., and Sharma, D.K. 2016a. Effect of organic and inorganic amendments in combination with halophilic bacteria on productivity of rice-wheat in sodic soils. *Bhartiya Krishi Anusandhan Patrika* **31**(3): 165-170.

Arora, S., Singh, Y.P., Vanza, M. and Sahni, D. 2016b. Bioremediation of saline and sodic soils through halophilic bacteria to enhance agricultural production. *J. of Soil and Water Cons. India* **15**(4): 302-305.

Arora S., Vanza, M. 2017. Microbial approach for bioremediation of saline and sodic soils. In: Arora S *et al.* (eds.), Bioremediation of Salt Affected Soils: An Indian Perspective, Springer International Publishing, Switzerland, pp. 87-100

Arora, S., Vanza, M., Mehta, R., Bhuva, C. and Patel, P. 2014b. Halophilic microbes for bio-remediation of salt affected soils. *African J. of Microbiol. Res.* **8** (33): 3070-3078.

Chhabra, R. 1996. *Soil Salinity and Water Quality*. Oxford & IBH Publishers. 282.

Chhabra, R. 2008. Recent advances in diagnosis of salt affected soils. In: Diagnosis and management of poor quality water and salt affected soils (eds. Lal *et al.*), CSSRI, Karnal, pp. 15-20.

Chhabra, R. and Abrol, I.P. 1977. Reclaiming effect of rice grown in sodic soil. *Soil Sci.***124**: 49-55.

Chhabra, R., Abrol, I.P., Dargan, K.S., and Gaul, B.L. 1980. Save on phosphatic fertilizers in the initial years of alkali soils reclamation. *Indian Farming*. **30**(3): 13-15.

Dagar, J.C., Tomar, O.S. and Kumar, Y. 2006. Cultivation of medicinal isabgol (*Plantago ovata*) in different alkali soils in semi-arid regions of northern India. *Land Degradation and Dev.* **17**: 275-283.

Dagar, J.C., Tomar, O.S., Kumar, Y and Yadav, R.K. 2004. Growing three aromatic grasses in different alkali soils in semi-arid regions of northern India. *Land Degradation and Dev* **15**: 143-151.

Dudal, R. and Bramao, D.L. 1965. Dark clay soils of tropical and subtropical regions. FAO Agricultural Development Paper No.8 FAO (Food and Agriculture Organization), Rome.

Dudal, R. and Purnell, M.F. 1986. Land Resources: Salt-affected soils. *Recl. and Revegetation Res*. **5**:1-9.

Gupta, R.D. and Arora, S. 2016. Salt affected soils in Jammu and Kashmir: Their management for enhancing productivity. *J. of Soil and Water Cons. India* **15**(3): 199-204.

Jalali, V.K., Pareek, N., Sharma, V., Arora, S. and Bhat, A.K. 2005. Relative efficiency of various doses of gypsum on reclamation of deteriorated sodic soil. *Environment and Ecology* **23**(2): 459-461.

Jalali, V.K., Pareek, N., Sharma, V., Arora, S. and Bhat, A.K. 2006. Characterization of salt affected soils of Ravi-Tawi command area of Jammu region. *Environment and Ecology* 24S(1): **181**-183.

Kanwar, J.S. and Bhumbla, D.R. 1969. Physico chemical characteristics of sodic soils of Punjab and Haryanaand their amelioration by the use of gypsum. *Agrokem.Talajt*, **18**:315-320.

Minhas, P.S., Singh, Y.P., Tomar, O.S., Gupta, R.K. and Gupta, R.K. 1997. Saline water irrigation for the establishment of furrow planted trees in north western India. *Agroforestry Systems*, **35**: 177-186.

NRSA and Associates. 1996. Mapping salt affected soils of India, 1:250,000 mapsheets, Legend. NRSA, Hyderabad.

Qadir, M., Oster, J.D., Schubert, S., Noble, A.D. and Sahrawat, K.L. 2007. Phytoremediation of sodic and saline-sodic soils. *Adv. in Agron.* **96**:197–247.

Rao, G.G., Nayak, A.K. and Chinchmalatpure, A.R. 2000. Dill (*Anethumgraveolens*): A potential crop for salt affected black soils. CSSRI Technical monograph 1, CSSRI, RRS, Anand.

Rao, G.G., Nayak, A.K. and Chinchmalatpure, A.R. 2003. Salvadorapersica: A life support of salt affected black soils. Technical bulletin 1/2003, CSSRI, RRS, Bharuch.

Rhoades, J.D. 1982. Reclamation and management of salt-affected soils after drainage. In *Proceedings of First Annual Western Provincial Conference, Soil Salinity*. Lethbridge, Alberta. 29 Nov. - 2 Dec. 1982 Pages 125-197

Sharma, R.C. and Mondal, A.K. 2006. Mapping of soil salinity and sodicity using digital image analysis and GIS in irrigated lands of the Indo-Gangetic Plain. *Agropedology* 16:71–76.

Sharma, V., Arora, S. and Jalali, V.K. 2012. Emergence of sodic soils under the Ravi-Tawi canal irrigation system of Jammu, India. *J. of the Soil and Water Cons. India* **11**(1): 3-6.

Singh, K.N. and Sharma, P.C. 2006. Salt tolerant varieties released for saline and alkali soils. Central Soil Salinity Research Institute, Karnal.

Singh, G., Abrol, I.P. and Cheema, S.S. 1994. Agroforestry techniques for the rehabilitation of salt lands. *Land Degradation and Rehabilitation* **5**: 223-242.

Singh, G., Dagar, J.C. and Singh, N.T. 1997. Growing fruit trees in highly alkali soils–a case study. *Land Degradation and Rehabilitation* **8**: 257-268.

Singh, G., Gupta, S.K., Sharma, D.P. and Tyagi, N.K. 2003. Reclamation and management of waterlogged salt affected soils. Technical Brochure, DOLR, MORD 02/2003, Department of Land Resources, Min. Rural Development, Govt. of India, p. 36.

Singh, G., Sharma, P.C., Ambast, S.K., Kamra, S.K. and Khosla, B.K. 2007. CSSRI: A Journey to Excellence (1969-2006), Central Soil Salinity Research Institute, Karnal, p-156.

Singh, N.T. 1992. Salt-affected soils in India. In: Land and Soil. T.N. Khoshoo and B.L. Deekshatulu (Ed).HarAnand Publications, New Delhi, India, pp. 65-100.

Singh, N.T. 1994. Land degradation and remedial measures with reference to salinity, alkalinity, waterlogging and acidity. In: Natural Resources Management for Sustainable Agriculture and Environment (ed. D.L. Deb), Angkor Publishers, New Delhi, pp. 109-124.

Singh, R.P., Setia, R., Verma, V.K., Arora, S., Kumar, P. and Pateriya, B. 2017. Satellite remote sensing of salt-affected soils: Potential and limitations. *J. of Soil and Water Cons., India* 16(2): 97-107.

Singh, Y.P., Singh, R., Sharma, D.K., Mishra, V.K. and Arora, S. 2016a Optimizing gypsum applications for amelioration of sodic soils to enhance growth, yield and quality of rice (*Oryza sativa* L.). *J. of the Indian Soc. of Soil Sci.* **64**(1): 33-40.

Singh, Y.P., Singh, G.B., Sharma, D.K. and Arora, S. 2017. Evaluation of different agro-forestry systems for their effect on alkali soils. *J. of the Indian Soc. of Soil Sci.* **65**(1): 42-47.

Singh, Y.P., Mishra, V.K., Sharma, D.K., Singh, G., Arora, S., Dixit, H. and Cerda, A. 2016b. Harnessing productivity potential and rehabilitation of degraded sodic lands through Jatropha based intercropping systems. *Agril. Ecosys. and Env.* **233**: 121-129.

Smedema, L.K. and Ochs, W.J. 1998. Needs and prospects for improved drainage in developing countries. *Irri. and Dranag. Sys.* **12**:359–369.

Szabolcs, I. 1979. Review of research on salt affected soils. Natural Resource Research, IS, UNESCO, Paris.

Tomar, O.S., Minhas, P.S., Sharma, V.K., Singh, Y.P. and Gupta R.K. 2003. Performance of 31 tree species and soil conditions in a plantation established with saline irrigation. *Fors. Ecol. and Mangt.* **177**:333-346.

15

Physio-molecular Mechanisms of Drought Tolerance in Crop

Shambhoo Prasad

Introduction

Drought is the most important abiotic factor that adversely affect growth and crop production. Drought stress can altered the normal physiological processes that influence one or a combination of biological factors for yield and yield attributing traits (Ashraf, 2010). The abnormal metabolism due to stress may reduce plant growth (Claves *et al.,* 2002). Production is limited by environmental stresses, according to different scholar estimates, only 10 per cent of the world's arable land is free from stress, in general, a major factor in the difference between yield and potential performance, environmental stresses. Drought is one of the most common environmental stresses that almost 25 per cent of agricultural lands for agricultural farm products in the world are limited (Loresto, 1976; Chaves, 2002). Drought occurrence in India is also very frequent. Drought occurred in past 1967, 1968, 1969, 1972, 1974, 1979, 1987, 2002, 2009 and had considerable impact on food grain production. For example, the drought of 1966-67 reduced overall food grains production by 19%. The drought of 1972-73 reduced the food grains production from 108.95 million tonnes to 95 million tonnes, causing a loss of about $ 400 million $. The 1987 drought in India damaged 58.6 million ha of cropped area affecting 285 million people (Ray *et al*., 2015). Drought in 2002 has reduced food grains production to 174 million tonnes from 212 million tonnes resulting in decline of 3.2% GDP (Rathore *et al*., 2009; Ray *et al*., 2015).

The water deficit reduces crop yield due to decrease in photosynthetic area, decreased radiation use efficiency and harvest index (Earl and Davis, 2003). Drought stress affect the water balance and plant metabolisms. Plants under drought conditions use various changes to tolerate stress conditions and increase drought tolerance which includes changes in whole plant, tissue, at physiological and molecular levels. Appearance of a single or a combination

**Corresponding author email: shambhoonduat@gmail.com*

of inherent changes determines the ability of the plant to stand under aridity conditions. Tolerant plant adjusts its water balance by osmotic adjustment, deep root systems and by transient leaf rolling.

Definition of drought

Drought is condition in which plant faces water shortage to meet the requirement of its growth and development. Drought categorize in two broad groups that is:

***Physical drought*:** It occurs due to lack of soil moisture and plant does not meet its water requirement for metabolic activity. It is related to a deficiency of precipitation over an extended period of time, usually for a season or more. This deficiency results in a water shortage for some activity, group, or environmental sector. Drought is also related to the timing of precipitation. Other climatic factors such as high temperature, high wind, and low relative humidity are often associated with drought.

Physiological drought: Water is available in soil but plant does not able to absorb due to physiological reasons. It may occur when the concentration of solutes in the soil water is equal to or higher than that in the root cells so water can not enter the plant by osmosis. This situation may be found in salt marshes and coastal mud flats. It also occurs during cold spells when there is an increase in the resistance of the root to water movement into the plant. Root resistance increases in response to low temperatures since the permeability of endodermal cells decreases rapidly below 5 °C.

The physical drought further divided into the following groups

Meteorological drought: Meteorological drought is defined on the basis of the degree of dryness, in comparison to a normal or average amount, and the duration of the dry period. Definitions of meteorological drought must be region-specific, since the atmospheric conditions that result in deficiencies of precipitation are highly region-specific. The variety of meteorological definitions in different countries illustrates why it is not possible to apply a definition of drought developed in one part of the world to another. For instance, the following definitions of drought have been given by the different country:

- United States (1942): < 2.5 mm of rainfall in 48 hours.
- Great Britain (1936): Fifteen consecutive days with daily precipitation < 0.25 mm.
- Libya (1964): When annual rainfall is < 180 mm.
- Bali (1964): A period of six days without rain.

- A drought year as a whole is defined by the Indian Meteorological Department (IMD) as a year in which, the overall rainfall deficiency is $< 10\%$ of the long period average value.

Data sets required to assess meteorological drought are daily rainfall information, temperature, humidity, wind velocity and pressure, and evaporation.

Agricultural drought: Agricultural drought links various characteristics of meteorological drought to agricultural impacts, focusing on precipitation shortages, differences between actual and potential evapo-transpiration, soil-water deficits, reduced groundwater or reservoir levels, and so on. Plant water demand depends on prevailing weather conditions, biological characteristics of the specific plant, its stage of growth, and the physical and biological properties of the soil. A good definition of agricultural drought should account for the susceptibility of crops during different stages of crop development. Deficient topsoil moisture at planting may hinder germination, leading to low plant populations per hectare and a reduction of yield. Data sets required to assess agricultural drought are soil texture, fertility and soil moisture, crop type and area, crop water requirements, pests and climate.

Hydrological drought: Hydrological drought refers to a persistently low discharge and/or volume of water in streams and reservoirs, lasting months or years. Hydrological drought is a natural phenomenon, but it may be exacerbated by human activities. Hydrological droughts are usually related to meteorological droughts, and their recurrence interval varies accordingly. Changes in land use and land degradation can affect the magnitude and frequency of hydrological droughts. Data sets required to assess hydrological drought are surface-water area and volume, surface runoff, stream flow measurements, infiltration, water-table fluctuations, and aquifer parameters.

Socio-economic drought: Socio-economic definitions of drought associate the supply and demand of some economic good with elements of meteorological, hydrological, and agricultural drought. It differs from the other types of drought in that its occurrence depends on the processes of supply and demand. The supply of many economic goods, such as water, forage, food grains, fish, and hydroelectric power, depends on the weather. Due to the natural variability of climate, water supply is ample in some years, but insufficient to meet human and environmental needs in other years. The demand for economic goods is increasing as a result of population growth and economic development. The supply may also increase because of improved production efficiency, technology, or the construction of reservoirs. When both supply and demand increase, the critical factor is their relative rate of change. Socio-economic drought is promoted when the demand for water for economic activities far exceeds the supply.

Drought effect on crop plants

In terms of plant physiology, dryness causes stress in plant growth 30-50%, yield reduction in drought stress due to low plant growth occurs as a result of the high evapotranspiration, temperature, high intensity of sunlight (Ghodsi *et al.*, 1998). High temperature caused by the drought stress increased the respiration, photosynthesis and enzyme activity in the plant. The drought under high light intensity, the light reaction of photosynthesis produces high free radicals of oxygen leading to plant death due to drought as well as oxidative stress. Absorption of nutrients from the upper soil horizon which is found in most foods gets reduced under drought (Bagheri, 2009). The prolong in drought conditions prolong accumulation of salts and ions in the upper layers of the soil around the root cause of osmotic stress and ion toxicity. The first response to stress is a biophysical response. In fact, with increasing drought stress, cell wall wizened and loose, with a decrease in cell volume, pressure decreases and the potential for the development of the cell, depending on the potential pressure decreases and growth is reduced. These factors reduces the size and number of leaves in plants (Bagheri, 2009). Leaf mesophyll cells become dehydrated due to drought. The amount of ABA increases in guard cell and fluxes out stored potassium and calcium from the cell and resultantly closure of stomata under drought stress condition (Bagheri, 2009).

Physio-molecular strategies of the plant for drought tolerance

Whole plants respond to drought through morphological, physiological, and metabolic modifications occurring in all plant organs. At the cellular level plant responses to water, the deficit may result from cell damage, whereas other responses may correspond to adaptive processes. Although a large number of drought-induced genes have been identified in a wide range of plant species, a molecular basis for plant tolerance to water stress remains far from being completely understood (Ingram and Bartels, 1996). The rapid translocation of ABA in shoots via xylem flux and the increase of ABA concentration in plant organs correlate with the major physiological changes that occur during plant response to drought (Zeevaart and Creelman, 1988).

Leaf rolling: Leaf rolling is one of the drought avoidance mechanisms to prevent water deficits during drought stress (O'Toole and Change, 1978). Loresto *et al.* (1976); Loresto and Chang (1981) have suggested leaf rolling as a criterion for scoring drought tolerance in tall and semi-dwarf rice cultivars. Varietal differences in leaf rolling and unrolling have been observed and it is correlated with internal water status of the leaf tissue. It is also related to the stomatal closure and decreases transpiration from rice leaves (O'Toole *et al.*, 1979). The degree of leaf rolling during water stress shows diurnal variation and is maximum at 1,300 hrs (O'Toole, 1982). Leaf rolling is induced by the loss of turgor and poor osmotic adjustment (Hsiao *et al.*, 1984). Delayed leaf rolling is an indication of turgor maintenance and a component of dehydration avoidance (Blum, 1988).

Rolling reduces effective leaf area and transpiration, and thus is a potentially useful drought avoidance mechanism in arid areas. Visual estimation of leaf rolling as an indicator of plant water status has been investigated in rice (*Oryza sativa* L.) (O' Toole and Moya 1978; Jones 1979) and wheat (*Triticum aestivum* L.) (Jones, 1979). In rice, both O'Toole and Moya (1978) and Jones (1979) found a strong correlation between visual leaf rolling score and leaf water potential. Greater rolling indicated lower leaf water potential. Leaf rolling score was poorly correlated with leaf water potential in wheat and rolling was confounded with leaf morphological differences (Jones, 1919).

Lemee (1954) concluded from a study of several kinds of grass of the Saharan region that rolling seems to occur only after extensive tissue water loss, and may be too late to save the leaf. This then raises the question of how to use leaf rolling as a selection criterion in drought-resistance improvement programs. Hsiao *et al.* (1984) concluded that selection for delayed leaf rolling in rice would increase yields through maintaining gas exchange and photosynthetic productivity.

Relative water content: Leaf relative water content (RWC) has been proposed as a more important indicator of water status than other water potential parameters under drought stress conditions (Lugojan and Ciulca, 2011). Schonfeld *et al.* (1988) showed that wheat cultivars having high RWC are more resistant against drought stress. Generally, it seems that osmoregulation is one of the main mechanisms maintaining turgor pressure in most of the plant species against water loss from soil and thus it causes the plant to continue water absorption and retain metabolic activities (Gunasekera and Berkowiz, 1992). Zlatev (2005) found that by exerting drought stress for 14 days and reaching soil potential to -0.9 Mpa, osmotic potential and turgor pressure in the first leaf of bean strongly was decreased. Ramos *et al.* (2003) stated that RWC of bean leaves under drought stress was significantly lesser than control. Lazacano-Ferrat and Lovat (1999) subjected bean plant to drought stress and after 10, 14 and 18 days of irrigation was withheld, they evaluated RWC of stem and found RWC was significantly lower compared with control plants. Gaballah *et al.* (2007) applied anti-transparent matters on two sesame cultivars named Gize 32 and Shanavil 3 and observed that by preventing water transpiration from leaves led to increasing in RWC in these cultivars.

Reactive oxygen species (ROS): Under drought and other abiotic stress, the plant faces oxidative stress due to high production of oxygen radicals that damage the cell membrane and cell homeostasis (Jacob and Haber, 1996). In order to cope with continuous ROS production plants have a battery of enzymatic and non-enzymatic antioxidants, which function as an extremely efficient cooperative system (Fig. 1). The major scavenging mechanisms include superoxide dismutase (SOD), enzymes and metabolites from the ascorbate-glutathione cycle and catalase (Noctor and Foyer, 1998). They are located throughout the different compartments of the plant cell, with the exception of catalase that is exclusively located in peroxisomes. SOD is the front-line enzyme in ROS attack since it rapidly scavenges superoxide, one of the first ROS to be produced, dismutating it to oxygen and H_2O_2. However, this reaction only converts one ROS to another, and H_2O_2 also needs to be destroyed since it promptly attacks thiol proteins. The major enzymatic cellular scavengers of H_2O_2 are catalase and ascorbate peroxidase (Noctor and Foyer, 1998; Noctor, 2000).

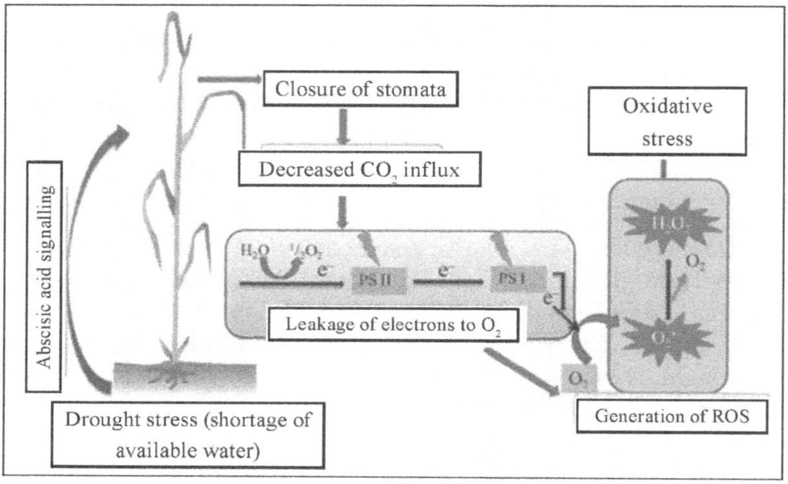

Fig. 1. Schematic representation of biosynthesis of ROS in plants under drought stress condition

Plants are natural producers of ROS and under normal conditions plants are continuously producing ROS. Unlike singlet oxygen and the hydroxyl radical whose production is kept at minimum levels (Asada and Takahashi, 1987; Foyer *et al.*, 1994) superoxide and H_2O_2 are synthesized at very high rates under normal conditions (Noctor and Foyer, 1998). One of the major cellular sites responsible for ROS production is the chloroplast (Foyer *et al.*, 1994). Plants can use the level of steady-state cellular ROS to monitor their intracellular level of stress (Miller, 2002). However, this steady-state level must be tightly regulated in order to prevent an oxidative burst by over accumulation of ROS, which would ultimately result in extensive cell damage and death (Dat *et al.*, 2000). Symptoms of oxidative damage (like lipid peroxidation) have been used to assess the increase in ROS production under drought stress. However, the lack of symptoms does not imply that increased ROS formation is not occurring. Instead, the lack of symptoms is likely to result on the concomitant increase in cellular antioxidant defenses.

Stomatal conductance: In the short term, stomatal closure to avoid excessive water losses occurs rapidly and is widely recognized as the primary effect of drought on carbon assimilation (Chaves *et al*., 2002). However, avoidance of excessive water losses occurs at the expense of reducing the CO_2 availability inside the leaf, and therefore limiting photosynthesis (Galmés *et al*., 2007a). Moreover, concomitant to stomatal closure, the leaf mesophyll conductance to CO_2 also decreases in the short term in response to water shortage, further worsening the CO_2 diffusion from the atmosphere to the sites of carboxylation (Flexas *et al*., 2008). On the opposite, metabolic impairment of the photosynthetic machinery only becomes relevant to limit CO_2 assimilation under moderate to severe drought stress intensity.

Reduced photosynthesis: Photosynthesis is an important process that is carried out in all living land plants. It involves the conversion of light energy into chemical energy. It is a primary process in plant productivity. The site of photosynthesis in plants is predominantly in the green leaf and the productivity of plants directly depends upon the chlorophyll bearing surface area, irradiance and their potential to utilize CO_2 (Baker, 1991; Chaves *et al*., 2002). Plant biomass production depends upon the amount of water use for growth as well as on water use efficiency (WUE). Productivity of crop plants may be increased by WUE and one of the major factors for enhanced WUE is net CO_2 assimilation rate (Glames *et al*., 2007a). Thus, the final biological or economical yield can be increased by increasing the net CO_2 assimilation rate (Nakashima and Suenaga, 2017).

Stomatal closure in response to drought is generally assumed to be the main cause of the drought. It decreases the photosynthetic activity, since stomatal closure increases CO_2 availability in the mesophyll. Moreover, the CO_2/O_2 ratio drops (Johanson *et al*., 1983; Hsiao *et al*., 1984) and photo respiration increases under drought (Noctor, 2000), whereas under severe drought, complete stomatal closure occurs, photosynthesis and photorespiration both decline, and thermal dissipation increases up to 90% of the total dissipation (Medrano *et al.*, 2000) and causes photo-inhibition or photodamage of PS-II. However, there is strong evidence that drought induced stomatal closure and also affects mesophyll metabolism. The stomatal limitation that causes inhibition in photosynthesis have been frequently described and reviewed by various scientists (Flexas *et al.*, 2002).

Stress hormones: Abscisic acid is believed to be the key hormone that mediates plant responses to adverse environmental stimuli since the level of ABA in plants usually increases during abiotic stress conditions, and elevated ABA can enhance plant adaptation to various abiotic stresses (Tuteja, 2007). Since, the first observation of ABA accumulation in drought stressed wheat (Wright,

1969), increased levels of endogenous ABA under drought stress conditions have been reported in many plant species which include maize (Wang *et al*., 2008), sorghum (Kannangara *et al*., 1983), rice (Henson, 1984), barley (Stewart and Voetberg, 1985; Thameur *et al*., 2011), soybean (Bensen *et al*., 1988), and wheat (Guoth *et al*., 2009).

Abscisic acid biosynthesis occurs in two places; it starts from plastids and ends in the cytosol. ABA in higher plants synthesized via the mevalonic acid-independent pathway also called indirect pathway. In this pathway, ABA is synthesized through cleavage of a C40 carotenoid precursor, followed by a two-step conversion of the intermediate xanthoxin to ABA via ABA aldehyde, which will be oxidized into ABA. Mutants defective in ABA biosynthesis have been isolated in many plant species from maize, tomato, tobacco, potato, barley and Arabidopsis (Xiong and Zhu, 2003). The primary mechanism of ABA biosynthesis pathway is shown in Fig. 2.

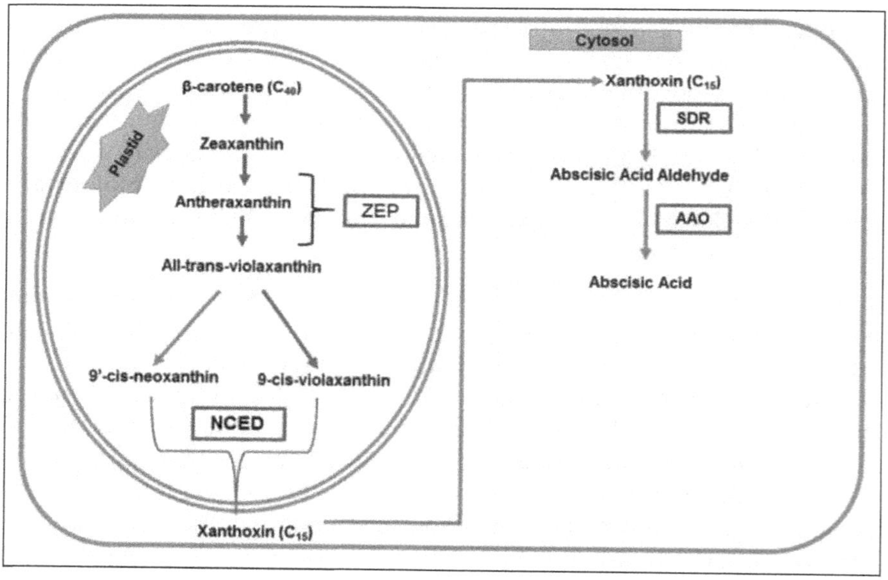

Fig. 2. Schematic representation of biosynthesis of ABA in plants (Modified from Mehrotra *et al.*, 2014)

Mechanism of drought tolerance adopted by plants

Understanding the mechanisms by which plants perceive environmental signals and transmit the signals to cellular machinery to activate adaptive responses is of fundamental importance to biology. Knowledge about stress signal transduction is also vital for continued development of rational breeding and transgenic strategies to improve stress tolerance in crops (Fig. 3).

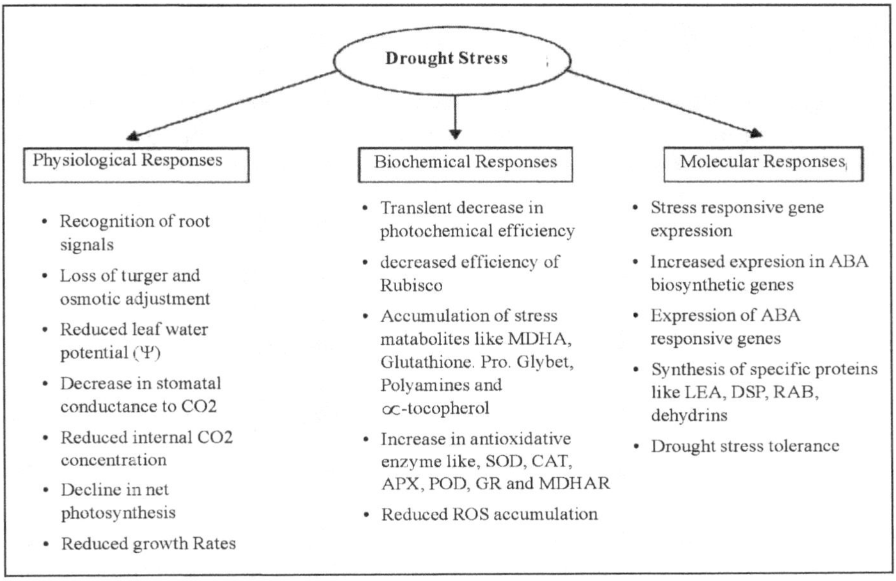

Fig. 3. Schematic representation of mechanism of drought tolerance in plants (Modified from Shao *et al.*, 2008)

Physiology of drought tolerance

Plants can be subjected to slowly developing water shortages (i.e. taking days, weeks, or months), or they may face short-term deficits of water (i.e. hours to days). In these situations, plants adapt by responding accordingly, minimizing water loss and maximizing water uptake (O'Toole and Cruz, 1980; Schonfeld *et al.,* 1988). Plants are more susceptible to drought stress during the reproductive stages of growth i.e., flowering and seed development. Therefore, the combination of short-term plus long-term responses allows plants to produce a few viable seeds. Some examples of short-term and long-term physiological responses include:

Short-term responses

- *In the leaf*: root-signal recognition, stomatal closure, decreased carbon assimilation.
- *In the stem*: inhibition of growth, hydraulic changes, signal transport, assimilation of transport
- *In the root*: cell-drought signaling, osmotic adjustment

Long-term responses

- *In the above-ground portion of the plant*: inhibition of shoot growth, reduced transpiration area, grain abortion, senescence, metabolic acclimation, osmotic adjustment, anthocyanin accumulation, carotenoid degradation, the intervention of osmoprotectants, ROS-scavenging enzymes.
- *In the below-ground portion of the plant:* Turgor maintenance, sustained root growth, increased root/shoot, increased absorption area.

Plants in naturally arid conditions retain large amounts of biomass due to drought tolerance and can be classified into four categories of adaptation:

- *Drought-escaping plants:* annuals that germinate and grow only during times of sufficient times of moisture to complete their life cycle.
- *Drought-evading plants:* non-succulent perennials which restrict their growth only to periods of moisture availability.
- *Drought-enduring plants:* also known as xerophytes, these evergreen shrubs have extensive root systems along with morphological and physiological adaptations which enable them to maintain growth even in times of extreme drought conditions.
- *Drought-resisting plants:* also known as succulent perennials, they have water stored in their leaves and stems for sparing uses.

The difference between drought avoidance and drought tolerance is given in table 1.

Table 1. Difference between drought avoidance and drought tolerance

Drought avoidance	Drought tolerance
Plants maintain favourable tissues water content.	Plants do not maintain favourable tissue water content.
It reduces photosynthesis and increase root development.	Better seed germination, seedling growth and photosynthesis.
In cereals, it operates during the vegetative phase.	In cereals, it operates during the reproductive phase.
Plants cannot withstand low tissue water content.	Plants can withstand low tissue water content.
It involves various morphological and anatomical features of the plant which reduce water loss through transpiration.	It generally involves those characters which support for better photosynthesis under drought conditions.

Structural adaptations: Many adaptations for dry conditions are structural, including the following:

- Adaptations of the stomata to reduce water loss, such as reduced numbers, sunken pits, waxy surfaces.

- Reduced number of leaves and their surface area.
- Water storage in succulent above-ground parts or water-filled tubers.
- Crassulacean acid metabolism (CAM metabolism) allows plants to get carbon dioxide at night and store malic acid during the day, allowing photosynthesis to take place with minimized water loss.
- Adaptations in the root system to increase water absorption.
- Trichomes (small hairs) on the leaves to absorb atmospheric water.

Molecular mechanisms of drought tolerance: In response to drought conditions, there is an alteration of gene expression, induced by or activated by transcription factors (TFs). These TFs bind to specific cis-elements to induce the expression of targeted stress-inducible genes, allowing for products to be transcribed that help with stress response and tolerance (Fig. 4). Some of these include dehydration-responsive element-binding protein (DREB), ABA-responsive element-binding factor (AREB), no apical meristem (NAM), Arabidopsis transcription activation factor (ATAF), and cup-shaped cotyledon (CUC). Much of the molecular work to understand the regulation of drought tolerance has been done in *Arabidopsis*, helping elucidate the basic processes of mechanisms of drought tolerance in crop plants (Varshaney *et al.*, 2018).

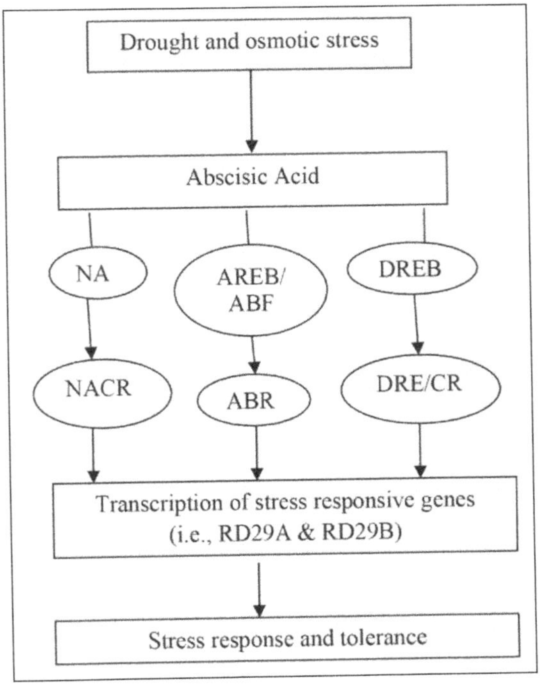

Fig. 4. Schematic representation of molecular mechanism of drought tolerance in plants

DREB TFs

DREB1/CBF TFs: DREB1A, DREB 1B, and DREB 1C are plant specific TFs which bind to drought responsive elements (DREs) in promoters responsive to drought, high salinity and low temperature in Arabidopsis. Over expression of these genes enhance the tolerance of drought, high salinity and low temperature in transgenic lines from arabidopsis, rice and tobacco.

DREB2 TFs: DREB proteins are involved in a variety of functions related to drought tolerance. For example, DREB proteins including DREB2A cooperate with AREB/ABF proteins in gene expression, specifically in the DREB2A gene under osmotic stress conditions. DREB2 also induces the expression of heat-related genes, such as heat shock protein. Over expression of DREB2Aca enhances drought and heat stress tolerance levels in Arabidopsis.

AREB/ABF TFs: AREB/ABFs are ABA responsive bZIP-type TFs which bind to ABA-responsive elements (ABREs) in stress-responsive promoters and activate gene expression. AREB1, AREB2, ABF3 and ABF1 have important roles in ABA signaling in the vegetative stage, as ABA controls the expression of genes associated with drought response and tolerance. The native form of AREB1 cannot target drought stress genes like RD29B in Arabidopsis, so modification is necessary for transcriptional activation (Fujita, 2013). AREB/ABFs are positively regulated by SnRK2s, controlling the activity of target proteins via phosphorylation. This regulation also functions in the control of drought tolerance in the vegetative stage as well as the seed maturation and germination (Fujita, 2005).

A number of transgenic have been identified, isolated, cloned and expressed in plants which are potential sources of resistance to abiotic stress (Table 2). The transferring of these genes into host significantly reduced the yield losses in drought stress regimes.

Table 2. Drought tolerant genes source and its mechanism of functions in plants

Plant genes/ Proteins	Host plants	Gene function
Beta gene	Cotton	Accumulates GB to increase drought tolerance in cotton crops.
	Wheat	Accumulates GB to increase drought tolerance in wheat plants.
Cox gene	Arabidopsis plant	Accumulates GB to reduce drought tolerance in Arabidopsis Plants.
	Tobacco	Accumulates GB to increase drought tolerance in Tobacco plants.
TPSP	Rice	Protection of plants
	Rice	Effective photosynthesis

ERA 1 Farnesyl Transferase	Canola Plant	Farnesyl Transferase downregulates and as a result ABA produces to prevent drought.
	Canola Plant	Reduced Stomatal pore conductance
p5csf129a Gene	Tobacco Plant	Increased Production of Proline amino acid.
	Various Plants	Increased turgor pressure in plants.
CDSP 32	Potato Plant	Protection of the Chloroplast from ROC's
	Potato Plant	Prevention of Oxidative Damage.
Other genes	Maize	Improved Photosynthetic machinery
	Tobacco	Increase in Stomatal pore conductance

Conclusion

Drought is one of the important abiotic stress that majorly effect the growth and yield of crop plants in India and many parts of the word. The decrease in yield due to drought varies from 20 to 70 per cent depending on time, duration, intensity of drought and growth stage of the crop. The mechanisms behind drought tolerance are complex and involve many pathways which allow plants to respond to specific sets of conditions at any given time. Some of these interactions include stomatal conductance, leaf rolling, chlorophyll degradation, relative water content and the intervention of osmoprotectants (such as sucrose, glycine, and proline) and ROS-scavenging enzymes. The molecular control of drought tolerance is also very complex and is influenced other factors such as environment and the developmental stage of the plant. This control consists mainly of transcriptional factors, such as dehydration-responsive element-binding protein (DREB) and abscisic acid (ABA)-responsive element-binding factor (AREB). The drought tolerant plant maintain its water balance by transient leaf rolling, high accumulation of compatible osmolytes, balance stomatal conductance and finally less reduction in yield and yield components over normal condition.

References

Asada, K. and Takahashi, M. 1987. Production and scavenging of active oxygens in chloroplasts. In: Kyle, D.J. Osmond, C.B., Arntzen CJ, eds. Photoinhibition. Amsterdam: Elsevier: 227-87.

Ashraf, M. 2010. Inducing drought tolerance in plants. *Recent Adv. in Biotech. Adv.* **28** (1): 169–183.

Bagheri, A. and Sadeghipour, O. 2009. Effects of salt stress on yield, yield components and carbohydrates content in four hullless barley (*Hordeum vulgare* L.) cultivars. *J. of Biological Sci.* **9**: 909-912.

Baker, N.R. 1991. A possible role for photosystem II in environmental pertubations of photosynthesis. *Physiol Plant.* **81**: 563-70.

Bensen, R.J., Boyer, J.S. and Mullet, J.E. 1988.Water deficit-induced changes in abscisic acid, growth, polysomes, and translatable RNA in soybean hypocotyls. *Plant Physiol.* **88**: 289–294.

Blum, A. 1988. Drought resistance. *In* plant breeding for stress environments. p. 43-77.

Chang, T.T. and Loresto, G.C. 1986. Screening techniques for drought resistance in rice. p. 108-129. *In* Approaches for incorporating drought and salinity resistance in crop plants. Ed. V. L. Chopra and R. S. Paroda. Oxford IBH, New Delhi.

Chaves, M.M., Pereira, J.S., Maroco, J., Rodrigues, M.L., Ricardo, C.P.P., Osorio, M.L., Carvalho, I., Faria T. and Pinheiro, C. 2002. How plants cope with water stress in the field: Photosynthesis and growth. *Ann. of Bot.* **89**: 907–916.

Dat, J., Vandenabeele, S., Vranová, E., Van Montagu, M., Inzé, D. and Van Breusegem, F. 2000. Dual action of the active oxygen species during plant stress responses. *Cell. Mol. Life Sci.* **57**: 779–795.

Earl, H.J. and Davis, R.F 2002. effect of drought stress on leaf and whole canopy radiation use efficiency and yield of maize. *Agron. J.* **95**(3): 688-696.

Flexas, J., Bota, J., Escalona, J.M., Sampol, B. and Medrano, H. 2002. Effects of drought on photosynthesis in grapevines under field condition: an evaluation of stomatal and mesophyll limitations. *Funct. Plant Biol.* **29**: 461–471.

Foyer, C.H. and Harbinson, J. 1994. Oxygen metabolism and the regulation of photosynthetic electron transport. In: Foyer, C.H. Mullineaux, P. eds. Causes of photooxidative stresses and amelioration of defense systems in plants. Boca Raton, FL: CRC Press: 1-42.

Foyer, C.H., Descourvières, P. and Kunert, K.J. 1994 Protection against oxygen radicals: An important defense mechanism studied in transgenic plants. *Plant Cell Environ.* **17**: 507-23.

Foyer, C.H., Descourvieres, P. and Kunert, K.J. 1994. Photo oxidative stress in plants. *Plant. Physiol.* **92**: 696-717.

François, T., Simonneau, T. and Muller, B. 2018. The physiological basis of drought tolerance in crop plants: a scenario-dependent probabilistic approach. *Annual Rev. of Plant Biol.* **69(1)**: 733–759.

Fujita, Y., Fujita, M., Satoh, R., Maruyama, K., Parvez, M.M., Seki, M. and Yamaguchi-Shinozaki, K. 2005. AREB1 is a transcription activator of novel ABRE-dependent ABA signaling that enhances drought stress tolerance in Arabidopsis. *The Plant Cell.* **17**: 3470–3488.

Fujita, Y., Yoshida, T. and Yamaguchi-Shinozaki, K. 2013. Pivotal role of the AREB/ABF-SnRK2 pathway in ABRE-mediated transcription in response to osmotic stress in plants. *Physiologia Plantarum.* **147**: 15–27.

Gaballah, M.S., Abou, B., Leila, H., El-Zeiny, A. and Khalil, S. 2007. Estimating the performance of salt stressed sesame plant treated with anti transpirants. *J. Appl. Sci. Res.* **3**: 811-817.

Galmés, J., Medrano, H. and Flexas, J. 2007. Photosynthetic limitations in response to water stress and recovery in Mediterranean plants with different growth forms. *New Phytologist.* **175**: 81–93.

Ghodsi, M.M., Nuzeri and A. Zarea-Fizabady. 1998. The reaction of new cultivars and Alite lines on spring wheat into drought stress, Collection of abstract articles of 5th Iranian agronomy and plant breeding conference, Karaj, Iran. 252p

Gunasekera, D. and Berkowitz, G.A. 1992. Evaluation of contrasting cellular-level acclimation responses to leaf water deficits in three wheat genotypes. *Plant. Sci.* **86**: 1-12.

Guoth, A., TariI, Galle, A., Csiszar, J., Pecsvaradi, A. and Cseuz, L. 2009. Comparison of the drought stress responses of tolerant and sensitive wheat cultivars during grain filling:changes in flag leaf photosynthetic activity, ABA levels, and grain yield. *J. Plant Growth Regul.* **28**: 167–176.

Henson, I.E. 1984. Effects of atmospheric humidity on abscisic acid accumulation and water status in leaves of rice. *Ann. Bot.* **54**: 569–582.

Hsiao, T.C., O'Toole, J.C., Yambo, E.B. and Turner, N.C. 1984. Influence of osmotic adjustment on leaf rolling and tissue death in rice (*Oryza sativa* L.). *Plant Physiol.* **75**: 338-341.

Hu, Honghong and Xiong, Lizhong. 2014. Genetic engineering and breeding of drought-resistant crops. *Annl. Rev. of Plant Biol.* **65(1)**: 715–741.

Hurd, E.A.1976. Plant breeding for drought resistance. Pages 317-353 in T. T. Kozlowski, ed. Water deficits and plant growth. Academic Press, New York.

Ingram, J. and Bartels, D. 1996. The molecular basis of dehydration tolerance in plants. *Annl. Rev. of Plant Phys. and Plant Mol. Biol.* **47**: 377-403.

Jakob, B. and Heber, U. 1996. Photoproduction and detoxification of hydroxyl radicals in chloroplasts and leaves in relation to photoinactivation of photosystems I and II. *Plant Cell Physiol.* **37**: 629-635.

Jones, H.G. 1979. Visual estimation of plant water status in cereals. *J. Agric. Sci.* 92: 83-89.

Kannangara, T., Seetharama, N., Durley, R.C. and Simpson, G.M. 1983. Drought resistance of sorghum bicolor: 6. Changes in endogenous growth regulators of plants grown across an irrigation gradient. *Can. J. Plant Sci.* **63**: 147–155.

Lazacano-Ferrat I. and Lovat, C.J. 1999. Relationship between relative water content, nitrogen pools, and growth of *Phaseolus vulgaris* L. and *P. acutifoolius* A. Gray during water deficit. *Crop. SCI.*, **39**: 467-475.

Lemee, G. 1954. L'economie de 1'eau chez quelques Gramineesvivaces du Sahara septentrional. *Vegetation.* **56**: 534-541.

Loresto, G.C. and Chang, T.T. 1981. Decimal scoring systems for drought reaction and recovery ability in rice screening nurseries. *Inter. Rice Res. Newsletter* **6(2)**: 9-10.

Loresto, G.C., Chang, T.T. and Tagumpay, O. 1976. Field evaluation and breeding for drought resistance. *Phil. J. Sci.* **1**: 36-39.

Lugojan, C. and Ciulca, S. 2011. Evaluation of relative water content in winter wheat. J. Hortic. *Fores. Biotechnol.* **15**:173–177.

Medrano, H., Bota, J., Abadia, A., Sampol, B., Escalona, J.M. and Flexas, J. 2000. Effects of drought on light-energy dissipation mechanisms in high-light-acclimated, field grown grapevines. *Funct. Plant. Biol.* **29**:1197–1207.

Mehrotra, R., Bhalothia, P., Bansal, P., Basantani, M.K., Bharti, V. and Mehrotra, S. 2014. Abscisic acid and abiotic stress tolerance-different tiers of regulation. *J. Plant Physiol.* **171**: 486–496.

Miller, G. Suzuki, N. Ciftci-Yilmaz, S. and Mittler, R. 2010. Reactive oxygen species homeostasis and signaling during drought and salinity stresses. *Plant Cell Environ.* **33**: 453–467.

Nakashima, K. and Suenaga, K. 2017. Toward the genetic improvement of drought tolerance in crops. *Japan Agril. Res. Quarterly: JARQ.* **51(1)**: 1–10.

Noctor, G. 2000. Oxygen processing in photosynthesis: Regulation and signalling. *New Phytol.* **146**: 359-388.

Noctor, G. and Foyer, C.H. 1998. Ascorbate and glutathione: Keeping active oxygen under control. *Annu. Rev. Plant Physiol Plant Mol. Biol.* **49**: 249-279.

O'Toole, J.C. and Moya, T.B. 1978. Genotypic variation in maintenance of leaf water potential in rice. *Crop Sci.* **18**: 873-876.

O'Toole, J.C. 1982. Adaptation of rice to drought prone environments. p. 195-213. *In* Drought resistance in crops with emphasis on rice. International Rice Research Institute, P.O. Box 933, Manila, Philippines.

O'Toole, J.C. and Cruz, R.T. 1980. Response of leaf water potential, stomatal resistance, and leaf rolling to water stress. *Plant Physiol.* **65**: 428-432.

O'Toole, J.C., Cruz, R.T. and Singh, T.N. 1979. Leaf rolling and transpiration. *Plant Sci. Lett.* **16**: 111-114.

Ramos, M.L.G., Parsons, R. Sprent, J.I. and Games, E.K. 2003. Effect of water stress on nitrogen fixation and nodule structure of common bean. *Pesq. Agropec. Brasilia.* **38**: 339-347.

Rathore, B.M.S., Sud, R., Saxena, V., Rathore, L.S., Singh, T., Rathore, V.G.S. and Ray, M.M. 2009. Drought condition and management study in India.

Ray, S.S., Sai, M.S. and Chattopathyay, N. 2015. Agriculture drought assessment: operational approach in India with special emphasis on 2012, In: High-impact weather events over the SAARC region. Springer, pp. 349-364.

Schonfeld, M.A., Johnson, R.C., Carwer, B.F., and Mornhinweg, D.W. 1988. Water relations in winter wheat as drought resistance indicators. *Crop. Sci.* **28**: 526-531.

Shao, H.B., Chul, L.Y., Jaleel, C.A. and Zhao, C.X. 2008. Water-deficit stress-induced anatomical changes in higher plants. *Comptes Rendus Biologies* **54(3)**: 215–225.

Stewart, C.R. and Voetberg, G. 1985. Relationship between stress-induced ABA and proline accumulations and ABA-induced proline accumulation in excised barley leaves. *Plant Physiol.* **79**: 24–27.

Thameur, A., Ferchichi, A. and López-Carbonell, M. 2011. Quantification of free and conjugated abscisic acid in five genotypes of barley (*Hordeum vulgare* L.) under water stress conditions. *S. Afr. J. Bot.* **77**: 222–228.

Tuteja, N. 2007. Abscisic acid and abiotic stress signaling. *Plant Signal. Behav.* **2**: 135–138.

Varshney, R.K., Roberto, T. and Francois, T. 2018. Progress in understanding drought tolerance: from alleles to cropping systems. *J Exp. Botany.* **69(13)**: 3175–3179.

Xiong, L. and Zhu, J.K. 2003. Regulation of abscisic acid biosynthesis. *Plant Physiol.* **133**: 29–36.

Zeevaart, J.A.D. and Creelmanm, R.A. 1988. Metabolism and physiology of abscisic acid. *Annu. Rev. Plant Physiol. Plant Mol. Biol.* **39**: 439–473.

Zlatev, S.Z. 2005. Effect of water stress on leaf water relations of young bean. *J. Cent. Eur. Agric.* **6**: 5-14.

16

Soil Biodiversity and Its Management for Sustainable Agriculture

Sanjay Arora

Introduction

Soil, a dynamic living matrix, is an essential part of the terrestrial ecosystem. It is a critical resource not only to agricultural production and food security but also to the maintenance of most life processes. Soils contain enormous numbers of diverse living organisms assembled in complex and varied communities ranging from the myriad of invisible microbes, bacteria and fungi to the more familiar macro-fauna such as earthworms and termites. The diversity in soils is several times higher than that above ground. Each hectare top soil contains approximately 1,000 kg of different fungi, 500 kg of bacteria, 750 kg actinomycetes and 150 kg of algae and many protozoa (Table 1). These diverse micro-organisms interact with one another and with the plants and animals in the ecosystem forming a complex system of biological activity. Environmental factors, such as temperature, moisture and acidity, as well as anthropogenic actions, in particular, agricultural practices affect soil biological communities and their functions to different extents. Diversity of soil micro-organisms has emerged in the past decade as a key area of concern for sustainable soil health and crop production. Besides, the well-being and prosperity of earth's ecological balance, the sustainability of agricultural production systems directly depends on the extent and status of microbial diversity of soil.

Table 1. Important soil microorganisms, their number and biomass in cultivated soil

Micro-organisms	Average population (per g soil)	Average biomass (kg/ha)
Bacteria	10^7-10^8	500
Fungi	10^5-10^6	1,000
Actinomycetes	10^6-10^7	750
Algae	10^3-10^4	150

Source: Adu and Oades (1978)

**Corresponding author email: aroraicar@gmail.com*

Soil is a dynamic and species-rich habitat containing all major groups of micro-organisms (Hagvar, 1998). The soil micro-community plays a vital role as the global element cycles and thus for life on earth because 60-90% of the whole terrestrial primary production is decomposed in the soil and furthermore, many waste products of human society are detoxified there (Giller, 1996). Micro-organisms in the soil form a part of the biomass and contribute to reserve the soil nutrients and are generally referred to as the microbial biomass. Microbial bio- mass regulates the transformation and storage of nutrients; these processes affect many nutrient cycling functions including soil fertility and soil organic matter (SOM) turnover (Horwath and Paul, 1994). There is a relationship between microbial diversity and soil functionality, by considering that 80-90% of the processes in soil are reactions mediated by microbes (Coleman and Crossley, 1996; Nannipeiri and Badalucco, 2003).

Soil micro-organisms are an integral part of agricultural ecosystems; and they play a critical role in maintaining soil health, ecosystem functions and crop production. Some of the important functions carried out by micro-organism in soil are:

- The activities of certain micro-organism such as fungi, algae and bacteria affect soil structure by encouraging soil aggregation. This directly affects various soil physical properties and determines vulnerability to soil erosion;
- Soil micro-organisms are central to decomposition processes and nutrient cycling in soil–plant–animal continuum. They, therefore, affect plant growth and productivity as well as the release of nutrients and pollutants in the environment, for example the leaching of nitrates into water resources;
- Certain soil organisms can be detrimental to plant growth by causing plant diseases. However, they can also protect crops from pest and disease outbreaks through biological control and reduced susceptibility;
- The activities of certain organisms determine the carbon cycle - the rates of carbon sequestration and gaseous emissions and soil organic matter transformation;
- Symbiotic relationships, especially of *Rhizobium* bacteria and Mycorrhiza, with crop plants play a key role in the uptake of nutrients and water, and contribute to the maintenance of soil porosity and organic matter content, through their growth and biomass;
- A group of soil microorganisms, referred as plant growth promontory rhizobacteria (PGPR), secrete plant growth hormones such as IAA, GA, cytokinin and enhance seed germination, root development and plant growth.

- Soil organisms can also be used to reduce or eliminate environmental hazards resulting from accumulations of toxic chemicals or other hazardous wastes (bioremediation).

Biodiversity

The term biodiversity refers to the variability and richness of species in an ecosystem. A species could broadly be defined as a collection of populations that may differ genetically from one another to a greater or lesser degree, but whose individuals are able to mate and produce offspring. Species are the most useful units for biodiversity research, and species diversity is the most useful indicator of biodiversity. At its lowest level, biodiversity depends on the sequences of genes in living organisms. Genes are composed of stretches of DNA and these sequences, along with the proteins encoded by the genes, are almost identical to their counterparts in other species. Thus, they are said to be highly conserved across species, and such commonalities (or even differences) are referred to as genetic diversity. The importance of genetic diversity is observed in the combination of genes within an organism (the genome), the variability in the proteins or traits (phenotype) that they produce, and their resilience, survival and function. Biodiversity has three interrelated elements: genetic, functional and taxonomic diversity (Fig. 1). Taxonomic diversity, i.e. the number of species, is an important part of an ecosystem's diversity and this is controlled by the genetic diversity, which may be greater than the number of recognized species. Several species may have the same functions, resulting in diverse functions called functional diversity. Biodiversity is, therefore the interaction of all these elements.

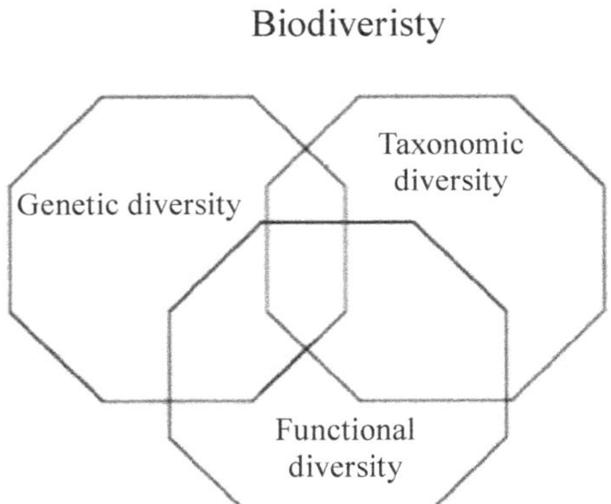

Fig. 1. Inter-related elements of soil biodiversity (*Source*: Biswas and Gawade, 2016)

Soil biodiversity

Soil biodiversity reflects the mix of living organisms in the soil. These organisms interact with one another and with plants and small animals forming a web of biological activity. Soil is by far the most biologically diverse part of Earth. The soil food web includes beetles, springtails, mites, worms, spiders, ants, nematodes, fungi, bacteria, and other organisms. These organisms improve the entry and storage of water, resistance to erosion, plant nutrition, and break down of organic matter. A wide variety of organisms provides checks and balances to the soil food web through population control, mobility, and survival from season to season.

Microbial diversity in soil

A variety of microorganisms are present in soil, though the space occupied by living organisms is less than 5% of the total space. Therefore, micro-organisms' remains confined to the hot spots i.e. aggregates with accumulated organic matter and rhizosphere. Each gram of soil has approximately 10^9 bacterial cells, 10^8 culturable bacteria, 10^4 bacterial species, 10^5 actinomycete cells and 10^5 fungal propagules. Bacteria in soil have varying mode of nutrition and can use sun light or chemical energy source. They can meet their carbon requirement from atmospheric CO_2 or organic compounds. Some take their nutrition from native organic matter or freshly added organic matter. The dominating soil bacteria include many species of *Arthrobacter, Pseudumonas, Bacillus, Xanthomonas, Clostridium, Azotobacter, Rhizobium* etc. The most dominating actinomycetes in soil is *Streptomyces.* Several genera of blue green algae, protozoa and algae are present in soil. All the soil micro-organisms have great diversity in their physiological activities and functions. On the basis of temperature requirement, they are thermophillic, mesophillic and psychrophillic. Some requires O_2 for respiration (aerobic) while others can survive in absence of O_2. The earlier taxonomic classification of soil micro-organisms is under process of tremendous change which can be noticed that root nodule rhizobia now has 36 species distributed among seven genera (*Allorhizobium, Azorhizobium, Bradyrhizobium, Mesorhizobium, Methlyorhizobium, Rhizobium, Sinorhizobium*). Plants play an important role in selecting and enriching the type of bacteria by constituents of their root exudates. The microbial community of wheat was studied extensively in IGP and it was observed that wheat genotype did not appreciably influence the total bacterial and pseudomonad populations. However, population was marginally different in rhizosphere and rhizoplane fractions. It was also reported that plants exert a strong influence on the composition of microbial communities in soil through rhizodeposition and the decay of litter and roots. The link between plant species and microbial communities in the rhizosphere soil is strict, being the result of co-evolution. Scientists suggested that competition in microbial communities of surface soils with prevalence of any microbial species was absent because the various microbial species inhabiting soil are spatially separated for most of the time. They assumed that the contact among microhabitats occur for a very short time immediately after rain, when water bridges are formed between the various soil particles and aggregates. Rapid drainage maintains the spatial isolation among the various microhabitats of soil. However, it does not take into account the mixing and transport by soil fauna and the stability of communities in biofilms at the interface between roots and soil which are not so strongly affected by wetting and drying. An alternative hypothesis to explain the large microbial diversity of surface soil is based on the presence of a greater variety and content of organic compounds than in deeper soil layer. This supports the diverse heterotroph-dominated microbial community in surface soil.

Microbial diversity and soil functions

Relationship between biodiversity and functions of terrestrial ecosystems is best described by the hump-shaped curve (Fig. 2), in which there is an increase in plant production (i.e. the function) with increasing biodiversity until a certain point is reached; then a further increase in biodiversity results in a decrease in plant production. Stability of ecosystem is another ecological concept and had strong relationship with biodiversity. Stability is defined as the property of an ecosystem to withstand perturbations. Stability includes both resilience (i.e. the property of the system to recover after disturbance) and resistance (i.e. the inherent capacity of the system to withstand disturbance). It is difficult to measure both resistance and resilience in soil. Generally microbe-mediated processes are the most sensitive to perturbations in the soil; for this reason the capacity of soil to recover from perturbations can be assessed by monitoring microbial activities. The links between microbial diversity and soil functioning, as well those between stability (resilience or resistance) and microbial diversity in soil, are unknown because, as stated above, it is difficult to measure microbial diversity. Therefore, soil functions are measured by determining the rates of microbial processes, without knowing the microbial species effectively involved in the measured process. According to O'Donnell *et al.* (2001), the central problem of the link between microbial diversity and soil function is to understand the relations between genetic diversity and community structure and between community structure and function. From a study it was found that soil fumigated with chloroform, with a much smaller microbial biomass than the corresponding non-fumigated soil, respired about the same amount of $^{14}C\text{-}CO_2$ from labelled straw as the non-fumigated soil. They also found that pollution with zinc affected respiration of non-fumigated and fumigated soils in the same way, indicating that the ratio between substrate C-to-microbial biomass C (larger in the fumigated than the non-fumigated soil) was not important. The ratio of CO_2 to ^{14}C-to-microbial biomass ^{14}C was linearly related to Zn pollution. It was reported that soil fumigated with chloroform, with greater microbial diversity, was more resistant and resilient than soil with a less diverse community to perturbations such as heating at 40 ^0C for 18 hrs or treatment with 500 mg Cu/g soil. However, they suggested that the observed effects were due to the physiological influence of $CHCl_3$ fumigation on the microbial community rather than differences in microbial diversity.

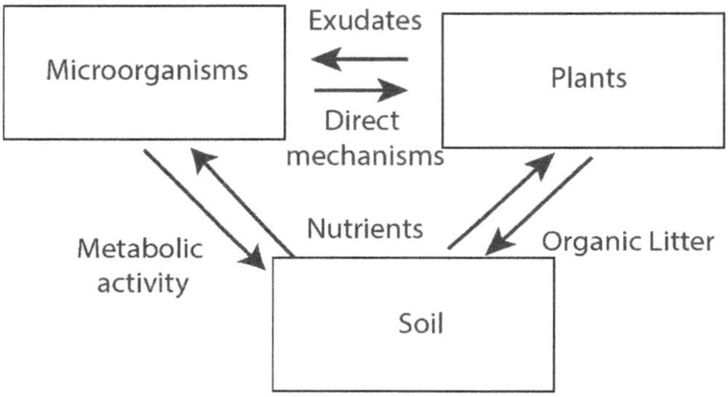

Fig. 2. Soil micro-organisms and functions (*Source*: Jacoby *et al*., 2017)

The fungi (pathogenic and non pathogenic) associated with the soil are known as soil borne fungi and play a fundamental role for the functioning of the soil ecosystem (Doron and Parkin, 1994, 1996; Hawksworth *et al.,* 1996). Due to their ability to decompose complex macro molecules like lignin or chitin they are essential for making the locked- up nutrients like C, N, P, S easily available. Moreover, the fungal mycelia play an important role for the stabilization of the soil because it binds soil aggregates and thus reduces erosion and helps to increase the water holding capacity (Kennedy and Gewin, 1997). Fungi are dominant in acid soils because an acidic environment is not suitable for the existence of either bacteria or actinomycetes, resulting in the monopoly of fungi for the utilization of organic substrates (Bolton *et al.,* 1993). In the frame of agriculture, the micro flora is of great significance because it has both beneficial and detrimental role in overall productions (Whitelaw, 2000). Farm practices including crop rotations and fertilizer or pesticide applications influence the nature and dominance of fungal species (Pelczar and Reid, 1972). Similar to Plant Growth Promoting Rhizobacteria (PGPR), some rhizosphere fungi able to promote plant growth upon root colonization are functionally designated as Plant-Growth-Promoting-Fungi (PGPF) (Hyakumachi, 1994 Srivastava and Sharma, 2014).

Role of soil microbial communities

Terrestrial ecosystem processes and functions are regulated by the activities of soil micro-organisms. Important ecological functions related to the microbial community activities are: litter decomposition, nutrient cycling, bioremediation, C sequestration, improving soil physical environment.

***Litter decomposition and nutrient availability*:** The most significant contribution of the soil microbes in terrestrial ecosystem is that of litter decomposition. By this process, the complex chemical compounds in dead leaves, roots, and other plant tissues are broken down into simpler chemical compounds, converting organically held nutrients into mineral forms available for renewed plant uptake. The nitrogen release from decomposition of organic residues is a prime example. Soil microbes also assimilate wastes from animals and other organic materials added to soils. As a by-product of their metabolism, microbes synthesize new compounds, some of which help to stabilize soil structure and others of which contribute to soil organic matter formation, and improve soil quality over time (Islam and Weil, 2000). To tackle the problem of burning of crop residues, use of microbial formulation for *in-situ* decomposition and nutrient recycling is feasible approach to be adopted. Investigations have shown that application of bio-formulation of microbial decomposers helps in decomposition of residues and enhances soil nutrient availability, soil health and succeeding crop yields (Arora *et al.*, 2019).

***Inorganic nutrient transformations and soil remediation*:** The transformation of inorganic nutrient elements and compounds is of great significance to the ecological functions of soil systems, including plant growth. Nitrate, sulfate, and to a lesser degree, phosphate ions are present in soil primarily due microbial activities. Bacteria and fungi assimilate some of the N, P and S in the organic materials they digest. Excess amounts of these nutrients may be excreted into the soil system in inorganic forms usable either by the bacteria and fungi themselves or by other organisms that feed on them. In this process, organically bound forms of N, P and S are converted into inorganic forms that are available to higher plants. Likewise, the solubility and availability of the other essential elements, such as Fe and Mn, are determined largely by microbial activities in soil. In well-drained soils, these elements are oxidized by autotrophic micro-organisms to their valence states, in which forms they are quite insoluble which keeps Fe and Mn mostly in non-toxic forms, even under fairly acidic conditions. Microbial oxidation also controls the potential for toxicity in soil contaminated with Se, As and Cr compounds.

Amongst different abiotic stresses, soil salinization, soil pH, acidification, drought and temperatures are the major restrictive factors in sustaining crop production. Soil salinity is a major issue for agriculture because high concentration of salt turns useful lands into unproductive areas. Both physical as well as chemical methods for reclamation of saline and sodic soils are not cost-effective and on the other hand, organic crop production is being promoted (Arora, 2019). The microbial strains available as bio-fertilizers for different crops do not perform effectively under salt stress and their activity decreases when used in salt affected soils due to osmolysis (Arora and Vanza, 2017; Arora, 2019). The soils of vast areas of IGPs in north India are sodic or saline-sodic. The halophilic plant growth promoting microbes have potential to ameliorate these soils (Arora and Singh, 2017; Sahay *et al*., 2018). The halophilic bacterial strains can help in recovery of salt affected soils by directly supporting vegetation growth thus indirectly increasing crop yields under salt stress conditions (Arora *et al*., 2014 a,b).

Biological N fixation: N is one of the prime nutrients for plant growth. The atmospheric N^2 is an inert gas, which cannot be used directly by higher plants. Biological N fixation is one of the most important microbial processes in soils. Actinomycetes in the genus *Frankia* fix major amounts of atmospheric N in forest ecosystems; *Cyanobateria* are important in flooded rice systems, wetlands and deserts; and *rhizobia* bacteria are the most important group for the fixation of gaseous inert N in agricultural soils (Brady and Weil, 2002). By far the greatest amount of N fixation by these micro-organisms occurs in root nodules or in other close associations with plants. Worldwide, enormous quantities of atmospheric N are fixed in soils annually by microbes into forms usable by higher plants (Tate, 1995). The free living bacteria also fixes nitrogen in rice-wheat system and the bio-formulation 'Halo-Azo' developed for the efficient N-fixers helps in bio-remediation of salt affected soils and enhances crop growth and yield under salt stress (Arora and Singh, 2018; Sarangi and Lama, 2018; Sahay *et al*., 2019).

Soil aggregate stability and carbon sequestration: Aggregate structure is a key biologically mediated soil property that results from close associations of fungal hyphae with plant roots, and release of low molecular weight organic acids associated with clays, metals, and micro-aggregates (Tisdall and Oades, 1982). In addition to C retention through efficient cell assimilation, extensive fungal hyphal associations with plant roots can jointly contribute to protect and accumulate C by enhancing macro-aggregation through physical enmeshing of soil micro-aggregates or releasing extracellular polysaccharide as binding agents or stabilization through complex association with metals as micro-aggregates (Oades, 1984; Chenu, 1989). The microbially derived decomposition products cement primary particles, organic debris, and micro-aggregates together to form and stabilize macro-aggregates (Jastrow and Miller, 2000). An enhanced physical protection of fragmented organic debris and microbial decomposition products within macro-aggregates is an important mechanism enabling the soil C accumulation (Jastrow and Miller, 2000). These mechanisms, in turn, increase the proportion of C in soil aggregates that is physically protected from microbial decomposition and protects soil from accelerated soil erosion (Islam and Weil, 2000).

Breakdown of toxic and xenophobic compounds: Many organic compounds toxic to plants or animals find their way into the soil. Some of these toxins are produced by soil microbes as metabolic by-products, some are applied as agrochemicals to kill pests and insects, and control weeds, and some others deposited in the soil because of environmental contamination. If these compounds accumulated unchanged, they would do enormous ecological damage. Some toxins are xenophobic compounds foreign to biological systems, and these may resist attack by commonly occurring microbes. Soil bacteria and fungi are especially important in helping maintain a nontoxic soil environment by breaking down toxic compounds.

Plant protection and diseases: Certain soil microbes protect higher plants from invasion by soil parasites and pathogens, but others attack plant roots. According to several recent research studies, many plants that are resistant to soil borne diseases have camouflaged rhizosphere (Brady and Weil, 2002; Imran and Shaukat, 2002). That is, although the rhizosphere of these resistant plants is enriched in microbial numbers, the types of microbes are more similar to those in the bulk soil than is the case for the rhizosphere of disease-susceptible plants. Having a rhizosphere microbial community similar to the community in the bulk soil may make the rhizosphere of these plants less identifiable to pathogens.

Beneficial rhizobacteria have an intriguing mode of action called induced systematic resistance, which helps plants from infection by diseases, insects or pests both above and below ground. In many cases studied so far on crops, the resistance-inducing organism has been a *Pseudomonas* or *Serratia* bacteria (Imran and Shaukat, 2002). This mechanism has been shown to effectively reduce damages by numerous fungal, bacterial, and viral pathogens and several leaf-eating insect pests (Imran and Shaukat, 2002).

Disease infestations occur in great variety and are induced by many different soil micro-organisms. Among the micro-organisms, bacteria, actinomycetes and fungi are responsible for most of the common soil borne diseases of crops. Included are wilts, damping-off, root rots, club root of cabbage, and blight of potatoes.

Impact of agricultural practices

Changes in environment affect both the number and kinds of soil organisms. Various agricultural management practices like tillage, cropping systems, fertilizer application, cultivation practices, soil organic amendments and pesticide application alter the microbial dynamics in the agro ecosystem. In a study on soil management practices, it was found that cluster analysis of community level physiological profile (BIOLOG) indicated much uniform microbial communities in organic farming than irrigated with salt water. The PCR-DGGE patterns of total soil DNA showed a moderate to significant variation among salt irrigated and organic soil samples. Tillage is practiced to destroy weeds, incorporate plant residue into the plough layer, destroy plant residues harbouring plant pathogens, and provide better soil structure for easy emergence of seedlings and for proper root growth. Labile organic reserves in soil generally decrease with cultivation and cropping and therefore, alter the microbial diversity and biomass in soil. These effects of tillage are largely confined to the surface (5-10 cm) soil layer. It has been reported that an increase in phosphatase and dehydrogenase activities and contents of moisture and, organic C and N in the surface (0 to 7.5 cm) of no-till soil compared to conventional tillage. However, at the 7.5 to 15 cm and 15 to 30 cm depths these trends were reversed and microbial populations, enzyme activities and moisture and organic C and N contents were the same or higher for conventional tillage than for no-till soil. Monoculture a practice of growing one crop year after year on the same piece of land is a result of newer innovations in the field of agriculture. This involves use of improved crop varieties, pesticides, inorganic fertilizer and has selective effect on microbial diversity. Wheat genotypes grown in IGPs did not significantly influence the total bacterial and pseudomonad population in soil, however, she reported distinct cluster of rhizosphere communities of wheat and mandua (*Eleusins coracana*) grown at Chaukhutia, Almora using 16S rDNA restriction profile. Crop rotation length, fallow substitute crops and fertilization also affect the microbial diversity in soil. Soil under continuous cropping has higher biomass C and N than soil under fallow and cropping + fallow soils. These variations could be explained in terms of substrate availability. While studying the effect of different cropping sequences on chemical and micro-biological properties it was reported maximum total N and organic matter content under early pigeonpea-lentil-greengram rotation. However, the population of bacteria was maximum under pigeonpea-maize-wheat rotation and that of fungi under maize-lentil rotation. Population of actinomycetes was not affected by different cropping sequences. Results of the long term experiment conducted with the application of N and P fertilizers indicated that total viable bacterial numbers increased with the addition of fertilizers and the highest population was detected in treatment with normal doses of NPK. However, selective influence of fertilizer application is observed on AMF species. *Acaulospora* sp.1 showed no change in spore number with fertilization whereas *Entrophosporas chenckii, Glomus mosseae, Glomous* sp.1, *Scutellospora fulgida* showed a declines in absolute number in response to fertilization. The effects of fertilizer applications on soil microbial biomass were due to an increase in root biomass, root exudates and crop residues thus providing increased substrate for microbial growth. Most pesticides applied at rates approximating those used in field applications caused only slight change in population and activities of micro-organisms, however, affected soil micro-organisms adversely at high rates of application. The effects of three post- emergence herbicides (2, 4-D, Picloram and glyphosate) on certain microbial variables was reported by scientists who concluded that the changes in microbial parameters measured as microbial numbers and soil respiration, occurred only at herbicide concentrations of much higher than that used for field application and the side effect of these chemicals were probably of little ecological significance.

GM crop and microbial diversity

Transgenic plants have been found to have significant effect on non-target soil micro-organisms. These crops may affect the soil microbial diversity directly by producing transgenic proteins and indirectly by mediating changes in plant proteins and root exudates composition. However, factors such as the composition and content of transgenic proteins in GM plant, the resistance of the proteins to degradation, soil physical, chemical and biological environment influence the accumulation and bio availability of the proteins in soil. It has been well established that flow of genes between organisms of different species occurs in nature. This is called horizontal gene transfer. Normally, special protection mechanisms prevent foreign genes, after entering a cell, to mix with its DNA. However, special genetic elements, so called vectors, are used in genetic engineering to overcome these so called species barrier mechanisms. It has been stated that this vector DNA may leak out into the soil and from there it may be transferred to other bacteria, may be a pathogenic to humans or playing role in soil fertility. Such gene transfer may brings the changes in soil micro-organisms with altered properties, that in the worst case may damage the soil fertilityormay promote the occurrence of new, dangerous and uncontrollable human pathogens.

Benefits of soil organisms

***Residue decomposition*:** Soil organisms decompose plant residue. Each organism in the soil plays an important role. The larger organisms in the soil shred dead leaves and stems. This stimulates cycling of nutrients. The larger soil fauna include earthworms, termites, pseudoscorpions, micro-spiders, centipedes, ants, beetles, mites, and springtails. When mixing the soil, the large organisms brings material to smaller organisms. The large organisms also carry smaller organisms within their systems or as "hitchhikers" on their bodies. Small organisms feed on the by-products of the larger organisms. Still smaller organisms feed on the products of these organisms. The cycle repeats itself several times with some of the larger organisms feeding on smaller organisms. Some larger organisms have a life span of two or more years. Smaller organisms generally die more quickly, but they also multiply rapidly when conditions are favourable. The food web is therefore quick to respond when food sources are available and moisture and temperature conditions are good (Fig. 3).

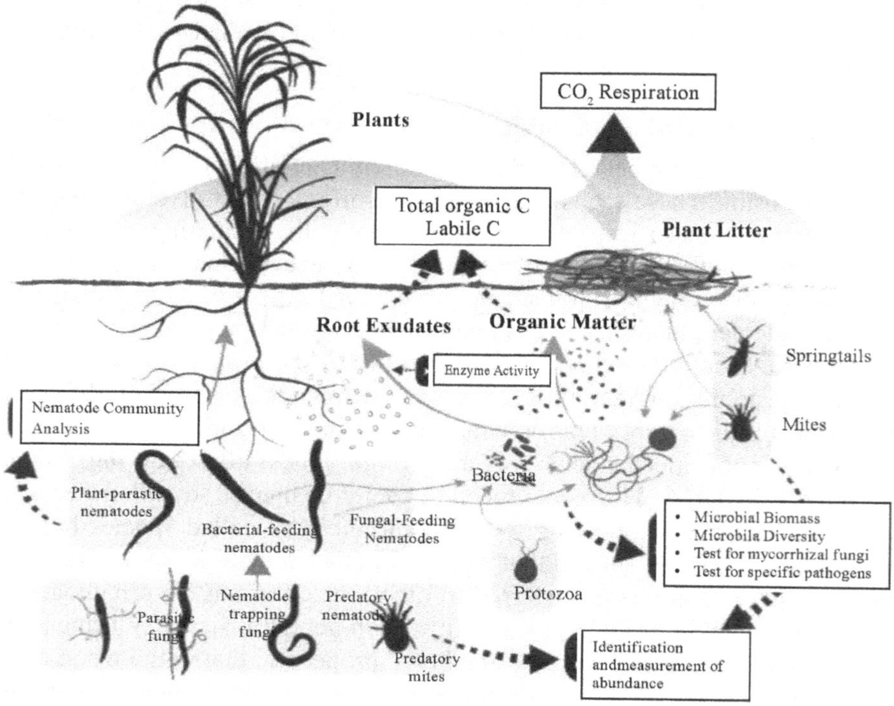

Fig. 3. Soil food web (*Source*: Brackin *et al.*, 2017)

Infiltration and storage of water: Channels and aggregates formed by soil organisms improve the entry and storage of water. Organisms mix the porous and fluffy organic material with mineral matter as they move through the soil. This mixing action provides organic matter to non-burrowing fauna and creates pockets and pores for the movement and storage of water.

Fungal hyphae bind soil particles together and slime from bacteria help hold clay particles together. The waterstable aggregates formed by these processes are more resistant to erosion than individual soil particles. The aggregates increase the amount of large pore space which increases the rate of water infiltration. This reduces runoff and water erosion and increases soil moisture for plant growth.

***Nutrient cycling*:** Soil organisms play a key role in nutrient cycling. Fungi often the most extensive living organisms in the soil, produce fungal hyphae. Hyphae frequently appear like fine white entangled thread in the soil. Some fungal hyphae (mycorrhizal fungi) help plants extract nutrients from the soil. They supply nutrients to the plant while obtaining carbon in exchange and thus extend the root system. Root exudates also provide food for fungi, bacteria, and nematodes.

When fungi and bacteria are eaten by various mites, nematodes, amoebas, flagellates, or ciliates, nitrogen is released to the soil as ammonium. Decomposition by soil organisms converts nitrogen from organic forms in decaying plant residues and organisms to inorganic forms which plants can use.

Management considerations

***Cultivation*:** The effects of cultivation depend on the depth and frequency of the cultivation. Tilling to greater depths and more frequent cultivations has an increased negative impact on all soil organisms. No-till, ridge tillage, and strip tillage are the most compatible tillage systems that physically maintain soil organism habitat and biological diversity in crop production.

***Compaction*:** Soil compaction reduces the larger pores and pathways, thus reducing the amount of suitable habitat for soil organisms. It also can move the soil toward anaerobic conditions, which change the types and distribution of soil organisms in the food web. Gaps in the food web induce nutrient deficiencies to plants and reduce root growth.

***Pest control*:** Pesticides that kill insects also kill the organisms carried by them. If important organisms die, consider replacing them. Plant-damaging organisms usually increase when beneficial soil organisms decrease. Beneficial predator organisms serve to check and balance various pest species.Herbicides and foliar insecticides applied at recommended rates have a small impact on soil organisms. Fungicides and fumigants have a much greater impact on soil organisms.

***Fertility*:** Fertility and nutrient balances in the soil promote biological diversity. Typically, carbon is the limiting resource to biological activity. Plant residue, compost, and manure provide carbon. Compost also provides a mix of organisms, so the compost should be matched to the cropping system.

***Cover crops and crop rotations*:** The type of crops that are used as cover or in crop rotations can affect the mix of organisms that are in the soil. They can assist in the control of plant pests or serve as hosts to increase the number of pests. Different species and cultivars of crops may have different effects on pests. However, the organisms and their relation to the crop are presently not clearly understood.

***Crop residue management*:** Mixing crop residue into the soil generally destroys fungal hyphae and favours the growth of bacteria. Since bacteria hold less carbon than fungi, mixing often releases a large amount of carbon as carbon dioxide (CO_2). The net result is loss of organic matter from the soil.

When crop residue is left on the soil surface, primary decomposition is by arthropod shredding and fungal decomposition. The hyphae of fungi can extend from below the soil surface to the surface litter and connect the nitrogen in the soil to the carbon at the surface. Fungi maintain a high C:N ratio and hold carbon in the soil. The net result is toward building the carbon and organic matter level of the soil. In cropping systems that return residue, macro-organisms are extremely important. Thus, it is important to manage the soil to increase their diversity and numbers.

Conclusion

Soil biological processes and soil biodiversity are central to the soil fertility and soil physical properties. Therefore, an understanding of microbial diversity is of paramount importance and requires immediate attention in order to maintain/ improve soil health. Our understanding of the links between microbial diversity and soil functions is poor because measurement of all soil micro-organisms is not easy; most are unculturable, even molecular techniques cannot detect all unculturable micro-organisms. In addition, the present assays for measuring microbial functions determine the overall rate of entire metabolic processes, such as respiration, or specific enzyme activities, without identifying the active microbial species involved in it. The central problem of the link between microbial diversity and soil function is to understand the relations between soil genetic diversity and microbial community structure and between community structure and soil function. The recent advances in RNA extraction from soil might permit us to determine active species in soil. Further, advances in understanding require determining the composition of microbial communities and microbial functions in microhabitats. The research priorities needed to address in future includes; 1) assessment of the total genetic diversity of soil, 2) to establish the link between microbial species and/or communities and soil functions, 3) to monitor changes in microbial communities due to agricultural practices and other perturbation and 4) identifying potential indicators of microbial diversity and soil quality.

References

Adu, J.K. and Oades, J.M. 1978. Utilization of organic materials in soil aggregates by bacteria and fungi. *Soil Biol. Biochem.* **10**: 117-122.

Arora, S. 2019. Bio-remediation and health management of salt affected soils. *Indian Farmg.* **69(11)**: 14-17.

Arora, S., Patel, P., Vanza, M. and Rao, G.G. 2014a. Isolation and characterization of endophytic bacteria colonizing halophyte and other salt tolerant plant species from coastal Gujarat. *Afri. J. of Microbiol. Res.* **8 (17)**: 1779-1788.

Arora, S., Vanza, M., Mehta, R., Bhuva, C. and Patel, P. 2014b. Halophilic microbes for bio-remediation of salt affected soils. *Afri. J. of Microbiol. Res.* **8**(33): 3070-3078.

Arora, S. and Singh, Y.P. 2017. Bioremediation of salt affected soils of Uttar Pradesh through halophilic microbes to promote organic farming. CSSRI Annual Report 2016–17. ICAR-CSSRI, Karnal, India, pp 123–127.

Arora, S. and Singh, Y.P. 2018. Bioremediation of salt affected soils of Uttar Pradesh through halophilic microbes to promote organic farming. CSSRI Annual Report 2017–18, ICAR-CSSRI. Karnal, India, pp 113–118.

Arora, S., Singh, Y.P. and Singh, A.K. 2019. Bio-augmenting crop residues degradation for nutrient cycling through efficient microbes to enhance productivity of salt affected soils. CSSRI Annual Report 2018-19, ICAR-CSSRI, Karnal, India, pp. 111-115.

Arora, S., Singh, Y.P., Vanza, M. and Sahni, D. 2016. Bioremediation of saline and sodic soils through halophilic bacteria to enhance agricultural production. *J. Soil Water Conser.* **15**(4): 302–305.

Arora, S. and Vanza, M. 2017. Microbial approach for bioremediation of saline and sodic soils. In: Arora S et al (eds) *Bioremediation of salt affected soils: An Indian perspective*. Springer International Publishing, Cham, pp 87–100.

Biswas, S. and Gawade, B.H. 2016. Soil biodiversity-types, methods of study and indices. *Biotech Articles* (https://www.biotecharticles.com/Authors/0/Dr-Sunanda-Biswas-37727.html).

Bolton, H.Jr., Fredrikson, J.K. and Elliot, L.F. 1993. Microbiology of the rhizosphere. In: F.B. Metting, Jr., Marcel Dekker (eds) *Soil Microbial Ecology,* Inc., New York, 27-63.

Brackin, R., Schmidt, S., Walter, D., Bhuiyan, S., Buckley, S. and Anderson, J. 2017. Soil biological health—What is it and how can we improve it? *Proc. Aust. Soc. Sugar Cane Technol.* **39**:141-154.

Brady, N.C. and Weil, R.R. 2002. Organisms and ecology of the Soil. In: *Nature and Properties of Soils*, 13th Ed., Prentice Hall, New Jersey, pp 449-497.

Chenu, C. 1989. Influence of a fungi polysaccharide, scleroglucan, on clay microstructures. *Soil Biol. Biochem.* **21**: 299-305.

Coleman, D.C. and Crossley, D.A. 1996. *Fundamentals of Soil Ecology.* Academic Press, London.

Doron, J.W. and Parkin, T.B. 1994. Defining and assessing soil quality. In: Doran, J. W. (ed.) *Defining Soil Quality for a Sustainable Environment.* SSSA Special Publication 35, Soil Science Society of America, Madison, 3-12.

Giller, P.S. 1996. The diversity of soil communities, the "poor man's tropical rainforest". *Biodiversity Conservation* **5**: 135-168.

Hagvar, S. 1998. The relevance of the Rio-Convention on biodiversity to conserving the biodiversity of soils. *Appl. Soil Ecol.* **9**: 1-7.

Hawksworth, D.L., Kirk, P.M., Sutton, B.C. and Pelger, D.N. 1996. *Ainsworth and Bisby's Dictionary of the Fungi.* 8th edition, CAB International, Wallingford.

Horwath, W.A. and Paul, E.A. 1994. Methods of soil analysis part 2. *Soil Sci. Soc. of America.*

Hyakumachi, M. 1994. Plant growth promoting fungi from turf grass rhizosphere with potential for disease suppression. *Soil Micro-organisms* **44**: 53-68.

Imran, A.S. and Shaukat, S.S. 2002. Mixtures of plant disease suppressive bacteria enhance biological control of multiple tomato pathogens. *Biol. Fertil. Soils* **36**: 260-268.

Islam, K.R. and Weil, R.R. 2000. Soil quality indicator properties in mid-Atlantic soils as influenced by conservation management. *J. Soil & Water Conser.* **55**: 69-78.

Jacoby, R., Peukert, M., Succurro, A., Koprivova, A. and Kopriva, S. 2017. The role of soil microorganisms in plant mineral nutrition-Current knowledge and future directions. *Front. Plant Sci.* **8**: 1617.

Jastrow, J.D. and Miller, R. 2000. Soil aggregation in the rhizosphere: optimal conditions for multiple mechanisms. The Ecol. Soc. Amer. 85th Ann. Meeting, Snowbird, UT, p. 21.

Kennedy, A.C. and Gewin, V.L. 1997. Soil microbial diversity: present and future considerations. *Soil Sci.* **162**: 607-617.

Nannipieri, P. and Badalucco, L. 2003. Biological processes. *In: Processes in the Soil-Plant System: Modelling Concepts and Applications,* D.K. Benbi & A. Nieder (eds). The Haworth Press, Binghamton, NY

Oades, J.M. 1984. Soil organic matter and structural stability: mechanisms and implications for management. *Plant Soil* **76**: 319-337.

Pelczar, M.J. Jr. and Aeid, A.D. 1972. *Microbiology.* Mc Graw-Hill Book Company, New Delhi pp. 936.

Sahay, R., Singh, A.K., Arora, S., Singh, A., Tiwari, D.K. *et al.* 2018. Effect of halophilic bioformulations on soil fertility and productivity of salt tolerant varieties of paddy in sodic soil. *Int. J. of Curr. Microbiol. and Appl. Sci.* **7**(9): 1174-1179.

Sarangi, S.K. and Lama, T.D. 2018. Evaluation of microbial formulations for crop productivity and soil health under coastal agro-ecosystem. ICAR-CSSRI Annual Report 2017–18, ICAR-CSSRI, Karnal. India, pp 165–166.

Srivastava, M.P. and Sharma, S. 2014. Potential of PGPR bacteria in plant disease management, New science, Germany, pp 81-96.

Tate, R.L. 1995. *Soil Microbiology.* John Wiley & Sons, Inc., New York, pp 147-170.

Tisdall, J.M. and Oades, J.M. 1982. Organic matter and water-stable aggregates in soils. *J. Soil Sci.* **33**: 141-163.

17

Impact of Climate Change in Crop Protection

Mukesh Sehgal, D.S. Srivastava and H. Ravindra*

Introduction

Climate change and agriculture are interrelated issues, both of which take place on a global scale. Global warming is projected to have significant impacts on conditions affecting agriculture, including fluctuation of temperature, carbon dioxide, glacial run-off, precipitation and the interaction of these factors (Sehgal *et al.*, 2006; Kalra *et al.*, 2008). These conditions determine the carrying capacity of the biosphere to produce enough food for the human population and for domesticated animals. The overall effect of climate change on agriculture will depend on the balance of these effects. Assessment of the effects of global climate changes on agriculture might help to properly anticipate and adapt farming to maximize agricultural production. Droughts, which have frequented different parts of India through the history, have been responsible for many famines, rural poverty and migration despite development of impressive irrigation potentials. Similarly, abnormal temperatures, high velocity winds and humidity during critical stages are known to significantly affect crop growth and development, pest incidences and epidemics, demand on irrigation resources and finally food production. Over the past few decades, the main-induced changes in the climate of the earth due to multifarious human activities linked to develop have become the focus of scientific and social attention. The predicted changes to agriculture vary greatly by region and crop. Findings for wheat and rice are reported here: The study found that increase in temperature (by about 2 °C) reduces potential grain yield in most places. Regions with higher potential productivity (such as northern India) were relatively less impacted by climate change than areas with lower potential productivity (the reduction in yields was much smaller). Climate change is also predicted to lead to boundary changes in areas suitable for growing certain crops. (Kalra *et al.*, 2002-2003). Reduction in yields as a result of climate change is predicted to be

**Corresponding author email: msehgalncipm@gmail.com*

more pronounced for rainfed crops (as opposed to irrigated crops) and under limited water supply situations because there are no coping mechanisms for rainfall variability. The difference in yield is influenced by baseline climate. In sub-tropical environments, the decrease in potential wheat yield ranged from 1.5 to 5.8%, while in tropical areas the decrease was relatively higher, suggesting that warmer regions can expect greater crop losses.

Climate change

Climate is changing naturally at its own pace, since the beginning of evolution of earth, 4-5 billion years ago, but presently it has gained momentum due to inadvertent anthropogenic disturbances. There is increase in the atmospheric temperature due to the increase levels of carbon dioxide (CO_2) and other greenhouse gases. The CO_2 concentration was 280±6 ppm between 1000 and 1750 A.D., and today, this value has become 370 ppm. Methane (CH_4), one of the important greenhouse gases contributing 18% global warming, has increased since pre-industrial time from 0.7 to currently 1.78 ppm and accounts for about 50% of global warming. Methane concentration in the atmosphere is presently increasing at 3% per year against 1.2% in the late 1970s. Nitrous oxide with its current concentration 310 ppm in the atmosphere is an important green house gas accounting for approximately 9% of the total green house effect and also responsible for the destruction of stratosphere ozone. Atmospheric concentration of nitrous oxide is increasing @ 0.22% per year. The emission of nitrous oxide is of serious concern because of its long atmospheric life time of 166± 16 years. Agricultural soils contribute 65% of anthropogenic N_2O emission.

The quantity of rainfall and its events has also become more uncertain. In certain place e.g. Maharashtra, Bihar, Rajasthan, Uttar Pradesh and Uttarakhand climatic extremes such as droughts, floods, timing of rainfall and snow melt have also increased.

The global mean annual temperatures at the end of 20th century are almost 7% carbon above those recorded at the end of the 19th century. The 1990s were, on an average, the warmest decade of the earth since instrumental measurement of temperature started in 1860s and the 1990 the warmest century during the last 1,000 years. All these changes have been ascribed primarily to the combustion of fossil fuel and land use changes. (Kalra *et al.*, 2005).

There has been a boost of agricultural research world over in understanding the effects of climate change, carbon sequestration in biosphere and in carbon trading potential of the land. In developing countries including India, however, there has been relatively less attention paid to this topic.

Existing global climate changes models sometimes are quiet contrasting, in generating the climate change scenarios for specific locations. The scenario needs to have higher spatial and temporal resolution, for subsequently working out the impact on agriculture.

Populations of insect-pests do not remain the same size in different years, but keep fluctuating around a mean equilibrium density. The mean equilibrium position can be defined as the long established average of pest density. Population dynamics result from interactions between the biotic potential of a pest species and environmental resistance. Environmental resistance comprises all the mortality factors encountered by a species in nature. Biotic mortality factors take the form of natural enemies such as predators, parasitoids, pathogens and inter specific or intra specific competition. Temperature, humidity, rainfall, wind, topography, photoperiod, etc. are the main abiotic factors usually regarded as key mortality factors.

Such components can directly influence pests by affecting their rate of development, reproduction, distribution, migration, adaptation and influence them indirectly through just plants, natural enemies and inter specific interactions with other insects. Hence, the wide variation in pest populations evident from year to year can be attributed to the intensity of different mortality factors.

Climate variability

Inter and intra seasonal climatic variability is quite large in this part of the world. The extent of variability in south-west and north-east monsoons fluctuate the crops productivity around the trend line, which shows positive slope due to technology advancement in various agro-ecological regions. Analysis with historic weather data sets, although indicate the rising temperature trend line are quite large. Sometimes these variations are confused with climate change.

Winter rains, in north-west India due to western disturbances, ranging up to 12% in the past years and received more spells of small amounts but is very much beneficial to the *rabi* crops, mainly wheat of this region. Rainfed wheat is dependent on this rainfall (Sharma *et al.,* 2005).

Understanding of the climate variability is difficult and most of the studies in this regard are with rains and to some extent temperature also. In our country, where most of the agriculture is under dryland and rainfed, the variability in the climatic parameters plays an important role in deciding the crop production. The climatic variability has not been related with crop response, though in some cases the empirical relations have been developed, which have the limitations of being location and time specific. Linking the climate variability with the productivity on regional scale can subsequently be linked with the climate changes scenarios to work out the impact on soil, crops, water resources and environment.

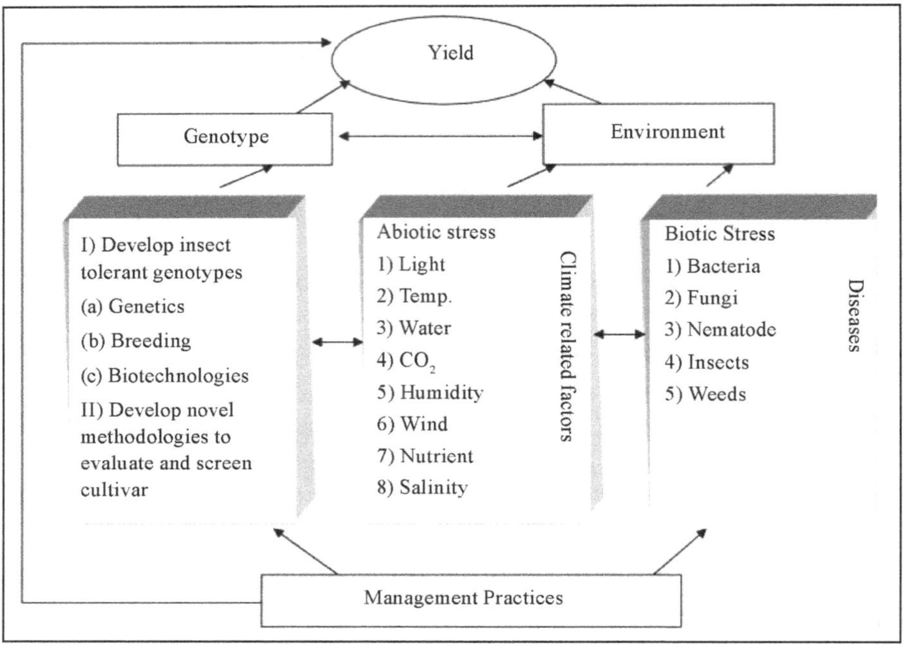

Source: Duli Zhao and Yangrui Li (2015)

Implications of physical factors for insect-pests

The importance of climate and weather events to the distribution of insects and their population dynamics has long been recognized. Insects are poikilothermic and are directly under the control of temperature for their growth. Light is not utilized directly by insects as an energy source, but it controls their daily behavioral rhythm. The movement of air masses determines the destiny of insect migration as a more or less regular phenomenon.

Insect mortality due to climatic conditions that are not generally identifiable gives, rise to the most frequently used concept of physiological death. Stilting (1993) concluded that climate was the reason for the failure of 30% of biological control programmers.

Temperature as the most conspicuous factor

Among various abiotic factors, the effect of temperature is most prominent, owing to the large body surface area relative to body volume of insects, which renders them susceptible to desiccation, thus restricting their distribution and abundance. Insects cannot regulate their body temperature and any change in the ambient temperature influences their survival, growth, development, fecundity and reproduction. Insects are found to be active in a temperature range within which their development rate increases with increases in ambient temperature. On either side of the favorable range, there are zones of inactivity in which metabolic activity decreases (ultimately leading to the death of the insect) to be restored when favorable conditions return.

Temperature has always been found to be important as a determinant for insect-pest dynamics; Campbell *et al.* (1974) defined physiological time as a measure of the amount of heat required over time, for an organism to complete various stages of development. The concept of 'degree days' is an effective tool in this regard. The minimum effective temperature, or the 'threshold temperature (also called the: base temperature) is the temperature below which insect development ultimately stops and the degree day can be defined as a unit of measures equal to one degree above the lower development threshold temperature over a 24 hrs. The time needed to reach a fixed accumulation associate with a recurring natural event usually varies from place to place and from year to year for different insect-pests according to temperature conditions. Degree days is represented by:

$$DD = (Tx + Ty)/2 - Ty$$

Where, Tx is the maximum day time temperature in °C,

Ty is the minimum day time temperature in °C.

The calculated degree day can be correlated with insect development, which can also help in finding the economic threshold level and in decision making. Laboratory studies of the effect of temperature on insect growth and development have mainly been conducted at different but constant temperatures. They have shown that insects remain active within a temperature range from 15 to 30-32 °C (Phadke and Ghai, 1994; also see Table 1) within this range, temperature increases the rate of development. In the case of the red cotton bug, the average duration of life cycle at constant temperature of 20 °C, 25 °C and 30 °C was found to be 61.3, 28.3 and 37.6 days respectively, while at 12.5 and 35 °C the pest did not develop (Bhatia and Kaul, 1966). Also, within the favorable limits, temperature growth rate increases were associated with an increasing trend in egg laying and egg hatching. Results are, however, not so clear for field experiments when pest populations are subjected to fluctuating temperatures as well as other factors that may be limiting (Kalra *et al.*, 2003 & 2004).

Table 1. Developmental time for each stage of *Thrips palmi* at six constant temperatures

Temperature (°C)	Developmental time				
	Egg	Ist instar larva	2nd instar larva	Pupa	Egg-to-adult
16	12.83±0.06 a	4.3±0.05 a	7.7±0.08 a	10.96±0.1 a	35.66±0.17 a
19	10.04±0.05 b	3.6±0.06 b	5.59±0.07 b	7.67±0.1 b	26.83±0.20 b
22	5.9±0.09 c	3.04±0.06 c	4.26±0.08 c	5.68±0.08 c	19.28±0.18 c
25	5.03±0.05 d	1.68±0.06 d	3.51±0.06 d	4.11±0.07 d	14.13±0.12 d
28	3.9±0.06 e	1.23±0.05 e	3.12±0.06 e	3.82±0.09 e	12.20±0.13 e
31	3.2±0.05 f	1.07±0.02 f	2.23±0.04 f	3.17±0.05 f	9.57±0.09 f

Source: Yadav and Chang (2014)

Temperature in conjunction with humidity

Studies conducted in Kerala during 1967-68 and 1971-72 revealed a positive correlation between gall midge infestation and mean rainfall, and a negative correlation between infestation and mean maximum temperature. Relative humidity and minimum temperature has no significant effect on infestation (Thomas, 1976). Long-term studies on peak emergence of gall midge indicated that temperature (28.63±1.03 °C), relative humidity (67.75±18.25%), sunshine hours (7.31±1.81) and rainfall (0-55 mm) were most favorable for higher emergence of adults.

Attempts have been made using the degree day concept to forecast outbreaks of rice yellow stem borer (*Schirpopga incertulas)* is important pest of rice in all the rice growing tracts of India (Johnson *et al.,* 1983). Using a minimum enumeration method, the cumulative degree days requirement for stem borers was calculated to be 653±29.4 °C (egg stage) and 615±26.8 °C (larval pupal stages) using threshold temperatures of 13 °C for egg and 16 °C for the larval pupal stages (Arnold, 1960; Baskerville and Emin, 1969; Krishnaiah *et al.,* 1997). There was a negative relationship between the population build up of stem borers and minimum temperature (Bhatnagar and Saxena, 1999).

Temperature and relative humidity are positively correlated with the incidence of infestation of sorghum Shoot fly (*Antherigona soccata*) (Hussaini and Rao, 1967; Dubey and Yadav, 1980). Peak emergence of adults also coincided with times of high rainfall (Ogwara, 1979; Chundurwar *et al.,* 1984). Similarly, in the case of the sorghum stem borer (*Chilo partellus*), temperature was found to be positively associated with incidence and the degree of infestation (Chelliah and Subramanian, 1972). Examples of studies involving weather factors include the following:

- The degree days concept is widely used in the USA and European countries for predicting the level of attack of pests, relating degree days with calendar dates for egg hatching and the optimum moment to destroy adult moths, so that they will not lay eggs that will produce fruit burrowing larvae within the optimum degree.

- During warm, rainy and springs, aphids multiply in pastures and migrate to larger areas, damaging crops on the way by inoculating viruses into them. During dry spring seasons, few aphids and viruses can be observed.
- In some advanced countries, integrated pest management is mainly based on meteorological information to regulate the intensity and cost of spraying of various non hazardous pesticides.
- Studies conducted in the tropics revealed that high atmospheric humidity, less sunshine and high temperature were favorable for the growth and multiplication of the rice gall midge (*Orseolia oryzae*).
- The lack of rainfall during monsoon seasons and a temperature range of 20-30 °C are optimum for stem borers. Potato beetle (*Leptinotarsa decemlineata*) is most active in temperature ranges from 16-27 °C and humidity of 70 per cent.
- The life cycle of locusts (*Acrididae* sp.) depends on various meteorological factors such as rainfall during the breeding period. Soil moisture and soil temperature are important during hatching, whereas wind and air temperature have a major influence during migration.

Scenarios predicted for the Indian subcontinent

Increase in temperature resulting from increase in greenhouse gases will have direct as well as indirect effects on various nutritional sources (Porter *et al.*, 1991; Cammell and Knight, 1992). Global warming will affect the growth and development of all organisms including insect-pests. Scenarios projected for the Indian subcontinent by the middle of the century suggest an increase in annual mean surface air temperature of 1.0 °C over land regions by the 2040s compared with the 1980s (Lal *et al.*, 1995). It is also assumed that there will be an increase of 0.1 °C in both minimum and maximum temperatures for each degree of latitude. Greater warming is thus expected at higher latitudes than in the tropics (Sutherst, 1991). Monsoon season warming will be less pronounced than in the winter months over the region. It is also predicted that rainfall will be highly erratic. There will be less frequent rains, but of greater intensity, leading to floods. A decrease of about 0.5 mm/day in summer monsoon rainfall is projected. The recurrence of climatic extremes such as droughts and floods has already increased in certain places. The snow cover is declining and there is some increase in sea level. It is expected that some of the coastal areas will be submerged, affecting land use patterns and the lifestyle of people living there.

Pest responses to climatic variability

If, after global warming, the ambient temperature remains within the favorable range for pests, the larger populations may be expected because insect species will complete more generations. However, if the ambient temperature rises above the favorable range, pest populations may be adversely affected.

The favorable temperature range for the rice stem borer (*Scirpophaga incertulas)* varies from 15.1-38 °C (Kalode, 1974). An examination of temperatures for New Delhi during temperature, the minimum temperature during June will become more favorable for pests, while the maximum temperature will become increasingly unfavorable. During the months from July to October, which constitute from 17.3 to 28.4 °C, while the maximum varied from 28.8 to 35.9 °C. Both minimum and maximum temperatures were thus well within the favorable range for this pest. With the presumed rise in temperature, there will still remain favorable. Therefore, with a rise in temperature the development rate of the pest will increase and greater numbers of generations can be expected. During the month of November, the minimum temperature, the development rate of the pest will increase and greater numbers of generation can be expected. Hence, any temperature rise during November will improve conditions for stem borer survival. The incidence of stem borers may be expected to rise with successive rises in temperature under New Delhi conditions. The expected reduction in rainfall may also increase the incidence of the pest. Pasalu *et al.* (1999) considered that high temperature and low rainfall could cause severe stem borer infestation. The percentage incidence of dead hearts and white heads were both negatively correlated with rainfall and minimum temperature and positively correlated with maximum temperature (Abraham *et al.*, 1972).

The favorable range of temperature for the brown plant hopper (*Nilaparvata lugens*) lies between 15 and 36 °C. From July to October, both minimum and maximum temperatures are favorable for the pest (Kalra *et al.*, 2008). With a rise in temperature, the minimum temperature will prove more favorable, while the maximum temperature may exceed the favorable range. There may not be any effect from a 1 or 2 °C rise in ambient temperature, as the beneficial effect of a rise in the minimum temperature may be nullified by an adverse effect of a rise in the maximum temperature. Similarly, the distribution and frequency of rainfall may affect the incidence of pests through changes in humidity levels, as well as directly. Small insects such as aphids, jassids and whiteflies are washed away by heavy rains, thereby reducing their incidence in crops. Aphid populations on wheat and other crops are adversely affected by rainfall and sprinkler irrigation (Daebeler and Hinz, 1977; Bakhetia and Sindhu, 1983; Chander, 1998). The army worm (*Mythimna separata*) reaches outbreaks proportions after heavy rains and floods. Lever (1969) analyzed the relationship between outbreaks of army worm and to a lesser extent, *Spodoptera mauritia* and rainfall from 1938 to 1965, observing that all but three outbreaks occurred when rainfall exceeded the average of 89 cm. The heavy incidence of army worm in New Delhi during the *kharif* (July-October, 1995) could be related more too maximum and minimum temperatures and rainfall as compared with previous years (Chander and Sharma, 1997).

Studies are needed to identify the threshold limit of development for insect-pests of all the major crops. There is a need to know the favorable range of temperatures, optimum temperature, zones of inactivity, lethal temperatures and favorable relative humidity. Apart from this, various empirical studies conducted on a regional basis on similar lines are needed.

Probable expansion in geographical range

Any increase in temperature is bound to influence the distribution of insects. It is predicted that a 1°C rise in temperature would enable species to spread 200 km northwards or 140 m upwards in altitude (Parry *et al.*, 1989). Areas that are not favorable at present due to temperature may become favorable with a rise in temperature. Minimum temperature rather than maximum temperature plays an important role in determining the global distribution of insect species (Hill, 1987), hence any increase in temperature will result in a greater ability to over winter at higher latitudes, ultimately causing a shift of pest intensity from south to north (EPA, 1989; Hill and Dyrnock, 1989). On the other hand, areas with an optimum temperature today may become less favorable with a rise in temperature.

As the species richness of insects tends to increase with temperature (Turner *et al.*, 1987), it is presumed that with an increase in temperature more species will be gained than lost. Over wintering survival and timing of the commencement of spring are important at higher latitudes (Adhikary *et al.*, 2008) leading to population build up of e.g., *Heliothis zea,* thereby endangering maize and soybean crops (EPA, 1989). It has been noted that insect populations under increasing temperatures move towards higher latitudes and elevation (Porter *et al.,* 1991; Sutherst, 1991). The spatial distribution of crops and cropping systems are also important in deciding insect-pests distribution and changes in these are imminent under climate change scenarios (Parry *et al.,* 1989; Carter *et al.,* 1991). The presence of over wintering sites, soil types and moisture availability can become deciding factors for the resurgence of insects in such areas (EPA, 1989).

Increased risk of invasion

A larger crop canopy and denser foliage will create more relative humidity, thereby making micro environments more favorable to pests. Increases in food quality, *i.e.* increase in the nitrogen content of plants due to high temperature can result in a sudden resurgence of population of pests. Moreover, under conditions of stress, plant defensive systems are less effective and they become more susceptible to pest attack (EPA, 1989).

Some important pests are long range migrants and move into crop areas where they cannot over winter successfully. However, with an increase in temperature, previously unfavorable areas may become suitable, thereby increasing survival. The cabbage butterfly (*Pieris brassicae*) migrates from the Himachal hills to the Punjab plain in winter and back to the hills in summer. The increase in temperature may affect this migration and extend distribution of the pest further north, lowering its activity in the plains.

Changes in over wintering success

With an increase in temperature, hibernation may be suspended earlier than usual, thereby initiating early pest activity. The paddy stem borer (*Scirpophaga incertulas*) becomes active in May or June with the availability of plants in rice nurseries. However, with the introduction of rice planting in April in Punjab and Haryana, which is prevalent today and with an increase in temperature, we may expect higher populations of pests. Similarly, the onset of hibernation may also be delayed with an increase in temperature, extending the period of activity of pests. On the other hand, the increase in temperature may reduce the activity of some pests in summer.

Natural enemy-pest interactions

Among insects, there is a distinction between the specialists and generalists, and their response to changing climatic conditions may be different. Generalist species have good dispersal capabilities and are best suited to tracking the changing climate, migrating with climatic shifts and taking the opportunity to feed on new parts of the range. Specialist species, even of the highly dispersive type, may not benefit as the plant species they feed on may not be present in the new region. All species will be under strong selection pressures, which may be different from those exerted when the climate is stable (Travis, 2000). This may also be applicable to natural enemies, which play decisive role in limiting the pest population in the newer areas. Bioclimatic studies on insect hosts and their natural enemies have confirmed potential physiological limitations for their geographic spread. Temperature is also found to exert a substantial influence on the growth and development of *Lydella jalisco*, a parasitoid of the mexican rice borer (*Eoreuma lifitini*). Its larval and pupal development is retarded with increasing temperature (Lauziere *et al.*, 2001). Climate shifts into the unfavorable zone would force all species to operate at or somewhere near the margins of their physiological range. Extinction might occur because of insufficient genetic diversity, but the species might survive in areas that become favorable as result of change. If climate change is slow, selection might enable the species to adapt. Otherwise the species may become extinct locally as conditions become more favorable for their competitors. This may occur at any tropic levels, but the effect may be greater at lower levels (Gutierrez, 2001).

Changes in population growth rates

If with an increase in temperature, the ambient temperature still remains favorable for the pest, then populations will grow quickly and higher pest populations will result. On the other hand, if with an increase in temperature, the favorable range is exceeded, then there may be reductions in the growth rate of pest populations. Each species has to be analyzed separately in the light of its favorable range and the range of temperature increases involved.

Changes in crop-pest synchrony

Crop-pest synchrony will almost inevitably be altered by changes in climate. During the *kharif*, we find heavy infestations of stem borer and shoot fly in maize and sorghum under north Indian cognitions; but *rabi* maize and sorghum are not affected by these pests. In southern parts, however, these pests are active during the *rabi* season. With an increase in temperatures, these pests may thus pose problems in both regions.

Effect on pest control factors

Pesticide use often leads to courses of pest resurgence and secondary pest outbreaks as well as adverse effects on ecosystems and human health (Vardan Bosch, 1978). The effects of weather on pest control operations are also significant. After application of chemicals at least 18-24 rain free hours are desirable if the chemicals are not to be washed away (Rao and Rao, 1996). Similarly, wind resulting in drifting problems is likely to cause damage.

Insect-pest complex: temporal and spatial changes due to crop management practices

Pest incidence is also affected by different crop management practices. The relationship between insects-pests and host has undergone many changes due to cropping systems, the emergence of alternative hosts, mono-cropping and the advent of new varieties and hybrids and changes in cultural agronomic and management practices. The availability of host plants becomes an important causal factor for insect outbreaks. Intercropping may give rise to increased parasite and predator populations through the availability of alternative prey, decreased colonization and reproduction of pests, chemical repellency, massing, feeding inhibition by odors from non-host plants, prevention of immigration and optimum synchrony in relation to pests and their natural enemies (Bhatnagar and Saxena, 1999).

The incidence of thrips, leaf hoppers and leaf miners was much higher in sole crop groundnut than in a groundnut/ cowpea system in Tamil Nadu (Chander and Garg, 1999). Alternative hosts also contribute to a large extent in sheltering insect-pests in the non-crop season. In the case of *Helicoverpa armigera,* over 98 wild and weed plants were recorded as alternative host plants. Lately, mono-cropping has emerged as major cause of increase in pests in plantation crops. Apart from all these, interacting between species, such as competition, predation and parasitism also contribute to pest damage.

Leaf folder incidence on rice and aphid incidence on wheat increased with an increase in nitrogen dosage (Chander and Garg, 1999). Nitrogen has been found to stimulate the incidence of several other pests.

The continuous flooding of rice fields intensifies the incidence of leaf and plant hoppers on the crop. Excluding water for a few days following flooding provides relief from them. Different cultivars planted simultaneously under the same ambient conditions may show different levels of inherent susceptibility or resistance towards pests. The cultivation of cultivars with monogenic or vertical resistance has also led to the development of biotypes in pests. Five biotypes of the gall midge (*Orseolia oryzae*) have been reported as being due to the cultivation of resistant cultivars of rice (Siddiqi, 1999). According to Banerjee (1964), the yellow stem borer, hispa and gundhi bug were major pests of paddy in most of the states, while gall midges, leaf folders, army worms, caseworms, grasshoppers, leaf and plant hoppers and thrips were either sporadic or minor pests prior to 1965. The leaf folder, leaf and plant hoppers and other pests, which are important pests today, became serious pests of paddy during the post green revolution years. The important pests of rice can be listed as the stem borer, gall midge (five biotypes), leaf folder, green leaf hopper, brown plant hopper, white backed plant hopper, cut worm, gundhi bug, hispa, thrips, mites and nematodes (Siddiqi, 1999). Continuous application of similar pesticides has resulted in the development of pesticide resistance. Broad spectrum pesticides have also killed natural enemies of some pests. The absence of both natural enemies and pest resistance provide the prelude for pest outbreaks.

Research priorities

Approaches to analysis of climate change on pests Sutherst (1991) developed the climatic matching GIS (Geographical Information System) software, Climate and Population modeling software CLIMEX, based on the innovative Fitzpatrick-Nix approach (Fitzpatrick and Nix, 1970). It has great utility in predicting the likely impact of climate change on the distribution of pests under present and future climate regimes. The algorithm has been used effectively to map the potential range of the Russian grain aphid. *Duaraphis noxia* and the pathogen *Phytophthpra cinnamonis* on *Quercus spp*. (Brasier and Scot, 1994). CLIMEX was used effectively to assess the favorableness of Australian environments for the Russian wheat aphid and its potential impact on Australian wheat yield (Hughes and Maywald, 1990).

Temperature summation and day length models have been used to predict that the northern boundary of *Plutella xylostella* would shift northwards by 300 km in Japan (Morimoto *et al.,* 1998). This temperature driven model can give rise to erroneous predictions, as observed arthropod community responses to global warming depend on the genetic flexibility of the population, the distribution of tolerant phenotypes, migration patterns, competition, and to some extent community and evolutionary dynamics (Heong *et al.,* 1995).

In addition to this, models developed for this purpose should incorporate a physiological approach based upon the precepts of biomass flow, which includes behaviors and ability to function and compete under changing climate scenarios. Such models can then be coupled with crop growth models to allow us to determine the effects of climate change that will ultimately affect crop production.

Resource base for the development of models

Any attempt to build up a system to assess the potential implications of climate and its success at regional levels depends entirely on its applicability. Questions arise as to how the developed methodology can be extended to entirely new domains. The approach, therefore, should be to extend the functioning of predictive systems on the basis of available technology. There are many empirical studies on insects that have been correlated with weather parameters. However, most of them are location based and area specific. These may, however, be of great help in exposing various facets pertaining to particular pests and crops.

Attempts have also been made to identify relationships between pest incidence and pesticide usage in relation to climate change and pest incidence (Chen and McCarl, 2001). Given the availability of data on pesticide usage, similar relationships could be identified, but again, applicability would be limited to a few specific parts of the country due to the wide disparity in pesticide usage on a national basis.

Approaches should be aimed at utilizing existing data generated by pest roving surveys such as that conducted by the Directorate of Plant Protection, Quarantine and Storage, India. This information is recorded for the entire country on weekly and monthly basis and could constitute key data for the work. The rapid pest roving survey may be further supplemented with results from empirical studies published by various agricultural universities.

Historical databases can be utilized for evaluating the insect-pests scenario over the short and long term to link annual climatic variability, climate change, extent of biophysical and socio-economic driving forces and change with the contemporary cropping systems. Demographic population analysis has diverse applications in predicting life history traits, analyzing population stability and structure, predicting outbreaks of pest species and examining the dynamics of colonizing or invading species (Mcpeek and Kalisz, 1993). Changes in pest complexes can be elucidated with the help of GIS and relational databases.

In the Indian context (as in many others), socio-economic factors need to be taken into consideration. Also, a recent major shift in research priorities in the area of climate change concerns the likely impact of climate change on the functioning of ecosystems and biodiversity. Long term policies are urgently needed to respond to medium term climate variability and changes in the frequency of extreme events so as to design strategies to tackle the problem on a regional basis. Given the large number of stakeholders in agriculture in a country such as India, there is a need for greater participation on the part of various workers to develop strategies based on available information on a regional scale, which can be effectively utilized by policy makers, farmers and researchers alike. Models used to predict climate change impacts on agriculture. The following models were developed to evaluate the impacts of changes in temperature and carbon dioxide on crops:

- INFOCROP, a generic growth model for various crops, was developed by IARI for optimal resource and agronomic management options.
- INFOCANE, a simple sugarcane growth model, was developed by IARI to measure effects on cane yield.
- Simple tea and coconut models were developed for tropical India and Sri Lanka.
- Pest damage mechanisms were coupled with INFOCROP for simulating the effect of pests. The use of this model meant that assessments could be made of the impact of climate change and its variability on incidence of pests for various crops.

- Interaction effects of climate changes (temperature rise, rainfall and radiation changes), with irrigation and nitrogen amounts, and agronomic management practices were established for various agro-ecologies. These were used to calculate the actual impact of climate change on agricultural production as well for suggesting agro techniques and resource management options for sustaining production in India.

Conclusion

Global warming has serious implications for our food security through its direct and indirect effects on crops, soils, livestock, fisheries and pests. At the same time, this is an issue with several social-economic-political implications. Assessment of the effects of global climate changes on agriculture might help to properly anticipate and adapt farming to maximize agricultural production. The study found that increase in temperature (by about 2 °C) reduces potential grain yield in most places. Regions with higher potential productivity (such as northern India) were relatively less impacted by climate change than areas with lower potential productivity (the reduction in yields was much smaller). In sub tropical environments, the decrease in potential wheat yield ranged from 1.5 to 5.8%, while in tropical areas the decrease was relatively higher, suggesting that warmer regions can expect greater crop losses. The importance of climate and weather events to the distribution of insects and their population dynamics has long been recognized. Insects are poikilothermic and are directly under the control of temperature for their growth. Light is not utilized directly by insects as an energy source, but it controls their daily behavioral rhythm. A larger crop canopy and denser foliage will create more relative humidity, thereby making micro environments more favorable to pests. Increases in food quality, *i.e.* increase in the nitrogen content of plants due to high temperature can result in a sudden resurgence of population of pests. Moreover, under conditions of stress, plant defensive systems are less effective and they become more susceptible to pest attack. The effects of weather on pest control operations are also significant. Approaches should be aimed at utilizing existing data generated by pest roving surveys such as that conducted by the Directorate of Plant Protection, Quarantine and Storage, India. Long term policies are urgently needed to respond to medium term climate variability and changes in the frequency of extreme events so as to design strategies to tackle the problem on a regional basis. Given the large number of stakeholders in agriculture in a country such as India, there is a need for greater participation on the part of various workers to develop strategies based on available information on a regional scale, which can be effectively utilized by policy makers, farmers and researchers.

References

Abraham, C.C., Thomas, B., Karunakaran, K. and Gopalakrishnan, R. 1972. Effect of planting seasons and the associated weather conditions of the incidence of the rice stem borer (*Tryporyza incertulas*). *Agril. Rech. J. of Kerala.* **10**(2): 141-151.

Adhikary, K.B., Shusheng, P. and Mark, P.S. 2008. Long-term moisture absorption and thickness swelling behaviour of recycled thermoplastics reinforced with Pinus radiata sawdust. Chem. Engg. J. **142** (2): 190-198.

Arnold, C.Y. 1960. Maximum-minimum temperatures as basis for computing heat units. *Procds. of the American Soc. of Hort. Sci.* **76**: 682-692.

Bakhetia, D.R.C. and Sindhu, S.S. 1983. Effect of rainfall and temperature on mustard aphid (*Lipaphis crysimi*). *Indian J. of Entom.* **45**(2): 202-205.

Banrjee, S.N. 1964. Paddy pests in entomology in India. Entomological society of India, New Delhi, pp 92-98.

Baskerville, G.L. and Emim, P. 1969. Rapid estimation of heat accumulation from maximum and minimum temperatures. *Ecol.* **50**: 514-517.

Bhatia, S.K. and Kaul, H.N. 1966. Effect of temperature on the development and oviposition of red cotton bug, Dysdercus koenigii and application of Pradhan's equation relating temperature to development. *Indian J. of Entom.* **28**(1): 45-54.

Bhatnagar, A. and Saxena, R.R. 1999. Environmental correlates of population buildup of rice insect pests through light trap catch. *Oryza.* **36**(3): 241-245.

Bowling, C.C. 1963. Effect of nitrogen levels on rice water weevil populations. *J. of Eco. Entom.* **56**(6): 826-827.

Brasier, C.M. and Scot, J.K. 1994. European oak decline and global warming; a theoretical assessment with special reference to the activity of *Phytophthora cinnamoni. Bulletin OEP.* 24: 221-232.

Cammell, M.H. and Knight, J.D. 1992. Effect of climatic change on the population dynamics of crop pests. *Adv. in Ecol. Rech.* **22**: 117-162.

Campbell, A., Frazer, B.D., Gilbert, N., Guitererrez, A.P. and Mackauer, M. 1974. Temperature requirements of some aphids and their parasites. *J. of App. Ecol.* **11**: 431-438.

Carter, T.R., Parry, M.L. and Porter, J.H. 1991. Climatic change and future agro-climatic potential in Europe. *Int. J. of Climatology.* **43**(1): 251-269.

Chander, S. 1998. Infestation of root and foliage ear head aphids on wheat in relation to predators. *Indian J. of Agril. Sci.* **68**(11): 754-755.

Chander, S. 1999. Effect of nitrogen and irrigation on the leaf folder, *Cnaphalocrosis medinalis* infestation in paddy. *Shashpa* **6**(1): 49-51.

Chander, S. and Garg., R.N. 1999. Effect of nitrogen and cultural practices on pests of rice and wheat under rice-wheat cropping system. *Annl. of Pl. Protec. Sci.* **7**(2): 159-162.

Chander, S. Garg, R.N. and Singh G. 1997. Termite infestation in relation to soil physical properties in wheat crop under rice-wheat cropping system. *Annl. of Agril. Res.* **18**(3): 348-350.

Chander, S. and Sharma, S.N. 1997. Infestation of armyworm (*Mythimna separate*) in paddy. *Annl. of Pl. Protec. Sci.* **5**(1): 108-121.

Chelliah, S. and Subramanian, A. 1972. Influence of nitrogen fertilization on the infestation by gall midge (*Pachydiplosis oryzae*) in certain rice varieties. *Indian J. of Entom.* **34**: 255-256.

Chen, C. and McCarl, B.A. 2001. An investigation of the relationship between pesticide usage and climate change. *Climate Change.* **50**: 475-487. .

Cheng, P.Y. 1990. Effect of abnormally high temperature and drought on the immigration of *Cnaphalocrosis medina lis* Guenee. *Insect Knowledge.* **27**: 195-197.

Chundurwar, R.D., Karanjkar, R.R., Karyankar, S.P. and Ali, M.A. 1984. Effect of weather parameters on incidence of sorghum shoot fly at Perbhani. *Sorghum News Letter.* **27**: 83.

Daebeler, F. and Hinz, B. 1977. Compensation of aphid injury to sugar beet by means of supplementary rain. *Arch. fur Phytopathologic und Pflanzenschutz.* **13**(3): 199-205.

Dhaliwal, G.S. and Arora, R. 1996. Principles of insect pest management. National agricultural technology information centre, Ludhiana 141 001. pp. 374.

Drake. 1994. Influence of weather and climate on agriculturally important insects. An Australian perspective. *Australian J. of Agril. Res.* **45**: 487-509.

Dubey, R.C. and Yadav, T.S. 1980. Sorghum shoot fly incidence in relation to temperature and humidity. *Indian J. of Entom.* **41**: 273-274.

Duli, Z. and Yangrui, L. 2015. Climate change and sugarcane production: Potential impact and mitigation strategies. *Int. J. of Agron.* **2**: 1-10.

E.P.A. 1989. The potential effects of global climate change on the United States. National Studies Review of the Report to Congress, Vol 2, US Environmental Protection Agency, Washington, DC

Fitzpatrick, E.A. and Nix, H.A. 1970. The climatic factor in Australian grassland ecology in More R.M., ed. Australian Grassland. Australian National University Press, Brisbane. pp. 3-26.

Garg, A.D. and Sethi, G.R. 1980. Succession of insect pests in *kharif* Paddy. *Indian J. of Entom.* **42**(3): 482-487.

Gutierrez, A.P. 2001. Crop ecosystem response to climate change. Pest and Population Dynamics, Division of Ecosystem Science, University of California.

Hill, D.S. 1987. Agricultural insect pests of temperate regions and their control. Cambridge University Press. Cambridge.

Hill, M.G. and Dyrnock. J.J. 1989. Impact of climate change agricultural/horticultural systems. DSIR Entomology Division submission to the New Zealand Climate Change Programme, Department of Scientific and Industrial Research, New Zealand.

Heong. K.L., Song, Y.H., Pimasamarn S., Zhang, R. and Bac. 1995. Global warming and rice arthropod community in Peng, S. Ingram, K.T. Nene, H.U. and Ziska L.H. Climate Change and Rice Symposium, Manilla, Philippines, March 1994, Springer Verlag, pp 326-335

Hughes, R.D. and Maywald, G.W. 1990. Forecasting the favorableness of Australian environment for the Russian wheat aphid, Duaraphis noxia (Homoptera; Aphididae) and its potential impact on Australian wheat yields. *Bull. of Entom. Res.* **80**: 165-175.

Hussaini, S.H. and Rao, P.V. 1967. Flowering behavior of the parents in CSH-1 and CSH- 2 Jowar hybrids under differential sewing's. *Sorghum News Letter.* **10**: 33-34.

IPCC. 1996. Inter-Governmental panel on climate for science of climate change in houghton. T. Meira Fillio, B.A. Callander, N. Harris, Kttenberg and Maskell. Climate Change, Cambridge: University Press, Cambridge.

Johnson, P.C., Mason, D. P., Radke, S. L. and Tracewski, K.T. 1983. Gypsy moth (*Lymantria dispar*) egg eclosion, degree day accumulation. *Envtl. Entom.* **12**: 929-932.

Kalode, M.B. 1974. Recent changes in relative pest status of rice insects as influenced by cultural, ecological and genetic factors, *Procd. of the Int. Rice Res. Conf.* 22-25 April, Los Banos, Philippines.

Kalra N. 2013. Climate Change Impacts on Agriculture in India. Indian Agricultural Research Institute, N Delhi.

Kalra, N., Chakraborty, D., Sharma, A., Rai, H. K., Jolly, M., Chander, S., Kumar, R., P., Bhadraray, S., Barman, D., Mittal, R. B., Lal, M. and Sehgal, M. 2008. Effect of increasing temperature on yield of some winter crops in northwest India. *Curr. Sci.* **94**(1): 82-89.

Kalra, N. Chander, S., Pathak, H. Aggarwal, P. K., Gupta, N. C., Sehgal, M. and Chakraborty, D. 2007. Impacts of climate change on agriculture. *Outlook on Agric.* **36**(2): 109-118.

Kalra, N., Chander, S., Misra, A. K. and Sehgal M. 2005. Linkage of agri-informatics with simulation to assess the productivity of crops. *Bioinformatics India* **3**(3): 45-52.

Kalra, N. Aggarwal, P. K., Chander, S., Pathak H., Choudhury, R. Chaudhury A., Sapra, R. L., Kumar, S., Hussain M. Z. and Sehgal, M. 2004. Correlation between the crop yields and climatic variability/change and Adaptation strategies. Proc. Vulnerability & Adaptation workshop on Agriculture, Forestry and Natural Ecosystems, 18-19 July 2003, Indian Institute of Science, Bangalore, India.

Kalra, N., Aggarwal, P. K., Chander, S., Pathak, H., Choudhary, R., Chaudhary, A., Sehgal, M., Rai, H. K., Soni, U. A., Sharma, A., Jolly, M., Singh, U. K., Ahmed, O. and Hussain, M. Z. 2003. Impacts of Climate Change on Agriculture. In Climate Change and India, Vulnerability Assessment and Adaptation (ed. Shukla, P. R. Sharma, S. K. Ravindranath, N. H., Garg, A. and Bhattacharya, S.). Universities Press.

Kalra, N., Aggarwal, P.K., Abrol Y.P., Pathak, H., Choudhury A., Chander, S., Sehgal M., Kumar, S., Soni U.A., Sharma S., Rai, H.K., Kumari, K., Sindhu, J., Chatterji, A., Hussain, M.Z. and Khan, S.A. 2002. Impact of climate variability and climate change on soil and crop productivity. *Sci. & Culture.* **68**(9-12): 233-243.

Kalra, N., Sehhgal, M., Soni, U.A. and Sharma, A. 2003. Impact of climatic variability and climate change on agriculture. In changing Horizon of Co-operative governance and restructuring of industry in India. M. Med trust Ambala. 24-29.

Kovitvadhi, K. 1972. Current research and future needs on rice insects problem in Thailand paper presented at a meeting of the Planning Group for Integrated Control Rice Pests in Southeast Asia, 9-12 May, Los Banos, Philippines.

Koyama, M.J. 1966. The relation between the outbreak of armyworm (Leucania separate) and richly nitrogenous manured cultivation of rice plant. *Japanese J. of Appl. Entom. and Zool.* **10**: 123-128.

Krishnaiah, N.V., Pasalu, I.C., Padmavathi, L., Krishnaiah, K. and Ram Prasad, A.S. 1997. Day degree requirement of rice yellow stem borer *Scirpophaga incertulus* (Walker). *Oryza.* **34**: 185-186.

Lal, M., Cubasch, U., Voss, R. and Waszkewitz, J. 1995. Effect of transient increase in greenhouse gases and sulphate aerosols on monsoon climate. *Curr. Sci.* **69**(9): 752-762.

Lauziere, 1. Setamou, M. Legasri, J. and Jones. 2001. Effects of temperature on life cycle of *Lydella jalisco* (Diptera:Tachinidae), a parasitoid of *Eoruma liftini* (Lepidoptera: Pyralidae). *Physical Chemistry and Ecology.* pp. 432-437.

Lever, R.J.W. 1969. Do armyworm follow the rain. *Wild Crops*. **21**: 351-352.

Mahadevan, N.R. and Chelliah, S. 1986. Influence of season and weather factors on occurrence of sorghum stem borer (*Chilo partellus*) Swinhoe in Tamil Nadu. *Trop. Pest Mangt.*, **32**: 212- 214.

Masters, G.J., Brown, V.K., Clarke, I.P., Whittaker, J.B. and Hollier, J.A. 1998. Direct and indirect effects of climate change on insect herbivores: Auchenorrhyncha (Homoptera). *Ecological Entomo.* **23**(1): 45-52.

Mcpeek, M.A., and Kalisz, S. 1993. Population sampling and bootstrapping in complex designs: demographic analysis in Scheiner, S.M. and Gurevitch, J. eds. *Designs and Analysis of Ecological Experiments,* Chapman and Hall, New York, pp. 232-252.

Morimoto, N., Imura, O. and Kura, P. 1998. Potential effects of global warming on occurrence of Japanese insects pests. *Appl. Entom. and Ecol.* **33**: 147-155.

Ogwara, K. 1979. Seasonal activity of sorghum shoot fly. *Entomological Exp. and Entom.* 26: 74-79.

Oya. S. and Suzuki, T. 1971. Studies on the population increase of green rice leafhopper Nephotettix cincticeps. II. The larval growth and oviposition on rice plants under controlled sunshine and nitrogenous fertilizer. *Procd. of the Assoc. of Plant Prot.,* Kokuriku, **19**: 45-49.

Padmanabhan, S.Y. and Chakrabarti, N.K. 1985. Diseases of rice, fungal diseases in *Rice Res. in India,* ICAR, New Delhi, pp. 459-517.

Partha Pratim Adhikary, Debashis Chakraborty, Naveen Kalra, C.B. Sachdev, A.K. Patra, Sanjeev Kumar, R.K. Tomar, Parvesh Chandna, Dhwani Raghav, Khushboo Agrawal, Mukesh Sehgal. 2008. *Australian J. of Soil Res.* **46**: 476-484

Parry, M.L., Carter, T.R. and Porter, J.H. 1989. The green house effect and the future of U.K. agriculture. *J. of the Royal Agril. Soc.* **150**: 120-121.

Pasalu, I.C., Krishnaiah, N.Y., Muralidharan, K., Mayee, C.D., Venkataraman, S., Rajarajeshwari, N.V.L. and Sarker, J. 1999. The power of climate components on the flare-up of diseases and pests in rice in Abrol, Y.P. and Gadgil, S. (Eds.) *Rice in a Variable Climate.* APC Publications, New Delhi.

Phadke, K.G. 1986. Ecological factors affecting aphid, *Lipaphis erysimi* (Kaltenbach) incidence of mustard crop. In: Agarwal, B.K. (Ed.) *Aphidology in India.* AR Printers, Calcutta, pp. 37-42.

Phadke, K.G. and Ghai, S. 1994. Effect of global warming on insect populations and crop damage. *Shashpa.* **1**(2):75-80.

Porter, J.H., Parry, M.L. and Carter, T.R. 1991. The potential effects of climatic change on agricultural insect-pests. *Agril. Fors. Metr.* **57**: 221-240.

Rao, B.V.R. and Rao, M.S. 1996. Weather effects on insect pests in Abrol, Y.P., Gadgils. and Pant G.B., eds, *Climate Variability and Agriculture,* Narosa, New Delhi, pp. 281-294.

Reguapthy, A., Palanisamy, S., Chanda M. N. and Gunathilagaraj, K. 1994. *A guide to crop pests,* Rajalakshmi Publications, Nagercoil.

Rodhe, A.L. 1990. A comparison of the contribution of various gases to the greenhouse effect. *Sci.* **248**: 1217-1219.

Sehgal, M., Das, S. Chander, S., Gupta, N.C. and Kalra, N. 2006. Climate studies and insect pests: Implications for the Indian context. *Outlook on Agric.* **35**(1): 33-40

Sharma, A.K., Kumar, J. and Nagarajan, S. 1998. Disease management strategy in wheat. *Indian Farmg.* 8(1): **51**-53; 64-69.

Sharma, A., Soni, U. A., Rai, H. K., Hussain, M. Z., Chander, S., Mishra, A. K., Sehgal, M. and Kalra, N. 2005. Estimation of solar radiation from temperature at different locations of India. *J. of Agrometeoro.* **7**(1): 14-20.

Siddiqi, E.A. 1999. Climate and rice production in India in Abrol. Y.P. and Gadgil, *Rice in a Variable Climate,* APC Publications, New Delhi.

Stilting, P. 1993. Why do natural enemies fail in classical biological control programs? *American Entom.* **39**: 31-37.

Subbarao, I.V. 1999. Soil and environmental pollution-a threat to sustainable agriculture. *J. of the Indian Soc. of Soil Sci.* **47**(4): 611-633.

Subramanian, V., Mani, M. and Guruswamy Roha, V.D. 1977. Effect of graded levels of nitrogen on the incidence of rice stem borer (*Tryporyza incertulas*). *Sci. and Cultiv.* **43**(5): 222-223.

Sutherst, R.W. 1991. Pest risk analysis and the green house effect. *Rev. of Agri. and Entom.*, **79**(11/12): 1177-118.

Thomas, B. 1976. Rice entomology. *News letter.* **4**(1): 33-34.

Travis, J.M.J. 2000. Climate change, dispersal and insect biodiversity (poster presented at ICE, 2000) Climate impacts group, department of ecology, plant ecology, Lund University, Lund, Sweden.

Turner, J.R.G., Gatehouse, C.M. and Corey, C.A. 1987. Does solar energy control organic diversity- butterflies, moths and British climate. *Oikos,* pp. 195-205.

Uthamasamy, S., Velu, V., Gopalan, M. and Ramanathan, K.M. 1983. Incidence of brown plant hopper Nilaparvata lugens on IR50 at graded levels of fertilization at Aduthurai. *Int. Rice Res. News letter,* **10**(1): 23.

Vardan, B. 1978. *Pesticide Conspiracy,* Doubleday, New York.

Vargas, R.I., Walsh, W.A., Kanehisa, D., Stark, H.D. and Nishida, T. 1999. Comparative demography of three Hawaiian fruit flies (Diptera: Tephritidae) at temperatures. *Annl. of the Entom. Soc. of America.* **93**(1): 75-81.

Yadav, R. and Chang, N.T. 2014. Effects of temperature on the development and population growth of the melon thrips (*Thrips palmi*) on eggplant (*Solanum melongena*). *J. of Ins. Sci.* **14**(1): 78.

18

Analysis of Field Experimental Data Using Statistical Calculator

D.S. Dhakre and D. Bhattacharya*

Introduction

Any computer that has Microsoft office excel in it can operate this statistical calculator. Operating system requirement is Windows 7 or its higher version. Usually looking at a data we become nervous and start thinking that how to analyse the data using computer programme, how to write the report of the research analysis etc. This newly developed statistical calculator will remove the fear of data analysis and make our life easy. This calculator will get the data analysis done within a very reasonable computing time.

This chapter introduces a new statistical calculator which provides a step-by-step guide to data analysis of the research data generated in agriculture, biological sciences and in other fields (Singh and Chaudhary, 1985; Chandel, 1998; Bhattacharya and Chowdhury, 2010). The process used to develop the program is quite involved with statistical theories and is not understandable to those who do not have any background in statistics and data analysis. Here we will explain the procedure adopted and the programme used, step-by-step so that it becomes easily comprehensible. We have provided examples in each spreadsheet for analysis. Data analysis with this statistical programme is not really difficult. According to the draft format required for analysis, you only need to arrange and enter the raw data. It does not require much information and knowledge about the formula used as well as the program developed to analyze the data (http://www.sbear.in/).

Methodology

Completely randomized design (CRD): Follow the steps described below to get the data analyzed and finding its outputs.

Step 1. Type *www.sbear.in* in any browser that will open the following web page (Fig. 1).

**Corresponding author email: dhakreds@gmail.com*

Fig. 1. Web page of society of bio-resource, environment and agricutural research

Step 2. Then click on statistical analysis which will open the page statistical calculator (Fig. 2) which is given below.

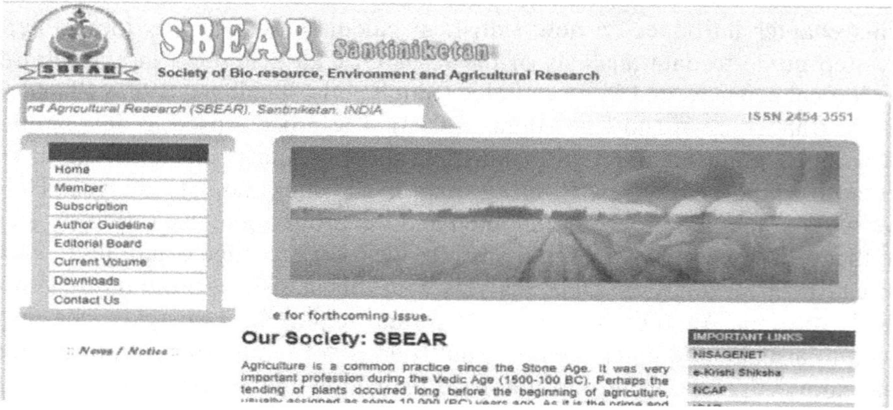

Fig. 2. Web page of statistical calculator

Step 3. Click on completely randmozed design (CRD) (Fig. 3).

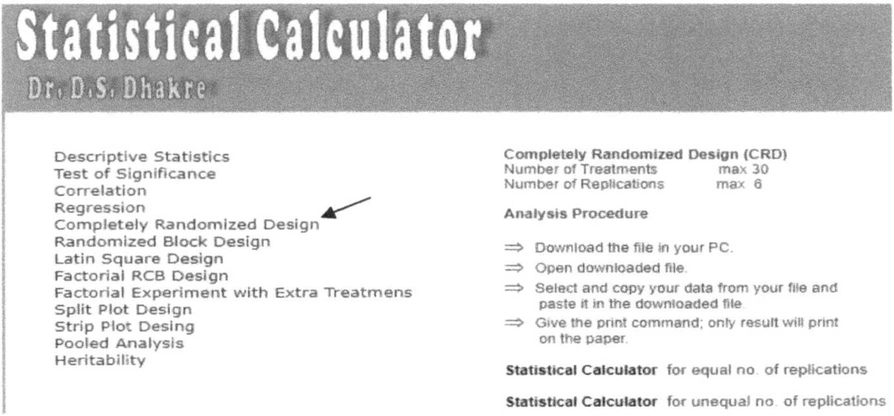

Fig. 3. Web page of CRD

Step 4. Here, you will get two statistical calculators, one for equal number of replications and other for unequal number of replications. Then click on any one of them according to your requirment and then that command will download the statistical calculator of completely randmozed design within a second. You can keep it in your computer or laptop forever. Next, you click on the downloaded of file which is an excel file, a data spreadsheet will open up. Here you can analyse upto 30 number of treatments with upto 6 number of replications. Now you can put your data directly in the given spreadsheet or copy from other sheet and paste it here.

Example 1. The following Table 1 give the yields in pound per plot, of five varieties of wheat after being applied to each of 4 plots, tested in a CRD. Carry out the analysis of a CRD for data.

Table 1. Yield of 5 wheat varieties

Varieties	Replications			
	R_1	R_2	R_3	R_4
A	8	8	6	10
B	10	12	13	9
C	18	17	13	16
D	12	10	15	11
E	8	11	9	8

Step 5. Put number of replications- 4, number of treatments-5 and name of the variable- yield of wheat in the following web page (Fig. 4).

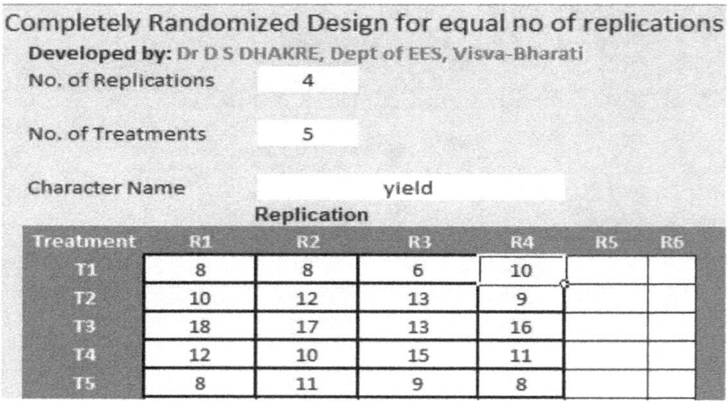

Completely Randomized Design for equal no of replications

Developed by: Dr D S DHAKRE, Dept of EES, Visva-Bharati

No. of Replications 4

No. of Treatments 5

Character Name yield

Treatment	R1	R2	R3	R4	R5	R6
	Replication					
T1	8	8	6	10		
T2	10	12	13	9		
T3	18	17	13	16		
T4	12	10	15	11		
T5	8	11	9	8		

Fig. 4. Web page of data spreadsheet

Step 6. Now you click on print command, then that command will start analysis of the data and produces the results in a printable format. If a printer is attached to the computer, then you can take a print out of the result, otherwise, you can view your result in a print preview mode (Fig. 5).

Output

ANOVA for yield of wheat

Source of variation	DF	SS	MS	F- value
Treatments	4	155.20	38.80	11.19**
Error	15	52.00	3.47	
Total	19	207.20		

** significant at the 0.01 level

Treatment	Mean
T1	8.00
T2	11.00
T3	16.00
T4	12.00
T5	9.00

Grand total	224.00	Root MSE	1.86
SE(m)	0.93	Grand Mean	11.20
SE(d)	1.32	Coefficient Variation	16.62%
Critical Difference	2.81	R Square	0.75

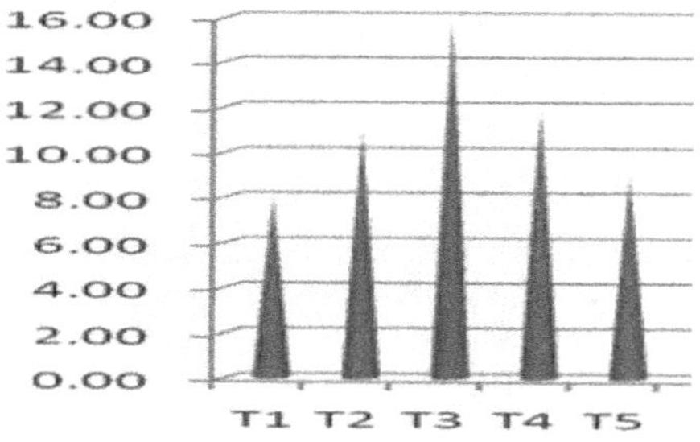

Fig. 5. Web page of ANOVA

Randomized block desing (RBD): Follow the steps described below to get the data analyzed and finding its outputs.

Step 1. Click on the RBD of statistical calculator web page (Fig. 6).

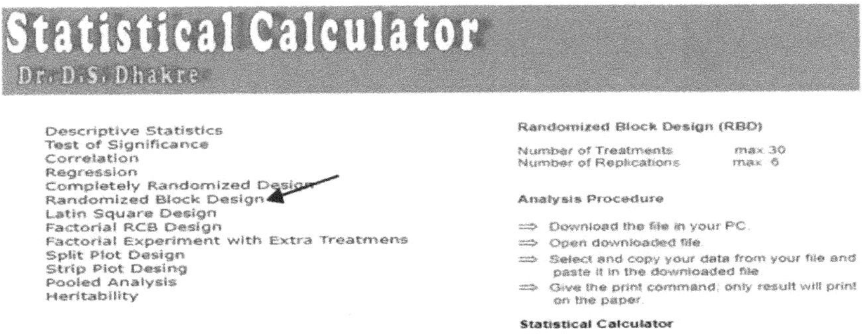

Fig. 6. Web page of RBD

Step 2. Next, after clicking on RBD, it will download the statistical calculator for RBD. You can keep it in your computer or laptop forever. After that, you click on the downloaded file which is an excel file, then a data spreadsheet will open up (Fig. 7). Here you can analyse upto 30 number of treatments with upto 6 number of replications. Now you can enter your data directly in the given spreadsheet or copy from other sheet and paste it here.

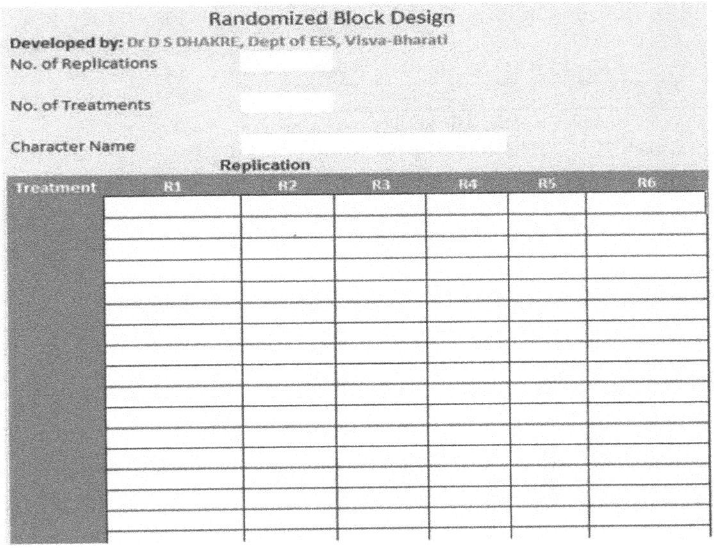

Fig. 7. Web page of data spreadsheet

Example 2. The following Table 2 gives the yields per plot of five varieties of wheat after being applied to each of 4 plots, tested in a RBD. Carry out the analysis of a RBD for data.

Table 2. Yield of 5 wheat varieties

Varieties	Replications			
	R_1	R_2	R_3	R_4
A	8	8	6	10
B	10	12	13	9
C	18	17	13	16
D	12	10	15	11
E	8	11	9	8

Step 3. Put number of replications-4, number of treatments-5 and name of the variable-yield of wheat in the following page.

Step 4. Now you click on print command and then that command will start analysis and produces the results in a printable format. If a printer is attached to computer, then you can take a print out of the results. Otherwise, you can see your result in a print preview mode. (Fig. 8).

Output

ANOVA for yield of wheat

Source of variation	df	SS	MS	F- value	p-value
Treatments	4	155.20	38.80	9.24	0.0012 **
Replications	3	1.60	0.53	0.13	0.9423
Error	12	50.40	4.20		

** Significant at the 0.01 level

Grand total	224.00	Root MSE	2.05
SE(m)	1.02	Grand Mean	11.20
SE(d)	1.45	Coefficient Variation	18.30%
Critical Difference	3.03	R Square	0.76

Treatment	Mean	
T1	8.00	
T2	11.00	
T3	16.00	Best treatment
T4	12.00	
T5	9.00	

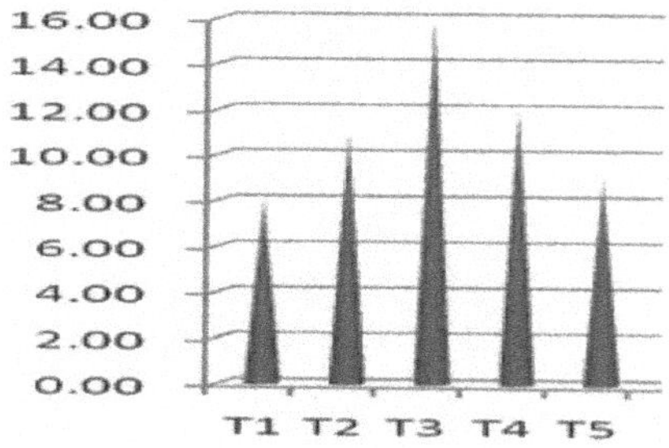

Fig. 8. Web page of ANOVA

Pooled analysis

Step 1. Click on pooled analysis items of the statistical calculator (Fig. 9).

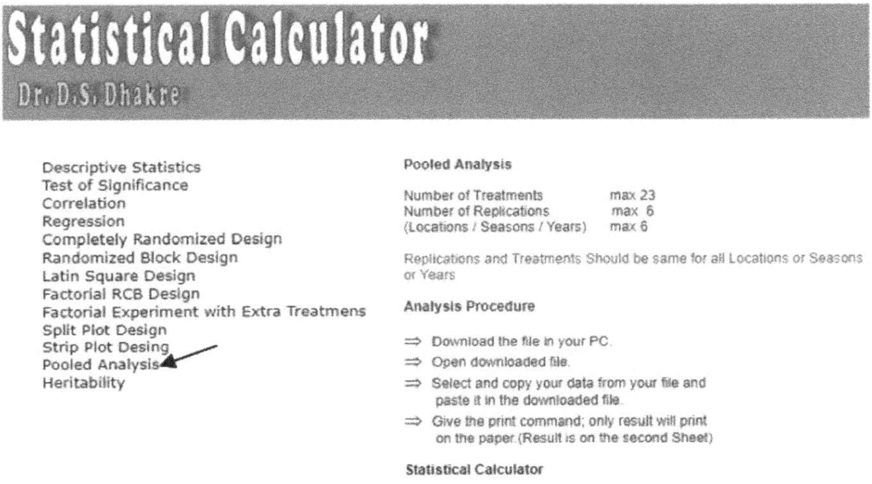

Fig. 9. Web page of pooled analysis

Step 2. Click on statistical calculator of pooled analysis, then that page will download the calculator within a second. You can keep it in your computer or laptop forever. Next, you click on the downloaded file which is an excel file and a data spreadsheet will open up. Here, you can analyse upto 23 number of treatments, upto 6 number of replications and upto 6 number of location/year/season. Now you can put your data directly in the given spreadsheet or copy from other sheet and paste it here.

Example 3. An experiment was conducted in RBD in three locations with 8 varieties and 4 replications. The data regarding yield of wheat (kg/plot) are given below in Table 3, 4 and 5. Analyze the data and draw your conclusions.

Table 3. Yield of wheat varieties (8) in location 1

	Location-1			
Variety	R_1	R_2	R_3	R_4
1	7.5	8.2	5.3	8.4
2	24.8	23.65	24.3	24.8
3	35	33.5	34.75	34
4	31.4	30	30.8	34.8
5	36	35.8	36.2	35.5
6	37.85	40	39.2	38.8
7	34.8	33.2	33	31.9
8	37.8	37.2	38.5	35.8

Table 4. Yield of wheat varieties (8) in location 2

	Location-2			
Variety	R_1	R_2	R_3	R_4
1	6.9	8	7.5	8.65
2	25.3	22.8	22.85	25.3
3	33.85	35	35.3	32.8
4	32.6	28.5	31.9	34.6
5	34.9	36.1	37.3	36
6	39	38.9	40.2	39.15
7	33.2	36	32.8	33.3
8	36.8	34.1	37.9	37

Table 5. Yield of wheat varieties (8) in location 3

	Location-3			
Variety	R_1	R_2	R_3	R_4
1	19.2	17.4	16	13.85
2	25.6	27.35	23.45	26.2
3	30.7	34.75	35.2	33.6
4	29.8	32	29.65	32.1
5	35.4	35.75	36.5	37.3
6	39.8	39.45	38.2	42.45
7	36.8	35.6	37.2	37.6
8	37.15	36.85	37.3	37.95

Step 3. Put number of replications-4, number of treatments-8, number of location-3 and name of locations (location 1, location 2, location 3) in the following page (Fig. 10).

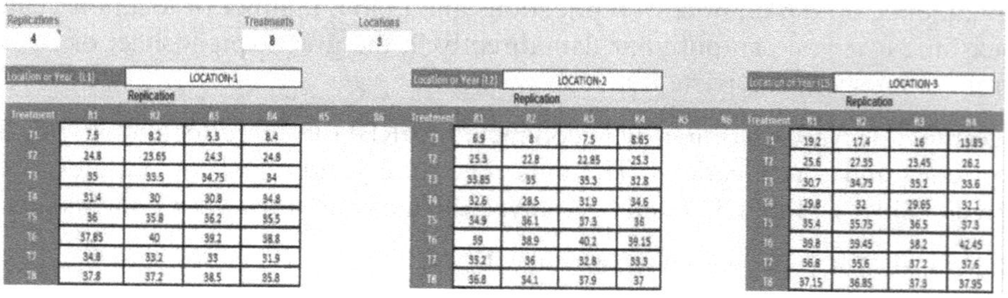

Fig. 10. Web page of data spreadsheet

Step 3. Now you click on print command and that will start analysis and produces the results in a printable format. If a printer is attached to computer, then you can take a print out of the results. Otherwise, you can see your result in a print preview mode (Fig. 11).

Output

ANOVA for Location-1

Source	DF	SS	MS	F-value	p-value
Treatments	7	2975.24	425.03	282.53**	0.00
Replications	3	1.06	0.35	0.24	0.87
Error	21	31.59	1.50		
Total	31	3007.90			
	Grand total =	972.75		Root MSE =	1.23
	SE (d) =	0.87		Grand mean =	30.40
	Critical difference =	1.80		CV =	4.03%
	R square =	0.99		SE(m) =	0.61

**Significant at the 0.01 level, ANOVA: analysis of variance, CD: critical difference, CV: coefficient of variance, DF: degree of freedom, SS: sum of square, MS: mean sum of square, SE(d): standard error of deviation, SE(m): standard error of mean

ANOVA for Location-2

Source	DF	SS	MS	F- value	p-value
Treatments	7	2918.18	416.88	193.18**	0.00
Replications	3	4.20	1.40	0.65	0.59
Error	21	45.32	2.16		
Total	31	2967.70			
	Grand total =	974.50		Root MSE =	1.47
	SE(d) =	1.04		Grand mean =	30.45
	Critical difference =	2.16		CV =	4.82%
	R square =	0.98		SE(m) =	0.73

ANOVA for Location-3

Source	DF	SS	MS	F- value	p-value
Treatments	7	1653.85	236.26	97.86**	0.00
Replications	3	4.97	1.66	0.69	0.57
Error	21	50.70	2.41		
Total	31	1709.519297			
	Grand total =	1028.15		Root MSE =	1.55
	SE(d) =	1.10		Grand mean =	32.13
	Critical difference =	2.28		CV =	4.84%
	R Square =	0.97		SE(m)=	0.78
	Location's	df	MSE		
	L1	21	1.50		
	L2	21	2.16		
	L3	21	2.41		

Fig. 11. Web page of ANOVA

According to Hartley's test the variances are equal. It means MSEs are homogeneous. If MSEs are heterogeneous, then we use Aitken's transformation to the data for pooled analysis (Table 6, 7, 8 and Fig. 12 and 13).

Table 6. Combined block design of experiment and ANOVA table with transformation

Source	DF	SS	MS	F-cal	p-value
Treatments	7	3808.95	544.14	544.14**	0.00
Locations	2	355.11	177.55	177.55**	0.00
Rep in loc	9	4.71	0.52	0.52	
Treat*loc	14	206.06	14.72	14.72**	0.00
Error	63	63.00	1.00		
Total	95	4437.83			
	SE	CD		GM=	22.06
Location	0.25	0.50		CV=	4.53
Treatment	0.41	0.82		Root MSE=	1
Trt*Loc	0.71	1.41		R Square =	0.99

Table 7. Combined block design experiment ANOVA table without transformation

Source	DF	SS	MS	F-cal	p-value
Treatments	7	7347.12	1049.59	518.17**	0.00
Locations	2	61.98	30.99	15.30**	0.00
Rep in loc	9	10.23	1.14	0.56	
Treat*loc	14	200.16	14.30	7.06**	0.00
Error	63	127.61	2.03		
Total	95	7747.10			
	SE	CD		GM=	30.99
Location	0.36	0.71		CV=	4.59
Treatment	0.58	1.16		Root MSE=	1.42
Trt*loc	1.01	2.01		R Square =	0.98

Table 8. Pooled mean (original) and (transformation) analysis

Treatment	Mean (Original)	Mean (Transformation)
T_1	10.58	7.32
T_2	24.70	17.59
T_3	34.04	24.29
T_4	31.51	22.49
T_5	36.06	25.71
T_6	39.42	28.08
T_7	34.62	24.60
T_8	37.03	26.42
L_1	30.40	24.78
L_2	30.45	20.73
L_3	32.13	20.68

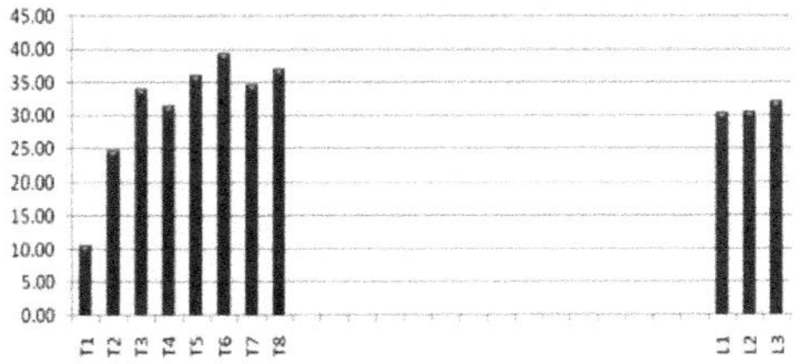

Fig. 12. Original values mean

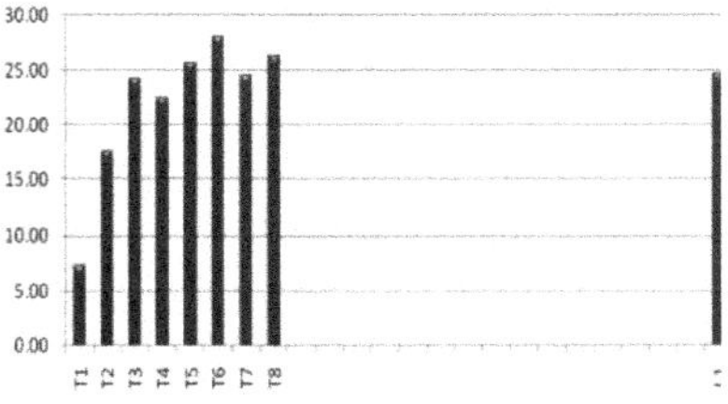

Fig. 13. Transformed values mean

Conclusion

The steps described in the statistical calculator are easy to follow and one will not have any difficulty in getting his/her data analysis done without much effort. Interpretation of the output, which is a very important but a shadow area in any analysis of data using software packages, has been made clear for the users. This newly developed statistical calculator will definitely remove the inhibition of the researchers in data analysis using appropriate statistical tools and techniques. The calculator can further be extended by incorporating more items in the calculator which will be our further research endeavour.

References

Bhattacharya, D. and Chowdhury, S.R. 2010. Statistics theory and practice, U.N. Dhur & Sons.

Chandel, S.R.S. 1998. A hand book of agricultural statistics, Achal Prakashan Mandir, Kanpur.

Singh, R.K. and Chaudhary, B.D. 1985. Biometrical methods in quantitative genetic analysis, Kalyani Publishers, Ludhiana